AF806136

Springer Climate

Series editor

John Dodson, Menai, Australia

Springer Climate is an interdisciplinary book series dedicated on all climate research. This includes climatology, climate change impacts, climate change management, climate change policy, regional climate, climate monitoring and modeling, palaeoclimatology etc. The series hosts high quality research monographs and edited volumes on Climate, and is crucial reading material for Researchers and students in the field, but also policy makers, and industries dealing with climatic issues. Springer Climate books are all peer-reviewed by specialists (see Editorial Advisory board). If you wish to submit a book project to this series, please contact your Publisher (elodie.tronche@springer.com).

More information about this series at http://www.springer.com/series/11741

Rais Akhtar • Cosimo Palagiano
Editors

Climate Change and Air Pollution

The Impact on Human Health in Developed and Developing Countries

Springer

Editors
Rais Akhtar
International Institute of Health Management and Research (IIHMR)
New Delhi, India

Cosimo Palagiano
Dipartimento Di Scienze Documentarie, Linguistico-Filologiche e Geografiche
Sapienza University of Rome
Rome, Italy

ISSN 2352-0698 ISSN 2352-0701 (electronic)
Springer Climate
ISBN 978-3-319-61345-1 ISBN 978-3-319-61346-8 (eBook)
DOI 10.1007/978-3-319-61346-8

Library of Congress Control Number: 2017952378

© Springer International Publishing AG 2018
This work is subject to copyright. All rights are reserved by the Publisher, whether the whole or part of the material is concerned, specifically the rights of translation, reprinting, reuse of illustrations, recitation, broadcasting, reproduction on microfilms or in any other physical way, and transmission or information storage and retrieval, electronic adaptation, computer software, or by similar or dissimilar methodology now known or hereafter developed.
The use of general descriptive names, registered names, trademarks, service marks, etc. in this publication does not imply, even in the absence of a specific statement, that such names are exempt from the relevant protective laws and regulations and therefore free for general use.
The publisher, the authors and the editors are safe to assume that the advice and information in this book are believed to be true and accurate at the date of publication. Neither the publisher nor the authors or the editors give a warranty, express or implied, with respect to the material contained herein or for any errors or omissions that may have been made. The publisher remains neutral with regard to jurisdictional claims in published maps and institutional affiliations.

Printed on acid-free paper

This Springer imprint is published by Springer Nature
The registered company is Springer International Publishing AG
The registered company address is: Gewerbestrasse 11, 6330 Cham, Switzerland

Foreword

Of all the effects which climate change is likely to induce, perhaps none is more complex, insidious, and capable of inflicting direct damage on people's health than increasing levels of air pollution. It is important to remember that even in the absence of climate change, air pollution is an increasingly serious health concern. This is particularly true in urban areas. Although developed regions such as California have seen decades of progress in decreasing air pollution through mechanisms such as catalytic converters on automobiles and stricter restrictions on emissions of particulate pollutants from sources such as diesel engines, the Los Angeles region still exceeded the federal health standard for ozone during 85 days in 2016. The current air pollution problems in developing megacities such as Beijing, Delhi, and Mexico City remain more somber. However, to focus only on large cities provides an incomplete picture of the problem at hand. According to data from the World Health Organization, the Iranian city of Zabol, with a population of less than 150,000 people, has the world's worst concentrations of PM 2.5 pollution due to dust generated by the desiccation of surrounding wetlands. Taken together, it has been estimated that globally air pollution contributes to some seven million premature deaths each year.

How anticipated climatic changes over the twenty-first century will effect air pollution is clearly of critical concern. However, it is a problem of great complexity with much local and regional variation. In some instances, warmer temperatures may attenuate local pollution by weakening atmospheric inversions. However, in the case of many large cities such as Los Angeles, higher temperatures promote increased rates of photochemical smog production. Decreased humidity may lessen atmospheric mixing. In semiarid regions, the increasing subsidence associated with stationary high pressure systems both decreases the potential of vertical dispersion of atmospheric pollutants and promotes landscape desiccation and the production of PM through fires and dust. There will be no simple global predictor for the influence of climate change on air pollution, nor one simple solution. One important and hopeful fact to bear in mind though is that as many of the sources of local air pollution, such as fossil fuels, are also drivers of climate change, efforts to decrease air pollutants will often contribute to decreasing climate change and vice versa.

With these challenges in mind, this volume is particularly timely and welcome. With chapters that span in geographic coverage from Europe to Africa and Asia and from Australia to North America and the Caribbean, the book provides a broad coverage of many different environmental and climatic settings. The range of cities, rural areas, and developed versus developing socioeconomic settings that are considered by the various authors is impressive as are the types of pollutants and health effects – including emissions from wildfires. In terms of science, the complex nature of climate change and its likely impacts on air pollution require just this type of broad analysis to begin appreciating its variability and the multifaceted challenges of mitigation. However, it is important to remember that the solutions for decreasing the toll of climate change and associated changes in air pollution will not be enacted by scientists but by policy makers. In this regard, it is good to see both explicit treatments of important policy initiatives such as the Paris Climate Agreement and the fact that considerations of policy and regulatory issues are woven into many of the chapters. The threats to human health posed by climate change and air pollution over the twenty-first century are daunting. However, seeing a large group of researchers from different countries and disciplines come together to produce this important compendium on the problem as it now stands and what we might anticipate in the future gives hope. It is by such international team efforts, from large-scale political agreements, such as the Paris Agreement, to focused research products, such as this book, that this problem can be tackled.

Los Angeles, California, USA — Glen M. MacDonald

Acknowledgment

In the process of writing, editing, and preparing this book, there have been many people who have encouraged, helped, and supported us with their skills, thoughtful evaluation of chapters, and constructive criticisms.

First of all, we are indebted to all the contributors of chapters from both developed and developing countries for providing the scholarly and innovative scientific piece of research to make this book a reality. We are also thankful to the reviewers who carefully and timely reviewed the manuscripts.

We are also grateful to Prof. Glen McDonald of the University of California, Los Angeles, for writing the foreword, which adds greatly to the book with his thoughtful insights.

Rais Akhtar thanks his family, wife, Dr. Nilofar Izhar; daughter, Dr. Shirin Rais; and son-in-law, Dr. Wasim Ahmad, who encouraged and sustained him in developing the structure of the book and editing tasks, and he is deeply grateful for their support and indulgence.

Cosimo Palagiano thanks his family, his daughters, Paola and Francesca Romana, who morally sustained him in the work; he also thanks Daniele Priori for the maps' retouch and Gianfredi Pietrantoni, who controlled the final editing of his chapter.

Finally and most essentially, we are deeply obliged to Springer and the entire publishing team, without whose patience, immense competence, and support, this book would not have come to fruition. We specially thank Dr. Robert K. Doe whose energizing leadership ensured that this book would indeed translate to reality.

We are also thankful to Ms. Anjana Bhargavan and Mr. Krishna Pandurangan for their constant guidance and cooperation during the preparation and review process of the manuscript. We are also grateful to Professor A.R. Kidwai for his useful suggestions.

Aligarh, India — Rais Akhtar
Rome, Italy — Cosimo Palagiano

Contents

About the Editors

Rais Akhtar is presently adjunct professor of the International Institute of Health Management Research, New Delhi. Formerly, he was national fellow and emeritus scientist (CSIR) at CSRD, Jawaharlal Nehru University, New Delhi, and visiting professor at the Dept. of Geology, AMU, Aligarh. He has taught at Jawaharlal Nehru University, New Delhi; the University of Zambia, Lusaka; and the University of Kashmir, Srinagar. He is recipient of a number of international fellowships including Leverhulme fellowship (University of Liverpool), Henry Chapman fellowship (University of London), visiting fellowship (University of Sussex), Royal Society fellowship (University of Oxford) and visiting professorship (University of Paris 10). He was lead author (1999–2007) on the Intergovernmental Panel Climate Change, which is the joint winner of the Nobel Peace Prize for 2007. He is a recipient of Nobel Memento. Professor Akhtar has to his credit 94 research papers and 17 books published from India, the United Kingdom, the United States, Germany and the Netherlands. His latest book is entitled *Climate Change and Human Health Scenario in South and Southeast Asia*, published in 2016 by Springer. Professor Rais Akhtar is member of the Expert Group on Climate Change and Human Health of the Ministry of Health and Family Welfare, Govt. of India.

Cosimo Cosimo is emeritus professor in geography at the Department of Documentary, Linguistic-Philological and Geographical Sciences, Sapienza University of Rome; co-chair of the Joint IGU/ICA Commission on Toponymy; and corresponding member of the Accademia Nazionale dei Lincei, the Accademia dell'Arcadia and the Istituto di Studi Romani. His main research interests are medical geography, geography of nutrition, history of cartography and toponymy. He was director of the Institute of Geography of Faculty of Letters and Philosophy and of both the Department of Territorial and Urban Planning and the Department of Geography of Sapienza University of Rome. For many years he has been in the Steering Committee of the Società Geografica Italiana and of the IGU Commission on **Health and Environment.** He has been member of the scientific board of the journals *Geography, Environment, Sustainability* and *Espacio y Tiempo*. He is director of the journal *Geografia*.

Part I
Introductory

Chapter 1
Climate Change and Air Pollution: An Introduction

Rais Akhtar and Cosimo Palagiano

Abstract Concern about air pollution has been known for thousands of years. Complaints about its effects on human health and the built environment were first voiced by the citizens of ancient Athens and Rome. Urban air quality, however, worsened during the Industrial Revolution, as the widespread use of coal in factories in Britain, Germany, the United States and other nations ushered in an "age of smoke" (Mosley, 2014). As urban areas developed, pollution sources, such as chimneys and industrial processes, were concentrated, leading to visible and damaging pollution dominated by smoke. This introductory chapter discusses about the impact of climate change on the level air pollution, and at same time highlights that Weather and climate play important roles in determining patterns of air quality over multiple scales in time and space, owing to the fact that emissions, transport, dilution, chemical transformation, and eventual deposition of air pollutants all can be influenced by meteorological variables such as temperature, humidity, wind speed and direction, and mixing height.The chapter quoted empirical studies on air pollution and impact on human health in both from developed and developing countries.

Keywords CO2 emissions • Ecosystems • Kolkata • Donald Trump • BRICS • Forest fires

According to Joseph Alcamo and Jørgen E. Olesen (2012), first of all we have to define the gap between common perception of what we mean by "climate" and its more scientific definition. In practice, climatologists in the first part of the twentieth century decided to use and the need for invariance in the conditions from one period to another. This led to the definition of 30-year climate norms, which started with

---

R. Akhtar (✉)
International Institute of Health Management and Research (IIHMR), New Delhi, India
e-mail: raisakhtar@gmail.com

C. Palagiano
Dipartimento Di Scienze Documentarie, Linguistico-Filologiche e Geografiche, Sapienza University of Rome, Rome, Italy
e-mail: cosimo.palagiano@uniroma1.it

© Springer International Publishing AG 2018

R. Akhtar, C. Palagiano (eds.), *Climate Change and Air Pollution*,
Springer Climate, DOI 10.1007/978-3-319-61346-8_1

the period covering 1901–1930. The latest climate norm is the period from 1961 to 1990. This period is also sometimes called the "climate normal period". With changing climates, one can question the applicability of 30-year periods in defining climate. Air pollution is considered the world's worst environmental risk. Though poor air quality and climate change are very different phenomena, both are closely related. The main sources of CO2 emissions – the extraction and burning of fossil fuels – are not only key drivers of climate change but also major sources of air pollutants. Furthermore, many air pollutants that are harmful to human health and ecosystems also contribute to climate change. Thus, initiating actions to reduce the pollution from fossil fuel burning will go a long way in improving air quality and addressing climate change (Bell et al. 2007). This line of argument has been further elaborated by Jacob and Winner who emphasized that "air quality is strongly dependent on weather and is therefore sensitive to climate change". Recent studies have provided estimates of this climate effect through correlations of air quality with meteorological variables perturbation analyses in chemical transport models (CTMs) and CTM simulations driven by the general circulation model (GCM) simulation of the twenty-first-century climate change (Jacob and Winner 2009). Evidence from modelling studies suggests that climate is likely to increase concentration of ozone, one of the leading urban air pollutants responsible for respiratory problems (Kris and McGregor 2008).

Having said that, it should have been stressed that "weather and climate play important roles in determining patterns of air quality over multiple scales in time and space, owing to the fact that emissions, transport, dilution, chemical transformation, and eventual deposition of air pollutants all can be influenced by meteorological variables such as temperature, humidity, wind speed and direction, and mixing height. There is growing recognition that development of optimal control strategies for key pollutants like ozone and fine particles now requires assessment of potential future climate conditions and their influence on the attainment of air quality objectives. In addition, other air contaminants of relevance to human health, including smoke from wildfires and airborne pollens and moulds, may be influenced by climate change" (Kinney 2008). In the study by Kinney, the focus was on the ways in which human health-relevant measures of air quality, including ozone, particulate matter, and aeroallergens, may be influenced by climate variability and change.

It is true. The major effect of the greenhouse effect is the sudden alternation of weather. The variability is a characteristic of the Mediterranean climate, but, during the last decades, such variability is more marked. Rainfall intensity and alternating high and low temperatures have strong impact on respiratory diseases, like influenza and pneumonia, which are very dangerous to the elder population. In the developed countries, the old people comprise the majority of the affected population. Today there is an increase in admission cost to hospitals than in the past. In addition the weather instability increases the number and the dangerousness of viruses and parasites responsible for various diseases.

Focusing on the impacts of climate change on air pollution, particularly ozone pollution, the Intergovernmental Panel on Climate Change (IPCC) has also clearly stressed that "pollen, smoke and ozone levels likely to increase in warming world, affecting health of residents in major cities. Rising temperatures will worsen air quality through a combination of more ozone in cities, bigger wild fires and worse pollen outbreaks, according to a major UN climate report. It is formed by the reaction with sunlight (photochemical reaction) of pollutants such as nitrogen oxides (NO2)" (Wynn 2014). Frequent forest fires in certain regions in Australia and in the state of California are examples of such events. The World Meteorological Organization (WMO) has now certified that 2016 was the warmest year.

With reference to human health implications, air pollution is currently the leading environmental cause of premature deaths. The findings of the World Health Organization (WHO) contend that air pollution is the world's biggest environmental health risk, killing 7 million people in 2012 (in comparison to 4 million deaths due to malaria and 3.1 million deaths of children under 5 due to malnutrition). Deteriorating air quality will mostly affect the elderly, children, people with chronic illness, and expectant mothers. Another report suggests that more than 5.5 million people die prematurely each year due to air pollution, with over half of those deaths occurring in China and India (Indian Express, Feb.13, 2016). Scientists have urged that in the face of future climate change, stronger emission controls are enforced to avoid worsening air pollution and the associated exacerbation of health problems, especially in more populated regions including megalopolises of the world encompassing both developing and developed countries. The American Lung Association's "State of the Air" report indicates that 166 million Americans are living in an environment with unhealthy ozone or particle pollution which induces health risks (Milman 2016, American Lung Association 2016). Another research highlights that "while the number of unhealthy polluted days has dropped in the past year, more than half of US population lives in areas with potentially dangerous air pollution, and about six out of 10 of the top cities for air pollution in the USA are located in the state of California" (McHugh 2016). Brazil, Russia, India, China, and South Africa (BRICS) have been drawing special attention due to the pollution emissions released into the atmosphere by their increasing number of industries and their exaggerated consumption of products (Cherni 2002).

In China alone, 1.2 million people die every year due to pollution. The estimated cost of environmental degradation in China is 9% of its gross domestic product (GDP), while it is 5.7% of its GDP for India (Zang 2015).

Another study by the researchers at the University of British Columbia in Canada revealed that about 1.4 million people in the South Asian nation and 1.6 million in its northern neighbour died of illnesses related to air pollution in 2013. The Indian and Chinese fatalities accounted for 55% of such deaths worldwide, the study said (Bhattacharya 2016).

This scenario has also been substantiated by the recently published *State of Global Air 2017* report. The report asserts that 92% of the world's population lives in areas with unhealthy air, and China and India together were responsible for over half of the total global attributable deaths. The study estimates that globally 2.7–3.4 million preterm births may be associated with PM2.5 exposure and South Asia is the worst hit, accounting for 1.6 million preterm births (Health Effects Institute 2017).

Referring to Africa, John Vidal asserts that air pollution is more deadly than malnutrition or dirty water. Vidal further elaborates that:

> "Africa's air pollution is causing more premature deaths than unsafe water or childhood malnutrition, and could develop into a health and climate crisis reminiscent of those seen in China and India. Governments in African countries are failing to address the links between air pollution and global warming. While most major environmental hazards have been improving with development gains and industrialisation, outdoor (or 'ambient particulate') air pollution from traffic, power generation and industries is increasing rapidly, especially in fast-developing countries such as Egypt, South Africa, Ethiopia and Nigeria" (Vidal 2016).

At the Paris Climate Conference in 2015, world leaders were urged to cut air pollution to save lives in poor countries. During the Paris climate summit, the World Health Organization said that tackling air pollution and global warming in tandem will reduce mortality in developing countries. However, even developed countries like Australia and California (USA) are not safe when rising temperature caused forest fires. A study published in the journal *Environmental Health Perspectives* in 2012 calculated that exposure to smoke from wildfires was already responsible for 339,000 premature deaths annually (Johnston 2012). Health impacts of wildfire occurrences have also been predicted in another review paper published in the *Environmental Health Perspectives*. The authors of the paper assert that wildfires are likely to increase in many parts of the world due to changes in temperature and precipitation patterns from global climate change. Wildfire smoke contains numerous hazardous air pollutants, and many studies have documented population health effects from this exposure (Reid et al. 2016). The air we breathe outdoors could be harming more people than ever, a new study suggests. Globally, more than 3 million people die prematurely each year from prolonged exposure to air pollution, according to the World Health Organization. By 2050, it could be 6.6 million premature deaths every year worldwide, a new study predicts. Chronic exposure to air pollution particles contributes to the risk of developing cardiovascular and respiratory diseases as well as lung cancer, WHO said. "The total number of deaths due to HIV and malaria is 2.8 million per year", said Jos Lelieveld, a professor at the Max Planck Institute for Chemistry in Germany and lead author of the study. "That's half a million less than the number of people who die from air pollution globally" (Ansari 2015). Residential energy emissions or domestic air pollution from fuels used for cooking and heating, especially in India and China, had the largest impact on deaths worldwide. In another 10 years, Delhi will record the world's largest number of premature deaths due to air pollution among all mega cities in the world. By 2025, nearly 32,000 people in Delhi will die solely due to inhaling polluted air. However, it will be another Indian city, Kolkata, that will record the highest number of such deaths by 2050 and Delhi will record the world's largest number of premature deaths due to air pollution (Sinha 2015).

The problems of climate change are not well considered by some people and governments. For example, US President Donald Trump does not believe in the damages caused by climate change to the environment and sadly reduced the

Environmental Protection Agency's current funding by more than 31%. President Trump announced on June 1, 2017, that he is withdrawing the United States from the landmark Paris climate agreement, an extraordinary move that puzzled America's allies and placed great hindrance in the global effort to address the warming planet.

References

Alcamo, Joseph and Olesen, Jorgen (2012) Life in Europe Under Climate Change, Wiley-Blackwell

American Lung Association (2016) State of the Air, 2016. www.lung.org

Ansari A (2015) Study: more than six million could die early from air pollution every year. CNN news, September 16

Bell ML, Goldberg R, Hogrefe C, Kinney PL, Knowlton K, Lynn B, Rosenthal J, Rosenzweig C, Patz JA (2007) Climate change, ambient ozone, and health in 50 US cities. Clim Chang 82:61–76

Bhattacharya S (2016) India and China have most deaths from pollution. Wall Street J, February 2016

Cherni JA (2002) Economic growth versus the environment: the politics of wealth, health and air pollution. Palgrave

Health Effects Institute (2017) First annual *state of global air* report, February 14, Boston

Jacob DJ, Winner DA (2009) Effect of climate change on air quality. Atmos Environ 43:51–63

Johnston FH (2012) Estimated global mortality attributable to smoke from landscape fires. Environ Health Perspect 120:695–701

Kinney PL (2008) Climate change, air quality and human health. Am J Pre Med 35(5):459–446

Kris KL, McGregor G (2008) Climate change, tropospheric ozone and particulate matter, and health impacts. Environ Health Perspect 116(11):1449–1456

McHugh J (2016) US air pollution. Worst cities for clean air are in California: report says, April 20. www.ibtimes.com/us-air-pollution-worst-cities-clean-air-are-california-report-says-23

Milman O (2016) More than half US population lives amid dangerous air pollution, report warns. The Guardian, April 20

Reid CE et al (2016) Critical review of health impacts of wildfires smoke exposure. Environ Health Perspect 124(9):1334–1343

Sinha K (2015) Delhi will record world's largest number of premature deaths due to air pollution. Times of India, September 17

Vidal J (2016) Air pollution more deadly in Africa than malnutrition or dirty water, study warns. The Guardian, London, October 20

Wynn G (2014) Climate change will hike air pollution deaths says UN study. Climate Home, March 28

Zang Q (2015) How much is pollution is costing China's economy? Asia, DW.com, May 18

Rais Akhtar is presently adjunct professor of the International Institute of Health Management Research, New Delhi.

Cosimo Palagiano is emeritus professor in geography at the Department of Documentary, Linguistic-Philological and Geographical Sciences, Sapienza University of Rome.

Chapter 2
Air Quality in Changing Climate: Implications for Health Impacts

Sourangsu Chowdhury and Sagnik Dey

Abstract Poor air quality is a leading risk factor for global disease. Two major pollutants – fine particulate matter ($PM_{2.5}$) and surface ozone – are also linked to climate change. A unified framework to quantify the morbidity and mortality burden from air pollution exposure was developed in Global Burden of Disease Study. 1500 and 2200 premature deaths from ozone and ambient PM2.5 exposure can be attributed to past climate change (from pre-industrial era to present day). For the future, air pollution exposure can be quantified by four Representative Concentration Pathways (RCPs) emission scenarios in a modelling framework. In addition to the role of climate change in modulating air quality in future, the changes in socio-economic and demographic condition of the future population are also expected to determine the burden due to air pollution. These may be quantified using the demographic and socioeconomic drivers used in formulating the Shared Socio-economic Pathways (SSP) scenarios. Combining the SSP and RCP scenarios in a scenario matrix framework would lead to the estimate of premature mortality burden for the future within an uncertainty range that can drive the policymakers to exercise adequate mitigation measures, which are expected to facilitate a healthier and climate secure society in future.

Keywords PM2.5 exposure • Ozone exposure • Changing climate • Premature mortality burden • RCP scenarios • SSP scenarios

Air Quality, Exposure and Health Impacts

Chronic exposure to $PM_{2.5}$ and ozone leads to cardiovascular and cardiopulmonary diseases and lung cancer and eventually premature death of millions of people worldwide (Cesaroni et al. 2014; Krewski et al. 2009; Pope et al. 2002; Chen et al. 2008). Some studies have depicted evidence of premature mortality due to diseases like neurological disorders and diabetes from exposure to ambient $PM_{2.5}$ (Gouveia

S. Chowdhury (✉) • S. Dey
Centre for Atmospheric Sciences, Indian Institute of Technology Delhi, Hauz Khas, New Delhi, India
e-mail: sourangsu@cas.iitd.ac.in; sagnik@cas.iitd.ac.in

© Springer International Publishing AG 2018

R. Akhtar, C. Palagiano (eds.), *Climate Change and Air Pollution*, Springer Climate, DOI 10.1007/978-3-319-61346-8_2

and Fletcher 2000; Bell et al. 2004). Though the problem is fast growing in the developing world (West et al. 2016), health impacts of air pollution have been documented in the developed countries even at very low air pollution exposure (Shi et al. 2015). $PM_{2.5}$ is emitted from various natural and anthropogenic sources and its spatio-temporal variation is modulated by meteorology and topography. Global burden of disease (GBD) effort (Lim et al. 2012; Murray 2015) establishes a unified framework to quantify the morbidity and mortality burden of air pollution globally. Studies showing evidence of mortality and morbidity due to diseases like chronic obstructive pulmonary diseases (COPD), ischemic heart diseases (IHD), stroke, lung cancer, diabetes and acute lower respiratory infection from $PM_{2.5}$ exposure are mostly limited to the developed countries. To address this issue, an integrated exposure-response (IER) function (Burnett et al. 2014) was developed for risk estimation by incorporating exposure spanning across ambient air pollution, household air pollution, passive smoking and active smoking (Burnett et al. 2014). This risk function enabled comparative assessment of the burden of diseases from air pollution across the world (Arnold 2014).

Exposure to ozone primarily affects the lungs causing short-term changes in lung function and escalates respiratory syndromes (Bell et al. 2004, 2005). Chronic long-term exposure to ozone may result in permanent impairment of the lungs, damage of the tissues lining the airways and development of pulmonary fibrosis (Lin et al. 2008; Jerrett et al. 2009; Li et al. 2016). Tropospheric ozone exposure not only results in impairment of human health but also damages vegetation with substantial reduction in crop yield and crop quality (Morgan et al. 2006; Avnery et al. 2011). In India wheat production is impacted the most due to exposure to ozone with an estimated loss of 3.5 ± 0.8 million tons followed by rice and other cereals (Ghude et al. 2014). On national scale, the yield loss due to ozone exposure is about 9.2% of the cereals required every year under the provisions of the recently implemented National Food Security Bill (2013) by the Government of India. Climate change can further exacerbate the current situation as it has been projected that ozone exposure will increase in the future (Horowitz 2006). This may lead to food shortage, which in turn can cause malnourishment impacting the health indirectly. A study by Jerrett et al. (2009) followed up 448,850 subjects as a part of the American Cancer Society Cancer Prevention Study II for 18 years and found that the relative risk (which may be defined as the ratio of probability of an event occurring in an exposed group to the probability of an event occurring in comparison with nonexposed group) of death from exposure to ground-level ozone due to respiratory causes with a 10 ppb increase in ozone concentration was 1.040 (95% CI 1.010–1.067). A global study (Anenberg et al. 2010) estimated that about 0.7 ± 0.3 million premature death/year can be attributed globally to ozone exposure. Another estimate (Silva et al. 2013) used ACCIMIP model simulations to determine exposure to ozone, the mortality attributed to exposure to ozone for past climate change (1850 to present day) was estimated to be around 1500 (−20,000 – 27,000) deaths/year. An India-based study (Ghude et al. 2016) has used a chemical transport model to estimate the exposure, and the resulting premature death due to chronic obstructive pulmonary diseases was estimated to be ~12,000 using the 2011 census data for the exposed population. Premature mortality (Mort) is generally estimated as a function of exposed population (*Pop*), relative risk (RR) and the background mortality (BM) rate (Anenberg et al. 2010; Murray 2015; Chowdhury and Dey 2016; Silva et al. 2016) and can be expressed as in Eq. 2.1. RR can be estimated as a function of exposure to pollutants (Pope et al. 2002; Burnett et al. 2014).

$$\mathrm{Mort} = \mathrm{Pop} \times \mathrm{BM} \times \frac{\mathrm{RR} - 1}{\mathrm{RR}} \tag{2.1}$$

Climate Change and Air Pollution

Since the Industrial Revolution, human activities have released huge amounts of carbon dioxide and other greenhouse gases (GHG) into the atmosphere, primarily from fossil fuel burning, to meet the energy demand of the growing population and industrial needs. Other activities like agricultural waste and solid fuel burning also contribute to climate-warming pollutants. Black carbon aerosol that is mostly emitted from incomplete combustion of fossil fuel, biofuel and biomass warms the atmosphere, which in turn influences the global and regional wind patterns, humidity and precipitation. Black carbon is also a major component of ambient $PM_{2.5.}$ Therefore, reducing black carbon has co-benefits to limit climate change and avert premature mortality burden. Changing meteorology under warming climate is expected to play an important role in modulating $PM_{2.5}$ by controlling its dispersion and life cycle due to changes in boundary layer depth, wind circulation pattern, precipitation frequency, relative humidity and temperature. Globally the climate is expected to become more stagnant in the future with weaker global circulation and decreasing frequency of mid-latitude cyclones (Daniel and Winner 2009). With increasing stagnation, the pollutants are expected to get piled near the surface thereby increasing the relative exposure. Increased humidity in the future can tend to influence local air quality at individual scale by diminishing ambient bio-aerosols (pollens, grains, spores and other aero-allergens) as they tend to clump together and become less respirable. Changes in precipitation pattern may also affect the aerosol scavenging. Wind speed and precipitation are projected to increase over India (Christensen et al. 2007; Menon et al. 2013) in the future under the warming climate. Although not much information is available about the projected mixing layer depth over India, it is expected that increasing temperature and wind speed will contribute towards expanding the mixing layer depth. It is implied that these projected meteorological factors in the future will contribute to escalated washout and ventilation. Thus we may expect that meteorology will partially help in reducing $PM_{2.5}$ exposure irrespective of the projected exposure strength in future.

Ozone is a secondary air pollutant formed in the atmosphere by photochemical processes in the presence of precursors like oxides of nitrogen (NO_x) and volatile organic compounds (VOC) which are mostly emitted by mobile vehicular sources (cars, trucks, etc.), industrial sources and natural sources like lightening, forest and grassland fires. In urban areas, power plants, industries, chemical solvents and vehicular emissions are the primary sources of the ozone precursors. In presence of sunlight, these precursors undergo chemical transformation to form ozone. The chemistry of ozone formation is temperature dependent and occurs in multiple number of steps. Methane (emitted primarily due to fossil fuel use, biomass burning, livestock farming, landfills and waste) which is one of the major components responsible for global warming is also one of the major components of VOC, but in urban settings, the non-methane volatile organic compounds (NMVOC) emitted generally outpace methane as the major component of VOC responsible for ozone formation. West et al. (2006) shows that reducing global anthropogenic methane emissions by 20% will avert around 30,000 premature deaths in 2030, and the cost-effectiveness of methane reduction is expected to be around $420,000 per avoided mortality (West et al. 2006). Thus it can be argued that mitigating methane emission can help to improve air quality globally bringing multiple benefits for air quality, climate, public health, agriculture and energy. With temperature projected to increase globally in the future (Daniel and Winner 2009), the ozone concentration is expected to escalate (Kinney 2008).

From Pre-industrial Era to Present Day

Since the pre-industrial era, human activities led to degradation of air quality across the globe. Measurements at various sites across the northern hemisphere indicate that surface ozone has increased by about fourfolds from 1860s to 2000s (Marenco et al. 1998). The change of surface concentration and exposure to $PM_{2.5}$ and ozone from the pre-industrial period to present can be attributed to multiple factors (Fang et al. 2013) – (a) changes in direct emissions of their constituents and precursors, (b) climate change induced changes in surface emissions, (c) the influence of increasing CH_4 concentration on tropospheric chemistry and (d) changes and transition in demographical features. Fang et al. 2013 have reported that global population-weighted $PM_{2.5}$ and O_3 have increased by about 8 ± 0.16 μg/m^3 and 30 ± 0.16 ppb, respectively, from the period 1860 to 2000 utilizing the Geophysical Fluid Dynamics Laboratory Atmospheric Model, version 3. Another study by Silva et al. (2013) used Atmospheric Chemistry and Climate Model Intercomparison Project (ACCMIP) group of models to conclude that global population-weighted $PM_{2.5}$ and ozone exposure increased by about 7.3 μg/m^3 and 26.5 ppb, respectively. Global mean concentrations of $PM_{2.5}$ and ozone in 1850 were estimated to be 11.4 μg/m^3 and 28 ppb, respectively, while the corresponding values in 2000 changed to 18.6 μg/m^3 and 54.5 ppb, respectively. Over the Indian landmass, mean concentrations of $PM_{2.5}$ and ozone increased from 14.3 μg/m^3 and 33.2 ppb, respectively, in 1850 to 22 μg/m^3 and 61.9 ppb, respectively, in 2000. The exposure over India and South Asia is generally underestimated by the global

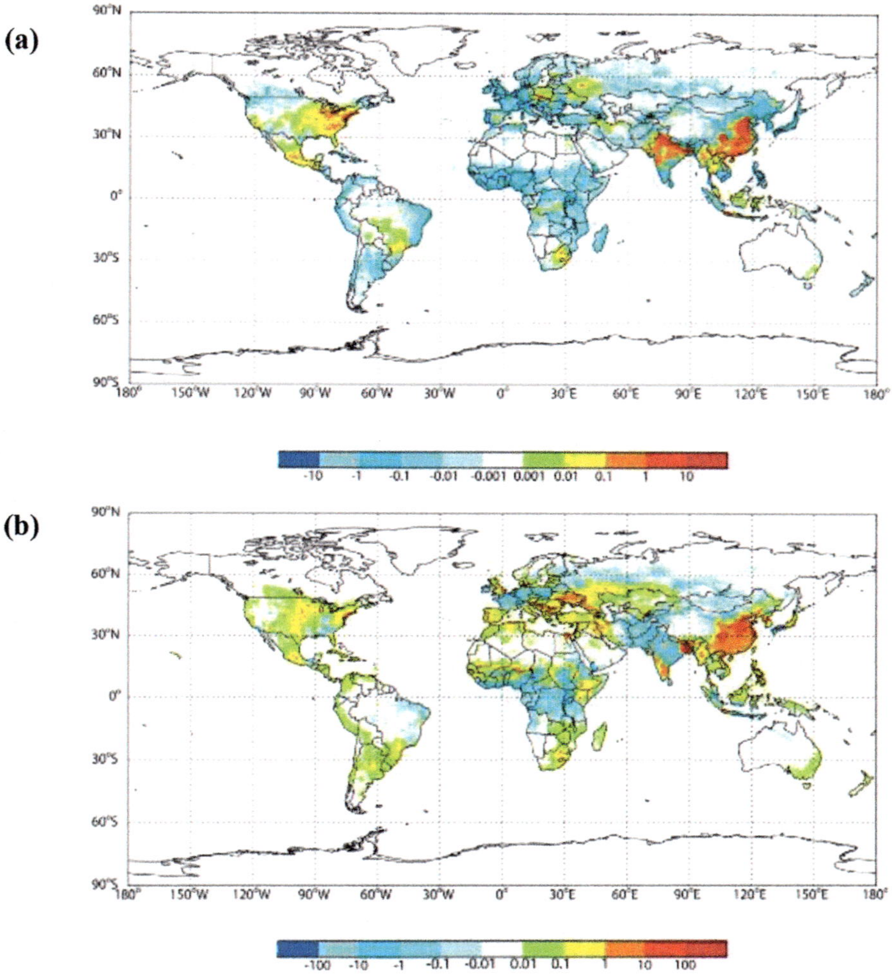

Fig. 2.1 Premature mortality attributed to past climate change in death/year (1000 km^2)$^{-1}$ for (**a**) ozone exposure (respiratory mortality) and (**b**) $PM_{2.5}$ exposure (cardiopulmonary diseases and lung cancer mortality (Adopted from Silva et al. 2013)

models (Pan et al. 2015). Figure 2.1 depicts the premature mortality that can be attributed to past climate change (1850–2000) due to $PM_{2.5}$ (a) and ozone (b) exposure, respectively. Past climate change (from pre-industrial era to present day) was estimated to cause ~1500 and 2200 premature deaths per year from ozone and $PM_{2.5}$ exposure, respectively.

Future Projections

Future air quality will be influenced by changes in emission and meteorology in warming climate. Changes in anthropogenic emissions are expected to dominate in the near future (Kirtman et al. 2013) and depend on various socio-economic factors such as demographic transition, economic growth, energy demand, technological choices, land use changes and implementation of policies regarding climate and air quality. Following a meeting on 8th September, 2008, a global initiative was taken for climate modelling under the fifth phase of the Coupled Climate Model Intercomparison Project (CMIP5) to enhance the earlier activities by incorporating 20 climate modelling groups from around the world (Taylor et al. 2012). These model simulations were targeted to focus on major gaps in understanding past and future climate changes described by Moss et al. (2010). The RCPs (summarized in Table 2.1) unlike the Special Report on Emission Scenarios (SRES) used for the earlier CMIP3 simulations include policy interventions and are built based on a range of projections of future socio-economic factors. These scenarios assume that certain policy actions will be taken to achieve certain emission targets. The labels of the four RCPs (RCP2.6, RCP4.5, RCP6 and RCP8.5) indicate a rough estimate of the radiative forcing at the end of the twenty-first century. Apart from the CMIP5 models, various chemical transport models (CTMs) like GEOS-Chem, WRF Chem and CMAQ models can also be utilized to estimate the concentration of the criterion pollutants. The CMIP5 and CTM model simulations for the projected concentration of various $PM_{2.5}$ components (viz. dust, black carbon, primary organic aerosols, secondary organic aerosols, sea salt and sulphate), VOC and NO_x in the future require data on emissions from RCP scenarios. The projected emissions of VOC and black carbon (among other constituents) over Asia are depicted in Fig. 2.2. One study (Silva et al. 2016) projected that global population-weighted ozone concentration in 2030 will change from present-day concentration by −2.5 to 15.2 ppb in

Table 2.1 Details about RCP scenarios

Scenario	Radiative forcing	Concentration (ppm)	Pathway	Model providing RCP
RCP8.5	>8.5 W/m^2 in 2100	>1370 CO_2 equiv. in 2100	Rising	MESSAGE
RCP6.0	~6 W/m^2 at stabilization after 2100	~850 CO_2 equiv. (at stabilization after 2100)	Stabilization without overshoot	AIM
RCP4.5	~4.5 W/m^2 at stabilization after 2100	~650 CO_2 equiv. (at stabilization after 2100)	Stabilization without overshoot	GCAM
RCP2.6	Peak at 3 W/m^2 before 2100 and the declines	Peak at ~490 CO_2 equiv. before 2100 and then declines	Peak and decline	IMAGE

The table is adopted from Moss et al. (2010)

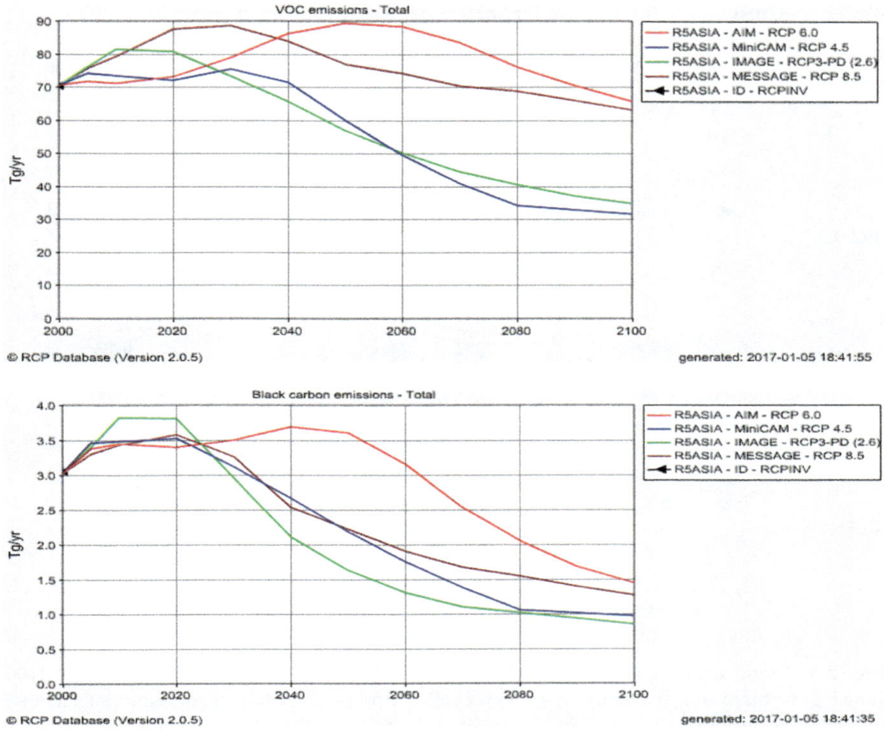

Fig. 2.2 Shows the emission of VOC (**a**) and black carbon (**b**) in future over Asia as projected by the RCP scenarios. These emissions go into the CMIP5 model simulations to determine the concentration of the pollutants in future decades. The figures are generated from the RCP scenario database hosted by IIASA

2030 across all the RCP scenarios using data from 14 ACCMIP models, while the change is projected to range from −11.7 to 13.6 ppb in 2100. They project an overall decrease of global population-weighted $PM_{2.5}$ exposure by 2100 ranging from −0.4 to −5.7 μg/m^3 across six ACCMIP models for all the RCP scenarios.

Projected Exposure to Ground-Level Ozone and Ambient $PM_{2.5}$

Very few studies have attempted to estimate the future exposure to ozone and $PM_{2.5}$. A recently published study (Madaniyazi et al. 2015) recognized the urgency to project premature mortality due to exposure to air pollutants in developing countries to facilitate implementation of policies. They also suggested that multi-model ensembles should be used to project the exposure to the air pollutants and

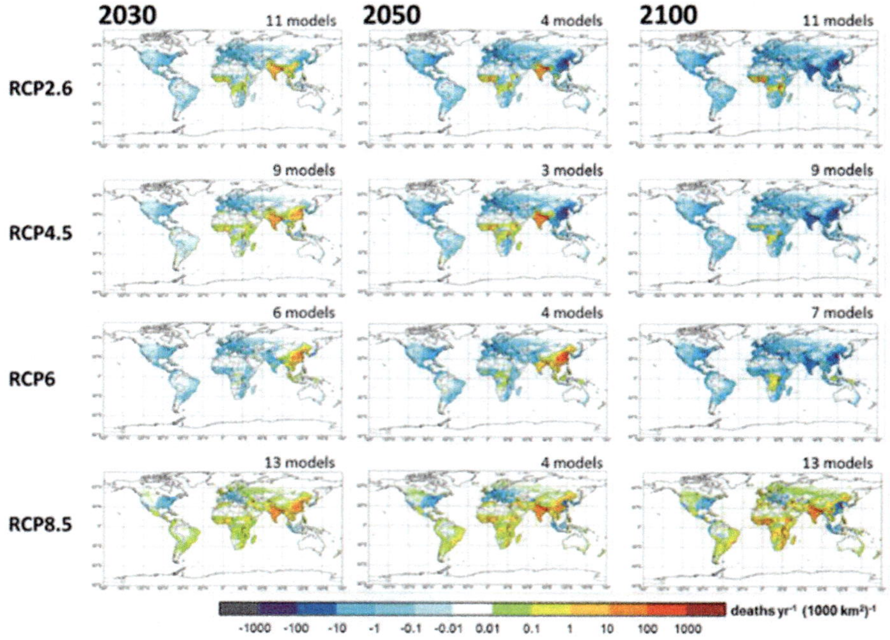

Fig. 2.3 Future ozone respiratory mortality for all RCP scenarios in 2030, 2050 and 2100, showing the multi-model average in each grid cell, for future air pollutant concentrations relative to 2000 concentrations (Adopted from Silva et al. 2016)

related excess premature mortality to better quantify the related uncertainty. Another study (Selin et al. 2009) used GEOS-Chem model to simulate future $PM_{2.5}$ exposure. Global ozone exposure is expected to increase by 6.1 ppb at the end of 2050, whereas the increase is expected to be about 24.4 ppb over India. They estimated that around 812,000 excess premature deaths per year can be attributed to exposure to ozone in 2050 as compared to 2000 due to changes in emission and climate. Recently, Silva et al. (2016) used an ensemble of ACCMIP models and projected global mortality burden due to ozone exposure to increase markedly from 382,000 (121,000–728,000) in 2000 to between 1.09 and 2.36 million deaths/year across all four RCPs in 2100. Figure 2.3 shows the projected premature mortality for three future decades. This study also identifies that change in premature ozone-related respiratory mortality/year in India in 2100 with respect to 2000 is projected to range from −230,000 to 292,000 across all the RCP scenarios.

The most unsettled issue regarding projection of aerosol concentration and $PM_{2.5}$ in future is whether $PM_{2.5}$ is expected to decrease or increase in future. Allen et al. (2016) projected that aerosol concentration is expected to increase in future, whereas Silva et al. (2016) projected that $PM_{2.5}$ exposure is expected to decrease relative to present-day exposure by the end of the century. Tagaris et al. (2009) used CMAQ modelling system to estimate 4000 premature mortality/year from $PM_{2.5}$ exposure in the USA due to climate change by the end of 2050. Tainio

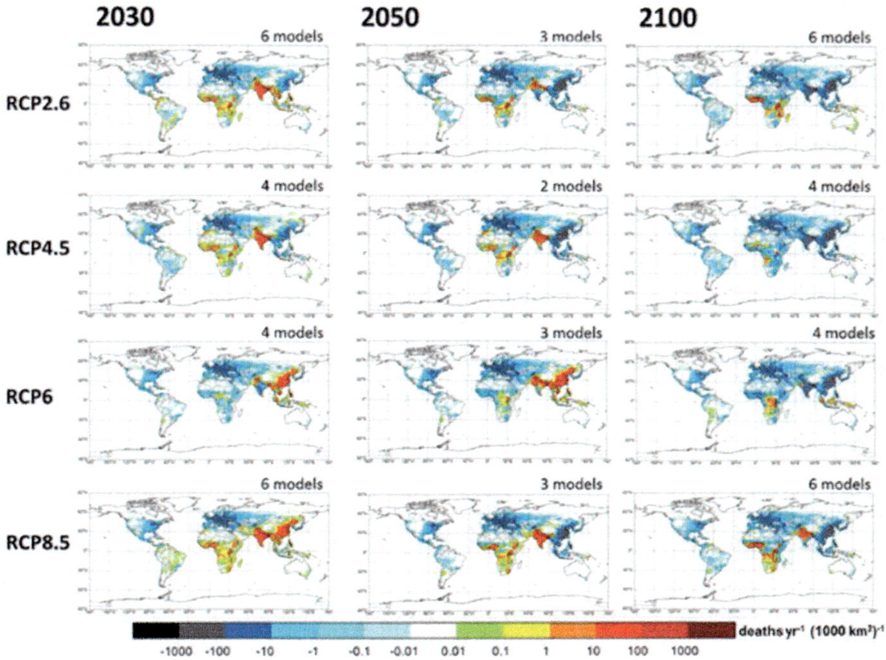

Fig. 2.4 Future mortality due to exposure to $PM_{2.5}$ for all RCP scenarios in 2030, 2050 and 2100, showing the multi-model average in each grid cell, for future air pollutant concentrations relative to 2000 concentrations (Adopted from Silva et al. 2016)

et al. (2013) used coupled RegCM and CAMx CTM to project a decrease in $PM_{2.5}$ exposure in future over Poland. Nawahda and Yamashita, (2012) projected $PM_{2.5}$ exposure to increase in future over East Asia using CMAQ modelling system which can be attributed to about 1,035,000 premature mortality by the end of 2020. Silva et al. (2016) projected that discounted exposure to $PM_{2.5}$ by the end of the century is expected to avert 1,310,000–1,930,000 premature mortality/year with respect to the estimated premature mortality using 2000 $PM_{2.5}$ exposure. Figure 2.4 depicts the projected premature mortality for three future decades (2030, 2050 and 2100).

Co-benefits of Reducing Air Pollutants

It is well established that human activities affect climate change, and as a consequence they are affected by climate change impacts (Smith et al. 2014). The focus to mitigate the concentration of warming climate-altering pollutants also holds the potential to benefit human health significantly. These co-benefits include health gains from strategies directed primarily at mitigation of climate change from policies implicated for health benefits (Haines et al. 2007; Smith and Balakrishnan 2009). In a nutshell, co-benefits are positive impacts on human health that arise from interventions to reduce the emission of climate-altering air pollutants. Co-benefits can be achieved in many ways (Smith et al. (2014) and references therein). For example, the reduction of co-pollutants from household solid fuel combustion will result in reduced exposure to air pollutants that are associated with diseases like chronic and acute respiratory illnesses, lung cancer, low birth weights and still births and tuberculosis. On the other hand, controlling household combustion of solid fuels will reduce emission of black carbon, CH_4, CO and other climate-altering air pollutants. Reduction in CH_4 and CO emission will also restrict the formation of tropospheric ozone. Cutting down the emission of health damaging co-pollutant from industries will reduce outdoor exposure to ambient air pollution and hence has the potential to avert large premature mortality. The benefits for climate include reduction in emission of climate-altering air pollutants like black carbon, CO and CH_4. Increased energy efficiency will reduce fuel demands and hence reduce emissions of climate-altering air pollutants. Health benefits of increased urban green space include reduced temperature and heat island effect, physiological benefits and better self-perceived health status. It also helps in partially reducing atmospheric CO_2 via carbon sequestration in plant tissues and soil. Increased urban greeneries will also facilitate deposition of climate-altering air pollutants emitted from various vehicular and industrial sources.

Few studies quantify the health and climate benefits of reducing climate-altering air pollutants. A study in India found that the benefits of hypothetically reducing solid fuel combustion in households by introducing clean cook stoves would help to avert about 2 million premature death and 55 million DALYs over the period of 10 years and reduction of 0.5–1 billion tons of CO_2 equivalent (Wilkinson et al. 2009). A study (Markandya et al. 2009) assessed the changes in emission of $PM_{2.5}$ and subsequent effects on human health that could result from climate change mitigation aimed to halve the GHG emission by 2050 from the electricity generation sector of India, China and European Union. In all these three regions, changes in modes of production of electricity to reduce CO_2 emission were associated with reduction in $PM_{2.5}$-related premature mortality.

Socio-economic Projection for Vulnerability Assessment

Certain group of population is more vulnerable and susceptible to air pollution than the others, like children, people with pre-existing heart and lung diseases, people with diabetes, outdoor workers and aged people (Balbus and Malina 2009; Makri and Stilianakis 2008). Socio-economic factors also influence the susceptibility towards air pollution exposure in terms of disproportionate exposure, coping capacities and access to health care (Makri and Stilianakis 2008). The most vulnerable population are the homeless with six times more odds to be morbid or die due to lungs or respiratory infections, asthma and cardiovascular and pulmonary diseases.

To assess the relationships between socio-economic development in response to climate change, the Integrated Assessment Modelling (IAM) and the Impacts, Adaptation and Vulnerability (IAV) group launched a set of scenarios that describe the future in terms of social and economic mitigation and adaptation challenges known as the Shared Socio-economic Pathways (SSP) (O'Neill et al. 2012; Ebi et al. 2014). This set of scenarios provides projections by age, sex and six levels of education for all the countries. The five SSP scenarios are a green growth strategy (SSP1), a middle of the road development pattern (SSP2), a fragmentation between the regions (SSP3), an increase in inequality across and within regions (SSP4) and a fossil fuel-based economic development (SSP5). To encompass a wide range of possible development pathways, the SSP are defined in terms of socio-economic challenges to mitigation and to adaptation (Fig. 2.5a). Figure 2.6 shows the

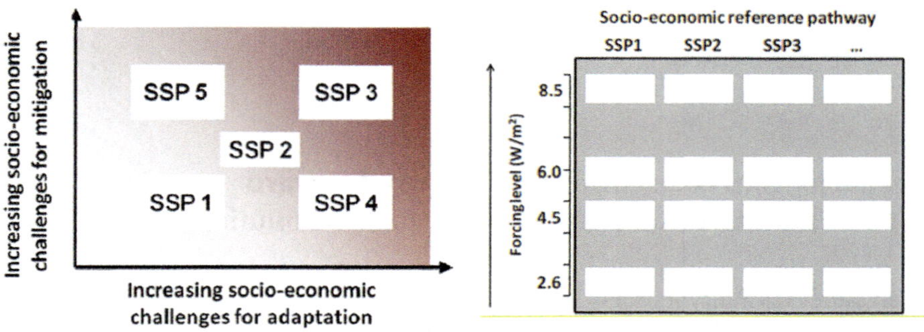

Fig. 2.5 (*Left*) The scenario space spanned by the SSP scenarios and (*right*) the scenario matrix architecture (Both figures are adapted from IPCC 2010)

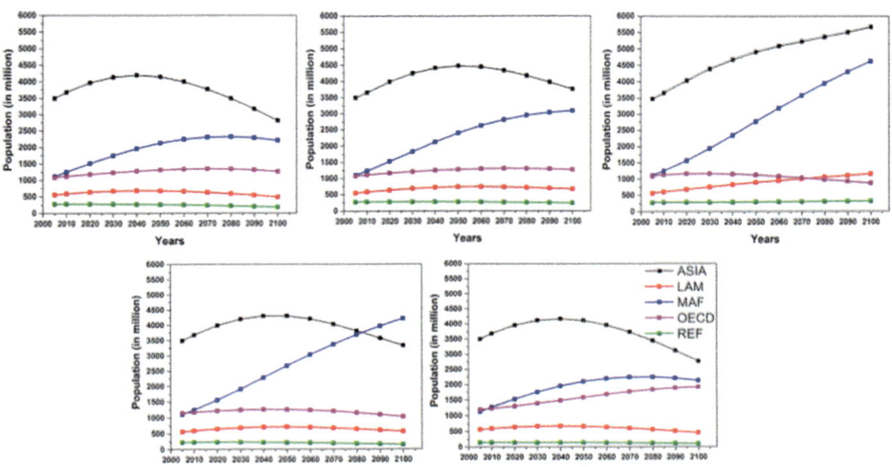

Fig. 2.6 Projected population used in developing the 5 SSPs'(numbered chronologically from a to b) for five world regions, namely, Asia, Latin American (LAM) countries, Middle East and Africa (MAF), OECD (OECD group of countries) and reforming economies (REF)

projected population used as a driver for developing each of the five SSP scenarios for five broad world regions, namely, Asia, Latin American (LAM) countries, Middle East and Africa (MAF), OECD (OECD group of countries) and reforming economies (REF).

SSP1 (Vuuren et al. 2016) considers the world to make definite progress towards sustainability, achieve development goals by cutting off resource intensity and dependency on fossil fuels. In SSP2 (Fricko et al. 2016) trends typical to recent decades are projected to continue, with some progress towards achieving development goals, historic reductions in resource and energy and slowly decreasing fossil fuel dependency. In SSP3 (Fujimori et al. 2016) scenario, the pathway assumed is opposite to sustainability which describes a world with stalled demographic transition. The SSP4 (Calvin et al. 2016) scenario predicts a very unequal world both within and across the countries, and the SSP5 scenario (Kriegler et al. 2016) envisions a world that stresses conventional development oriented towards economic growth.

Scenario Matrix Architecture: A Way Forward for Estimating Future Burden Due to Air Pollution

Premature mortality burden from air pollution depends on concentration of pollutant, exposed population and background mortality rate. The air pollutant concentration in the future can be projected by climate models following RCP emission scenarios, while the socio-economic factors like population and background mortality rate can be quantified following SSP scenarios. Therefore, the burden of disease in the future is expected to be quantified by RCP-SSP scenario matrix (4×5). This scenario matrix architecture (Fig. 2.5b) can be used in different ways for scientific and policy analyses. For example, analysts can compare consequences under the same climate scenario (RCP driven) across all socio-economic scenarios ("what is the effect of future socio-economic conditions on the impacts of a given climate change"). An assessment of the effect of future socio-economic conditions on the effectiveness and costs of a suite of mitigation and adaptation measures to combat climate change would allow comparing the differences between the SSP scenarios across a single RCP scenario (IPCC 2010; O'Neill et al. 2012). For example, population distribution from a particular SSP scenario can be combined with the exposure to $PM_{2.5}$ or ozone under different climate change scenarios to estimate the premature mortality burden that can be expected for a particular population if the world follows different climate change pathways in future.

Concluding Remarks

Despite continuous efforts to restrict various sources of pollutants by the government, air pollution remains as one of the major environmental hazards in India (Dey and Di Girolamo 2011). GBD study shows large premature mortality from air pollution exposure in India. However, burden estimates at regional level needs to be adjusted for the local condition. Our recent study (Chowdhury and Dey 2016) estimated about 800,000 premature adult deaths per year by adjusting for the heterogeneity in background mortality rate as a function of socio-economic development represented in terms of gross domestic product. Following example will clarify the importance of local adjustment. Delhi has the highest ambient $PM_{2.5}$ exposure in India, but at the same time, its GDP is also one of the highest in the country. If the baseline mortality of Delhi is not adjusted and instead a single India-specific value is considered, the premature mortality burden of Delhi is overestimated. Similarly, the burden would have been underestimated in regions that are less developed and have higher background mortality than all-India average. Therefore, identification of vulnerable regions based on premature mortality burden and prioritization of mitigation measures in these regions should be facilitated by such analysis.

The exposure to $PM_{2.5}$ and ozone has been increasing over India in the last decade (Saraf and Beig 2004; Dey and Di Girolamo 2011; Dey et al. 2012) resulting in increasing number of the population being pitched at risk of dying prematurely. What intrigues the policymakers is whether exposure to air pollution will continue to increase in the future. To project future premature mortality from air pollution exposure for India or any other country, the scenario matrix framework will be useful, because it will enable to isolate the relative roles of meteorological, demographic and epidemiological changes on the projected burden. Such strategic knowledge will provide the government adequate information to formulate policy to mitigate air pollution and develop climate change resilient society.

Acknowledgement Financial support from the Department of Science and Technology, Government of India, through a research grant (DST/CCP/NET-2/PR-36/2012(G)) under the first phase of the network program of "climate change and human health" is acknowledged.

References

Allen RJ, Landuyt W, Rumbold ST (2016) An increase in aerosol burden and radiative effects in a warmer world. Nat Clim Change 6:269–274. doi:10.1038/NCLIMATE2827

Anenberg SC, Horowitz LW, Tong DQ, West JJ (2010) An estimate of the global burden of anthropogenic ozone and fine particulate matter on premature human mortality using atmospheric modeling. Environ Health Perspect 118:1189–1195. doi:10.1289/ehp.0901220

Arnold C (2014) News|science selections disease burdens associated with PM 2.5 exposure how a new model provided global estimates. Environ Health Perspect 13:981277

Avnery S, Mauzerall DL, Liu J, Horowitz LW (2011) Global crop yield reductions due to surface ozone exposure: 1. Year 2000 crop production losses and economic damage "and" Global crop yield reductions due to surface ozone exposure: 2. Year 2030 potential crop production losses and economic damage under. Atmos Environ 45:2297–2309. doi:10.1016/j.atmosenv.2012.12.045

Balbus John M, Malina Catherine M (2009) Identifying vulnerable subpopulations for climate change health effects in the United States. J Occup Environ Med 51:33–37. doi:10.1097/JOM.0b013e318193e12e

Bell ML, Samet JM, Dominici F (2004) Time-series studies of particulate matter. Annu Rev Public Health 25:247–280. doi:10.1146/annurev.publhealth.25.102802.124329

Bell ML, Dominici F, Samet JM (2005) A meta-analysis of time-series studies of ozone and mortality with comparison to the National Morbidity, mortality, and air pollution study. Epidemiology 16:436–445. doi:10.1001/jama.292.19.2372.Ozone

Burnett RT, Arden Pope C, Ezzati M et al (2014) An integrated risk function for estimating the global burden of disease attributable to ambient fine particulate matter exposure. Environ Health Perspect 122:397–403. doi:10.1289/ehp.1307049

Calvin K, Bond-lamberty B, Clarke L et al (2016) The SSP4: A world of deepening inequality. Glob Environ Chang. doi:10.1016/j.gloenvcha.2016.06.010

Chen H, Goldberg MS, Villeneuve PJ (2008) A systematic review of the relation between long-term exposure to ambient air pollution and chronic diseases. Rev Environ Health 23:243–297. doi:10.1515/reveh.2008.23.4.24

Chowdhury S, Dey S (2016) Cause-specific premature death from ambient PM2.5 exposure in India: estimate adjusted for baseline mortality. Environ Int 91:283–290. doi:10.1016/j.envint.2016.03.004

Christensen JH, Hewitson B, Busuioc A, et al (2007) Regional climate projections. In: Climate change 2007: the physical science basis. Contribution of Working Group I to the Fourth Assessment Report of the Intergovernmental Panel on Climate Change

Daniel JJ, Winner DA (2009) Effect of climate change on air quality. Atmos Environ 43:51–63. 10.1016/j.atmosenv.2008.09.051

Dey S, Di Girolamo L (2011) A decade of change in aerosol properties over the Indian subcontinent. Geophys Res Lett 38:1–5. doi:10.1029/2011GL048153

Dey S, Di Girolamo L, van Donkelaar A et al (2012) Variability of outdoor fine particulate (PM2.5) concentration in the Indian subcontinent: a remote sensing approach. Remote Sens Environ 127:153–161. doi:10.1016/j.rse.2012.08.021

Ebi KL, Hallegatte S, Kram T, et al (2014) A new scenario framework for climate change research: background, process, and future directions. pp 363–372. doi:10.1007/s10584-013-0912-3

Fang Y, Naik V, Horowitz LW, Mauzerall DL (2013) Air pollution and associated human mortality: the role of air pollutant emissions, climate change and methane concentration increases from the preindustrial period to present. Atmos Chem Phys 13:1377–1394. doi:10.5194/acp-13-1377-2013

Fricko O, Havlik P, Rogelj J et al (2016) The marker quantification of the Shared Socioeconomic Pathway 2: a middle-of-the-road scenario for the 21st century. Glob Environ Chang. doi:10.1016/j.gloenvcha.2016.06.004

Fujimori S, Hasegawa T, Masui T et al (2016) SSP3: AIM implementation of shared socioeconomic pathways. Glob Environ Chang. doi:10.1016/j.gloenvcha.2016.06.009

Ghude SD, Jena C, Chate DM et al (2014) Reductions in India's crop yield due to ozone. Geophys Res Lett 41:5685–5691. doi:10.1002/2014GL060930

Ghude SD, Chate DM, Jena C et al (2016) Premature mortality in India due to PM 2.5 and ozone exposure. Geophys Res Lett 43:4650–4658. doi:10.1002/2016GL068949

Gouveia N, Fletcher T (2000) Time series analysis of air pollution and mortality: effects by cause, age and socioeconomic status. J Epidemiol Community Health 54:750–755. doi:10.1136/jech.54.10.750

Haines A, Smith KR, Anderson D et al (2007) Policies for accelerating access to clean energy, improving health, advancing development, and mitigating climate change. Lancet 370:1264–1281. doi:10.1016/S0140-6736(07)61257-4

Horowitz LW (2006) Past, present and future concentrations of tropospheric ozone and aerosols: methodology, ozone evaluation, and sensitivity to aerosol wet removal. J Geophys Res Atmos. doi:10.1029/2005JD006937
IPCC (2010) IPCC workshop on socio-economic scenarios. IPCC Work Socio-Economic Scenar Victor's Residenz-Hotel, Berlin, Ger 1-3 November 2010, pp 1–61
Jerrett M, Burnett RT, Pope CA, et al (2009) Long-term ozone exposure and mortality. N Engl J Med 360:1085–1095. doi:10.1056/NEJMoa0803894
Kinney PL (2008) Climate change, air quality, and human health. Am J Prev Med 35:459–467. doi:10.1016/j.amepre.2008.08.025
Kirtman B, Power SB, Adedoyin J, et al (2013) Near-term climate change: projections and predictability
Kriegler E, Bauer N, Popp A, et al (2016) Fossil-fueled development (SSP5): an energy and resource intensive scenario for the 21st century. doi:10.1016/j.gloenvcha.2016.05.015
Li C, Balluz LS, Vaidyanathan A et al (2016) Long-term exposure to ozone and life expectancy in the United States, 2002 to 2008. Medicine (Baltimore) 95:e2474. doi:10.1097/MD.0000000000002474
Lim SS, Vos T, Flaxman AD, et al (2012) A comparative risk assessment of burden of disease and injury attributable to 67 risk factors and risk factor clusters in 21 regions, 1990-2010: a systematic analysis for the global burden of disease study 2010. Lancet 380:2224–2260. doi:10.1016/S0140-6736(12)61766-8
Lin S, Liu X, Le LH, Hwang SA (2008) Chronic exposure to ambient ozone and asthma hospital admissions among children. Environ Health Perspect 116:1725–1730. doi:10.1289/ehp.11184
Madaniyazi L, Guo Y, Yu W, Tong S (2015) Projecting future air pollution-related mortality under a changing climate: progress, uncertainties and research needs. Environ Int 75:21–32. doi: 10.1016/j.envint.2014.10.018
Makri A, Stilianakis NI (2008) Vulnerability to air pollution health effects. Int J Hyg Environ Health 211:326–336. doi:10.1016/j.ijheh.2007.06.005
Marenco A, Gouget H, Nedelec P et al (1998) Evidence of a long-term increase in tropospheric ozone from Pic du Midi data series: consequences: positive radiative forcing. J Geophys Res Atmos 99:16617–16632
Markandya A, Armstrong BG, Hales S et al (2009) Public health benefits of strategies to reduce greenhouse-gas emissions: low-carbon electricity generation. Lancet 374:2006–2015. doi:10.1016/S0140-6736(09)61715-3
Menon A, Lvermann A, Schewe J, Lehmann JFK (2013) Consistent increase in Indian monsoon rainfall and its variability Geoscientific across CMIP-5 models. Dynam Earth Syst:287–300. doi:10.5194/esd-4-287-2013
Morgan PB, Mies TA, Bollero GA et al (2006) Season-long elevation of ozone concentration to projected 2050 levels under fully open-air conditions substantially decreases the growth and production of soybean. New Phytol 170:333–343. doi:10.1111/j.1469-8137.2006.01679.x
Moss RHRH, Edmonds JAJA, Hibbard KAKA et al (2010) The next generation of scenarios for climate change research and assessment. Nature 463:747–756. doi:10.1038/nature08823
Murray C (2015) Global, regional and national comparative risk assessment of 76 behavioural, environmental, occupational and metabolic risks or clusters of risks in 188 countries 1990–2013: a systematic analysis for the GBD Background. Lancet 6736:1–27. doi:10.1016/S0140-6736(15)00128-2
Nawahda A, Yamashita K (2012) Evaluation of premature mortality caused by exposure to PM 2.5 and ozone in East Asia:2000, 2005, 2020, pp 3445–3459. doi:10.1007/s11270-012-1123-7
O'Neill BC, Carter TR, Ebi KL, et al (2012) Workshop on the nature and use of new socioeconomic pathways for climate change research core writing team acknowledgments. Meet Rep Work Nat Use New Socioecon Pathways Clim Chang Res:1–37
Pan X, Chin M, Gautam R et al (2015) A multi-model evaluation of aerosols over South Asia: common problems and possible causes. Atmos Chem Phys 15:5903–5928. doi:10.5194/acp-15-5903-2015

Pope A, Burnett RT, Thun MJ, et al (2002) Lung cancer, cardiopulmonary mortality, and long-term exposure to fine particulate air pollution
Saraf N, Beig G (2004) Long-term trends in tropospheric ozone over the Indian tropical region. Geophys Res Lett 31:1–5. doi:10.1029/2003GL018516
Selin NE, Wu S, Nam KM et al (2009) Global health and economic impacts of future ozone pollution. Environ Res Lett 4:44014. doi:10.1088/1748-9326/4/4/044014
Shi L, Zanobetti A, Kloog I, et al (2015) Low-concentration PM and mortality: estimating acute and chronic effects in a population-based study. Environ Health Perspect:46–52. doi:10.1289/ehp.1409111
Silva RA, West JJ, Zhang Y et al (2013) Global premature mortality due to anthropogenic outdoor air pollution and the contribution of past climate change. Environ Res Lett 8:34005. doi:10.1088/1748-9326/8/3/034005
Silva RA, West JJ, Lamarque JF et al (2016) The effect of future ambient air pollution on human premature mortality to 2100 using output from the ACCMIP model ensemble. Atmos Chem Phys 16:9847–9862. doi:10.5194/acp-16-9847-2016
Smith KR, Balakrishnan K (2009) Mitigating climate, meeting MDGs, and moderating chronic disease: the health co-benefits landscape. "Commonwealth Health Ministers" Update
Smith KR, Woodward A, Al E (2014) Human health: impacts, adaptation, and co-benefits. In: Field CB, Barros VR, Dokken DJ (eds) Climate change: 2014 impacts, adaptation, and vulnerability. Part A: global and sectoral aspects. Contribution of working group II to the fifth assessment report of the intergovernmental panel on climate change, pp 709–756
Tagaris E, Liao K, Delucia AJ, Deck L (2009) Potential impact of climate change on air pollution-related human health effects. Environ Sci Technol:4979–4988
Tainio M, Juda-rezler K, Reizer M, et al (2013) Future climate and adverse health effects caused by fine particulate matter air pollution: case study for Poland, pp 705–715. doi:10.1007/s10113-012-0366-6
Taylor KE, Stouffer RJ, Meehl GA (2012) An overview of CMIP5 and the experiment design. Bull Am Meteorol Soc 93:485–498. doi:10.1175/BAMS-D-11-00094.1
Van Vuuren DP, Stehfest E, Gernaat DEHJ, et al (2016) Energy, land-use and greenhouse gas emissions trajectories under a green growth paradigm. Glob Environ Chang:1–14. doi:10.1016/j.gloenvcha.2016.05.008
West JJ, Cohen A, Dentener F et al (2016) What we breathe impacts our health: improving understanding of the link between air pollution and health. Environ Sci Technol 50:4895–4904. doi:10.1021/acs.est.5b03827
West JJ, Fiore AM, Horowitz LW, Mauzerall DL (2006) Global health benefits of mitigating ozone pollution with methane emission controls. Proc Natl Acad Sci U S A 103:3988–3993. doi:10.1073/pnas.0600201103
Wilkinson P, Smith KR, Davies M et al (2009) Public health benefits of strategies to reduce greenhouse-gas emissions: household energy. Lancet 374:1917–1929. doi:10.1016/S0140-6736(09)61713

Sourangsu Chowdhury received his MSc degree in atmospheric sciences from the University of Calcutta in 2012. He is a doctoral candidate at the Centre for Atmospheric Sciences, IIT Delhi. His primary research interest is to quantify the impact of particulate matter on human health with a focus on India.

Sagnik Dey is an associate professor at the Centre for Atmospheric Sciences, IIT Delhi. His research interest is to understand air quality, climate change and human health connection. He is a science team member of the NASA MAIA mission. He has published more than 60 peer-reviewed articles with an h-index of 22.

Chapter 3
International Conferences on Sustainable Development and Climate from Rio de Janeiro to Paris

Giovanni De Santis and Claudia Bortone

Abstract To cope with the problems caused by global warming whose effects began to be felt in the second half of the twentieth century, 21 summits have been held in order to identify the causes and the measures to be taken for a sustainable solution to the problem. This article reviews the results obtained in the various summits, highlighting both their positive and negative aspects and emphasizing the close relationships between climatic and territorial conditions. This approach is inevitable given the disastrous consequences that would result if the current trend of climate change were to escape human control, at least for that part of it caused by human activities.

We examine the current state of affairs by studying the causes that led to such a situation, the seriousness of which the major powers seem unable to accept nor find acceptable solutions that would reduce the dangers. A decisive role has been played by increased pollution in its many forms (agriculture, industry, domestic heating, traffic, etc.) caused by the use of fossil fuels that have led to an impressive increase in greenhouse gas emissions, with inevitable repercussions on the increase in the global temperature of the planet. Numerous global conferences have been held with the explicit aim of setting up the necessary safeguards, whose results to date have not, unfortunately, led to final decisions but to mere declarations of willingness to resolve the issue. All this has had and has an immediate feedback in further health-related issues, due to an increase in diseases closely related to environmental pollution, as well as the growing desertification of many areas resulting in a reduced quality of life.

Keywords Climate change • Global warming • Desertification • Greenhouse gases • The Conference of the Parties (COP) • Diseases

---

G. De Santis (✉) • C. Bortone
University of Perugia, Perugia, Italy
e-mail: giovanni.desantis@unipg.it; claudia91@formialex.it

© Springer International Publishing AG 2018

R. Akhtar, C. Palagiano (eds.), *Climate Change and Air Pollution*,
Springer Climate, DOI 10.1007/978-3-319-61346-8_3

The Reality of Global Warming: Causes and Consequences

> Global warming and many of the phenomena observed in recent decades didn't occur for hundreds, sometimes thousands, of years. The atmosphere and the oceans are heated, the stock of snow and glaciers has decreased, the sea level has risen and the level of greenhouse gases has increased. The human influence on climate is obvious. This is evidenced by the increased concentration of greenhouse gases and radiation in the atmosphere, by the increased heating and more intense climate variability. (...) It is highly probable that the influence of man has been the dominant cause of global warming since the middle of the last century. (...) The constants emissions greenhouse gas will result in a further increase in temperature and changes in all conditions of weather. Limiting climate change will require a substantial and sustained reduction of greenhouse gas emission[1].

These observations have led to a growing awareness (even if this is not the case for all countries) of the need to adopt structural measures for regulating pollutant emissions that were causing rises in temperature and, subsequently, climate change. This increase is connected to the so-called greenhouse effect, a natural phenomenon which regulates the ability of the atmosphere to deal with the energy from the sun, by means of a translucent membrane that 'traps' the sun's rays. Specifically, sunlight passes through the layer formed of greenhouse gases that envelops the entire planet and heats it; at the same time, however, part of the heat imprisoned can then be dispersed into the atmosphere, thus obtaining the climate balance which regulates life on Earth.

Fundamental components of this phenomenon are the greenhouse gases such as water vapour, carbon dioxide (CO2), methane (CH4), nitrous oxide (N2O) and ozone (O3), which regulate the Earth's temperature, just like a greenhouse. The effect becomes irreversible with the continuous increase of greenhouse gases in the atmosphere that tend to thicken the layer, preventing the required heat loss, which then has a negative impact on human activity. It is well established that humans have a growing influence on Earth's climate and on temperature with the use of fossil fuels, which add huge amounts of greenhouse gases to those naturally present in the atmosphere, leading to continual global warming. One need only mention carbon dioxide, a greenhouse gas produced primarily by human activity and responsible for 63% of global warming. Its concentration in the atmosphere exceeds 40% of the level recorded at the beginning of the industrial age. To that must be added other greenhouse gases that, even if in smaller amounts, have the power to generate large amounts of heat, so much so that, for example, methane is responsible for 19% of man-made global warming and nitric oxide for 6%.

The main causes, then, of rising temperatures lie in the burning of fossil fuels and deforestation, since the destruction of vegetation, which helps regulate the climate by absorbing carbon dioxide from the atmosphere, puts the trapped CO_2 back into the atmosphere. No less damaging is the development of livestock breeding which produces large amounts of methane and the increasing use of

---

[1]As strongly denounced by the Intergovernmental Panel on Climate Change (IPCC) in the Summary for Policy Makers of the fifth report, published in October 2013.

nitrogen fertilizers in agriculture which produce emissions of nitric oxide. Another negative factor can be found in the use of chlorofluorocarbons (CFCs), now banned, and the use of hydrofluorocarbons (HFCs)[2], which are being gradually eliminated. What is certain is that the current global average temperature is 0.85 °C higher compared to the end of the nineteenth century and each of the past three decades, since the first surveys, in 1850, have been hotter than its predecessor.

Global climate experts believe that human activities are almost certainly the main cause of the rising temperatures observed since the second half of the twentieth century. An increase of 2 °C above the pre-industrial era temperature is considered by scientists as the threshold beyond which there is a real risk of dangerous and potentially catastrophic environmental changes occurring globally. For this reason, after lengthy discussions, the international community has accepted the necessity of keeping global warming below 2 °C.

Before the industrial revolution, human beings released very little gas into the atmosphere; but today factors such as population growth, resulting in more demand for food, the use of fossil fuels and deforestation are, little by little, changing the level of greenhouse gases in the air and causing an excess in quality and quantity of substances that are not compatible with the protection of the planet. The result is the alteration of the delicate climatic balance which regulates the global temperature of the Earth. Climate change, already underway with devastating effects, will have significant implications for human health and the integrity of the environment, by strongly influencing agriculture, the availability of water, biodiversity, energy and the economy. There is an urgent need to at least interrupt this process and then drastically reduce emissions. If the planet continues to overheat, the first result would be an increase in sea levels, resulting from the thawing of terrestrial glaciers and ice caps, in part already taking place, leading to the submersion of coastal areas and settlements. No less important is the drying out and desertification of many temperate areas, with a loss of biodiversity of flora and fauna, and with the invasion of species of plants and animals typical of tropical areas, in addition to the increased frequency of extreme events caused by the collision of cold and warm currents.

At a health level, this could mean the onset of diseases now relegated to the southern hemisphere; it seems, in fact, that climate change could encourage the spread of tropical diseases like malaria and dengue fever, as the mosquitoes that carry the disease shift northward, where the temperature is on the rise. In addition, the increase in temperature favours the biological pollution of water, leasing to the proliferation of invasive plant and animal organisms.

[2]Refrigerant gases.

The Need for Action at the Global Level

From November 30 to December 12, 2015, there were held in Paris the 21st session of the Conference of the Parties (COP) arising from the United Nations Framework Convention on Climate Change (UNFCCC) and the 11th session of the Conference of the Parties (CMP11) on the activation of the Kyoto Protocol. After long and exhausting negotiations, a new universal and legally binding agreement was reached on climate change, given the urgency of initiating actions to limit global warming. To this end, various governments pledged to keep the global average temperature increase below 2 °C compared to pre-industrial levels, through national plans of action aimed at reducing their emissions, and to communicate their achievements and results every 5 years. However, in order to reduce significant disparities, the EU and countries with advanced development (CAD) will continue to provide funding for developing countries (DC) to reduce their emissions and to become more resilient to the effects of climate change.

On the basis of this agreement by which 195 countries are committed to reducing polluting emissions from the next September, October 14, 2016, can be considered a decisive date for the health of the planet. In fact, in the conference at Kigali in Rwanda, the UN member country subscribers to the 1987 Montreal Protocol on phasing out of CFC emissions endorsed the ban on production and use of HFCs, which are equally responsible for the greenhouse effect. The elimination of HFCs will be divided in to three stages: the first involving industrialized countries, who by 2019 will have to achieve a 10% reduction in the emission of these gases; the second will affect China, countries of South America and developing countries (DC), whose reduction will start from 2024, while the third will be India, Pakistan, Iran, Iraq and the Gulf countries from 2028 because their economies need longer timescales. The importance of this agreement lies mainly in the fact that HFCs have become the third element responsible for the greenhouse effect, after carbon dioxide and methane, so much so that it is estimated that this accord will mean a reduction of global warming of 0.5 °C by the end of the century.

If we return to our examination of the various conferences relating to measures for climate change reduction, it was only in 1979, when the first World Climate Conference was held in Geneva, that the issue was recognized as urgent, as a result of the many criticisms and appeals from the scientific world on the changes that might have long-term effects both on humans and the environment. The Conference ended with a statement addressed to all world leaders 'to foresee and prevent potential man-made changes in climate that might be adverse to the well-being of humanity'. The Conference also set up the World Climate Program (WCP) under the direct responsibility of the World Meteorological Organization (WMO), the United Nations Environment Program (UNEP) and the International Council of Scientific Unions (ICSU). From the late 1980s, there followed numerous intergovernmental conferences on climate change (Villach, 1985; Toronto, 1988; Ottawa, 1989; Tata, 1989, The Hague, 1989; Noordwiik, 1989; Cairo, 1989; Bergen, 1990;

and the second World Climate Conference (November 1990, Geneva)), but no binding decisions were arrived at that could be accepted by all the states.

Meanwhile, in 1990, the Intergovernmental Panel on Climate Change (IPCC), set up by UNEP and WMO, published its first report on climate and on the serious transformations taking place, while the UN General Assembly approved the conducting of negotiations for the draft of an international treaty. Despite the constant meetings, often unnecessary, and the continuous interjections by scientists and ecologists, it was only in June 1992 that talks began at a global level, with the World Conference on Environment and Development in Rio de Janeiro.[3] At this meeting, the member countries of the United Nations signed several documents committing them to sustainable development, including the United Nations Framework Convention on Climate Change (UNFCCC). By signing this agreement, governments undertook the adoption of programmes and measures aimed at the prevention, control and mitigation of the effects of human activity on the planet. In particular, the objective of the Convention is to (art. 2) 'stabilize greenhouse gas concentrations in the atmosphere at a level that would prevent dangerous anthropogenic interference with the climate system'. It also established a body called the

[3]It should be mentioned that on this occasion was held the fist meeting of the United Nations Conference on Environment and Development (UNCED), better known as Agenda 21, which is the programme of action by the international community (states, governments, NGOs, private sector) in the area of environment and development for the twenty-first century. It is a complex document which starts from the premise that human societies cannot continue to increase the economic gap between countries and between the classes of the population within them, increasing poverty, hunger, disease and illiteracy and causing the continuing deterioration of the ecosystems that are responsible for the maintenance of life on the planet. The Agenda 21 document is divided into four thematic sections that are detailed in the respective chapters: (1) social and economic areas: poverty, health, environment, demographics, production, etc. (2) Conservation and management of resources: atmosphere, forests, deserts, mountains, water, chemicals, waste, etc. (3) Strengthening the role of the most significant groups: women, youth, NGOs, ethnic groups, farmers, trade unions. (4) Methods of implementation: finances and institutions. To achieve these objectives, after the Rio Conference, several initiatives and projects were launched, and various governments outlined plans for the sustainable development of their countries, based on the specific existent conditions and environmental and social issues. Concerning the status of implementation of the commitments of Agenda 21 at a global level the UN Conference, 'Rio + 10' was held in August 2002 in Johannesburg (South Africa), on sustainable development, whose resolutions were signed by the governments of 183 countries. Among these documents is the 'United Nations Framework Convention on Climate Change', which commits governments to promote, through coordination with all the actors of the territory, an action plan for improving the quality of life and social and economic development in harmony with the environment. It was also hoped that all countries would undertake the consultative process with their populations and seek consensus on a Local Agenda 21 by 1996: 'Every local authority has to open a dialogue with its citizens, with associations and with private companies and adopt a Local Agenda 21. Through consultation and consensus building, local authorities can learn from the local community and businesses and can acquire the information necessary for the formulation of the best strategies. The consultation process can raise the awareness of families on issues of sustainable development. The programs, policies and laws passed by the local administration could be evaluated and amended on the basis of the new plans thus adopted. These strategies could also be used to support the proposals and to access local, regional, national and international funding' (article 28 of Agenda 21).

'Conference of the Parties (COP)', which was entrusted with the crucial task of implementing the general commitments contained in the Convention itself. This led to the calling of numerous conferences listed below:

Rio de Janeiro, Brazil 1992 followed by:

COP-1, Berlin Mandate 1995
COP-2, Geneva, Switzerland 1996
COP-3, the Kyoto Protocol on Climate Change 1997
COP-4, Buenos Aires, Argentina 1998
COP-5, Bonn, Germany 1999
COP-6, The Hague, Netherlands 2000
COP-6 bis, Bonn, Germany 2001
COP-7, Marrakesh, Morocco 2001

World Summit on Sustainable Development (WSSD), Johannesburg, South Africa 2002

COP-9, Milan, Italy 2003
COP-10, Buenos Aires, Argentina 2004
COP-11, Montreal, Canada 2005
COP-12, Nairobi, Kenya 2006
COP-13, Bali, Indonesia 2007
COP-14, Poznan, Poland 2008
COP-15, Copenhagen, Denmark 2009
COP-16, Cancun, Mexico 2010
COP-17, Durban, South Africa 2011
COP-18, Doha, Qatar 2012
COP-19, Warsaw, Poland 2013
COP-20, Lima, Peru 2014
COP-21, Paris, France 2015

As the present study was being drafted, the latest Conference (COP-22) opened in Marrakesh on October 8, 2016, with the clear intention to give full and formal launching of the Treaty of Paris, since the resolutions subscribed therein were approved by the governments of more than 100 countries whose share of pollution exceeds 70% of the greenhouse gases released into the atmosphere. Finally, the election on November 8, 2016, of US President Donald Trump threatens to undermine the decisions so painstakingly reached, according to a statement withdrawing America's adhesion to the agreements made.

Various Summits and the Measures to Be Taken for a Sustainable Solution to the Problem

Since 1990, the Intergovernmental Panel on Climate Change (IPCC), set up in 1988 by the UN and consisting of two bodies, the World Meteorological Organization (WMO) and the United Nations Environment Program (UNEP), has highlighted the risk of global warming due to increased greenhouse gas emissions and their effect on climate, mainly caused by the use of fossil fuels. Officially, from this point onwards, governments and transnational organizations began to take into account the innumerable problems and the serious damage that global warming could create for territories and societies. This is a valid issue of concern for mankind and brings with it the need, worldwide, to issue new guidelines on environmental protection, whose objectives and priorities should reflect local conditions and the degree of development of each individual state. This desire for immediate and consistent action in single situations is the common thread that unites, despite numerous differences, the 21 conferences that have taken place from 1992 to 2015.

In fact, it must be stressed that although numerous alarms have arrived from the entire scientific world, not all conferences have led to concrete results; many of these summits led to no decisions whatsoever and even brought out the hostility of some countries, including major world powers, often lined up on opposite sides. The only, disastrous, result has been the aggravation of climatic conditions, because of the power dynamics of the various countries who have demanded different intervention policies and specifications for each area, in order to avoid the application of consistent regulations, which often proved inadequate. Leaving aside the meetings whose results were purely formal, we will attempt to give a brief history of those which obtained positive results, indicating the choices made and agreements reached.

Having said that, we must start with the 'Earth Summit', the United Nations Framework Convention on Climate Change, held in Rio de Janeiro, in 1992. The moral substrate of the agreement is governments finally becoming aware of climate change and of the influence of human activities on such change, as well as the desire to protect the climate system of the planet, although full scientific certainty has not yet been reached on the causes and effects of the phenomenon.

However, the summit was unable to impart a value or a legally binding commitment to the agreement nor the need to set a mandatory limit on emissions by individual states. Nevertheless, its importance should not be underestimated because the countries involved were obliged to provide regular reports on policies chosen for implementing reduction measures and promoting adaptation to climate change. This obligation led to the subsequent Conferences of the Parties (COP) of Berlin (1995) and Geneva (1996), the results of which, though not formally binding, have the merit of encouraging more accurate and specific research, with which to identify the most appropriate action for each state as indicated by the Berlin Mandate. Since the effects of climate change were becoming increasingly evident, during the Geneva Summit, a regulatory plan was developed to be tested and officially approved at COP-3.

This meeting was held in Kyoto in 1997, and the important 'Kyoto Protocol on Climate Change'[4] was signed, in which, for the first time, 38 countries, including both industrialized and developing nations, formally pledged to reduce emissions of six types of greenhouse gases. The agreement, taking into account the social, economic and environmental conditions of the signatory states, carefully measured reduction measures, to 5.2% of the emissions of 1990, the year of the first IPCC report, to be implemented in 2008–2012. These agreements were also analysed in the COP-4 in Buenos Aires in 1998, while at the conference in Bonn in 1999, guidelines were drawn up to outline the relations and communications between member states to further the study of flexible mechanisms, such as the Joint Implementation[5] and the Clean Development Mechanism (CDM)[6], in addition to identifying the capacity building of individual states. However, the general interests of the various governments are not always identical, as demonstrated by the conflicts characterized the COP-6 (2000) of The Hague, marked by clashes between the USA and the EU, so that in 2001, the political and financial problems left unresolved at COP-6 bis in Bonn had to be addressed anew, just 4 months after the withdrawal of the USA from the ranks of the signatory countries of the Kyoto Protocol.

Five years after the Kyoto Protocol on Climate Change, the conditions for its implementation had to be decided, and during the COP-7 (2001), the 'Marrakesh Accords' were signed, to guarantee compliance with the agreed stipulations and the reporting of each firmatory's activities. The 'Marrakesh Ministerial Declaration' was also signed for the World Summit on Sustainable Development scheduled for 2002 in Johannesburg, with the intent of determining progress 10 years after the Earth Summit, whose importance was reaffirmed also during the COP-8 (2002,

---

[4]The Protocol commits the industrialized countries and those with economies in transition (Eastern European countries) to reduce (5 % in the period 2008–2012) GHG emissions capable of altering the natural greenhouse effect. Greenhouse gases covered by the Protocol are carbon dioxide, methane, nitrous oxide, hydrofluorocarbons, perfluorocarbons and sulphur hexafluoride. Unfortunately, not all states have acceded to the Protocol: the USA, responsible for 30 % of the total emissions from developed countries, signed but then refused to ratify the Treaty. For newly industrialized countries, the Protocol does not provide for any reduction target. China, India and other developing countries have been exempted from obligations because they are not considered among the 'historical' major emitters of greenhouse gases (i.e. those that remain in the atmosphere for about a century and which are the cause of climate change). The non-member countries are responsible for 40 % of global emissions of greenhouse gases.

[5]Joint Implementation (JI): If two industrialised countries that have signed a commitment to do so produce a plan to reduce greenhouse gas emissions, the investing country is accredited the emission rights of the host country. The investing country may then produce a larger quantity of greenhouse gases, which will be equivalent to the reduction obtained in the host country.

[6]Unlike JI projects, in the Clean Development Mechanism (CDM) projects, partners are developing countries that have not signed PSA reduction commitments. In this case, therefore, emissions rights are not transferred but created. The investing country may emit greater amounts of greenhouse gases without the host country having to reduce its total emissions.

New Delhi). In Milan (COP-9) in 2003, the Special Fund on Climate Change and the Fund for Less Developed Countries were set up, and the rules and methods for including agroforestry activities in the CDM outlined, objectives that were resumed at Buenos Aires (COP-10, 2004) and broadened to include issues such as development and technology transfer and sustainable use of territory, as well as in addition to identifying the specific needs of individual countries.

The Summit in Montreal (COP-11, 2005) had an important part to play, since 7 years after the Kyoto Protocol on Climate Change, the countries that had approved the protocol committed themselves to determining specific tasks to be implemented after the 2012 deadline. This consideration for the future was also a point of discussion at COP-12 (Nairobi, 2006) which included the 'work programme on impacts, vulnerability and adaptation' and the 'Nairobi Framework', aimed at providing additional support for developing countries, and also the 'Compliance Committee of the Kyoto Protocol on Climate Change', which made it fully operational. COP-13 (Bali, 2007) was of fundamental importance for climate balance, with the 'Bali Road Map', namely, an international long-term agreement for combating climate change that would involve the entire world political system, entrusting the control and organization to a specific working group, to ensure a long-term cooperative action (AWG-LCA); these actions were further expanded at the COP-14 (Poznan, 2008).

At the Copenhagen Conference (COP-15, 2009), there were strong political tensions; interventions in favour of the poorest countries to allow them to reach the technological levels needed for the use of renewable energy sources were met by the choice to limit the increase in global warming to no more than 2 °C. Despite the worsening global climate conditions and related issues affecting many areas of the planet, at the Cancun Conference (COP-16, 2010), it was decided to limit the amount of thermal reduction no longer at 2 °C but at least 1.5 °C. The role that technological development could play in achieving the required objectives was barely recognized by the establishment of the Adaptation Committee and the Technological Mechanism, which included within it the Technology Executive Committee (TEC) and the Climate Technology Centre and Network (CTCN).

Given the need to implement the commitments of the Kyoto Protocol (2013–2020), in 2011, the Durban COP-17 set up the 'Ad Hoc Working Group on the Durban Platform for Enhanced Action' (ADP), with the task of 'developing a protocol with the force of law, according to the Convention, applicable to all parties', which was then modified during COP-18 (Doha 2012) and COP-19 (Warsaw 2013) so as to close the gap between pre-2020 commitments and the scares results already obtained. At COP-20 in Lima (2014) discussion focused on the results of the fifth assessment report presented by the IPCC, which indicated the increased reliability of scientific evidence regarding climate change and its cause-and-effect dynamics, since the early 1990s. It was considered necessary, also, to implement the sanctions mechanism introduced at the Conference in Warsaw (2013), on financial compensation to be paid by countries who caused damage related to climate change. Discussion was also held on awareness and education about gender difference and of a different approach than that of the INDC[7], combining top-down and bottom-up, in other words integrating the decisions taken by the COP with the voluntary choices of individual governments and keeping in mind the transparency of any action.

We then arrive at the Conference in Paris that also hosted the 11th session of the meeting of the Parties to the Kyoto Protocol of 1997, with the commitment, after nearly 20 years, of reaching a legally binding global agreement that transcended any political tension and/or financial claim. With the Paris accord, countries agreed to reduce greenhouse gas emissions 'as soon as possible' and according to voluntary parameters. However, despite this commitment, the salvation of the planet remains uncertain, since the agreement will come into force only after it has been ratified by at least 55 states, responsible for 55% of total CO2 emissions (with respect to 1990), caused mainly by the USA (19%), China (11.9%), Japan (9.4%), Germany (3.9%), India (3.4%), Africa (3.2%), South America (2.7%), Canada (1.8%), Italy (1.8%), the UK (2.5%) and Oceania (1.3%).

In conclusion, it is clear that these numerous conferences have only partially changed the current state of affairs. In 25 years of work, the 22 conferences (including Marrakesh in October 2016) achieved very little, paradoxically given the seriousness of the problems to be faced. Successes such as the signing of the Kyoto Protocol were made possible only through compromises with minimal and unsatisfactory end results. Certainly there is no denying that some COP have made concrete policy choices, such as the Warsaw COP-19, which helped increase awareness that without specific, competent organs, no change could be contemplated. Also, mention can be made of the setting up of commissions and special bodies such as the Adaptation Committee and the Technology Mechanism, inside which, at COP-16, were formed the Technology Executive Committee (TEC) and the Climate Technology Centre and Network (CTCN).

Agreements, such as the 'Nairobi Framework', aimed at providing additional support to developing countries (COP-12), or the 'Bali Road Map' of COP-13, called on all the world's political powers to implement rapid action on climate change. Unfortunately, it is clear that every effort made to limit or reduce the effects of climate change has disappeared under the constant pressing political and economic influence exerted by the individual states involved. Even the COP-21 in Paris, despite the good intentions of actually beginning the battle against climate change, was a race against time to reach the key conditions for the implementation of the treaty, which was approved and signed in the COP-22 (Marrakesh). However, this could prove to be ineffective owing to the anti-ecological stance adopted by the new US presidency in the field of environmental protection. It is clear, therefore, that it is necessary to discuss and define a new framework of environmental protection that could lead to a state of affairs very different from the present one.

[7]Intended Nationally Determined Contributions (INDC): contributions to the global reduction of greenhouse gases that the nations intended to give on a voluntary basis by means of 'clear and transparent plans'.

Effects and Repercussions on Health

Climate change has already today radical effects on human health and will have even more in the future, because of its great influence on various factors such as food, water, cleanliness, health care and the control of infectious diseases, resulting in increased mortality and morbidity, especially among the elderly and the poor. While the most serious risks are expected in cities in the middle and high latitudes, warmer winters will probably reduce cold-related deaths in some countries. In contrast, heat waves will tend to affect our cardiovascular and respiratory system, due to periods of extreme heat suddenly becoming more frequent and close together, or even real weather inversions, which can prevent the dispersion of pollutants. These, added to emissions caused by fires, radically worsen air quality in many cities. The quality of water resources will also be at risk as their quantity is reduced, as already happens in many countries where clean drinking water is becoming more and more depleted, undermining the quality of life of the natives, but above all, further weakening the already poor health-care systems in the most disadvantaged areas. It will become imperative to take action against the increasing concentrations of bacteria and other microorganisms responsible for many of the new outbreaks of disease in Africa, India and Southeast Asia, where the scarcity of clean water forces people to use other sources of low quality, often at risk, such as polluted rivers. The result is a massive increase in diseases such as dysentery, cholera, blindness and infectious diseases generally, which can reach epidemic proportions following a further deterioration in climatic conditions. Heat waves, floods, cyclones and droughts, in fact, cause death and disease, the migration of entire populations, epidemics and serious psychological problems, and while scientists remain uncertain about how climate change will affect the frequency of tornadoes and hurricanes, they have no doubts when foreseeing that some regions will be victim of floods and droughts.

Coastal flooding is also on the increase, due to rising water levels, with serious damage to the already disadvantaged local economies. The increase in phenomena connected to climate change has substantial and multiple consequences for human health, both directly and indirectly, which can arise both in the short and long term. It has been estimated that around 150,000 deaths occurred worldwide in 2000, according to a recent study by the World Health Organization, and by 2040 the figure could reach around 250,000 deaths a year.

Among the major risks that threaten health are extreme weather events, as mentioned above, and deaths due to heat waves, and floods are expected to increase. In fact, different types of extreme weather events affect different regions: for example, heat waves are a problem especially in Southern Europe and the Mediterranean but also, to a lesser extent, in other regions. Suffice it to say that, according to estimates, the heat wave of 2003 caused more than 70,000 deaths in 12 European countries, especially among the older members of the population who were more vulnerable to disease. It is predicted that by 2050, heat waves will cause more than 120,000 deaths per year in the EU, generating costs of 150 million euro if appropriate measures to cope with the situation are not taken. These estimates are higher not only because of rising temperatures and the increased frequency of heat waves but also because of the changes taking place in European demographics: in fact, currently about 20% of EU citizens are over 65, and it is estimated that by 2050, they will number about 30% of the total population. High temperatures, often associated with air pollution, can cause respiratory problems and cardiovascular diseases, especially among children and the elderly, and lead to premature deaths.

These climate changes also affect communicable diseases, because changes in the local microclimate can result in the spread of insects that act as vectors, and temperature changes facilitate or inhibit the proliferation of bacterial or parasitic species. There are many ways in which communicable diseases can spread, and these are usually divided into four simple categories:

Water-borne diseases are those of faecal-oral transmission, like cholera or various forms of diarrhoea. Cholera is still endemic in some countries, notably in Bangladesh and other poor countries, and is also showing changes in its distribution, since the increase in temperature of the sea and inland waters encourage the proliferation of the cholera bacteria. *Water-based* diseases are those in which a parasite lives part of its life cycle in the water, as in the case of schistosomiasis. There are signs that this disease is also spreading outside its traditionally endemic areas, for example, in some areas of China, and this is a grave cause of concern, since the parasite is carcinogenic and causes tumours in the bladder and liver.

According to the traditional classification, *water-washed* diseases are those in which the causative agents are routinely eliminated if elementary rules of hygiene are followed; examples include scabies and trachoma. Here the crucial problem is the availability of water for washing, and therefore the desertification of large areas of the planet is a major cause for concern. Finally, *water-related* diseases are those where the carrier, and not the parasite itself, has a cycle involving water. The most obvious example is malaria, carried by anopheles mosquito and linked to the presence of stagnant water. Malaria is perhaps the transmissible disease most studied in relation to climate change, and there is evidence of its spreading outside the areas where it is endemic. It is important to note that the change of distribution of communicable diseases as a result of climate change is not a phenomenon that involves only the low-income countries, although these will be the most affected. The risk will also affect economically evolved countries, so much so that we are nowadays witnessing the emergence of infectious diseases in Europe which are not related only to migration but also to changes of climate or the interaction between these two phenomena. This problem of interaction is of particular concern and preoccupation to epidemiologists, because the concomitant and partially linked phenomena of mass migrations and climate change can together have important and unpredictable effects.

Every inhabitant of the planet should have access to sufficient quantities of good quality water, uncontaminated and not stagnant. We know that this is not the case. By 2025, nearly half of the world's population will be faced with extreme water shortages, and drinking water quality is declining in many parts of the world. Fifty percent of wetlands have been lost, with their flora and fauna, while at the same time, 70% of available water reserves are used for irrigation. There is no denying that there is also a strong component of social inequality, not only for the obvious fact that those without access to water of good quality are poor but also because the rich are responsible for colossal waste such as the irrigation of golf courses in very dry areas such as Kuwait or Qatar. Apart from diseases directly related to scarce good quality water, drought is itself the cause of various diseases. In large areas of China where drought is becoming an acute problem, respiratory diseases are rife. This is due to the fact that in cities particle pollution is on the increase, while in rural areas, dust storms are more frequent and disastrous because of soil erosion. There are also indirect risks, mainly due to the deterioration and contamination of the environment, such as pollutants from industrial processes or waste water and sewage, which carried by floods could lead to the contamination of drinking water and agricultural land or even reach and contaminate rivers, lakes and seas and enter the food chain. The same applies to the forest fires caused by high temperatures and drought (or often set alight intentionally), which damage property and increase air pollution.

Finally, the expected changes in the distribution of vector-borne diseases will also have important consequences for human health. The higher temperatures, milder winters and wetter summers are colonizing large areas where insects, vectors of disease, survive and multiply, allowing the proliferation of diseases like Lyme disease, dengue fever or malaria in new regions whose natural habitat was not previously conducive to their development and to their transmission. Seasonal variations, in which some seasons seem to start earlier and last longer, may have negative consequences for human health, especially for people suffering from allergies, which are on the rise globally, with the possible risk of asthma attacks brought on by the combined exposure to different allergens at the same time. All this might also lead to an increased pressure on health facilities, intensifying financial commitment in rich countries, while the situation in developing countries would become even more untenable.

The risks associated with climate change are also long term: changes in temperature and precipitation will probably affect the food production capacity of territories now exploited by agriculture. Their general massive reduction, combined with the problem of unequal distribution of resources, would not only exacerbate the problem of malnutrition but also trigger other consequences, such as mass migratory movements, political instability as well as an increase in food prices at a global scale. Climate change is a factor to consider when it comes to food security and access to food, something that can aggravate existing social and economic problems. Finally, while the health services of the developed countries are generally more predisposed to face the inevitable consequences of climate change, those more economically disadvantaged could suffer huge setbacks, because individual events such as floods, droughts, long-lasting heat waves or a drastic reduction of food resources will continue to exert increasing pressure on health services in affected areas.

Acknowledgements The present study is the result of the joint work of the two authors; in the drafting of the text, De Santis dealt with §§ 1, 2 and 4 and Bortone § 3. We would like to thank Prof. Mike George Riddell for valuable advice regarding the drafting of the text in English.

References

AA.VV (2013) Processi naturali e intervento umano. II Mulino, Bologna, p 357
Akhtar R (ed) (2016) Climate change and human health. Scenario in South and Southeast Asia. Springer, Cham, p 290
Bargagli A. e Michelozzi M.P. (2011), Clima e salute. Come contrastare i rischi immediati e a lungo termine delle ondate di calore, Roma, II Pensiero Scientifico Editore, 160.
Chelazzi G (2013) L'impronta originale. Storia naturale della colpa ecologica. Einaudi, Milano
Contarini E (2010) Qualcuno ha parlato di clima? Milano, Discanti, p 144
Dominguez Bascòn P. et Munoz Péres A. (1998), Clima e salute in Andalusia. Studio dell'incidenza dell'onda di calore del luglio 1995 sulla base dei ricoveri al pronto soccorso, C. Palagiano, G. De Santis e D. Castagnoli (a cura di), Metodi di raccolta dati e percezione della salute, "Atti del Sesto Seminario Internazionale di Geografia medica (Roma, 4–6 dicembre 1997)", Perugia, Edizioni Rux, 391–404.
Flannery T (2015) Una speranza nell'aria. I cambiamenti climatici e la sfida che siamo chiamati ad affrontare. Corbaccio Editore, Milano, p 240
Gelbspan R (1998) Clima Rovente. Baldini & Castoldi, Milano, p 307
Giddens A (2015) La politica del cambiamento climatico. II Saggiatore, Milano, p 287
Kandel R (1999) L'incertezza del clima. Einaudi, Milano, p 145
Lanza A (2000) II cambiamento climatico. II Mulino, Bologna, p 128
Lanza A (2006) Lo sviluppo sostenibile. II Mulino, Bologna, p 121
Lehner B, Döll P et al (2006) Estimating the impact of global change on flood and drought risks in Europe: a continental, integrated assessment. Clim Chang 75(3):272–299
Navarra A, Pinchera A (2002) II clima. Laterza, Roma-Bari, p XIV–229
Provenzale A. (a cura di) (2013), II mutamento climatico. Processi naturali e intervento umano, Bologna, II Mulino, 357
Räisänen J, Hansson U et al (2004) European climate in the late 21st century: regional simulations with two driving global models and two forcing scenarios. Clim Dyn 22:242–248
Ruddiman W. (2015), L'aratro, la peste, il petrolio. L'impatto umano sul clima, Università Bocconi Milano, pp 246
Smith L.C. (2011), 2050. II futuro del nuovo nord, Milano, Einaudi, pp 381
Von Weizsacker E.U., Lovins A.B. e Lovins H.L. (2009), Fattore 4. Come ridurre l'impatto ambientale moltiplicando per quattro l'efficienza della produzione, Milano, Edizioni Ambiente, 2009, pp. 337
Wigley T (2005) The climate change commitment. Science 307(5716):1766–1769
Wandel J (2006) Adaptation, adaptive capacity and vulnerability. Glob Environ Chang 16 (3):282–292
www.eea.europa.eu/it/.../cambiamento-climatico-e-salute-umana

Giovanni De Santis is a professor of human geography and geography, environment and health at the University of Perugia (Italy). Giovanni's scientific studies has developed through the analysis of various geographic aspects, highlighting in particular the existing relationship between man and environment, such as the following:

- The population and demographic dynamics
- Localization of inhabited areas
- Issues related to medical geography in particular
- The distinctive features and certain environmental reflections from tourism and circulation in general
- Some specific aspects of agricultural, historical and cartographic geography
- The problems and the implications related to environment sustainability

Giovanni is a member of the committee of the Commission on Health and Environment of the IGU-UGI.

Claudia Bortone holds two master's degrees: one in philosophy and a second in civil economics at the University of Perugia. She is a scientific associate to geography magazines in the production of scientific papers. With excellent knowledge of the English language, she has enriched her education with university activities of tutoring and teaching collaboration.

Chapter 4
COP21 in Paris: Politics of Climate Change

Rais Akhtar

Abstract An attempt has been made to discuss various dimensions of Paris Climate Agreement, its likely impact on the levels of global warming, and various voices in favour of and against the climate deal, and the US withdrawal from the Paris Agreement under President Donald Trump. Since USA backed out from the Paris Agreement, China and India are bound to re-visit their commitments on emissions, and the future of our planetary world looks bleak.

Keywords COP21 • Temperature increase • Air pollution • India • Renewable energy • Donald Trump

Introduction

The Paris Climate Agreement emerged successful with a narrow escape from disaster as it ran into overtime. As differences persist between the USA and emerging economies, the President Barack Obama used his authority to save American interests. The most important push to this climate deal was not the perception and understanding of climate change impact among participating countries, but a phone call from President Barack Obama to Chinese and Brazilian presidents and the Indian Prime Minister on the last day, i.e., 11 of December, of the conference, which led to the signing of this "historic" agreement. Had President Obama been so powerful politically and internationally, the Copenhagen Summit in 2009 would have been successful. There is further scope for research as to what pressure tactics as well as assurances were extended by the USA to emerging economies of China and India.

R. Akhtar (✉)
International Institute of Health Management and Research (IIHMR), New Delhi, India
e-mail: raisakhtar@gmail.com

© Springer International Publishing AG 2018

R. Akhtar, C. Palagiano (eds.), *Climate Change and Air Pollution*,
Springer Climate, DOI 10.1007/978-3-319-61346-8_4

Paris Climate Agreement

The Paris Climate Agreement, when 197 countries committed to keeping the global temperature rise "well below" the limit of 2 °C above preindustrial levels, came into force on 4 November 2016. By June 2017, 151 of 197 countries ratified the Agreement. The Paris Climate Agreement has just forced everyone to quickly switch to natural gas or nuclear in the power generation sector for the next 10–15 years. Because of that the price of natural gas would become high when demand is high that will have a positive influence on the renewable energy sector. Personally, I am crossing my fingers to believe that there will be some technological breakthrough in the energy production in our near future (combination of fuel cell and solar—use solar to generate hydrogen or thermo-exchange members that can capture waste heat under low-temperature difference). The policy that the world is currently working on is just to slow down the climate change and hope to avoid extreme climate or nonreversible devastating disaster in our planetary system (Lam 2016).

Politics of Climate Change

As for India, the newspaper headlines concerning Paris climate conferences varied from "Creators of climate change must cut emissions" to "Nations whose rise was powered by fossil fuels must bear more burden" attributed to the Prime Minister of India. At the same time, a group of developing nations comprising India, China, and others stated the global climate deal must produce a clear climate finance road map and ensure that the rich nations bear a heavier burden. Contrary to this, the Paris Climate Agreement reveals that the "Least developed countries and Small Island Developing States have special circumstances" that are eligible for provision of support. It is evident that both China and India are not eligible for any adaptation and mitigation support. The Guardian reported on 13 December: "When US officials realised Paul Oquist, Nicaragua's delegate, planned to deliver a fiery speech denouncing the deal, Secretary of State John Kerry and Raúl Castro, the Cuban leader, telephoned Managua to make sure that Oquist spoke after the agreement was adopted, when it would in effect be too late". Thus the US involvement in the shaping and architecture of the Paris climate deal was significant. Nigel Purvis has rightly called the White House's COP21 goals: less climate idealism, more political realism. The International Business Times remarked that COP21 Paris climate talks have failed by letting the rich off the hook. The Guardian reported on 12 December 2015 that James Hansen, an Adjunct Professor at Columbia University and known as the father of climate change awareness, calls Paris talks "a fraud." Of course the idea of financial support to a certain category of nations, particularly the least developing and island nations cannot be ignored. In this connection Stephen Dinan of the Washington Times,—Sunday, 29 November 2015, quoted Ugandan Foreign Minister Sam Kutesa who was explicit earlier this year when asked what it would take for developing countries to sign up for the emerging US-led climate deal: "Money." Thus the issues of equity and common but differentiated responsibilities (CBDR) were laid to rest with this agreement. Why were the USA and other developed countries eager to conclude a climate deal? Baseless arguments have been made by developed countries that developing countries including India and China will be the worst sufferer from climate change impacts. In a recent example of pressurizing India to accept developed countries' analysis that India may be hotter by 8°C and lose $200 billion per year (Hindustan Times, 16 July 2015), forgetting the devastation caused by European heat waves that killed 70,000 Europeans in 2003. In the ten global ranking of heat wave mortality, European heat wave mortality was at the top, followed by Russian heat wave, and US heat wave mortally figured at third, fourth, seventh, eighth and ninth positions. India's heat wave mortality in 2003 was placed at number six. In one of my papers, I argued that not only India and China but even developed countries—the USA, the UK and other nations of Europe—are vulnerable to climate change. Katrina (2005), Sandy (2012) and Harvey (2017) hurricanes had devastated the USA, while flooding in Europe and forest fires in Australia and recently in California are examples that show that Western countries are even more vulnerable. The last week of December 2015 had been a great disaster for England and southern USA as flooding devastated these regions. The huge blizzard which pounded the eastern coast of eastern Virginia (USA) during the fourth week of January 2016 has broken all records. WMO confirms 2016 as the hottest year on record, about 1.1 °C above preindustrial era.

In my view the developed countries, particularly the USA were adamant to conclude the Paris Climate Agreement in their favour, as the Americans and other developed nations realized that they are more vulnerable to climate change impacts.

Indian Context

Regarding the use of coal for energy, the reality is that each and every country uses its own resource for power generation. Australia, Germany, and India possess rich coal reserves. Therefore, these and other countries with rich coal reserves use it mostly for its power generation. As the meeting of COP21 in Paris concluded in a climate agreement, in my opinion India failed to take the stand based on the Kyoto Protocol that states "common but differentiated responsibilities", clearly meaning that the West must first reduce their emissions substantially. In one of the papers published in 2010 from Brussels, I have clearly stated the association between country's GDP and CO2 emissions (Akhtar 2010). Thus, high emissions are a must for development for developing countries. In Paris P.M. Modi has rightly asserted that "Climate change is a major global challenge. But it is not of our making" (Hindustan Times, 1st December, 2015) and "Nations whose rise was powered by fossil fuels must bear more burden industrialized countries" (Hindustan Times, 1st December, 2015). At the earlier meeting of the G8+5 in Heiligendamm in July 2007, the former Indian Prime Minister also indicated that we are determined to see that India's per capita emissions never exceed the per capita emissions of the industrialized countries.

During US Election 2016

Since India has taken a logical stand on emission reduction, and in the USA, the congress has rejected Obama's efforts to reduce GHG emissions, it seems unlikely that a Paris climate treaty will be approved by the Republican-dominated congress. Both Donald Trump and Ted Cruz, candidates for Republican nomination for Presidential elections in the USA, are against the Paris Climate Agreement. "I don't believe in climate change," Trump said flatly, while Ted Cruz doesn't believe in man-made climate change or Science behind it (quoted from The Atlantic, 9 December 2015). It seems the likely that if the Republicans wins the US Presidential election, the USA might pull out of the Paris Climate Agreement as they did when the Kyoto Protocol accord was signed. However, "President Obama's special envoy for climate change has warned Republican presidential hopefuls, including Donald Trump and Ted Cruz that any attempt to scrap the Paris Climate Agreement would lead to a "diplomatic black eye" for the US" (The Guardian, 16 February 2016).

Donald Trump: President of the USA

After election victory, Donald Trump met Al Gore who shared the 2007 Nobel Peace Prize with the IPCC; later he met with William Happer, a Princeton professor of physics who has been a prominent voice in questioning whether we should be concerned about human-caused climate change. It should be noted that in the 2015 senate testimony, Happer argued that the "benefits that more [carbon dioxide] brings from increased agricultural yields and modest warming far outweigh any harm". "While not denying outright that increasing atmospheric carbon dioxide levels will warm the planet, he also stated that a doubling of atmospheric carbon dioxide would only cause between 0.5 and 1.5 degrees Celsius of planetary warming (Mooney 2016). The most recent assessment of the United Nations' Intergovernmental Panel on Climate Change puts the figure much higher, at between 1.5 degrees and 4.5 degrees C". Scott Pruitt, the Oklahoma attorney general who has been a longtime adversary of the Environmental Protection Agency (EPA), has been named as the head of this agency and a close friend to the fossil fuel industry. Pruitt wrote that the debate on climate change is "far from settled", adding: "Scientists continue to disagree about the degree and extent of global warming and its connection to the actions of mankind" (Sidahmed 2016). Rex Tillerson who was the CEO of Exxon, a company that funded climate change denial for years, has been nominated Secretary of State. Worried Obama, just 2 days before Donald Trump took over the presidency, transferred $500 m to the Green Climate Fund in an attempt to protect the Paris climate deal (Slezak 2017).

In the historical Indian context, Paul Baran in his book *The Political Economy of Growth* (1957, New York) states that the colonial drain was a mercantilist concept—India's loss of economic resource and their transfer to Britain was a consequence of her political subordination. The coming of the British rule in India had broken up pre-existing self-sufficient agricultural communities and forced a shift to the production of export crops, which distorted the internal economy (Baran 1957). The resources from African and South Asian colonies were used to develop industrial base of Liverpool and Manchester. Baran also suggests that about 10% of India's gross national product was transferred to Britain each year in the early decades of the twentieth century. In light of the above, India failed to assert the Kyoto Protocol principle of "common but differentiated responsibility" between developed and developing nations, for gaining access to green technology and finance for both adaptation and mitigation.

The Paris Climate Agreement entered into force on 4 November 2016, 30 days after the date on which at least 55 parties to the convention accounting in total for at least an estimated 55% of the total global greenhouse gas emissions deposited their instruments of ratification, acceptance, approval or accession with the depositary.

Conclusion

On the first day of Trump's presidency, and shortly after inauguration on 20th January, 2017, the White House website was scrubbed of most climate change references. Instead, highlighted at the top of the issue list is the "America First Energy Plan," which talks about the need to roll back former president Barack Obama's far-reaching climate regulations, known as the Climate Action Plan. Trump had also appointed several most prominent climate change deniers, including Secretary of State in his team.

After about five months, President Trump announced on 1st June 2017 that he is withdrawing the United States from the landmark Paris climate agreement, an extraordinary move that puzzled America's allies and placed great hindrance in the global effort to address the warming planet. US joins only Syria and Nicaragua on climate accord 'no' list However, China, European Union, and. India have vowed to support Paris climate agreement, despite Trump's decision to withdraw from this landmark accord.

Nevertheless, future seems not encouraging and the Paris Climate Agreement may be dead following the decision by Donald Trump to withdraw from the Paris climate treaty. An Australian politician has said that though Australia has ratified the Paris Climate Agreement, "US withdrawal means Paris is cactus." As opined by Sneed, among numerous pledges made during Trump election campaign, include "cancelling" American involvement in the Paris climate accord, reviving the coal industry and rolling back federal environmental regulations. If Trump follows through, scientists say it could have a profound long-term effect on the planet" (Sneed 2017). Since USA backed out from the Paris Climate Agreement, China and

India are bound to re-visit their commitments on emissions, and the future of our planetary world looks bleak. Reffering to the US policy on climate change under Donald Trump and the hurricane Harvey that devastated Texas in late August, 2017, Mark Lynas has justly noted " we all have a duty to confront denial and speak out. If we fail, the Harveys, Katrinas and Sandys of the future will be even worse than the storms we experience today. And in future, as now, each subsequent climate disaster will just be "news". Surely we can do better than that" (Lynas 2017) Because of such grim scenario Stephen Hawking "has warned that Donald Trump's decision to withdraw from the Paris Climate Agreement on climate change could "push the Earth over the brink" and lead to a point where global warming is "irreversible" (The Independent 2017).

References

Akhtar R (2010) CO2 emission reduction and the emerging socio-economic development in developing country: a case study of India. In: Dapper MJ, Swinne D, Ozer P (eds) Developing countries facing global warming: a post-Kyoto assessment. Royal Academy of Overseas Sciences, Brussels, pp 15–26

Baran PA (1957) The political economy of growth. Monthly Review Press, New York

Lam N (2016) Personal correspondence

Lynas M(2017) Now we have a moral duty to talk about climate change, CNN, August 31

Mooney C (2016) Trump meets with Princeton physicist who says global warming is good for us, Washington Post, January, 13, www.washingtonpost.com

Sidahmed M (2016) Climate change denial in the Trump cabinet: where do his nominees stand? The Guardian, London, December 15. https://www.theguardian.com › Environment › Climate change scepticism

Slezak M (2017) Barack Obama transfers $500m to Green Climate Fund in attempt to protect Paris deal, The Guardian, London, January 18

Sneed A (2017) Trump day 1: global warming's fate, Scientific American, Climate, January 20

The Independent (2017) Stephen Hawking has warned that Donald Trump's decision to withdraw from the Paris Agreement on climate change could "push the Earth over the brink" and lead to a point where global warming is "irreversible", July 4

Rais Akhtar is presently adjunct professor of the International Institute of Health Management Research, New Delhi.

Part II
Case Studies: Developed Countries/Regions

Chapter 5
Climate Change Impacts on Air Pollution in Northern Europe

Ruth M. Doherty and Fiona M. O'Connor

Abstract The impacts of climate change on air pollution are discussed in the context of Northern Europe. Europe as a whole benefits from a wealth of data and statistics from the European Environment Agency and the European Monitoring and Evaluation Programme that also considers long-range transboundary air pollution and its own EU air quality standards. In this region projected future air quality levels are determined not only by climate change impacts affecting the regional to local-scale air pollution but also by climate drivers and phenomena that change hemispheric background pollution levels. This chapter reviews the impacts on air pollution in Northern Europe associated with projections of greenhouse gas emissions and emissions of pollutant primary species and precursors for the future, produced for the Intergovernmental Panel for Climate Change (IPCC). Studies relating these air pollution impacts to future changes in air pollution-related mortality and morbidity for Europe are also presented.

Keywords Air pollution • Climate change impacts • Ozone • Particulate matter • Human health • Northern Europe

Introduction: Health Effects and EU Air Pollution Levels in the 2000s

The European Environment Agency (EEA), who provides independent information on the environment to inform policy and decision-making across the EU, states that the three air pollutants that most significantly affect human health are particulate matter (PM), nitrogen dioxide (NO_2) and ground-level ozone (O_3) (EEA 2016). PM

R.M. Doherty (✉)
School of GeoSciences, University of Edinburgh, Scotland, UK
e-mail: ruth.doherty@ed.ac.uk

F.M. O'Connor
Met Office, Hadley Centre, Exeter, UK
e-mail: fiona.oconnor@metoffice.gov.uk

© Springer International Publishing AG 2018

R. Akhtar, C. Palagiano (eds.), *Climate Change and Air Pollution*,
Springer Climate, DOI 10.1007/978-3-319-61346-8_5

exposure, both short term (acute) and long term (chronic), is associated with all-cause and, in particular, cardiovascular and respiratory disease and mortality (WHO 2013a, b). PM has been measured for the last decade or so in Europe mainly as PM_{10} (PM_{10} (diameter <10 μm), often referred to as coarse PM) and more recently $PM_{2.5}$ (particle diameter <2.5 μm) often referred to as fine PM (e.g. in the UK, $PM_{2.5}$ measurements at regular monitoring sites have been available typically from the late 2000s). The health effects are thoughts to be greater for the smaller size particles due to their ability to penetrate more deeply into the thoracic and respiratory systems. Short-term exposure to O_3 is also associated with cardiovascular and respiratory mortality. The evidence base for long-term effects due to O_3 exposure is increasing, but this evidence is mainly from North American studies (COMEAP 2015). For NO_2 there has been much debate about whether effects are caused by NO_2 itself or by co-pollutants emitted by the same sources, notably traffic (COMEAP 2015). However, evidence now suggests independent effects of short-term exposure to NO_2 – associated with respiratory and cardiovascular outcomes – whilst for long-term exposure, a causal relationship is suggested (WHO 2013a; US EPA 2015).

The European Union (EU) has developed an extensive legislation establishing health-based standards and objectives for these and other air pollutants (http://ec.europa.eu/environment/air/quality/standards.htm). There are legally binding and target values for annual average $PM_{2.5}$, 24 h and annual average PM_{10}, NO_2, and a target value for maximum daily 8-h mean O_3. For 24-h mean PM_{10} and NO_2 35 and 18 exceedances respectively are allowed per year under these limit values. For O_3, the target values require no more than 25 exceedances averaged over 3 years.

However, despite substantial emission controls that have improved air quality for some pollutants, the percentage of the EU population exposed to air pollutant concentrations higher than the EU limit or target values (as given above) is between 8 and 30% (EEA 2015, Table ES.1). There are several underlying reasons:

(a) O_3 and some components of $PM_{2.5}$ are secondary pollutants, i.e. they are formed in the atmosphere from primary precursor emissions. Hence, besides precursor emissions, there are meteorological and transport factors as well as chemical transformation and deposition processes that determine their ambient concentrations.
(b) In addition, O_3 and some PM components are relatively long lived such that long-range transport of O_3 or PM pollution from outside the EU contributes significantly to regional EU levels (EEA 2015).
(c) For PM a further complication is the natural components due to dust; sea salt that cannot be regulated contributes to both $PM_{2.5}$ and PM_{10} levels (although more of these emissions are in the larger size fractions).
(d) O_3 chemistry is non-linear, and titration of O_3 by NO occurs when NO_x levels are high. This has led to increases in O_3 concentrations in the highly urbanised areas in the EU, including Belgium, Germany, the Netherlands and the UK (Bach et al. 2014; EEA 2015).

A key question is how will air pollution levels change further in Northern Europe under future emission policies and as a result of climate change? The following sections address this question by considering first future scenarios for greenhouse gas emissions and their impacts on climate as well as pollutant primary and precursor emissions developed for Intergovernmental Panel on Climate Change (IPCC) assessment reports (section "Future IPCC scenarios of climate and pollutant precursor emissions change"), in addition to outlining the impacts of climate change on air pollution (section "Climate change impacts on air pollution"). The impacts of IPCC climate and combined climate and emissions scenarios on O_3 and $PM_{2.5}$ pollution for Northern Europe are discussed in sections "Climate change impacts on air pollution: IPCC future climate scenarios" and "Air pollution Impacts from Combined Future IPCC climate and emissions scenarios", respectively. Section "Health effects of air pollution under climate change and combined emissions and climate change" present a synthesis of health impacts related to future changes in air pollutant concentrations which is followed by discussion and conclusions (section "Discussion and conclusions").

Future IPCC Scenarios of Climate and Pollutant Precursor Emissions Change

The Special Report on Emissions Scenarios (SRES; Nakicenovic et al. 2000) provided projections of future emissions of greenhouse gases including primary pollutants and pollutant precursor species for the IPCC third and fourth assessments in 2001 and 2007. These SRES emissions projections were based on a diverse future in terms of demographic, economic, population and technological driving factors and comprised four main families: A1, A2, B1 and B2. Global climate models (GCMs) using the SRES scenarios projected an average across the GCMs (termed "ensemble" average) change in global mean temperature for the years 2090–2099 compared to 1980–1999 of 1.4–6.3 °C (Meehl et al. 2007).

For the latest IPCC Fifth Assessment Report in 2013, a new series of emissions scenarios for greenhouse gases and pollutant precursors were developed on the basis of a radiative forcing value at the top of the atmosphere in 2100, termed Representative Concentration Pathways (RCP) scenarios. There are four RCPs covering a range of net radiative forcing projections: RCP2.6 (vanVuuren et al. 2011), RCP4.5 (Thomson et al. 2011), RCP6.0 (Masui et al. 2011) and RCP8.5 (Riahi et al. 2012). Unlike the SRES scenarios, some RCP scenarios considered aspects of climate mitigation and stabilisation. Figure 5.1 depicts projected temperature changes with time for the different RCP scenarios. By the end of the twenty-first century, the increase of global mean surface temperature is projected to be 0.3–1.7 °C for RCP2.6 and 2.6–4.8 °C for RCP8.5 compared to 1986–2005 (Collins et al. 2013).

Regionally in Europe, near-term (2016–2035 relative to 1986–2005) projections of ensemble mean changes in mean temperature over Northern Europe are typically

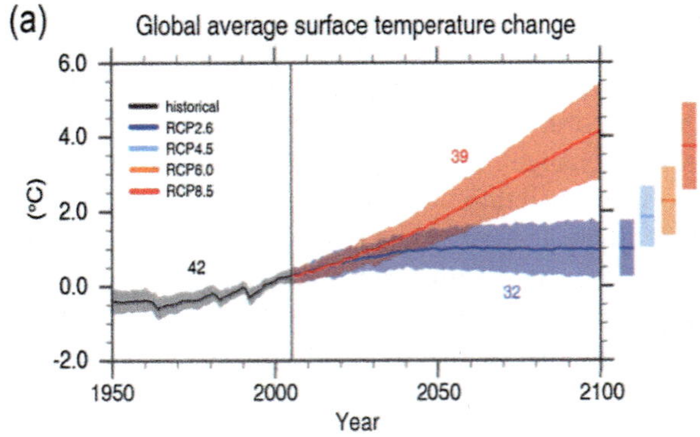

Fig. 5.1 CMIP5 time series from 1950 to 2100 of global annual mean surface temperature relative to the 1986–2005 time period. The projections out to 2100 are based on RCPs 2.6 and 8.5. The shading represents one standard deviation, and the number of models is given in the same colour. The projected global annual mean temperature change for 2081–2100 relative to 1986–2005 and the associated standard deviations for the 4 RCPs are shown as coloured vertical bars to the right of the figure. This is a reproduction of Fig SPM.7 in (IPCC 2013)

between 0–0.9 °C in winter and 0.6–1.2 °C in summer, whilst ensemble mean summer precipitation changes by −30 to +10%, and winter precipitation changes by −5 to 15% (see Figure 11.18, Kirtman et al. 2013). Near the end of the twenty-first century under the RCP 4.5 scenario, the median value across an ensemble of GCMs for area mean temperature is projected to increase (relative to 1986–2005) in Northern Europe by 2.7 °C and by 2.2 °C in summer and 3.4 °C in winter, respectively (Christensen et al. 2013). Across the GCMs, the increase in winter temperature is likely to be greater in Northern Europe compared to Central and Southern Europe and the converse for summer mean temperature (Christensen et al. 2013). The GCM ensemble median value for area mean annual precipitation in Northern Europe under RCP4.5 is projected to increase by 8% in 2081–2100, with summer precipitation increasing by 5% and winter precipitation increasing by 11% (Christensen et al. 2013). Across the GCMs annual mean precipitation is likely to increase in Northern Europe (Christensen et al. 2013).

All RCP scenarios assume aggressive abatement measures (Fiore et al. 2012). In particular NO_x emissions are reduced by ~50% in 2100 (from ~80 Tg N yr.-I to 30–50 Tg yr.-1) compared to 2000 levels, and black carbon (BC) emission also reduce by a similar percentage (Fig. 5.2). These measures generally result in large decreases in pollutant precursor species globally (Fig. 5.2; Fiore et al. 2012). However, ammonia (NH_3) increases in all scenarios due to increased agricultural-related emissions and methane (CH_4) that more than doubles in 2100 compared to 2005 (Fig. 5.2; Fiore et al. 2012).

However, as air pollution controls are not the primary focus of the SRES or the RCP emissions, the diversity in the ranges of precursor pollutant emission changes are somewhat small across the different RCP scenarios (Garcia-Menendez et al.

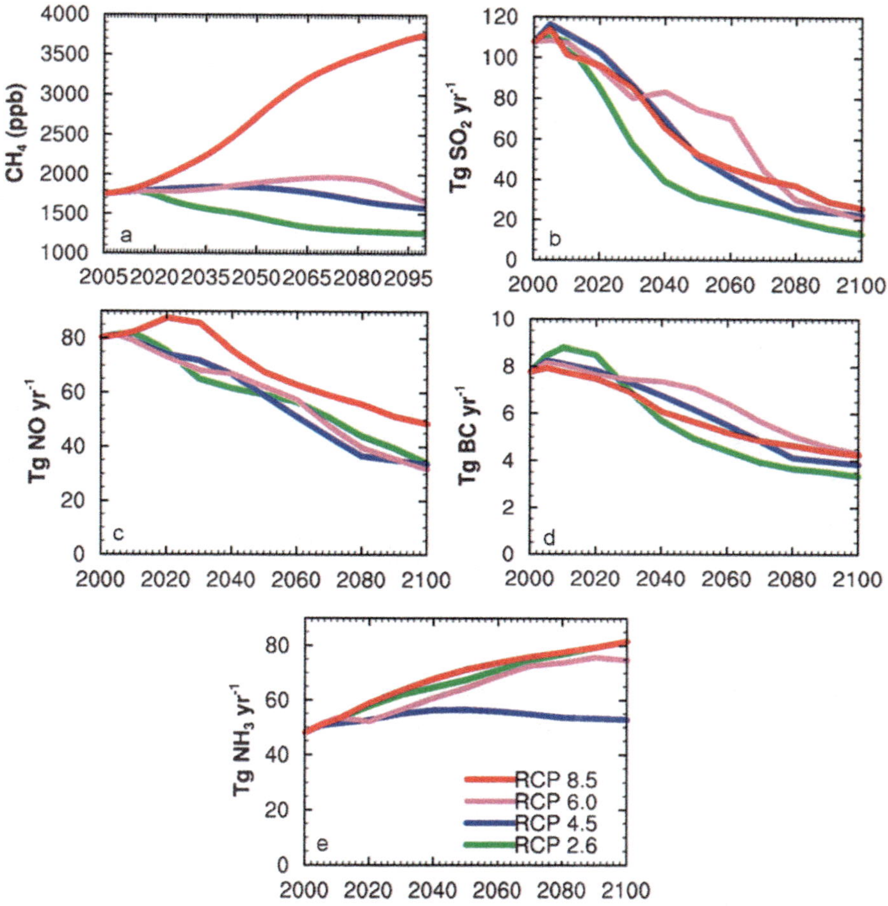

Fig. 5.2 Future evolution of (**a**) CH_4 abundance and selected global emissions of air pollutants and precursors, (**b**) SO_2, (**c**) NO, (**d**) BC, and (**e**) NH_3, from anthropogenic plus biomass burning sources combined, under the RCP scenarios (Reprinted from Fiore et al. (2012))

2015), except for the RCP8.5 scenario with very high levels of methane emissions as described above.

Coupled climate-chemistry models have been used to simulate the impacts of these IPCC SRES and RCP climate scenarios, resulting from changes in greenhouse gas emissions. Studies relating to SRES/RCP climate scenarios for Northern Europe are discussed below. The impacts of pollutant precursor emissions changes as well as climate change from the SRES/RCP scenarios have also been studied and are outlined in section "Air pollution impacts from combined future IPCC climate and emissions scenarios" following the discussion of the impacts associated with IPCC climate projections. First chemistry-climate change interactions are outlined below.

Climate Change Impacts on Air Pollution

Changes in mean temperature affect chemical reaction rates that influence production and loss rates of gaseous pollutants and hence affect local and regional pollution levels. Notably, higher temperatures increase the decomposition rate of peroxyacetyl nitrate (PAN), a reservoir species for nitrogen oxides (NO_x=NO+NO_2), reducing NO_2 or O_3 production following long-range transport but increasing local NO_2 and O_3 levels (Jacob and Winner 2009; Doherty et al. 2013). PM pollution is also impacted by temperature. However since PM comprises many different components, the overall impact is difficult to discern (Dawson et al. 2013; Garcia-Menendez et al. 2015). For example, higher temperatures enhance chemical reaction rates that lead to increased oxidation of sulphur dioxide (SO_2), which can condense to form sulphate aerosol – a major component of PM. Higher temperatures can reduce the partitioning of nitrate into the aerosol phase and hence reduce nitrate aerosol levels – another major component of PM in Northern Europe (e.g. in the UK; Yin and Harrison 2008; Harrison et al. 2012) and also some organic aerosol species (Fiore et al. 2012). Change in temperature also influences natural emissions of O_3 and PM precursors, e.g. wildfire emissions, emissions of isoprene – a biogenic volatile organic compound (VOC) – and emissions of methane from wetlands (O'Connor et al. 2010). One study focusing on agricultural areas in Europe also suggested that natural emissions of NO_x from soils increased slightly with higher temperature (Forkel and Knoche 2006). Regional O_3 and PM levels in Northern Europe can be impacted by transport of these emissions from elsewhere. For example, large parts areas in Northern Europe were impacts by PM pollution from forest fires in Spain and Portugal during the 2003 heatwave in Europe (Hodzic et al. 2007).

Changes in mean precipitation amount as well as frequency impact wet deposition that removes pollutants and in particular PM from the atmosphere. PM levels decrease in areas where increased precipitation frequency is simulated and vice versa (Fang et al. 2011; Penrod et al. 2014; Allen et al. 2016). Cloud amount also influences the magnitude of incoming solar radiation and hence photolysis rates that influence gaseous pollutants. Several studies for Europe link increased summer O_3 concentrations to enhanced NO_2 photolysis rates in turn caused by reduced cloud amount (Meleux et al. 2007; Katragkou et al. 2011). Depending on the spatial extent of the changes in rainfall regional and local PM pollution may be influenced by climate-induced changes in precipitation.

Besides changes in mean temperature and precipitation, changes in other mean climate variables notably humidity and boundary layer mixing height also impact on air pollution levels (see Table 1; Fiore et al. 2012). Higher humidities occur under climate change as the warmer atmosphere hold more moisture, lead to greater O_3 destruction in low NOx regions and hence can impact regional O_3 levels transported across the oceans into Northern Europe (Colette et al. 2015). Changes in the height of the boundary layer as well as wind speed exert a major control on the mixing and dispersion of local pollution. However, the impacts of climate change on these meteorological variables associated with dispersion that are influenced strongly by local-scale features are highly uncertain.

In addition to change in mean climate, mean air pollution levels and in particular episodes associated with exceedances of air quality standards will be affected by changes in climate extremes such as heatwaves as well as climate phenomena that influence climate extremes and regional air pollution. For Northern Europe the most relevant climate phenomena are the North Atlantic Oscillation (NAO), extra-tropical cyclones or storms and blocking high-pressure systems (Christensen et al. 2013). These phenomena typically impact pollution transport at the regional scale.

The NAO represents the intensity of the pressure gradient and thus the wind strength and direction across the North Atlantic across Northern Europe (e.g. Hurrell 1995). The positive NAO phase (strong pressure gradient) is associated with pollution transport from North America to Northern Europe, whilst the negative NAO phase (weak pressure gradient) is related to slower transport of pollution from Eastern Europe to Western Europe primarily Central and Southern Europe (Christiados et al. 2012; Pausata et al. 2012). Pollution export from Europe has also been associated with the NAO (Eckhardt et al. 2003; Duncan and Bey 2004). The response of the NAO to climate change is uncertain, although recent analysis suggests a small increase in positive NAO values in winter; however, model-to-model uncertainty is large (Christensen et al. 2013).

The strength and phase of the NAO are related to the relevant storm track pathways across the North Atlantic and into Europe. A poleward shift of the North Atlantic storm track has previously been linked to climate change and a tendency towards a positive NAO phase (Yin 2005), but recent studies suggest a weak extension of the storm track towards Europe, and a small reduction in frequency globally (Ulbrich et al. 2009; Christenesen et al. 2013). In addition, in relation to O_3, mid-latitude cyclones are also associated with transport of ozone-rich air from the lower stratosphere to the troposphere (Neu et al. 2014). This transport leads to elevated O_3 levels in the upper-mid troposphere which may or may not reach the surface. Under climate change stratosphere-troposphere exchange is predicted to increase as a result of an enhanced Brewer-Dobson circulation (Butchart and Scaife 2001). These changes in horizontal and vertical transport may impact the dispersion of pollution to and from Northern Europe. However there is much uncertainty across GCM projections of climate-driven changes in this mid-latitude cyclone pathways, frequency and intensity, all of which may potentially impact pollution transport.

Blocking high-pressure systems are associated with low wind speeds and are slow moving and have been associated with PM pollution in winter (e.g. Webber et al. 2017) and are often the cause of heatwaves in summer. Over Europe, a multi-model GCM study suggested a decrease in winter and summer blocking frequency (Masato et al. 2013) under climate change. Warmer mean temperature leading to drier soils have been suggested as a mechanism to enhance heatwave impacts over Europe, e.g. through suppression of O_3 deposition leading to higher O_3 levels (Vautard et al. 2005; Solberg et al. 2008; Emberson et al. 2013). The relationship between blocking – which is a large-scale feature – and stagnation, which reflects local wind speeds pollution, is however complex. Hence the overall relationship between blocking and air pollution episodes remains uncertain (Kirtman et al. 2013). Horton et al. (2014) report annual mean changes in stagnation under RCP8.5 using an air stagnation index, but these were not significant over Northern Europe. However, several studies suggest that the extreme temperature experienced during the 2003 heatwave in Europe will become the average summertime mean temperatures by around 2050 based on SRES scenarios (Schär et al. 2012; Stott et al. 2004). The passage of mid-latitude cyclones followed by blocking high-pressure systems has been shown to be a means of O_3 pollution transport whereby pollution transported to the mid-troposphere descends to the surface either in dry air streams embedded within the cyclone (Brown-Steiner and Hess 2011; Lin et al. 2012) or with subsidence to the surface that occurs through the subsequent passage of a high-pressure system (Knowland et al. 2015).

Climate Change Impacts on Air Pollution: IPCC Future Climate Scenarios

Climate change impacts on O_3 pollution have been widely investigated in the literature. Figure 5.3, based on a synthesis of the literature for different IPCC climate projections, depicts the global and regional including European annual mean surface O_3 response to climate change in 2030 compared to 2000 (green bars) (Kirtman et al. 2013). The change in annual mean regionally averaged surface O_3 due to changes in pollutant primary and precursor emissions from SRES (blue) and RCP (red) are also shown for comparison. The green solid bar ranges show the multi-model standard deviation of the annual mean reflecting model-to-model differences in regionally averaged O_3, whilst the dashed line shows the spatial variation in the annual across Europe for one model (Fiore et al. 2012; Kirtman et al. 2013). In general, O_3 changes and variability in the annual mean area average values are larger as a result of changes in emissions alone compared to climate change alone (Fig. 5.3).

The decrease in global and European average O_3 is due to higher water vapour and temperatures affecting background O_3 levels mainly in unpolluted regions. However, as discussed above, higher temperatures can lead to local O_3 increases in highly polluted regions – here an increase during the peak pollution season of 2–6 ppbv for Central Europe is depicted from a study by Forkel and Knoche, (2006). This increase in surface O_3 due to climate change, which occurs in polluted regions, is termed the "climate penalty effect" and has been reported in observational as well as modelling studies (Wu et al. 2008b; Bloomer et al. 2009; Rasmussen et al. 2013; Collette et al. 2015). Most recently, Collette et al. (2015) addressed the question of whether the ozone climate penalty was robust across Europe by performing a meta-analysis based on data from 11 studies covering different time periods and climate scenarios – mainly the SRES scenarios used from simulations

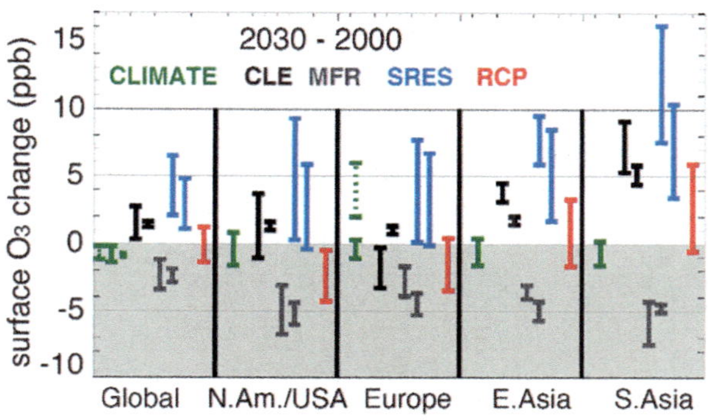

Fig. 5.3 Reprinted from Chap. 11, IPCC WG1 Fifth Assessment Report, figure 11.22, adapted from Fiore et al. (2012). Changes in surface O_3 (ppb) between year 2000 and 2030 driven by climate alone (CLIMATE; *green*) or emissions alone following CLE (*black*), MRF (*grey*), SRES (*blue*) and RCP (*red*) emission scenarios. Bars represent multi-model standard deviation (for further details, see Kirtman et al. (2013))

using both global and regional climate-chemistry models. In agreement with the results in Fig. 5.3, the near-term (2010–2040) changes in surface O_3 were found to be small for all regions in Northern Europe.

Figure 5.4 depicts the change in summertime surface O_3 between 2041 and 2070 based on the SRES A1B scenario and historical levels and significance levels (assessed using a student t-test) that are used to depict robustness when two-thirds of the models agree either on the significance on the change or its non-significance. Over Northern Europe the climate penalty ranges from around 1 ppbv in France and mid-Europe, but with a lack of model agreement, to a decrease or climate benefit over Scandinavia and the British Isles up to 1 ppbv. Considering the range of climate scenarios, for France and mid-Europe region, average median O_3 increases for this period reach up to 5 ppbv and up to ~7 ppbv for 2080–2100 (see Fig. 3, Collette et al. 2015) which occurs under the SRES A2 scenario. For the British Isles and Scandinavia, there are consistently small changes – typically decreases in 2040–2070 of ~1 ppbv (as shown in Fig. 5.4). For 2070–2100, over Scandinavia, most climate scenarios yield further small decreases in O_3, whilst for the British Isles simulations performed with the RCP 8.5 scenario show a larger median decrease (~1.5 ppbv) but an increase of a similar magnitude using the SRES A2 climate scenario. Overall, this meta-analysis demonstrates that surface O_3 changes are significant across Europe with a latitudinal gradient showing a O_3 climate penalty for large parts of continental Northern Europe and a climate benefit further north in the vicinity of the North Atlantic. It is likely that the decreases in the northernmost regions are associated with regional O_3 decreases due to higher humidities leading to higher O_3 destruction as discussed in section "Climate change impacts on air pollution". Previous studies of climate change impacts on surface O_3 using high-resolution regional models have shown typical results to those described above for Northern Europe in terms of spatial patterns, although the magnitude of change varies with metric (Collette et al. 2013; Langner et al. 2012a, b). In a

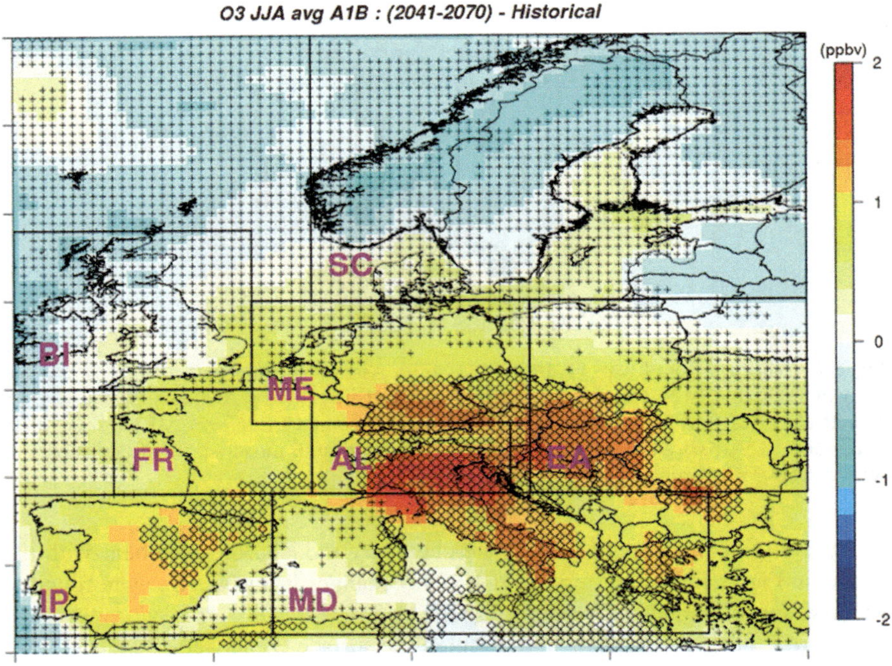

Fig. 5.4 Anomaly of average JJA ozone (ppbv) under the A1B scenario by the middle of the century (2041–2070) according to nine models for 144 simulated years. At each grid point, the shading is the average of the nine model ensembles, each model response being the average change between future and present conditions (see Table 1 for the exact years corresponding to present conditions for each model). A diamond sign (respectively a plus sign) is plotted where the change is significant (respectively not significant) for two-third of the models so that the absence of any symbol indicates the lack of model agreement. Subregions used in Fig. 5.3 are displayed on the map with the following labels: *AL* Alps, which includes Northern Italy, *BI* British Isles, *EA* Eastern Europe, *FR* France, *IP* Iberian Peninsula, *MD* Mediterranean, *ME* mid-Europe, *SC* Scandinavia (Reprinted from Collette et al. (2015), ERL)

regional European multi-modelling study using the SRES A1B climate scenario, Langner et al. (2012a) report that in 2100 in Northern Europe climate change leads to reductions of 0–3 ppbv for both mean and daily maximum O_3 in summer. Langner et al. (2012b) suggest that climate change has greater impact on episodic O_3 (they examine the 95th percentile of hourly O_3) than on longer-term (mean and daily maximum O_3) summer averages.

Climate change impacts on PM are much less certain, as discussed in section "Climate change impacts on air pollution", due to its multiple components being influenced by a number of climate factors, often acting in opposite directions, leading to cancelling effects. $PM_{2.5}$ concentrations are expected to decrease in regions where precipitation increases enhance wet removal (Kirtman et al. 2013). However, there is a lack of consensus on other climate-driven factors leading to low confidence in the overall impact of climate change on $PM_{2.5}$ distributions (Kirtman et al. 2013). One regional modelling study over Europe reported the geographical patterns of the impact of climate on surface summer PM levels appeared much less robust than for O_3 (Collette et al. 2013). However, most recently, a PM climate penalty has been suggested (Garcia-Menendez et al. 2015; Allen et al. 2016). A PM climate penalty simulated in 2100 in the Eastern United States was attributed to enhanced sulphate concentrations due to faster and greater SO_2 oxidation with higher temperature (Garcia-Menendez et al. 2015). A recent multi-model study using the RCP 8.5 climate scenario suggested that climate change may increase the aerosol burden and surface PM concentrations, through a reduction in large-scale precipitation over northern mid-latitude land regions (Allen et al. 2016). To date there is no emerging consensus on a PM climate penalty for Europe.

As discussed in section "Climate change impacts on air pollution", climate change can affect climate phenomena that can impact air pollution transport and episodes. In particular changes in mid-latitude storm track pathways and frequency affect large-scale pollution transport (Wu et al. 2008; Barnes and Fiore 2013), and large-scale blocking may affect local stagnation and heatwave episodes. Modelling studies generally suggest increases in the frequency and duration of extreme O_3 pollution events, but there is considerable uncertainty in spatial patterns of these events and their drivers (Forkel and Knoche 2006; Fiore et al. 2012; Kirtman et al. 2013). Overall, it is suggested that the peak pollution levels will increase in polluted regions due to higher temperatures associated with stagnation episodes (Fiore et al. 2012; Kirtman et al. 2013), but further work to improve understanding on the linkage between climate change impacts on blocking, stagnation and pollution events is needed.

Air Pollution Impacts from Combined Future IPCC Combined Climate and Emissions Scenarios

The majority of studies on air pollution impacts in the future consider both climate change and emission change. Typically these studies, especially those that use global-scale models, use compatible scenarios for future emissions of pollutant species and precursors and for greenhouse gases emissions that generate climate scenarios such as the SRES or RCP scenarios. As such, the joint effect of emission and climate change under the four RCPs scenarios on O_3 and PM air quality averaged over Europe is shown in Fig. 5.5.

The differing annual mean surface O_3 response across Europe (as well as globally) with an increase in RCP 8.5 as compared to decreases in other three RCP scenarios is clear. In 2100, under RCPs 2.6, 4.5 and 6.0, there is a reduction in annual mean surface O_3 between -5 and -20 ppbv compared to 2005. This decrease is primarily due to a reduction in NO_x and VOCs precursor emissions (Fig. 5.2; see also Fig. 1 Cionni et al. 2011). These changes are larger than those

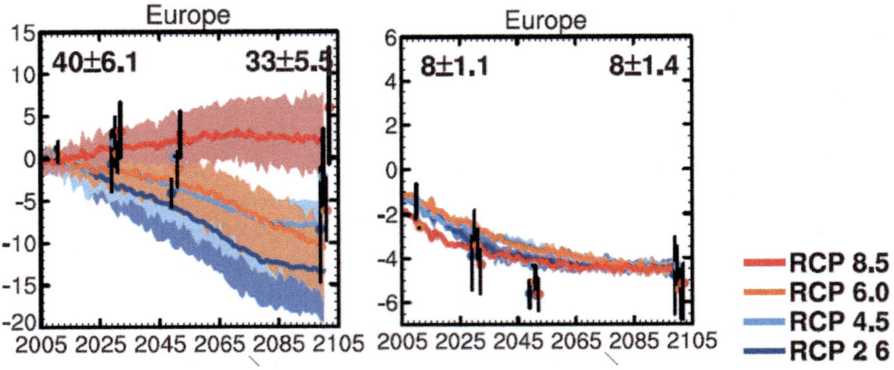

Fig. 5.5 Projected changes in annual mean surface (*left*) O_3 (ppbv) and (*right*) $PM_{2.5}$ (μg m-3) from 2000 to 2100 following the RCP scenarios (8.5 *red*, 6.0 *orange*, *light blue* 4.5, 2.6 *dark blue*) averaged over Europe (land). *Coloured lines* show the average, and shading denotes the full range of four chemistry-climate models, and *coloured dots* and bars represent the average and full range of ~15 ACCMIP models (Taken from Fiore et al. (2012) as used in Kirtman et al. (2013). The European panels are extracted from Figures 11.23a and 11.23b (Kirtman et al. 2013))

discussed due to climate change alone in 2100 (section "Climate change impacts on air pollution: IPCC future climate scenarios"). The increase in annual mean surface O_3 (~2 ppbv) under RCP8.5 reflects primarily the large increase in CH_4 emissions and outweighs the impact of reductions in other O_3 precursor species (Fig. 5.2). The corresponding changes in European average annual mean $PM_{2.5}$ concentrations are similar across the four RCP scenarios. All scenarios lead to a decrease in $PM_{2.5}$ compared to present day of ~4–5 μg m-3. The reductions in $PM_{2.5}$ generally follow reductions in SO_2 emissions (Fig. 5.2) and primary organic carbon emissions (Fiore et al. 2012; Kirtman et al. 2013). However, as noted in section "Future IPCC scenarios of climate and pollutant precursor emissions change" NH_3 emissions increase over time which led to higher ammonium aerosol. Increased ammonium alongside reduced SO_2 emissions may lead to relatively higher ammonium nitrate aerosol levels (Kirtman et al. 2013). Overall, it appears that the emission changes generally are the main drivers of changes in annual mean O_3 and $PM_{2.5}$. These impacts are either augmented or reduced by the impact of climate change. Conversely, changes in peak levels of pollution during O_3 or PM episodes can well be largely driven by changes in climate affecting climate phenomena.

Several higher-resolution regional modelling studies for Europe have also highlighted the dominance of pollutant primary and precursor emissions changes over climate change in driving future changes in O_3 and PM levels (Langner et al. 2012a; Coleman et al. 2013; Collette et al. 2013; Lacressonnière et al. 2014). Using the SRES A1B climate scenario together with the RCP 4.5 for pollutant precursor emissions, Langner et al. (2012a) found lower summertime daily maximum surface O_3 of around 9 ppbv in Northern Europe in 2100. Similar findings were reported by Coleman et al. (2013) who noted that changes in meteorology over the North Atlantic region became more influential over time. In contrast, using the RCP 8.5

scenario, Lacressonnière et al. (2014) reported increased surface O_3 over NW Europe in both the 2030s and 2050s due to the large and unique CH_4 emissions increase under this scenario. As discussed in section "Climate change impacts on air pollution", natural emissions of pollutant precursors may be impacted by climate change. Hence climate-driven increases in natural NO_x emissions from lightning and soils have the potential to offset anthropogenic NO_x emission reductions (Kim et al. 2015).

Health Effects of Air Pollution Under Climate Change and Combined Emissions and Climate Change

A very limited number of studies have linked climate change impacts on air pollution to changes in human health; most of these studies have been global studies (Fang et al. 2013, Silva et al. 2016) or for the USA (e.g. Knowlton et al. 2004; Bell et al. 2007; Tagaris et al. 2009; Post et al. 2012; Garcia-Menendez et al. 2015). These studies have been focussing on chronic or long-term exposure.

One study by Fang et al. (2013) examined global $PM_{2.5}$ and O_3 mortality associated with climate change under the SRESA1B climate scenario. They found that $PM_{2.5}$ levels increased in 2081–2100 relative to 1981–2000 over most major emission regions due to reduced precipitation – in agreement with Allen et al. (2016) (see section "Climate change impacts on air pollution: IPCC future climate scenarios")-, except in parts of Northern Europe where $PM_{2.5}$ levels decreased (Fang et al. 2013). Across Europe annual premature mortality associated with chronic exposure to $PM_{2.5}$ increased by ~1% with an additional 3300 deaths (95% confidence interval, CI, of 2200–4400). Years of life lost (YLL) increasing by 1% and by approximately 17,000 (95% CI, 11,000–22,000) years. For O_3, again, a mixed response was found with increases in continental Northern Europe and decreases in Scandinavia (Fig. 2, Fang et al. 2013) in agreement with previous studies (see section "Climate change impacts on air pollution: IPCC future climate scenarios"). This led to an overall increase in annual premature mortality due to respiratory disease from chronic O_3 exposure across Europe of 0.6% with an additional annual 300 deaths (95% CI, 100–500), with YLL increasing by 0.5% or 5800 years lost (95% CI, 3000–8600). This study assumed a constant population. A regional European modelling study using the SRES A2 climate scenario also estimated annual premature mortalities due to exposure to ozone in the 2030s and 2050s. O_3-related mortality and morbidity increased over most of Europe but decrease over the northernmost Nordic and Baltic countries, with the largest change being a 34% increase over Belgium (Orru et al. 2013). This study highlighted the results described above that the effects of climate change on ozone concentrations could differentially influence mortality and morbidity across Europe (Orru et al. 2013).

The health impacts of combined emissions of precursor pollutants change and climate change under the four RCP scenarios have recently been investigated by Silva et al. (2016). These authors additionally considered future population projections and changes in baseline mortality rates over time. Premature $PM_{2.5}$ mortality for Europe was reduced by 137,000–176,000 annual deaths across the RCP scenarios in the 2030s compared to the 2000s. Corresponding annual avoided mortalities for Europe were 187,000–200,000 deaths for the 2050s and 103,000–112,000 deaths for the 2100s, respectively. As discussed in section "Air Pollution impacts from combined current and future IPCC climate and emissions scenarios." These decreases were driven by reductions in primary and precursor emissions in the future (Fig. 5.2). O_3-related respiratory premature mortality for Europe also decreased with −880 to −8,870 annual avoided deaths in 2030 and −440 to −9760 annual avoided deaths in 2050. The lower limits were associated with the RCP8.5 scenario whereby increasing CH_4 emissions offset the impacts of reductions in other O_3 precursor species emissions. In 2100, under RCP8.5 premature mortality increases by 2,390 annual deaths, whilst premature mortalities are reduced by −24,900 to −44,600 annually for the other three RCP scenarios. Overall, the changes in emissions and population, rather than climate change, are the main drivers of change in $PM_{2.5}$ and O_3 pollution-related health effects.

Discussion and Conclusions

The impacts of climate change on air pollution have been summarised and discussed in relation to climate and emission scenarios produced for recent IPCC assessments. There is much literature outlining the effects of climate change on surface O_3 air pollution, and a number of studies focus on Europe. The effects of climate change on surface PM pollution are less well documented, in part due to the complexity and uncertainties in quantifying the combined effect of climate change on PM arising from the net change in its different PM components, but new studies are emerging. Northern Europe is influenced by several climate phenomena in relation to pollution transport (and photochemistry) that may in turn be influenced by climate change: the NAO, mid-latitude cyclones and blocking high-pressure systems that can be associated with heatwaves in summer. These changes in climate affect background pollution as well as regional/local pollution episodes in Northern Europe.

Numerous studies have examined the impacts of climate change based on IPCC climate scenarios. Over Europe the robustness of a climate penalty has been discussed. Surface O_3 concentrations are generally projected to increase under climate change in continental Northern Europe (a climate penalty) but decrease (a climate benefit) further north over the British Isles and Scandinavia. This leads to a latitudinal gradient in the surface O_3 response and consequent health impacts due to climate change across Northern Europe and shown in both global and regional modelling studies. For PM, climate penalties and benefits have also been suggested to occur across Northern Europe. Further studies in this region are needed to understand (a) the local precipitation response to climate change levels and (b) - temperature-driven changes in precursor oxidant gases and their partitioning and their respective influences on surface PM levels and thereby on PM-related mortality and morbidity. There is much uncertainty and few studies on the impacts of climate change on air pollution episodes.

When projections of primary and precursor pollutant species are considered in combination with changes in climate, generally the impacts of emission changes outweigh those due to climate change when considering annual and summertime mean air pollution levels.

Overall, a key uncertainty is the range of projected changes in surface O_3 and PM across different models when driven by the same climate scenarios. Health impacts, in relation to chronic exposure to PM and O_3, are also uncertain in several aspects. In particular, there are uncertainties in risk estimates associated with different health outcomes for O_3 and PM exposure, and how these risk estimates are modified due to temperature or multi-pollutants. In addition, daily baseline mortality and morbidity rates may not remain constant in the future.

In terms of linkages between climate change and air quality policies, the latest RCP scenarios highlight the potential for climate and air pollution control policy scenarios to act in tandem, whereby reductions in methane and black carbon have benefits for air quality as well as climate change (UNEP 2011; Shindell et al. 2012).

References

Allen RJ, Landuyt W, Rumbold ST (2016) An increase in aerosol burden and radiative effects in a warmer world. Nat Clim Change 6:269–274. doi:10.1038/nclimate2827

Bach H, Brandt J, Christensen JH, Ellermann T, Geels C, Hertel O, Massling A, Nielsen HØ, Nielsen O-K, Nordstrøm C, Nøjgaard JK, Skov H, Chatterton T, Hayes E, Laxen D, Irwin J, Longhurst J, Pelsy F, Zamparutti T (2014) Services to assess the reasons for non-compliance of ozone target value set by directive 2008/50/EC and potential for air quality improvements in relation to ozone pollution: final report. ECORYS, Rotterdam

Barnes EA, Fiore AM (2013) Surface ozone variability and the jet position: implications for projecting future air quality. Geophys Res Lett 40:2839–2844. doi:10.1002/grl.50411

Bell ML, Goldberg R, Hogrefe C, Kinney P, Knowlton K, Lynn B, Rosenthal J, Rosenzweig C, Patz J (2007) Climate change, ambient ozone, and health in 50 US cities. Clim Chang 82 (1–2):61–76. doi:10.1007/s10584-006-9166-7

Bloomer BJ, Stehr JW, Piety CA, Salawitch RJ, Dickerson RR (2009) Observed relationships of ozone air pollution with temperature and emissions. Geophys Res Lett 36:L09803. doi:10.1029/2009GL037308

Brown-Steiner B, Hess P (2011) Asian influence on surface ozone in the United States: a ... chemistry, seasonality, and transport mechanisms. J Geophys Res 116:D17309. doi:10.1029/2011JD015846. 1

Butchart N, Scaife AA (2001) Removal of chlorofluorocarbons by increased mass exchange between the stratosphere and troposphere in a changing climate. Nature 410:799–802. doi:10.1038/35071047

Christensen JH, Krishna Kumar K, Aldrian E, An S-I, Cavalcanti IFA, de Castro M, Dong W, Goswami P, Hall A, Kanyanga JK, Kitoh A, Kossin J, Lau N-C, Renwick J, Stephenson DB, Xie S-P, Zhou T (2013) Climate phenomena and their relevance for future regional climate change. In: Stocker TF, Qin D, Plattner G-K, Tignor M, Allen SK, Boschung J, Nauels A, Xia Y, Bex V, Midgley PM (eds) Climate change 2013: the physical science basis, Contribution of Working Group I to the Fifth Assessment Report of the Intergovernmental Panel on Climate Change. Cambridge University Press, Cambridge, UK, pp 1217–1308. doi:10.1017/CBO9781107415324.028

Christoudias T, Pozzer A, Lelieveld J (2012) Influence of the North Atlantic Oscillation on air pollution transport. Atmos Chem Phys 12:869–877. doi:10.5194/acp-12-869-2012

Cionni I, Eyring V, Lamarque JF, Randel WJ, Stevenson DS, Wu F, Bodeker GE, Shepherd TG, Shindell DT, Waugh DW (2011) Ozone database in support of CMIP5 simulations: results and corresponding radiative forcing. Atmos Chem Phys 11:11267–11292. doi:10.5194/acp-11-11267-2011

Coleman L, Martin D, Varghese S, Jennings SG, O'Dowd CD (2013) Assessment of changing meteorology and emissions on air quality using a regional climate model: impact on ozone. Atmos Environ 69:198–210

Colette A, Bessagnet B, Vautard R, Szopa S, Rao S, Schucht S, Klimont Z, Menut L, Clain G, Meleux F, Curci G, Rouïl L (2013) European atmosphere in 2050, a regional air quality and climate perspective under CMIP5 scenarios. Atmos Chem Physics 13:7451–7471. doi:10.5194/acp-13-7451-2013

Colette A, Andersson C, Baklanov A, Bessagnet B, Brandt J, Christensen JH, Doherty R, Engardt M, Geels C, Giannakopoulos C, Hedegaard G, Katragkou E, Langner J, Lei H, Manders A, Melas D, Meleux F, Rouïl L, Sofiev M, Soares J, Stevenson DS, Tombrou-Tzella M, Varotsos KV, Young P (2015) On the robustness of the ozone climate penalty in Europe. Environ Res Lett 10:084015. doi:10.1088/1748–9326/10/8/084015

Collins M, Knutti R, Arblaster J, Dufresne J-L, Fichefet T, Friedlingstein P, Gao X, Gutowski WJ, Johns T, Krinner G, Shongwe M, Tebaldi C, Weaver AJ, Wehner M: Long-term climate change: projections, commitments and irreversibility. In: Climate Change 2013: the physical science basis. Contribution of Working Group I to the Fifth Assessment Report of the Intergovernmental Panel on Climate Change [Stocker TF, Qin D, Plattner G-K, Tignor M, Allen SK, Boschung J, Nauels A, Xia Y, Bex V, Midgley PM (eds) Cambridge University Press, Cambridge/New York 2013

COMEAP (2015) Quantification of mortality and hospital admissions associated with ground-level ozone, UK Department of Health Committee on the Medical Effects of Air Pollution. ISBN 978-0-85951-776-8, https://www.gov.uk/government/collections/comeap-reports.

Dawson JP, Bloomer BJ, Winner DA, Weaver CP (2013) Understanding the meteorological drivers of U.S. particulate matter concentrations in a changing climate. Bull Am Meteorol Soc 95(4):521–532

Doherty RM, Wild O, Shindell DT, Zeng G, MacKenzie IA, Collins WJ, Fiore AM, Stevenson DS, Dentener FJ, Schultz MG, Hess P, Derwent RG, Keating TJ (2013) Impacts of climate change on surface ozone and intercontinental ozone pollution: a multi-model study. J Geophys Res Atmos 118. doi:10.1002/jgrd.50266

Duncan BN, Bey I (2004) A modeling study of the export pathways of pollution from Europe: seasonal and interannual variations (1987–1997). J Geophys Res-Atmos 109:D08301. doi:10.1029/2003JD004079

Eckhardt S, Stohl A, Beirle S, Spichtinger N, James P, Forster C, Junker C, Wagner T, Platt U, Jennings SG (2003) The North Atlantic Oscillation controls air pollution transport to the Arctic. Atmos Chem Phys 3:1769–1778. doi:10.5194/acp-3-1769-2003

EEA (2015) Air quality in Europe – 2015 report. EEA Report No 5/2015., European Environment Agency. ISBN: 978–92–9213-701-4. http://www.eea.europa.eu/publications/air-quality-in-europe-2015

Emberson LD, Kitwiroon N, Beevers S, Büker P, Cinderby S (2013) Scorched Earth: how will changes in the strength of the vegetation sink to ozone deposition affect human health and ecosystems? Atmos Chem Phys 13:6741–6755. doi:10.5194/acp-13-6741-2013

Fang Y, Fiore AM, Horowitz LW, Gnanadesikan A, Held I, Chen G, Vecchi G, Levy H (2011) The impacts of changing transport and precipitation on pollutant distributions in a future climate. J Geophys Res 116(D18):D18303. doi:10.1029/2011jd015642

Fang Y, Mauzerall DJ, Liu J, Fiore AM, Horowitz (2013) Impacts of 21st century climate change on global air pollution-related premature mortality. Clim Chang 121:239–253. doi:10.1007/s10584-013-0847-8

Fiore AM, Naik V, Spracklen DV, Steiner A, Unger N, Prather M et al (2012) Global air quality and climate. Chem Soc Rev 41:6663–6683

Forkel R, Knoche R (2006) Regional climate change and its impact on photooxidant concentrations in southern Germany: simulations with a coupled regional climate-chemistry model. J Geophys Res 111:D12302

Garcia-Menendez F, Saari RK, Monier E, Selin NE (2015) U.S. air quality and health benefits from avoided climate change under greenhouse gas mitigation. Environ Sci Technol 49 (13):7580–7588. doi:10.1021/acs.est.5b01324

Harrison RM, Dall'Osto M, Beddows DCS, Thorpe AJ, Bloss WJ, Allan JD, Coe H, Dorsey JR, Gallagher M, Martin C, Whitehead J, Williams PI, Jones RL, Langridge JM, Benton AK, Ball SM, Langford B, Hewitt CN, Davison B, Martin D, Petersson KF, Henshaw SJ, White IR, Shallcross DE, Barlow JF, Dunbar T, Davies F, Nemitz E, Phillips GJ, Helfter C, Di Marco CF, Smith S (2012) Atmospheric chemistry and physics in the atmosphere of a developed megacity (London): an overview of the REPARTEE experiment and its conclusions. Atmos Chem Phys 12:3065–3114. doi:10.5194/acp-12-3065-2012

Hodzic A, Madronich S, Bohn B, Massie S, Menut L, Wiedinmyer C (2007) Wildfire particulate matter in Europe during summer 2003: meso-scale modeling of smoke emissions, transport and radiative effect. Atmos Chem Phys 7:4043–4064

Horton DE, Skinner CB, Singh D, Diffenbaugh NS (2014) Occurrence and persistence of future atmospheric stagnation events. Nat Clim Chang 4:698–703. doi:10.1038/NCLIMATE2272

Hurrell JW (1995) Decadal trends in the North Atlantic Oscillation and relationships to regional temperature and precipitation. Science 269:676–679

IPCC (2013) In: Stocker TF et al (eds) Climate change 2013: the physical science basis, contribution of Working Group I to the Fifth Assessment Report of the Intergovernmental Panel on Climate Change. Cambridge University Press, Cambridge

Jacob DJ, Winner DA (2009) Effect of climate change on air quality. Atmos Environ 43:51–63

Katragkou E, Zanis P, Kioutsioukis I, Tegoulias I, Melas D, Krüger BC, Coppola E (2011) Future climate change impacts on summer surface ozone from regional climate-air quality simulations over Europe. J Geophys Res. 116(D22307) doi:10.1029/2011JD015899.

Kim MJ, Park RJ, Ho CH, Woo JH, Choi KC, Song CK, Lee JB (2015) Future ozone and oxidants change under the RCP scenarios. Atmos Environ 101:103–115

Kirtman B, Power SB et al (2013) Near-term climate change: projections and predictability. In: Stocker TF et al (eds) Climate change 2013: the physical science basis. Contribution of Working Group I to the Fifth Assessment Report of the Intergovernmental Panel on Climate Change. Cambridge University Press, Cambridge, UK

Knowland KE, Doherty RM, Hodges KI (2015) The effects of springtime mid-latitude storms on trace gas composition determined from the MACC reanalysis. Atmos Chem Phys 15:3605–3628. doi:10.5194/acp-15-3605-2015

Knowlton K, Rosenthal JE, Hogrefe C, Lynn B, Gaffin S, Goldberg R, Rosenzweig C, Civerolo K, Ku JY, Kinney PL (2004) Assessing ozone-related health impacts under a changing climate. Environ Health Perspect 112:1557–1563

Lacressonnière G, Peuch VH, Vautard R, Arteta J, Deque M, Joly M, Josse B, Marcal V, Saint-Martin D (2014) European air quality in the 2030s and 2050s: impacts of global and regional emission trends and of climate change. Atmos Environ 92:348–358

Langner J, Engardt M, Andersson C (2012a) European summer surface ozone 1990-2100. Atmos Chem Physics 12:10097–10105

Langner J, Engardt M, Baklanov A, Christensen JH, Gauss M, Geels C, Hedegaard GB, Nuterman R, Simpson D, Soares J, Sofiev M, Wind P, Zakey A (2012b) A multi-model study of impacts of climate change on surface ozone in Europe. Atmos Chem Phys 12:10423–10440

Lin M et al (2012) Transport of Asian ozone pollution into surface air over the western United States in spring. J Geophys Res 117:D00V07. doi:10.1029/2011JD016961

Masato G, Hoskins BJ, Woollings TJ (2013) Winter and summer northern hemisphere blocking in CMIP5 models. J Clim 26:7044–7059. doi: http://dx.doi.org/10.1175/JCLI-D-12-00466.1

Masui T et al (2011) An emission pathway for stabilization at 6 Wm-2 radiative forcing. Clim Chang 109:59–76

Meehl GA, Stocker TF, Collins WD, Friedlingstein P, Gaye AT, Gregory JM, Kitoh A, Knutti R, Murphy JM, Noda A, Raper SCB, Watterson IG, Weaver AJ, Zhao Z-C (2007) Global climate projections. In: Climate Change 2007: the physical science basis. Contribution of Working Group I to the Fourth Assessment Report of the Intergovernmental Panel on Climate Change [Solomon S, Qin D, Manning M, Chen Z, Marquis M, Averyt KB, Tignor M, Miller HL (eds) Cambridge University Press, Cambridge/New York

Meleux F, Solmon F, Giorgi F (2007) Increase in summer European ozone amounts due to climate change. Atmos Environ 41(35):7577–7587. doi:10.1016/j.atmosenv.2007.05.048

Nakicenovic N et al (2000) Special report on emission scenarios (SRES) report of Working Group III of the Intergovernmental Panel on change. Cambridge University Press, Cambridge, UK

Neu JL, Flury T, Manney GL, Santee ML, Livesey NJ, Worden J (2014) Tropospheric ozone variations governed by changes in stratospheric circulation. Nat Geosci 7:340–344. doi:10.1038/ngeo2138

O'Connor FM et al (2010) Possible role of wetlands, permafrost, and methane hydrates in the methane cycle under future climate change: a review. Rev Geophys 48:RG4005. doi:10.1029/2010RG000326

Orru H, Andersson C, Ebi KL, Langner J, Astrom C, Forsberg B (2013) Impact of climate change on ozone-related mortality and morbidity in Europe. Eur Respir J 41:285–294. doi:10.1183/09031936.00210411

Pausata FSR, Pozzoli L, Vignati E, Dentener FJ (2012) North Atlantic Oscillation and tropospheric ozone variability in Europe: model analysis and measurements intercomparison. Atmos Chem Phys 12:6357–6376. doi:10.5194/acp-12-6357-2012

Penrod A, Zhang Y, Wang K, Wu SY, Leung L (2014) Impacts of future climate and emission changes on US air quality. Atmos Environ 89:533–547

Post ES et al (2012) Variation in estimated ozone-related health impacts of climate change due to modeling choices and assumptions. Environ Health Perspect 120(11):1559–1564

Rasmussen D, Hu J, Mahmud A, Kleeman MJ (2013) The ozone-climate penalty: past, present, and future. Environ Sci Technol 47:14258–14266

Riahi K, Dentener F, Gielen D, Grubler A, Jewell J, Klimont Z, Krey V, McCollum D, Pachauri S, Rao S, van Ruijven B, van Vuuren DP, Wilson C (2012) Energy pathways for sustainable development. In: Schär C, Vidale PL, Lüthi D, Frei C, Häberli C, Liniger MA, Appenzeller C (eds) Global energy assessment: toward a sustainable future. The role of increasing temperature variability in European summer heatwaves. Nature 427:332–336

Shindell D, Kuylenstierna JCI, Vignati E, van Dingenen R, Amann M, Klimont Z et al (2012) Simultaneously mitigating near-term climate change and improving human health and food security. Science 335:183–189. doi:10.1126/science.1210026

Silva RA, West JJ, Lamarque J-F, Shindell DT, Collins WJ, Dalsoren S, Faluvegi G, Folberth G, Horowitz LW, Nagashima T, Naik V, Rumbold ST, Sudo K, Takemura T, Bergmann D, Cameron-Smith P, Cionni I, Doherty RM, Eyring V, Josse B, MacKenzie IA, Plummer DS, Righi M, Stevenson DS, Strode S, Szopa S, Zeng G (2016) The effect of future ambient air pollution on human premature mortality to 2100 using output from the ACCMIP model ensemble. Atmos Chem Phys 16:9847–9862. doi:10.5194/acp-16-9847-2016

Solberg S, Hov O, Sovde A, Isaksen ISA, Coddeville P, De Backer H, Forster C, Orsolini Y, Uhse K (2008) European surface ozone in the extreme summer 2003. J Geophys Res 113(D07307), doi:10.1029/2007JD009098.

Stott PA, Stone DA, Allen MR (2004) Human contribution to the European heatwave of 2003. Nature 432:610–614

Tagaris E, Liao K-J, DeLucia AJ, Deck L, Amar P, Russell AG (2009) Potential impact of climate change on air pollution-related human health effects. Environ Sci Technol 43(13):4979–4988. doi:10.1021/es803650w

Thomson AM et al (2011) RCP4.5: a pathway for stabilization of radiative forcing by 2100. Clim Chang 109:77–94

Ulbrich U, Leckebusch GC, Pinto JG (2009) Extra-tropical cyclones in the present and future climate: a review. Theor Appl Climatol 96:117–131

UNEP/WMO (2011) UNEP/WMO integrated assessment of black carbon and tropospheric ozone. http://www.unep.org/dewa/Portals/67/pdf/BlackCarbon_report.pdf

US EPA (2015) Integrated science assessment for oxides of nitrogen – health criteria (Second External Review Draft) United States Environmental Protection Agency. http://cfpub.epa.gov/ncea/isa/recordisplay.cfm?deid=288043

Vautard R, Honoré C, Beekmann M, Rouil L (2005) Simulation of ozone during the august 2003 heat wave and emission control scenarios. Atmos Environ 39:2957–2967

van Vuuren DP et al (2011) RCP2.6: exploring the possibility to keep global mean temperature increase below 2 degrees C. Clim Chang 109:95–116

Webber CP, Dacre HF, Collins WJ, Masato G. (2017) The dynamical impact of Rossby wave breaking upon UK PM10 concentration. Atmos Chem Phys 17:867–881. doi:10.5194/acp-17-867-2017

WHO (2013a) Review of evidence on health aspects of air pollution – REVIHAAP project: technical report. World Health Organisation, Copenhagen. 2013

WHO (2013b) Health risks of air pollution in Europe –HRAPIE project. New emerging risks to health from air pollution – results from the survey of experts. World Health Organization, Regional Office for Europe, Copenhagen

Wu S, Mickley LJ, Leibensperger EM, Jacob DJ, Rind D, Streets DG (2008) Effects of 2000-2050 global change on ozone air quality in the United States. J Geophys Res 113. doi:10.1029/2007JD008917

Yin JH (2005) A consistent poleward shift of the storm tracks in simulations of 21st century climate. Geophys Res Lett 32:4

Yin J, Harrison RM (2008) Pragmatic mass closure study for PM1.0, PM2.5 and PM10 at roadside, urban background and rural sites. Atmos Environ 42:980–988

Ruth M. Doherty is a professor of atmospheric sciences at the University of Edinburgh, UK. She has 20 years' experience in modelling air pollution and climate at global, regional and urban scales, linked to health effects, and authored over 65 papers. She is a member of the UNECE Task Force on Hemispheric Transport of Air Pollution and leads their climate change research programme.

Fiona M. O'Connor is a climate scientist working on atmospheric chemistry, with a particular interest in methane, chemistry-climate interactions, air quality and climate system feedbacks. Fiona's work aims to gain a better understanding of the sources and sinks of atmospheric methane, the interannual variability of methane emissions and atmospheric concentrations and the potential feedbacks in the climate system which may affect future concentrations of atmospheric methane. She is also interested in modelling and understanding the role of short-lived atmospheric trace gases, such as ozone and methane, in climate change. A key aspect of her work is developing and running the UK Chemistry and Aerosol (UKCA) model on a global scale and on a decadal-to-centennial timescale. She is also working towards implementing UKCA in a regional climate model, so that the interactions between climate and air quality on a more regional scale can be explored.

Chapter 6
The Impact of Climate Change and Air Pollution in the Southern European Countries

Cosimo Palagiano and Rossella Belluso

Abstract The extension of Europe from the North to the South, i.e., from Knivskjellodden in Magerøya Island in Norway at 71° 11′ 08″ N and the Isola delle Correnti in Sicily at 36°38′44″N, is of 35°, with a difference in latitude of about 35°. We pass from the Arctic Ocean to the Mediterranean, with a significant difference in temperature and rainfall. In addition the population density varies from 15.5 inhabitants per sq.km to about 196 inhabitants per sq.km, about 13 times more. The climate parameters and the distribution of population have considerable importance in air pollution and in its variation.

In the Southern European countries, which we consider in this chapter, the car traffic and the solar irradiation have a great impact on the pollution, together with the industrial pollution.

Keywords Europe • Air pollution • Ambient • EU Ambient Quality Directive • Human health • Urban quality of life

Introduction

In this chapter, we consider the climate change and air pollution of the Southern European countries. The European countries which belong to the European Union are 28, including the United Kingdom, before the referendum on Brexit. The chapter will take in consideration only the major Mediterranean countries, such as Portugal, Spain, France, Italy, and Greece.

First of all, we should consider the most evident effects of the climate change all over Europe, summing up in the Table 6.1. According to Alcamo and Olesen (2012, p. 209), the Mediterranean region is one of the most vulnerable areas of Europe

C. Palagiano (✉)
Dipartimento Di Scienze Documentarie, Linguistico-Filologiche e Geografiche, Sapienza University of Rome, Rome, Italy
e-mail: cosimo.palagiano@uniroma1.it

R. Belluso
Sapienza University of Rome, Rome, Italy
e-mail: rossella.belluso@uniroma1.it

© Springer International Publishing AG 2018

R. Akhtar, C. Palagiano (eds.), *Climate Change and Air Pollution*, Springer Climate, DOI 10.1007/978-3-319-61346-8_6

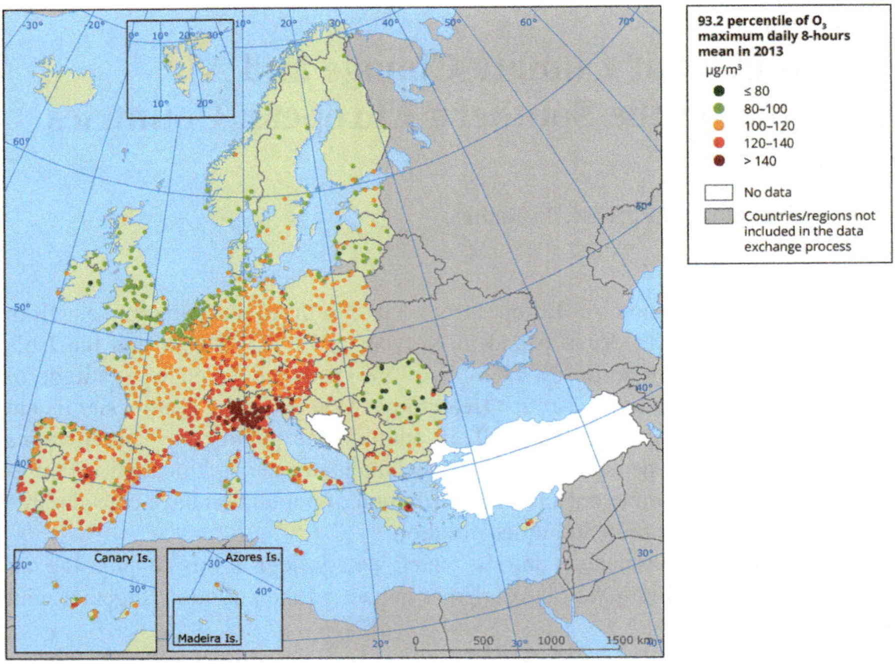

Fig. 6.1 93.2 percentile of O3 maximum daily 8-h mean in 2013 in the EU-28. This map shows highest values of ozone in in Po Valley and generally in all industrialized areas of Europe, particularly in Germany, Catalonia, and mainly along the coasts of the Western Mediterranean. In particular the problems of the Po Valley are that the atmospheric instability due to the presence of the thermal inversion promotes chemical and physical chemical reactions (Pinna 1989, p. 25 f.). Among the pollutants, a prevailing place is taken by ozone, due, as we said above, to the splitting of oxides and combination of their oxygen with atmospheric oxygen. Notes: the graph is based, for each member state, on the 93.2 percentile of maximum daily 8-h mean concentration values, corresponding to the 26th highest daily maximum of the running 8-h mean. For each country, the lowest, highest, and median values (in $\mu g/m^3$) at the stations are given. The rectangukes give the 25 and 75 percentiles. At 25% of the stations, levels are below the lower percentile; at 25% of the stations, concentrations are above the upper percentile. The target value threshold set by the EU legislation is marked by the *heavy line* (Source: EEA (2013b))

because it is "threatened by a combination of warmer and drier conditions leading to longer and more frequent droughts, aggravated water scarcity, declining crop productivity, and higher fie risks." We could add to these problems also the increasing variability and instability of the weather, with drought and floods alternatively. Phenomena like these strongly damage crops and houses and favor the diffusion of vector-borne diseases (malaria, dengue, chikungunya, and West Nile virus) (Alcamo and Olesen 2012, p. 57 ff,).

The Commission of the European Union publishes some tables and thematic maps also on the air pollution.

The most interesting map (Fig. 6.1) considers the distribution of ozone (O3) in Europe. This map shows the major intensity of ozone in Mediterranean Europe, due to the solar heating. But ozone is very dangerous because it is produced by the sun, which breaks off the oxygen from the pollutants discharged from the cars and joins it to the atmospheric oxygen. For example, in NO2, SO2, etc., the oxygen (O2) joins to the O and turns into O3, the ozone, exactly.

The target value applied by EU member states from 1 January 2010 is that the threshold should not be exceeded at a monitoring station on more than 25 days per year, determined as a 3-year average starting from 2010. The long-term objective does not exceed the threshold level at all (EEA Report|No 5/2015; EEA (2013b), p. 25) (Tables 6.1 and 6.2), (Figs. 6.2 and 6.3).

The Ambient Air Quality Directive (EU 2008) sets limit values for both short-term (24-h) and long-term (annual) PM10 concentrations, whereas values for only long-term PM2.5 concentration have been set (Table). The short-term limit value for PM10 is the limit value for PM10 that is most often exceeded in Europe.

The annual PM10 limit value is set at 40 µg/m^3. The deadline for member states to meet the PM10 limit values was 1 January 2005. The deadline for meeting the target value for PM2.5 (25 µg/m^3) was 1 January 2010, and the deadline for meeting the exposure concentration obligation for PM2.5 (20 µg/m^3) was 2015. The Air Quality Guidelines (AQGs) set by WHO are stricter than the EU air quality standards for PM and have the aim to achieve the lowest concentrations possible. The PM2.5 annual mean guideline corresponds to the lowest levels beyond which total cardiopulmonary and lung cancer mortality have been shown to increase (with >95% confidence) in response to long-term exposure to PM2.5 (WHO 2006a).

Table 6.1 Air quality standards for O3 as defined in EU Ambient Air Quality Directive and WHO AQG

EU Air Quality Directive			WHO AQG
Averaging period	Objective and legal nature	Concentration	
Maximum daily 8-h mean	Human health and target value	120 µg/m^3, not to be exceeded on more than 25 days per year averaged over 3 years	100 µg/m^3
AOT40 accumulated over May to July	Vegetation target value	18,000 (µg/m^3).h averaged over 5 years	
Maximum daily 8-h mean	Human health long-term objective	120 µg/m^3	
Accumulated over May to July	Vegetation long-term objective	6000 (µg/m^3 AOT).h	
1 h	Information threshold	180 µg/m^3	
1 h	Alert threshold	240 µg/m^3	

Note: AOT 40, accumulated O3, exposure over a threshold of 40 ppb. It is the sum of the differences between hourly concentrations

>80 µg/m^3 accumulated over all hourly values measured between 8.00 and 20.00 Central European Time

Source: EU 2008; WHO 2006a, 2008

Table 6.2 EU and WHO AQG directive

		EU Air Quality Directive		WHO AQG
Size fraction	Averaging period	Objective and legal nature and concentration	Comments	
PM_{10}	1 day	Limit value: 50 µg/m^3	Not to be exceed on more than 35 days per year	50 µg/m^3
PM_{10}	Calendar year	Limit value: 40 µg/m^3		20 µg/m^3
$PM_{2.5}$	1 day			25 µg/m^3
$PM_{2.5}$	Calendar year	Target value: 25 µg/m^3		10 µg/m^3
$PM_{2.5}$	Calendar year	Limit value: 25 µg/m^3	To be met by 1 January 2015 (until then, margin of tolerance)	
$PM_{2.5}$		Exposure concentration Obligation, 20 µg/m^3	2015	
$PM_{2.5}$		Exposure reduction target, 0–20% reduction in exposure (depending on the average exposure indicator in the reference year) to be met by 2020		

In 2013, the PM2.5 concentrations were higher than the target value at several stations in Bulgaria, the Czech Republic, Italy, and Poland, as well as one station in France, Macedonia, Kosovo, Romania, and Slovakia (Figs. 6.4 and 6.5).

Transport, energy, industry, commerce, institutions, household, agriculture, and waste are the major sectors contributing to emissions of air pollutants in Europe. The transport sector has considerably reduced its emissions over the past decade, with the exception of BaP (benzo[a]pyrene) emissions, which have increased by 9% in the EU-28 and 60% in EEA-33 countries from 2004 to 2013. BaP is the result of incomplete combustion at temperature between 300 °C (572 °F) and 600 °C (1112 ° F). The ubiquitous compounds can be found in coal tar, tobacco smoke, and many foods, especially grilled meats.

The commercial, institutional, and households fuel combustion sector dominates the emissions of primary PM2.5 and PM10, BaP and CO in the EU-28 in 2013. Some countries use household wood and other biomass combustion for heating, thanks to government incentives/subsides. In addition they have the perception that it is a "green" opportunity.

Industry considerably reduced its air pollutant emissions between 2004 and 2013, with the exception of BaP emissions. It still largely uses Pb, As, Cd, NMVOC (non-methane volatile organic compound), Ni, primary PM, SOx, and Hg emissions. Although the industrial BaP emissions are of only the 5% of the total BaP emissions of EU-28, they may affect population exposure in the vicinity of the industrial sources.

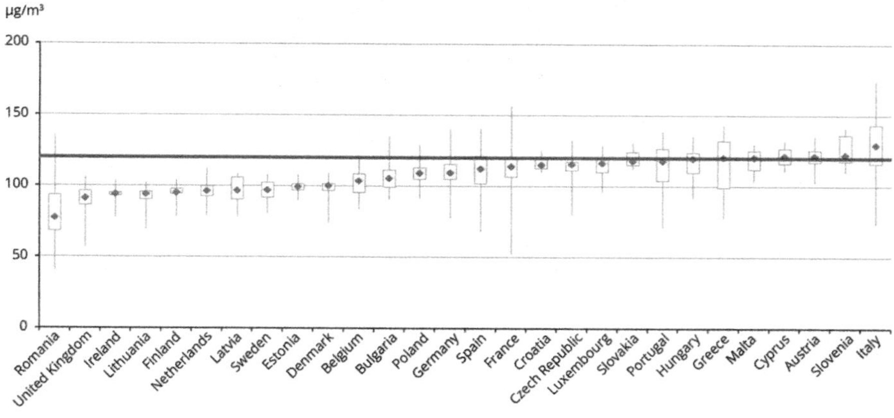

Fig. 6.2 As we can see in Fig. 6.2, the O3 target value was exceeded more than 25 times in 2013 in 18 of the EU countries, which are Austria, Bulgaria, Croatia, Cyprus, the Czech Republic, France, Germany, Greece, Hungary, Italy, Luxemburg, Malta, Poland, Portugal, Romania, Slovakia, and Spain. At least ten of these countries belong to South Europa, but only Germany and Poland are entirely at a latitude of 50° and over, if we can consider 50° as rough line of geographical separation between the North and the South of Europe. Another source of air pollution is particulate matter. The Ambient Air Quality Directive (EU 2008) sets limit value for both short-term (24-h) and long-term (annual) PM10 concentration, whereas values for only long-term PM2.5 concentration have been set. The Air Quality Guidelines (AQGs) set by WHO are stricter than the EU air quality standard for PM. The PM2.5 annual mean guideline corresponds to the lowest levels beyond which total, cardiopulmonary and lung cancer mortality have been shown to increase (with >95% confidence) in response to long-term exposure to PM2.5 (WHO 2006a) (Source: EEA (2013b))

From 2004 to 2013, energy production and distribution considerably reduced their emissions, even if they are a large source of primary PM, SO, Hg, Ni, and NOx emissions.

In agriculture sector the air pollutants have least decreased in EU-28. Its NH3 emissions have decreased from 2004 to 2013, thanks to the European policies, with the exception of NH3.

Agriculture is the most important source of PM10. In addition agriculture emits the 50% of total CH4 emissions in the EU-28.

The waste sector contribution to the total emissions of air pollutants is relatively small, with the exception of CH4.

The Impact of Pollutants on Human Health

The ozone molecule is extremely reactive, able to oxidize many cellular components, including amino acids, proteins, and lipids.

At a concentration of 0.008–0.02 ppm (15–40 g/mc), the smell can already be detected; 0.1 ppm causes irritation of the eyes and throat for its action against the

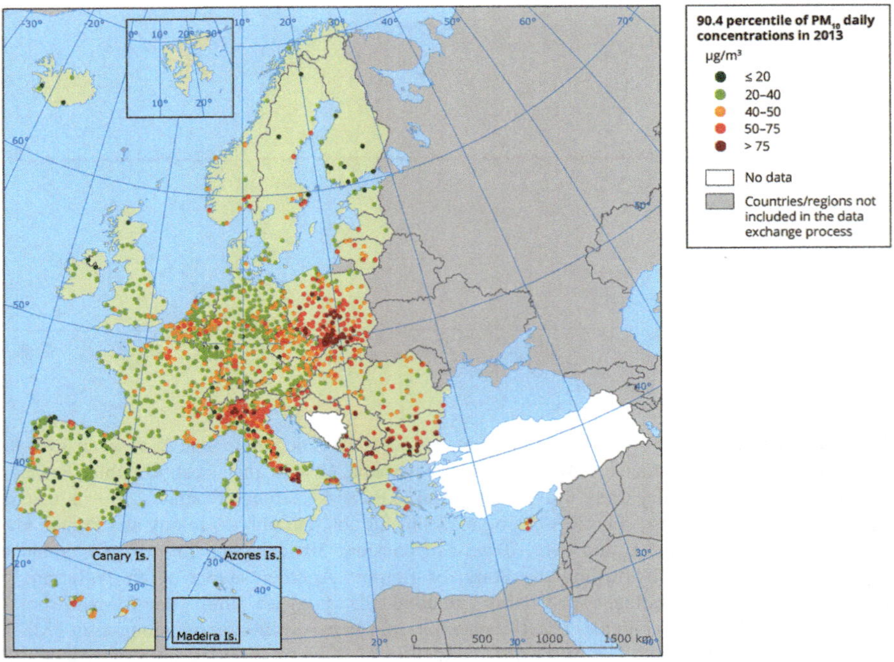

Fig. 6.3 9.4 percentile of PM10 daily concentration in 2013 in EU-2

mucous membranes. Higher concentrations cause irritation of the respiratory tract, coughing, and a tightness in the chest which makes breathing difficult. The most sensitive subjects, such as asthmatics and the elderly, may be subject to asthma attacks even at low concentrations. At a concentration of 1 ppm, it causes headaches and to 1.7 ppm can produce pulmonary edema.

In the presence of other photochemical oxidants, sulfur dioxide and nitrogen dioxide, ozone action is always enhanced to synergistic effect. High concentrations may result in death.

BaP can affect the nervous, immune, and reproductive systems. In addition BaP's metabolites are mutagenic and highly carcinogenic (Kleiböhmer 2001, pp. 99–122; Denissenko et al. 1966, pp. 430–2; Le Marchand et al. 2002, 205–14; Truswell 2002, pp. p. 19–24; Sinha et al. 2005).

Atmospheric particulate matter – also known as particulate matter (PM) or particulates – is microscopic solid or liquid matter suspended in the Earth's atmosphere.

Its toxicity depends on both physical characteristics (particle size) and the chemical composition that consists predominantly of organic compounds of carbon and oxides of toxic elements. In recent years, attention has focused on the finer fractions of particulate matter and in particular on the PM10 (aerodynamic diameter $<$ of 10 µm) and PM2.5 (aerodynamic diameter $<$2.5 m). The particles that are dispersed in the air have a diameter from 0.1 to 10 µm. They can penetrate into the

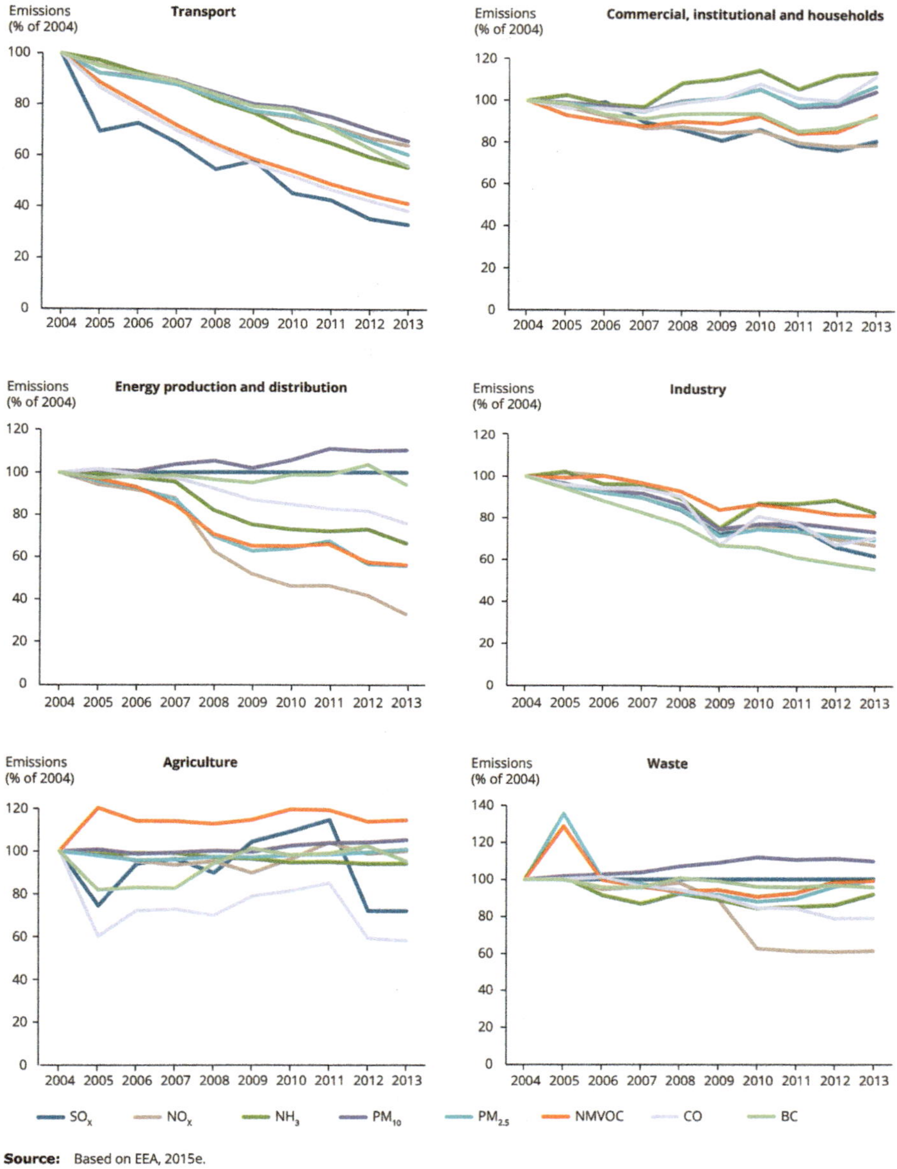

Fig. 6.4 Development in EU-28 emissions 2004–2013 of SOx, NOx, NH3 (ammonia), PM2.5, NMVOC (non-methane volatile organic compound), CO, and BC

respiratory tract, reaching the epithelium of the bronchioles and alveoli and then the blood. Particulate matter can produce acute and chronic effects. Sensitive population (elderly, children, asthmatics) may be affected by lung inflammation, respiratory, and cardiovascular diseases. Chronic effects include an increase in lower

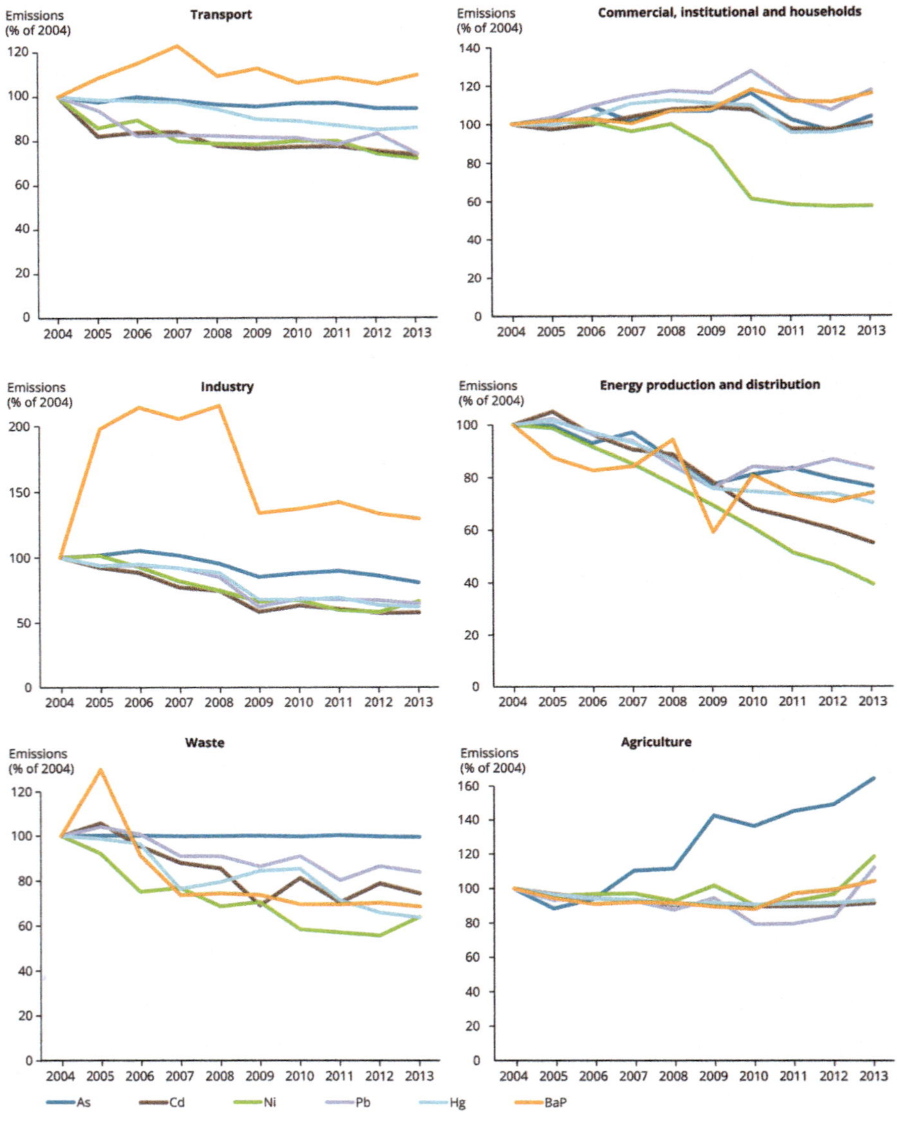

Source: Based on EEA, 2015e.

Fig. 6.5 Development in EU-28 emissions of As, Cd, Ni, Pb, Hg, and BaP (*bottom*), 2004–2013 (% of 2004 levels) (Source: EEA (2013b))

respiratory diseases (chronic obstructive pulmonary disease and reduced lung function, heart disease, and lung cancer).

NOx includes all nitrogen oxides and mixtures of their chemical compounds. The nitrogen oxides are produced from any combustion, from those of the automobile engine, of the wood-burning fireplaces and by power plants. Monoxide (NO) and nitrogen dioxide (NO2) are both potentially dangerous to human health. NO2 is particularly toxic to the eyes and the respiratory system and may contribute to chronic bronchitis, asthma, and emphysema.

Traffic Emergency

The traffic emergency due to unsustainable increase in the number of cars cannot be resolved by the production of less polluting vehicles but by their reduction (Table 6.3).

The Commune of Rome, with its 693.7 cars per 1000 inhabitants, has the not-enviable primacy of the cars per 1000 inhabitants in Italy. The reduction of traffic is an urgent problem to solve. Many communal administrations face the problem closing some streets and squares to the private traffic or establishing the walking Sundays. But such measures are absolutely useless, because the air pollution does not decrease. A possible solution can be the diffusion of the electric and hybrid cars (Table 6.4).

The health impact assessment presents, for each pollutant, the population-weighted concentration, the estimated number of YLL (years of life lost), and the years of life lost per 100,000 inhabitants. In total, in the 40 countries assessed, 4,804,000 YLL are attributed to PM2.5 exposure, and 828,000 YLL and 215,000 YLL are attributed to NO2 and O3 exposure, respectively. In the EU-28, the attributed YLL to PM2.5, NO2, and O3 exposure are 4,494,000, 800,000, and 197,000, respectively. In the South Europe which we have considered, Bulgaria, Croatia, Cyprus, France, Greece, Italy, Malta, Portugal, Romania, Slovenia, Spain, Albania, Andorra, Bosnia and Herzegovina, Macedonia, Monaco, Montenegro, San

Table 6.3 Cars and buses per 1000 inhabitants in some EU countries

Countries	Cars per 1000 inhabitants	Buses per 1000 inhabitants
Austria	515.3	1.1
Belgium	478.7	1.5
Finland	507.3	2.3
France	499.9	1.4
Germany	502.3	0.9
United Kingdom	496.9	1.5
Ireland	440.7	–
Italy	608.1	1.6
Netherland	473.5	0.7
Spain	493.4	1.4
Sweden	467.7	1.5
Average	544.2	1.3

Table 6.4 Years of life lost (YLL) attributable to PM2.5, O3, and NO2 exposure in 2012 in 40 European countries and the EU-28. SOMO35 means sum of means over 35 ppb (of daily maximum 8-h O3 concentrations)

	$PM_{2.5}$			O_3			NO_2		
Countries	Annual mean	YLL	$YLL/10^5$ inhabitants	SOMO35	YLL	$YLL/10^5$ inhabitants	Annual mean	YLL	$YLL/10^5$ inhabitants
Austria	14.8	65,400	776	5419	3800	46	18.81	7000	83
Belgium	15.8	99,500	894	2050	2100	19	23.41	24,200	218
Bulgaria	24.9	141,500	1937	5960	5900	81	16.38	7100	97
Croatia	16.8	46,900	1099	7143	3200	74	14.89	500	12
Cyprus	25.0	8000	729	8369	500	47	9.42	0	0
Czech Republic	18.8	116,300	1106	4806	4700	44	17.14	0	31
Denmark	10.0	31,400	562	2662	1300	24	12.90	500	10
Estonia	7.9	7000	532	2310	300	24	10.30	0	0
Finland	7.1	20,800	385	1650	700	14	10.12	0	0
France	14.7	508,900	778	3635	21,100	32	18.71	89,900	137
Germany	13.3	645,200	802	3357	25,100	31	20.63	112,400	140
Greece	19.2	116,700	1.057	9378	9200	84	15.46	13,900	126
Hungary	18.9	141,900	1431	6.342	7700	77	16.57	8000	81
Ireland	8.1	14,400	315	1479	500	11	10.76	0	0
Italy	18.9	652,200	1095	7328	40,500	68	25.30	237,300	399
Latvia	12.4	19,900	976	3103	800	40	13.65	1000	50
Lithuania	12.9	25,100	839	3358	1000	35	9.88	0	0
Luxemburg	12.6	2800	524	2561	100	16	21.79	600	122
Malta	12.4	2300	551	8022	300	64	15.61	0	0
Netherlands	13.7	112,700	673	1949	2700	16	23.26	31,000	185
Poland	23.9	560,400	1472	4045	16,100	42	16.72	20,000	53
Portugal	9.9	59,900	570	4240	4000	38	15.45	5200	49
Romania	20.8	279,700	1395	3967	9900	49	16.22	16,600	83

Slovakia	20.5	65,400	1209	6103	21,900	63	15.97	600	12
Slovenia	17.7	19,900	967	7092	1700	61	16.65	300	17
Spain	11.9	274,100	586	5850	7200	47	17.88	63.600	136
Sweden	7.2	35,200	370	2233	1700	18	12.49	100	1
United Kingdom	11.9	420,800	661	1183	7200	11	23.32	156,900	246
Albania	21.1	24,500	854	8760	2300	81	16.33	1200	42
Andorra	15.9	600	838	8058	100	71	14.74	0	0
Bosnia and Herzegovina	18.5	41,200	1,074	7322	2700	71	14.90	900	23
Macedonia	29.2	32,200	1560	8472	30	8	9.00	0	0
Iceland	4.7	600	181	1242	1800	89	19.13	2300	111
Lichtenstein	10.2	200	546	5132	20	43	20.59	0	94
Monaco	18.2	300	957	6979	20	62	24.34	100	213
Montenegro	18.7	6800	1093	8584	600	93	15.47	200	36
Norway	7.2	16,400	327	2182	800	16	13.38	2000	39
San Marino	16.7	300	978	6048	20	56	17.65	0	0
Serbia	24.3	140,200	1557	6844	7000	77	18.61	11,500	127
Switzerland	12.6	46,500	582	4990	3100	39	21.58	10,200	128
Total		480,400	895		215,000	40		828,000	154
EU-28		449,400	898		197,000	39		800,000	160

Source: EEA (2013b)

Marino, and Serbia registered 19,695, 1011, and 1826 YLL/100,000 inhabitants for impact of PM2.5, O3, and NO2, respectively (Table 6.5).

Premature deaths occur when a person dies before the standard age expectancy of a country and gender. These deaths can be preventable if their cause can be eliminated. Years of life lost (YLL) is determined by an estimated average years that a person would have lived if he (she) had not died prematurely.

The South European countries have charged in total of 214,480 premature deaths attributable to PM2.5, 7328 to O3, and 29,737 to NO2 exposure, respectively, that is of 53.22, 45.73, and 41.30 in percent of the values of EU-28, respectively. It depends on the total population of the South European countries. The South European countries have a total population of about 237,771,436 inhabitants, which is about the 28.57% of the total European population. But the South EU-28 population of 227,287,593 inhabitants is of 44.56% of the total EU-28 population.

The highest numbers of YLL from PM2.5 are observed in the countries with the largest populations (France and Italy), but if we consider YLL per 100,000 inhabitants, we can observe the largest impacts in the central and eastern European countries, which have also the highest concentrations.

Regarding O3, the countries with the largest impacts are Italy, Spain, and France. The highest rate of YLL per 100,000 inhabitants is presented by the countries in the Western Balkans and Italy.

The largest health impact attributable to NO2 exposure is in the hot-spot regions, as Italy (Po Valley).

The Benefits

The benefits can be real if the EU countries slow down the use of polluting fuels and initiate a serious program of nonpolluting fuels, such as solar, wind, etc., energy in all economic and domestic sectors. The photovoltaic system is increasing in Europe. We hope that this will be prevailing in the immediate future. According to Alcamo and Olesen (2012), p. 253, "Wind energy is one of the fastest growing energy technologies in the world, with Europe leading the world with 69 percent of total capacity: Wind energy now satisfies about 5% of the EU's total electricity demand…It accounts for over 5% of electricity usage in five countries (Germany, Ireland, Portugal, Spain, and Denmark)… In terms of total capacity, the leaders in the EU are Germany with over 22,000 MW and Spain with more than 15,000 MW installed capacity."

The photovoltaic technology is growing at a tremendous rate: total installed capacity of photovoltaic panels in the EU was 1542ì1 MW at peak performance, led by Germany (1103), Spain (341), and Italy (50) (Enery.eu – Europe's energy portal: www.energy.eu/#renewable).

The capacity of thermal solar collectors in the EU is 14,289 MW led by Germany (2301), Austria (1987), and France (812).

Table 6.5 Premature deaths attributable to PM2.5, O3, and NO2 exposure in 2012 in 40 Europe countries and in the EU-28

Countries	$PM_{2.5}$	O_3	NO_2
Austria	6100	320	660
Belgium	9300	170	2300
Bulgaria	14,100	500	700
Croatia	4500	270	50
Cyprus	790	40	0
Czech Republic	10,400	380	290
Denmark	2900	110	50
Estonia	620	30	0
Finland	1900	60	0
France	43,400	1500	7700
Germany	59,500	2100	10,400
Greece	11,100	780	1300
Hungary	12,800	610	720
Ireland	1.200	30	0
Italy	59,500	3300	21,600
Latvia	1800	60	90
Lithuania	2300	80	0
Luxemburg	250	10	60
Malta	200	20	0
Netherlands	10,100	200	2800
Poland	4460	110	1600
Portugal	5400	320	470
Romania	25,500	720	1500
Slovakia	5700	250	60
Slovenia	1700	100	30
Spain	25,500	1800	5900
Sweden	3700	160	10
United Kingdom	37,800	530	14,100
Albania	2200	140	270
Bosnia and Herzegovina	3500	4	0
Macedonia	3000	130	210
Iceland	100	2	0
Lichtenstein	20	1	3
Monaco	30	2	7
Montenegro	570	40	20
Norway	1700	70	200
San Marino	30	2	0
Serbia and Kosovo	13,400	550	1100
Switzerland	4300	210	950
Total	432,000	17,000	75,000
Total EU-28	403,000	16,000	72,000

Source: EEA (2013b)

By 2030, the potential annual production of "environmental compatible" biomass could be 40 MtOE from the waste and forest residues and municipal waste. This amount to 15–16% of Europe's projected primary energy production in 2030 (Alcamo and Olesen 2012, p. 259).

The contribution of hydroelectric facilities to overall electricity production is substantial, providing 17% of global electricity and 17.9% of electricity in the EU.

Renewable energies are not limited to wind, sun of falling water, but also include the heat of the earth and the movement of oceans. The sources can make a contribution to reducing CO2 emissions, but their long-term potential for producing electricity is not as great as for other renewable sources (Alcamo and Olesen 2012, pp. 254–261).

Conclusive Remarks

According to the last data of the EEA (European Environment Agency), 467,000 deaths occur in UE yearly (2012–2014) from air pollution (particulate, ozone, nitrogen dioxide, benzo(a)pyrene, and sulfur dioxide). Despite the air quality in Europe is improving, air pollution remains the biggest risk environmental factor for human health and lowers the quality of life. The goal is to reduce the smog and water pollution effects on 50% of the population by 2030. In Italy there are 63,630 victims of the particulate, 21,040 of nitrogen dioxide, and 3380 for ozone. Alarming also German data, der United Kingdom and France. Among the top places for victims of smog, there are also Poland and Spain. Looking for a sustainable mobility is possible. Smog deteriorates the quality of life especially in big cities, affecting 60% of population.

The introduction of gas methane cars or electrical partly or entirely could mitigate the emission of harmful gases. However, the responsibility is not only collective but also individual.

References

Ainsworth EA, Yendrek CR, Sitch, Collins W, Emberson LD (2012) The effects of tropospheric ozone on net primary productivity and implication or climate change. Annu Rev Plant Biol 63:637–661

Alcamo J, Olesen JE (2012) Life in Europe under climate change. Wiley-Blackwell, Chichester

Bach H, Brandt J, Christensen JH, Ellermann T, Geels C, Hertel O, Massling A, Nielsen HØ, Nielsen O-K, Nordstrøm C, Nøjgaard JK, Skov H, Chatterton T, Hayes E, Laxen D, Irwin J, Longhurst J, Pelsy F, Zamparutti T (2014) Services to assess the reasons for non-compliance of ozone target value set by directive 2008/50/EC and potential for air quality improvements in relation to ozone pollution: final report. ECORYS, Rotterdam

Dahl A, Gharibi A, Swietlicki E, Gudmundsson A, Bohgard M, Ljungman A, Blomqvist G, Gustafsson M (2006) Traffic-generated emissions of ultrafine particles from pavement–tire interface. Atmos Environ 40(7):1314–1323

De Vries W, Hetteling J-P, Posch M (eds) (2015) Critical loads and dynamic risk assessments: nitrogen, acidity and metals in terrestrial an aquatic ecosystem, Environmental Pollution Series, vol 25. Springer, Dordrecht

Denissenko MF, Pao A, Tang M, Pfeifer GP (1996) Preferential formation of benzo[a]pyrene adducts at lung cancer mutational hotspots in P53. Science 274(5286):430–432

Grice S, Stedman J, Kent A, Hobson M, Norris J, Abbott J, Cooke S (2009) Recent trends and projections of primary NO2 emissions in Europe. Atmos Environ 43(13):2154–2167

Harmens H, Mills G (eds) (2012) Ozone pollution: impacts on carbon sequestration in Europe. ICPVegetation Programme Coordination Centre, Centre for Ecology and Hydrology, Bangor

Héroux ME, Anderson HR, Atkinson R, Brunekreef B, Cohen A, Forastiere F, Hurley F, Katsouyanni K, Krewski D, Krzyzanowski M, Künzli N, Mills I, Querol X, Ostro B, Walton H (2015) Quantifying the health impacts of ambient air pollutants: recommendations of a WHO/Europe project. Int J Public Health 60:619–627

Holland MR, Jones HH, Berdowski J, Bleeker A, Visschedijk AJH (2001) Economic evaluation of air quality targets for PAHs. AEAT/ENV/R0593, AEA Technology Environment, Culham Science Centre, Abingdon. http://ec.europa.eu/environment/enveco/air/pdf/pah_report.pdf

Karlsson PE (2005) Economic assessment of the negative impacts of ozone on the crop yield and forest production. A case study of the estate Östads Säteri in Southwestern Sweden. Ambio 34(1):32–40

Kleiböhmer W (2001) Polycyclic aromatic hydrocarbon (PAH) metabolites, environmental analysis (volume 3 of Handbook of Analytical Separations). Elsevier

Kumar P, Pirjola L, Ketzel M, Harrison RM (2013) Nanoparticle emissions from 11 non-vehicle exhaust sources – a review. Atmos Environ 67:252–277

Lave LB, Seskin EP (1973) An analysis of the association between U.S. mortality and air pollution. J Am Stat Assoc 68(342):286. doi:10.1080/01621459.1973.10482421

Le Marchand L, Hankin JH, Pierce LM et al (2002) Well-done red meat, metabolic phenotypes and colorectal cancer in Hawaii. Mutat Res 506–507:205–214. doi:10.1016/s0027-5107(02)00167-7

Lim SS, Vos T, Flaxman AD, Danaei G, Shibuya K, Adair-Rohani H et al (2012) A comparative risk assessment of burden of disease and injury attributable to 67 risk factors and risk factor clusters in 21 regions, 1990–2010: a systematic analysis for the Global Burden of Disease Study 2010. Lancet 380(9859):2224–2260

Mills G, Harmens H (eds) (2011) Ozone pollution: a hidden threat to food security. ICP Vegetation Programme Coordination Centre, Centre for Ecology and Hydrology, Bangor

Pieralice E, Triolo IL, Earth O (2011) Prove tecniche di emissioni. Gli effetti sanitari del PM 10 sulla salute e le scelte modali nelle città italiane. Roma, Isfort, Competenze e risorse per la mobilità

Pinna M (1989) Inquinamento atmosferico in Italia. Note climatico-ecologiche. In: Palagiano C. (ed) 1989Geografia della salute in Italia. F. Angeli, Milano

Pope CA 3rd, Dockery DW (2006) Health effects of fine particulate air pollution: lines that connect. J Air Waste Manage Assoc 56:709–742

Pufelete M, Battershill J, Boobis A, Fielder R (2004) Approaches to carcinogenic risk assessment for polycyclic aromatic hydrocarbons: a UK perspective. Regul Toxicol Pharmacol 40 (1):54–66

Querol X, Alastuey A, Pandolfi M, Reche C, Pérez N, Minguillón MC, Moreno T, Viana M, Escudero M, Orio A, Pallarés M, Reina F (2014) 2001–2012 trends on air quality in Spain. Sci Total Environ 490:957–969

Rexeis M, Hausberger S (2009) Trend of vehicle emission levels until 2020 – prognosis based on current vehicle measurements and future emission legislation. Atmos Environ 43 (31):4689–4698

Seinfeld J, Spyros P (1998) Atmospheric chemistry and physics: from air pollution to climate change, 2nd edn. Wiley, Hoboken, p 97

Sinha R, Kulldorff MJ, Gunter M, Strickland P, Rothman N (n.d.) Dietary benzo[a]pyrene intake and risk of colorectal adenoma. DOI: 10.1158/1055–9965.EPI-04-0854 Published August 2005

Slootweg J, Maximilian P, Hettelingh J (eds) (2010) Progress in the modelling of critical thresholds and dynamic modelling, including impacts on vegetation in Europe, CCE status report 2010. Coordination Centre for Effects, Bilthoven

Truswell, AS (2002) Meat consumption and cancer of the large bowel. Eur J Clin Nutr 56 (Suppl 1). doi:10.1038/sj.ejcn.1601349. PMID 11965518

Institutional Publications

EC (2005) Communication from the Commission to the Council and the European Parliament, thematic strategy on air pollution. European Commission, Brussels. COM(2005) 0446 final 21.9.2005. http://eurlex.europa.eu/LexUriServ/LexUriServ.do?uri=CELEX:52005DC0446:EN:NOT

EC (2013a) Commission staff working document accompanying the communication on a revised EU strategy on air pollution proposal for a revision of directive 2001/81/EC on national emission ceilings for certain atmospheric pollutants. Proposal for a legislative instrument on control of emissions from medium combustion plants – impact assessment, SWD (2013) 531, European Commission, Brussels. http://ec.europa.eu/environment/archives/air/pdf/Impact_assessment_en.pdf

EC (2013b) The clean air package. European Commission, Brussels. http://ec.europa.eu/environment/air/clean_air_policy.htm

EEA (2010) The European environment state and outlook 2010. European Environment Agency

EEA (2012b) NATURA 2000, version 2012 rev. 1′. https://sdi.eea.europa.eu/internal-catalogue/srv/eng/search?uuid=c773f1a8–16ed-4f7a-91aa-f7475105f805

EEA (2012a) Particulate matter from natural sources and related reporting under the EU air quality directive in 2008 and 2009, EEA technical report No 10/2012. European Environment Agency

EEA (2013a) Environment and human health, joint EEA-JRC report, EEA report no 5/2013. European Environment Agency

EEA (2013b) Air quality in Europe – 2015 report

EEA (n.d.) Report no 9/2013. European Environment Agency

EEA (2013c) The impact of international shipping on European air quality and climate forcing, EEA technical report no 4/2013. European Environment Agency

EEA (2014a) Past and future exposure of European freshwater and terrestrial habitats to acidifying and eutrophying air pollutants, EEA technical report no 11/2014. European Environment Agency

EEA (2014b) Air quality in Europe – 2014 report, EEA report no 5/2014. European Environment Agency

EEA (2014c) Focusing on environmental pressures from long-distance transport. TERM 2014: transport indicators tracking progress towards environmental targets in Europe, EEA report no 7/2014. European Environment Agency

EEA (2015a) Air quality e-reporting database. http:/ www.eea.europa.eu/data-and-maps/data/aqereporting

EEA (2015b) Exposure of ecosystems to acidification, eutrophication and ozone (Indicator CSI 005). European Environment Agency

EEA (2015c) European Union emission inventory report 1990–2013 under the UNECE convention on Long-range Transboundary Air Pollution (LRTAP), EEA technical report no 8/2015. European Environment Agency

EEA (2015d) Exceedance of air quality limit values in urban areas (Indicator CSI 004). European Environment Agency

EEA (2015e) National emissions reported to the convention on Long-range Transboundary Air Pollution (LRTAP Convention). http://www.eea.europa.eu/dataand-maps/data/data-viewers/air-emissions-viewerlrtap

ETC/ACC (2009) Traffic and air quality contribution of traffic to urban air quality in European cities, Hak, C., Larssen, S., Randall, S., Guerreiro, C., Denby, B. and Horálek, J., ETC/ACC Technical Paper 2009/12

ETC/ACC (2010) Status and trends of NO2 ambient concentrations in Europe, Guerreiro, C., Horálek, J., de Leeuw, F., Hak, C., Nagl, C., Kurfürst, P. and Ostatnicka, J, ETC/ACC Technical Paper 19/2010.ETC/ACM, 2011, Evaluation of current limit and targetvalues as set in the EU Air Quality Directive, de Leeuw, F.and Ruyssenaars, P., ETC/ACM Technical Paper 2011/3

ETC/ACM (2015a) European air quality maps of PM and ozone for 2012 and their uncertainty, Horálek, J., de Smet, P., Kurfürst, P., de Leeuw, F. and Benešová, N., ETC/ACMTechnical Paper 2014/4

ETC/ACM (2015b) Mapping ambient concentrations of benzo(a)pyrene in Europe – population exposure and health effects for 2012, Guerreiro, C., Horálek, J., de Leeuw, F. and Couvidat, F., ETC/ACM Technical Paper 2014/6

EU (1992) Council Directive 92/43/EEC of 21 May 1992 on the conservation of natural habitats and of wild fauna and flora (OJ L 206, 22.7.1992, p. 7–50). http://eur-lex.europa.eu/legal-content/EN/ TXT/?uri=CELEX:31992L0043

EU (2001) Directive 2001/81/EC of the European Parliament and of the council of 23 October 2001 on national emission ceilings for certain atmospheric pollutants (OJ L 309, 27.11.2001, p. 22–30). http://eurlex. europa.eu/LexUriServ/LexUriServ.do?uri=OJ:L:2001: 309:0022:0030:EN:PDF

EU (2002) Decision No 1600/2002/EC of the European parliament and of the council of 22 July 2002 laying down the Sixth Community Environment Action Programme (OJ L 242, 10.9.2002, p. 1–15). http://eur-lex.europa.eu/LexUriServ/LexUriServ. do?uri=CELEX:32002D1600:EN:NOT

EU (2004) Directive 2004/107/EC of the European Parliament and of the council of 15 December 2004 relating to arsenic, cadmium, mercury, nickel and polycyclic aromatic hydrocarbons in ambient air (OJ L 23, 26.1.2005, p. 3–16). http://eur-lex.europa.eu/LexUriServ/LexUriServ.do?uri=OJ:L:2005:023:0003:0016:EN:PDF

EU (2008) Directive 2008/50/EC of the European Parliament and of the council of 21 May 2008 on ambient air quality and cleaner air for Europe (OJ L 152, 11.6.2008, p. 1–44). http://eur-lex.europa.eu/ LexUriServ/LexUriServ.do?uri=OJ:L:2008:152:0001:0044:EN:PDF

EU (2009) Directive 2009/30/EC of the European Parliament and of the council of 23 April 2009 amending Directive 98/70/EC as regards the specification of petrol, diesel and gas-oil and introducing a mechanism to monitor and reduce greenhouse gas emissions and amending Council Directive 1999/32/EC as regards the specification of fuel used by inland waterway vessels and repealing Directive 93/12/EEC (OJ L 140, 5.6.2009, p. 88–113). http://eur-lex.europa.eu/LexUriServ/LexUriServ.do?uri=OJ:L:2009:140:0088:0113: EN:PDF

EU (2013) Decision No 1386/2013/EU of the European Parliament and of the Council of 20 November 2013 on a General Union Environment Action Programme to 2020 'Living well, within the limits of our planet, OJ L 354, 20.12.2013, p. 171–200. http://eur-lex.europa.eu/legal-content/EN/TXT/PDF/?uri=CELEX:32013D1386& from=EN
IARC (2012) Chemical agents and related occupations – volume 100 F – a review of human carcinogens. International Agency for Research on Cancer (IARC) Monographs on the Evaluation of Carcinogenic Risks to Humans, Vol. 100F. http://monographs.iarc.fr/ENG/Monographs/vol100F/mono100F.pdf
IARC (2013) Outdoor air pollution a leading environmental cause of cancer deaths. Press Release No 221. http:// www.iarc.fr/en/media-centre/iarcnews/pdf/pr221_E. pdf
OECD/IEA (2013) Nordic energy technology perspectives: pathways to a carbon neutral energy future. Organisation for Economic Co-operation and Development/International Energy Agency, Paris
UN (2012) World population prospects: the 2010 revision. United Nations, Department of Economic and Social Affairs, Population Division, New York
UNECE (1979) Convention on long-range transboundary air pollution. United Nations Economic Commission for Europe, Geneva. http://live.unece.org/fileadmin/DAM/env/lrtap/full%20text/1979.CLRTAP.e.pdf
UNECE (2011) Manual on methodologies and criteria for modelling and mapping critical loads & levels and air pollution effects, risks and trends – chapter 3: mapping critical levels for vegetation. United Nations Economic Commission for Europe, Geneva. http://icpvegetation.ceh.ac.uk/manuals/mapping_manual.html
WHO (2000) Air quality guidelines for Europe. World Health Organization, Regional Office for Europe, Copenhagen. http://www.euro.who.int/__data/assets/ pdf_file/0005/74732/E71922.pdf
WHO (2005) Effects of air pollution on children's health and development – a review of the evidence. World Health Organization, Regional Office for Europe, Copenhagen
WHO (2006a) Air quality guidelines. Global update 2005. Particulate matter, ozone, nitrogen dioxide and sulphur dioxide. World Health Organization, Regional Office for Europe, Copenhagen
WHO (2006b) Health risks of particulate matter from long-range transboundary air pollution. World Health Organization, Regional Office for Europe, Copenhagen
WHO (2008) Health risks of ozone from long-range transboundary air pollution. World Health Organization, Regional Office for Europe, Copenhagen
WHO (2010) WHO guidelines for indoor air quality: selected pollutants. World Health Organization, Regional Office for Europe, Copenhagen
WHO (2013a) Review of evidence on health aspects of air pollution – REVIHAAP project, technical report. World Health Organization, Regional Office for Europe, Copenhagen
WHO (2013b) European health for all database [online database]. WHO Regional Office for Europe, Copenhagen. http://data.euro.who.int/hfadb/
WHO (2013c) Health risks of air pollution in Europe – HRAPIE project. New emerging risks to health from air pollution – results from the survey of experts. World Health Organization, Regional Office for Europe, Copenhagen
WHO (2014a) Burden of disease from ambient air pollution for 2012 – summary of results. World Health Organization, Geneva. http://www.who.int/phe/health_topics/ outdoor air/databases/AAP_BoD_results_March2014.pdf
WHO (2014b) Ambient (outdoor) air quality and health, fact sheet no 313, updated March 2014. World Health Organization. http://www.who.int/mediacentre/ factsheets/fs313/en/
WHO (2004) Health risks of particulate matter from long-range transboundary air pollution. World Health Organization, Regional Office for Europe, Copenhagen

Cosimo Palagiano is emeritus professor in geography at the Department of Documentary, Linguistic-Philological and Geographical Sciences, Sapienza University of Rome.

Rossella Belluso holds a master's degree in geography and a PhD in economic geography at Sapienza University of Rome, Department of European, American and Intercultural Studies. Her research interests are mainly addressed to the study of wine-and-food folk traditions and cultures, to international migrations to Italy (she is member of the relevant PRIN 2008 project) and to climate change. She is member of the multidisciplinary project sponsored by the Italian Ministry for Cultural Heritage in cooperation with the Società Geografica Italiana. She is member, since 2008 to present, of the international project *Education for Rural People*, sponsored by the FAO of the UN, as well as of the FAO *World Food Day* project. She has published about 70 articles, 20 reviews and more than 20 notes on major Italian and international geographical magazines.

Chapter 7
Canada: Climate Change, Air Pollution and Health

Stefania Bertazzon and Fox Underwood

Abstract Canada is a very large country with a very sparse population. As one of the highest-latitude countries in the northern hemisphere, it is exposed to extreme effects of climate change. Many of these effects have an impact on air quality. Canada is also one of the world's largest economies, with its wealth tightly linked to natural resource extraction. This resource dependency has led to a remarkable awareness of the potentially negative consequences of a resource-based economy on the environment, climate change, and air quality and, hence, to a tension between economic development and environmental protection. Canada has the ability to invest significantly in the monitoring and modelling of air quality. In translating this knowledge to the medical community and the general public, health risks related to air pollution could be mitigated and better health could be promoted. However, monitoring efforts should focus far more on the spatial dimension, in addition to the temporal one, owing to the great expanse of Canada's geography.

Keywords Geography • North • Spatial variability • Climate change • Air modelling • Health

An Overview of Canada and Its Geography

Canada is the second largest country in the world, with a land surface of almost 9 million square kilometres and a population of 36 million (Statistics Canada 2017a). For comparison, Canada is almost as large as Europe, albeit with a population approximately one-twentieth the size. Most of Canada lies north of the 49° parallel, and most people live in the ten provinces, which lie approximately between the 49° and the 60° parallel. Approximately 80% of the Canadian population has been classified as living in urban areas (Statistics Canada 2011). The remaining 20% of the population live in rural and remote areas. As medical services are clustered in a few large urban areas, health care is generally accessible to only urban and near-urban populations and is fairly inaccessible to rural and remote

S. Bertazzon (✉) • F. Underwood
Department of Geography, University of Calgary, Calgary, AB, Canada
e-mail: bertazzs@ucalgary.ca

© Springer International Publishing AG 2018
R. Akhtar, C. Palagiano (eds.), *Climate Change and Air Pollution*,
Springer Climate, DOI 10.1007/978-3-319-61346-8_7

populations. Northern residents (i.e. those living above the 60° parallel), in particular, may only be able to drive on certain ice roads, provided the weather is cold: in warmer temperatures, ice roads may be unusable or unsafe (Natural Resources Canada 2014). Consequently, to travel long distances for care, air travel may be the only option and only if residents can afford it.

While primary industry is necessarily co-located with natural resources, manufacturing and related industries are typically located on the eastern sides of urban areas in North America (Bailie and Beckstead 2010). Combined with prevailing west and north winds throughout the country, the pattern of polluting facilities being located on the eastern sides of urban areas tends to have positive effects on urban air, as most residential areas lie upwind from noxious emissions. Unsurprisingly, rural residents and residents of minor urban centres do not necessarily benefit from this locational advantage. Further, indigenous communities are often located in rural areas and reserves and, likewise, may experience greater exposure to industrial pollution related to primary sector activities.

Typical Canadian cities feature sprawling suburbs characterized by single-family dwellings and low population density and a tendency to expand into surrounding towns and villages. These pervasive urban dynamics are known as urban sprawling, a dynamic and growing phenomenon in many North American metropolitan and urban centres. Urban sprawling is associated with a variety of environmental issues, ranging from the consumption of rural and natural areas, to biodiversity reduction, to the need for extensive utility lines such as electricity and gas. Most notably, this spatial pattern leads to increasingly low population density and relative scarcity of services in newer communities, where residents are subject to long daily commutes that predominantly take place in private vehicles carrying only a single occupant. As a result, commuting and residential traffic tends to be heavy and aggravated by heavy commercial traffic, which is also present in urban areas. Further, because of the country's northern location, traffic can be slow and difficult during the long winter season, when heavy snow and wind often result in treacherous road conditions and frequent minor accidents and, therefore, increased traffic-related air pollution. Traffic is among the main sources of air pollution in Canadian cities.

Canada's economy is largely based on the resource sector, where, over the last few decades, the oil and gas sectors have gained prominence with intense exploitation of oil deposits and oil sands, particularly in the northern portion of the province of Alberta. Exploration and extractive activities tend to be associated with high levels of air and water pollution, especially as surface deposits are depleted and extraction inevitably occurs at greater depths. Pollution associated with resource activities, particularly oil extraction, tends to more directly affect rural and remote populations, as well as minor urban centres and indigenous communities.

Geography and the Spatial Turn in the Health Sciences

With its vast territory and large variability in resources, climate, and population distribution, Canada is truly a place where space matters. Indeed, due to its need to manage expansive land parcels, Canada is the country where the first geographic information system, Canada geographic information system (CGIS), was realized by Roger Tomlinson in the 1960s (Tomlinson 1968). Yet, even in a country of this size, researchers do not always think spatially. Goodchild and Janelle (2004) have argued that a spatial turn, begun perhaps a decade ago in the social sciences, is now invoking a similar turn in the health sciences as well (Richardson et al. 2013). Yet, "People die each year because no one bothers to properly analyse disease and death data for unusual localised concentrations" (Openshaw 1997). Indeed, "spatial is special" (Anselin 1989): environmental exposures, residential location, economic activities, and transportation routes occur in space, at different scales, and with peculiar regional and local characteristics. They interact with each other, as well as with climate change, mass migration, and population genetics.

Canadian health research is generally aware of the spatial dimension of health and health care when it comes to health service research, distance, and geographic accessibility. Back in 2002, a report commissioned by the Canadian government noted that "Canadians want and expect to have access to health care services when and where they need them" and that "concerns also exist about *timely access to existing services, particularly in rural and remote areas*" (Romanow 2002). Spatial epidemiology is another field where the use of spatial reasoning and methods has become a routine. Paradoxically, when it comes to air pollution and climate, many researchers seem to forget all about space, spatial thinking, and spatial methods. Air pollution is measured with painstaking frequency and regularity over time—but only at sparse and irregular locations in space. Disregarding space when measuring air quality is just the tip of the iceberg: we tend to be unaware of the variability of air pollution over space even though we are aware of the variability of weather conditions within our cities.

Climate Change and Effects of Air Pollution and Health

Due to its high latitude and large land area, Canada is exposed to severe impacts of climate change. For example, changes in winter cyclonic patterns in recent years have been associated with dichotomous patterns, with mild winters in Western Canada in contrast to harsh seasons and abundant precipitation in Eastern Canada. These patterns in turn affect air circulation at the continental as well as urban scale, impacting air and pollution transport.

Moreover, it is becoming accepted that climate change is associated with greater variability and greater exposure to severe atmospheric events. These events do not necessarily have direct impacts on air quality, but their indirect impacts in terms of cost and human health are very significant. Owing to its northern location, Canada has been experiencing increasing severe atmospheric events in the spring, with more sudden and faster snowmelt, and in the fall, with early-season blizzards. In both Atlantic and Pacific coastal regions, these events have been accompanied by remnants of tropical storms. The most recent examples are the floods of autumn 2016 in southern British Columbia on the Pacific coast, which occurred almost simultaneously with devastating floods in Atlantic Canada, particularly in Nova Scotia and Newfoundland and Labrador, and is associated with the end tail of Hurricane Matthew after the hurricane's devastating effects on Haiti and parts of the United States.

The city of Calgary, in southern Alberta, had not experienced severe flooding until the summer of 2005, when century-long record-breaking river levels occurred due to unusually high rainfall. While relatively minor damage was experienced by large numbers of homes in many parts of the city (e.g. flooding of basements and lower floors), severe damage was experienced in some of the oldest residential communities, which are located near major riverbeds. Calgarians may have hoped that the 2005 floods were an isolated event, but the city experienced much worse flooding in the summer of 2013. The meteorological event originated in the uphill Rocky Mountain regions, where the winter had been characterized by heavy snowfall, which remained on the ground into the summer. With warming temperatures in the month of June, heavy rainfall in the mountains triggered the movement of large masses of snow that began to slide into the rivers, resulting in devastating floods in downstream communities, namely, Canmore, High River, and Calgary. This flood was named Canada's most costly natural disaster, costing between 5 and 6 billion dollars in damages (Milrad et al. 2015).

Factors that contributed to the flood included higher than normal snowfalls in the mountains, excess amount of precipitation during the early spring in the Bow and Elbow River watershed, and a wet spring that left soils saturated with no room to absorb additional precipitation (Eccles et al. 2017). At its peak discharge rate, the Bow River was flowing at an estimated 1700 m^3/s (Milrad et al. 2015). Several inner-city communities in Calgary were evacuated, resulting in displacement of over 100,000 people throughout the region. Recent studies (Eccles et al. 2017) suggest that the floods of 2005 and 2013 were also associated with contamination of rural drinking water, which is supplied from private water wells. Among the consequences of the floods was a drastic loss in market value of some of the oldest and more attractive residential communities. Calgary was founded a century ago at the intersection of the Bow and Elbow rivers. At the time, with the limited local knowledge of a recently settled land, the location was considered safe.

Early-season blizzards bring abundant snow precipitation over short periods, accompanied by strong winds as early as October and, recently, even in September. Some of the impacts of these events are not substantially different from normal winter events, such as reduced visibility and treacherous driving and walking conditions. However, early-season blizzards are worsened by the unpreparedness of the general population; for example, snow tires have not yet been mounted on vehicles, while pedestrians walk wearing lighter, less sturdy shoes. This leads to increased probability of traffic accidents and falls and injuries. More severe and emblematic consequences occur when these events occur before the early season, such as the event that hit Calgary on September 10, 2014 when the trees still had full foliage. According to the City of Calgary (2014), about half of all trees were damaged, with more severe damage to the larger and more mature trees of older communities.

Aside from the impact of tree loss on urban air quality, one of the major consequences of this September storm was the damage to aerial electricity lines caused by falling branches, which left some 30,000 families, particularly in older communities, without electricity. Canadians rely heavily on electricity: while residential heating is typically fuelled by natural gas, some components of furnaces and thermostats require electricity. Cooking depends on electricity too, as the vast majority of cooking stoves are electric. The lack of electricity therefore sparked an emergency in the cold conditions, with residents lighting wood-burning stoves and fireplaces. Wood fires are considered a major source of particulate matter air pollution in the region.

Forest fires are natural events in the mountain regions and the boreal forest of Canada. However, forest fires frequently result from so-called prescribed burns: burns that are mandated under forest management programmes to control pests. Climate change may be associated with increased frequency of forest fires, both of natural and man-made origin, with man-made fires being associated with greater occurrence of pests (e.g. the pine beetle). Forest fires release large quantities of smoke into the air, leading to severe loss of visibility and lower air quality. For example, in the summer of 2015, smoke drifting from fires in Washington State caused more than ten times the annual average of fine particulate matter ($PM_{2.5}$) in the air throughout southern Alberta, blotting the sky with grey clouds and leading health authorities to issue alerts and air quality advisories. The following year, a devastating forest fire burned for over a month in the summer of 2016 in Fort McMurray (northern Alberta), spreading over 590,000 hectares of forest and destroying approximately 2400 homes. This forest fire became the costliest disaster in Canadian history, surpassing the 2013 Calgary floods of only 3 years before. The causes of the fire have not been determined, nor has a connection with climate change been positively established. In contrast, El Niño has been considered as a probable contributing factor to a mild and dry fall and winter and to an even more unusually warm spring season.

The Canadian northern climate is characterized by long winters, with relatively short spring and summer seasons, which are the prime blooming seasons. Yet the changing climate in recent years, along with increasing climate variability, has seen shorter springs and more rapid transitions to summer. For instance, the ragweed season has increased by 1 month in parts of Canada (Ziska et al. 2011). Changes in the seasons may also bring about new changes, with unpredictable impacts on allergy and asthma sufferers.

Air Quality Monitoring

As a wealthy country, Canada can rely on financial resources and an established infrastructure that enables timely and accurate air quality monitoring. This infrastructure is important for monitoring present air quality conditions and may become increasingly important as the climate continues to change, leading to less predictable patterns and levels of air pollution. There are several agencies managing air quality monitoring at various levels (e.g. federal, provincial, and local). As an example, the Calgary Region Airshed Zone (2017) manages a network of passive and continuous air quality monitoring stations, providing monthly and hourly data on a suite of common air pollutants such as particulate matter ($PM_{2.5}$), ozone (O_3), and volatile organic compounds (VOCs). Environment Canada issues hourly Air Quality Health Index (AQHI) updates. The index, a composite of pollutants that are known to be noxious for human health, can be accessed easily and readily alongside the weather forecast (Environment Canada 2010).

The AQHI is issued hourly for each of the three continuous monitoring stations in Calgary. While the AQHI is clearly a useful indicator, it may not be truly representative of local air quality and associated health risk, due to the complexity of Canada's geography and the great geographic spread of its population. Consider, again, the city of Calgary alone. In 2016, the city of Calgary had a population of 1,239,220 over a land area of 825 km^2 (Statistics Canada 2017b)—a far greater area than three monitoring stations can hope to represent in their sensor readings. Calgary has expanded over a relatively flat prairie where space was once abundant. Over the years, its expansion model has been increasingly characterized as urban sprawl, dominated by single-family dwellings. This urban dynamic pattern has resulted in a very low population density: 1329 persons per square kilometre. This residential pattern in turn results in long commuting journeys: in 2006, more than 75% of commuters in the metropolitan area commuted by personal vehicle, over 60% of them carrying only a single occupant (Bailie and Beckstead 2010). In 2011, the average commute time was 25 min, which is above the national average (Statistics Canada 2016). In 2006, Calgary had the second largest average commute distance in Canada (after Toronto) and the greatest 5-year increment among the largest Canadian cities (Bailie and Beckstead 2010). Of the total greenhouse emissions in 2006, 70% was attributed to buildings, 27% to transportation, and 3% to waste (Bailie and Beckstead 2010). The land use pattern is also relatively segregated, with distinct residential, commercial, and industrial zones. In this complex urban landscape, air pollution is affected by localized emission patterns. Conversely, population health risk is affected by the emission pattern, as well as by the residential pattern.

The AQHI is issued frequently, but at a very coarse spatial resolution. The AQHI is calculated based on the values of three air pollutants: particulate matter ($PM_{2.5}$), ozone (O_3), and nitrogen dioxide (NO_2). Of these three pollutants, PM and O_3 are considered regional pollutants; that is, their variation over space is relatively moderate and their value is fairly constant regionally. Conversely, NO_2 tends to display large variations over space as a function of industrial emissions, residential emissions (heating in the winter), and traffic-related emissions (motorized vehicles). Further, air pollution levels are affected by meteorological factors, which include primarily wind, along with temperature, humidity, and air pressure. Notably, regional air pollutants, such as PM and O_3, exhibit lesser spatial variation, but they still exhibit a spatial pattern, especially in complex urban environments with localized emissions, and their dynamics may be aggravated by wind and other meteorological variables.

Empirical research is based on recorded land use variables, such as traffic volumes, residential heating, population density, and wind speed and direction. Therefore, analytical models are flexible and can serve as tools to estimate and predict changes in air pollution in the presence of climate change (Bertazzon et al. 2015). Such models can help to address what-if questions to simulate climate change scenarios and estimate the resulting variations in air pollution. Perhaps the most important component of empirical work is the knowledge translation (i.e. communicating results to local, community stakeholders and to health practitioners) in an effective framework in order to mitigate the impact of air pollution on human health. Owing to its relative development and wealth, and relatively low pollution, Canada is in a strong position to lead the way in promoting extensive spatial monitoring and modelling of air pollution.

Walking and Breathing

Canada is a relatively young country, with few major cities dating back a few centuries, while many, such as Calgary, have been around for just over a century. Compared with the oldest cities of Europe, Asia, and North Africa, Canada's cities have had a much shorter history and experienced their major development during the automobile era. This urban pattern, related to the urban sprawl discussed earlier, has led to largely car-dependent urban environments, where it is often difficult to carry out daily activities without the aid of a private vehicle.

Obesity has been linked to a lack of balance between calorie intake (nutrition) and calorie consumption (exercise). In recent years, health researchers and professionals, along with city managers, politicians, and urban planners, have come to promote more walking, cycling, and public transit use in daily activities, as opposed to solely driving in private vehicles. However, in this effort to promote walking, it has been observed that walking may be problematic in North American cities, given that they were designed for driving. Indeed, walking may be extremely unsafe where there are no sidewalks, where pedestrians are forced to cross high-traffic roads without proper pedestrian crossings, where lighting is poor (especially in the winter), or in neighbourhoods where crime rates are high. Moreover, distance plays an important role, as people can comfortably walk over a certain threshold, but having to cover large distances would require so much time and effort as to impact work or other routine activities. Finally, walking is also seen as providing a certain experience; therefore, some neighbourhoods and roads provide a better experience than others do. Walking along a street with shop windows, trees, benches, well-maintained buildings and well-maintained yards provides a positive experience, whereas walking along a path that requires frequent busy-road crossings, rundown neighbourhoods and narrow or bumpy or uneven sidewalks provides a less positive walking experience. All these elements have come to form the concept of *walkability*, and walkability indices have been defined to rank neighbourhoods in North American cities based on their geographical accessibility and distance from amenities. That is, the *walkscore* of a neighbourhood increases with the sum of amenities that can be reached by walking or that are located within a walking distance of that neighbourhood.

We shall argue that when we walk outdoors, we breathe ambient air. Walking in an urban environment exposes us to varying levels of air pollution, yet air quality is rarely considered a dimension of walkability. Nonetheless, variations in pollution levels within urban environments can be large. Consider the vast urban and metropolitan area of Calgary where, in addition to the wide spatial extension, elevation ranges by 300 metres from highest to the lowest elevation, and winds can be strong and can vary in direction within a day, thereby quickly moving air masses. Moreover, most pollution sources are localized, notably over the industrial park, the airport, the railway yard, and along the railway and major roads. In winter, more traffic occurs on local roads, as people tend to walk less and drive more, while residential zones exhibit higher pollution levels due to residential heating. Accordingly, several traffic-related pollutants exhibit a spatial pattern with pollutant concentration declining rapidly as distance from roads increases; therefore, walking within a few hundred metres of a major road leads to higher pollution exposure than if we were to walk further away from the road. Truly, *where* we walk can make a difference in the quality of the air we breathe.

As climate changes and tends to become warmer overall, Canadian cities will become naturally more walkable, as the ambient air will be more pleasant and the danger of ice and treacherous sidewalks will decrease. We can expect weather patterns to change over a range of scales, including local circulation over urban and metropolitan areas. We can expect greater variability that can make it harder to choose where to walk and breathe less polluted air. A spatial analytical approach to the study of air quality and walkability can only benefit our understanding of urban air pollution, reducing our exposure to noxious pollutants, for the benefit of our health.

Conclusion

This chapter has presented an overview of Canada's geographic position related to climate change and air pollution. In northern territories, residents are far from health care, while pollution from Canada's strong primary industry affects more rural, remote, and indigenous residents than those living in urban areas. Conversely, within urban areas, sprawling development has lowered walkability and given strong incentive for people to drive, often in single-occupancy vehicles, giving rise to greater air pollution from traffic in large cities. At the same time, Canada's long winters have led to higher pollution from residential heating. With changes to spring and summer, ragweed season may now be longer in parts of Canada. Finally, more severe weather has begun to appear in the form of serious fires, floods, and snowstorms.

Climate change exposes Canada to major changes and therefore major risks. Many of these risks involve increased air pollution and greater harm to our health. However, as a wealthy country concerned with the effects of air quality on human health, Canada could potentially put the proper infrastructure and research in place—with a strong emphasis on expanding air monitoring over space—to understand these effects and mitigate their impact on human health, as its climate changes. Written by geographers, this chapter was centred on a geographic perspective to air quality, because air quality varies over space, and often scientists are entirely engrossed with temporal variability.

References

Anselin L (1989) What is special about spatial data? Alternative perspectives on spatial data analysis. Presented at the Spatial Statistics: Past, Present, and Future Symposium, Institute of Mathematical Geography, Syracuse University. Retrieved from https://deepblue.lib.umich.edu/handle/2027.42/58237

Bailie A, Beckstead C (2010) Canada's coolest cities. Retrieved 13 Mar 2017, from http://www.pembina.org/pub/2021

Bertazzon S, Johnson M, Eccles K, Kaplan G (2015) Accounting for spatial effects in land use regression for urban air pollution modelling. Spat Spatio-temporal Epidemiol 14-15:9–21

Calgary Region Airshed Zone (2017) What we monitor. Retrieved 13 Mar 2017, from http://craz.ca/monitoring/what-we-monitor/

City of Calgary (2014, Oct 16) Recovering from "Snowtember" 2014. Retrieved 13 Mar 2017, from http://www.calgary.ca/CSPS/Parks/Pages/Tree-recovery-and-restoration.aspx

Eccles KM, Checkley S, Sjogren D, Barkema HW, Bertazzon S (2017) Lessons learned from the 2013 Calgary flood: assessing risk of drinking water well contamination. Appl Geogr 80:78–85. https://doi.org/10.1016/j.apgeog.2017.02.005

Environment Canada (2010, Jan 7) Environment and climate change Canada – air – National Air Pollution Surveillance Program (NAPS) – Home. Retrieved 13 Mar 2017, from http://www.ec.gc.ca/rnspa-naps/

Goodchild, M. F., Janelle, D. G. (2004) Spatially integrated social science. Oxford University Press

Milrad SM, Gyakum JR, Atallah EH (2015) A meteorological analysis of the 2013 Alberta flood: antecedent large-scale flow pattern and synoptic–dynamic characteristics. Mon Weather Rev 143(7):2817–2841. https://doi.org/10.1175/MWR-D-14-00236.1

Natural Resources Canada (2014 June 17) Canada in a changing climate: sector perspectives on impacts and adaptation. Retrieved 13 Mar 2017, from http://www.nrcan.gc.ca/environment/resources/publications/impacts-adaptation/reports/assessments/2014/16309

Openshaw, S (1997) GIS in health and crime analysis. Retrieved 13 Mar 2017, from http://www.geog.leeds.ac.uk/presentations/97-2/sld025.htm

Richardson DB, Volkow ND, Kwan M-P, Kaplan RM, Goodchild MF, Croyle RT (2013) Spatial turn in health research. Science 339(6126):1390–1392. https://doi.org/10.1126/science.1232257

Romanow, R (2002, Jul 1) Building on values: the future of health care in Canada: CP32-85/2002E-IN – Government of Canada Publications. Retrieved 13 Mar 2017, from http://www.publications.gc.ca/site/eng/237274/publication.html

Statistics Canada (2011 Feb 4) Population, urban and rural, by province and territory (Canada). Retrieved 13 Mar 2017, from http://www.statcan.gc.ca/tables-tableaux/sum-som/l01/cst01/demo62a-eng.htm

Statistics Canada (2016 Sep 15) Commuting to work. Retrieved 13 Mar 2017, from https://www12.statcan.gc.ca/nhs-enm/2011/as-sa/99-012-x/99-012-x2011003_1-eng.cfm

Statistics Canada (2017a Feb 8) Focus on geography series, 2016 census – census subdivision of Calgary, CY (Alberta). Retrieved 13 Mar 2017, from https://www12.statcan.gc.ca/census-recensement/2016/as-sa/fogs-spg/Facts-csd-eng.cfm?LANG=Eng&GK=CSD&GC=4806016

Statistics Canada (2017b Feb 8) The daily—population size and growth in Canada: key results from the 2016 Census. Retrieved 13 Mar 2017, from http://www.statcan.gc.ca/daily-quotidien/170208/dq170208a-eng.htm?HPA=1

Tomlinson, R. (1968) A geographic information system for regional planning. Department of Forestry and Rural Development, Government of Canada. Retrieved from https://gisandscience.files.wordpress.com/2012/08/1-a-gis-for-regional-planning_ed.pdf

Ziska, L., Knowlton, K., Rogers, C., Dalan, D., Tierney, N., Elder, M. A., … Frenz, D. (2011). Recent warming by latitude associated with increased length of ragweed pollen season in central North America. Proc Natl Acad Sci, 108(10), 4248–4251. https://doi.org/10.1073/pnas.1014107108

Stefania Bertazzon A health geographer, Stefania Bertazzon has worked extensively on spatial modelling of air pollution. She authored over 60 peer-reviewed publications, many on quantitative modelling, air pollution, and public health. She actively engages with health researchers, hospitals, public health officers; she is a member of the Calgary Region Airshed Zone PM-O3 management committee. Her research was featured in major Canadian media.

Fox Underwood is a geographic research analyst working in a supporting role in the areas of air pollution and digestive disease epidemiology. She prepares maps and geographic variables to study digestive diseases; in particular, the inflammatory bowel diseases of Crohn's disease and ulcerative colitis. She has also carried out data collection for two substantial air pollution monitoring campaigns in Calgary.

Chapter 8
Climate Change, Forest Fires, and Health in California

Ricardo Cisneros, Don Schweizer, Leland (Lee) Tarnay, Kathleen Navarro, David Veloz, and C. Trent Procter

Abstract Wildland fire is an important component to ecological health in California forests. Wildland fire smoke is a risk factor to human health. Exposure to smoke from fire cannot be eliminated, but managed fire in a fire-prone ecosystem for forest health and resiliency allows exposure to be mitigated while promoting other ecosystem services that benefit people. The California Sierra Nevada is a paragon of land management policy in a fire-prone natural system. Past fire suppression has led to extreme fuel loading where extreme fire events are much more likely, particularly with climate change increasing the length of fire season and the probability of extreme weather. We use the California Sierra Nevada to showcase the clash of increased development and urbanization, past land management policy, future scenarios including climate change, and the intertwining of ecological health and human health. Fire suppression to avoid smoke impact has proven to be an unreliable way to decrease smoke-related health impacts. Instead ecological beneficial fires should be employed, and their management should be based on smoke impacts at monitors, making air monitoring the foundation of fire management actions giving greater flexibility for managing fires. Tolerance of smoke impacts from restoration fire that is best for forest health and resiliency, as well as for human health, is paramount and preferred over the political expediency of reducing smoke impacts today that ignores that we are mortgaging these impacts to future generations.

Keywords Wildland fire smoke • Climate change • Public health • Air quality • Policy • Ecological health

---

R. Cisneros (✉) • D. Veloz
Health Science Research Institute, University of California, Merced, CA 95340, USA
e-mail: rcisneros@ucmerced.edu

D. Schweizer
Health Science Research Institute, University of California, Merced, CA 95340, USA

USDA Forest Service, Pacific Southwest Region, Vallejo, CA 94592, USA

L. (Lee) Tarnay • K. Navarro • C.T. Procter
USDA Forest Service, Pacific Southwest Region, Vallejo, CA 94592, USA

© Springer International Publishing AG 2018

R. Akhtar, C. Palagiano (eds.), *Climate Change and Air Pollution*,
Springer Climate, DOI 10.1007/978-3-319-61346-8_8

Introduction

In this chapter, we discuss forest fires in California ecosystems and the subsequent human health impacts via smoke exposure and the implications of climate change and past suppression policy. Projections under likely climate change scenarios demonstrate that the area burned from wildfires and the length of the wildfire season will continue to increase in the western United States and in many other parts of the world. Wildfires emit many air pollutants of concern for public health. Wildland fire is an important component to ecological health in California Forests. Large tracts of this land have been set aside for ecological protection and are adjacent to areas of high population living in poor air quality such as the Central Valley. Much of the attention is given to the Sierra Nevada (see Fig. 8.1) which covers about 25% of California' land area and supplies more than 60% of the developed water. Approximately 63% of the Sierra Nevada are federally protected public lands including nine national forests, three national parks, and two national monuments containing 20 designated wilderness areas. The intersection of this fire-prone ecosystem and large amounts of anthropogenic emissions creates a unique setting to understand natural process function and ecological health, the implications of climate change to that system, and the consequences to human health from forest management in an already anthropogenically polluted environment.

Intact functioning ecosystems are essential to defend against climate change (Martin and Watson 2016, p. 123), but one-tenth of global wilderness has been lost in the past two decades (Watson et al. 2016, pp. 2–3). Fire is a natural process integral to California Forests and shrub lands, including the Sierra Nevada, determining vegetation distribution and structure (Kilgore 1981, pp. 58–89; Swetnam 1993, pp. 887–888; Swetnam 2009, pp. 133–140). Smoke is an inevitable consequence. Native American tribes have a long history attributing fire and smoke to successful landscape management including active and widespread use of fire to increase desired results (Levy 2005, p. 305). As C.H. Merriam chief of the US Division of Biologic Survey wrote in 1898 of smoke and visibility in the Sierra Nevada (Cermak 2005, p. 17):

> Few see more than the immediate foreground and a haze of smoke which even the strongest glass is unable to penetrate.

Numerous other accounts attest to much more smoke being encountered during European settlement of the Sierra Nevada. Wildland fire smoke in California in the late nineteenth century was said to "choke up the atmosphere" and with any increase "…our farmers will be able to cure bacon and ham without the aid of a smokehouse" but "Nobody seemed to care; it was all public land, and what is everybody's business is nobody's business" (Cermak 2005, pp. 15–17). There was a largely indifferent attitude to wildland fire and smoke among the mountain residents during European settlement of the Sierra Nevada with a general belief that it was an essential process (Cermak 2005, pp. 9–18).

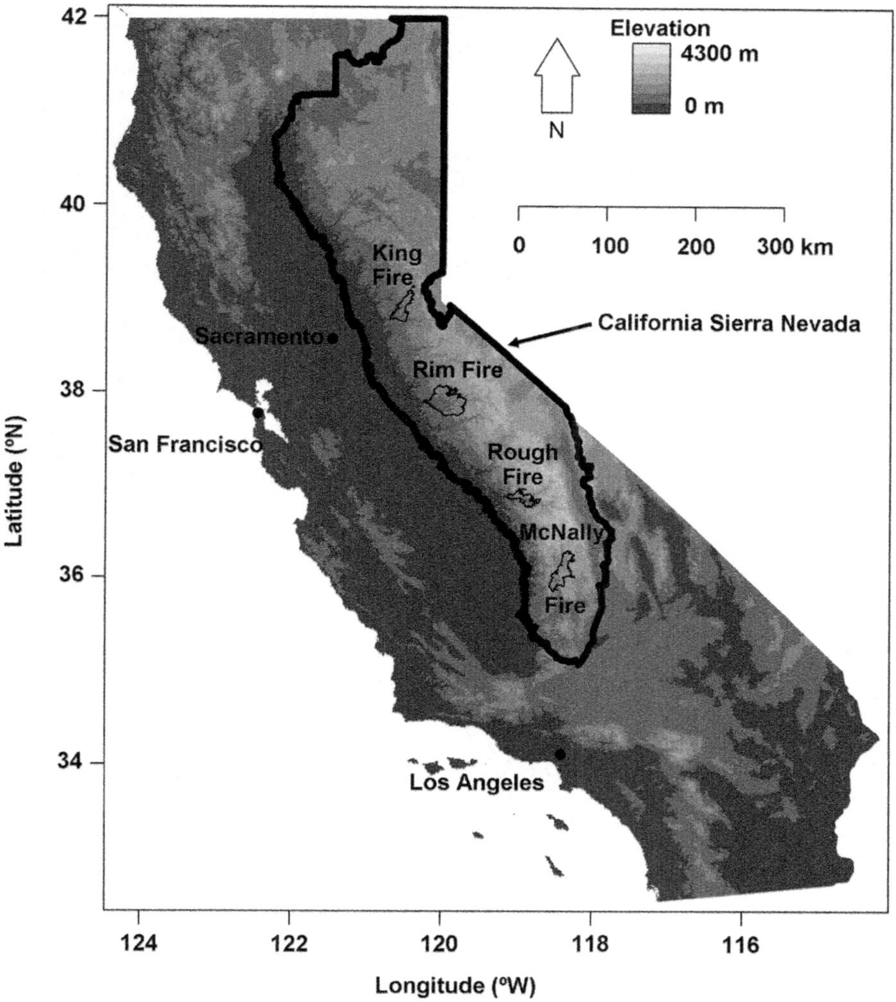

Fig. 8.1 Location map

The benefits of wildland fire slowly moved out of favor as land was developed and industrialized (Cermak 2005, pp. 19–20). Suppression of wildland fire became the normal management action in the United States toward the end of the nineteenth century. This fire suppression policy, dating back 150 years, has created western forests with an abundant fuel loading problem (Steel et al. 2015, pp. 8–10). Suppression policy largely transferred smoke exposure to a later date. Smoke impacts were effectively removed during the early years of suppression. Climate change and extreme fuel loading are combining to create an environment where full suppression is no longer an option. The fuel loading problem has become an air pollution emissions problem leading to human health exposure with only the

Fig. 8.2 The 1940s era; US Forest Service public campaign likens forest fires to death and destruction

question of whether emissions will be released in uncharacteristically high-intensity wildfire or through active management of wildland fire for ecological benefit.

Emissions from large wildfires analyzed using satellite data document air quality impacts (Langmann et al. 2009, pp. 112–113) with smoke toxicity (Wegesser et al. 2009, pp. 894–895) and the negative impacts of large wildfires on human health being published (Tham et al. 2009 p. 72) creating concern for exposure and public health (see Fig. 8.2). Extreme suppression policy beginning in the early 1900s has led to generations unaccustomed to smoke impacts from fire-adapted forests that historically have burned much more frequently. The impacts to human health from wildland fire emissions (smoke) must be understood and incorporated into any discussion, management, or public health policy.

Climate Change and Forest Fires in California or in the Western United States

In California, increased fuels and a changing climate are creating a post-suppression era where large high-intensity destructive wildfires (megafires) are becoming more common (see Table 8.1). Wildland fire in California and throughout much of the American west is expected to increase in activity in the post-suppression era with increased large fire frequency, longer duration fire, and a longer fire season (Westerling et al. 2006, p. 940; Flannigan et al. 2013, p. 847). The frequency of wildfires is projected to increase in many parts of the western United States due to alterations of temperature and precipitation patterns related to climate change that lead to increases in spring and summer temperatures and earlier spring snowmelt (Westerling et al. 2006, p. 940). The increased wildfire activity in California involves both climate and land management practices (Westerling et al. 2006, p. 943). Land management practices are discussed below in the forest fire policy section.

California is one of the most biologically and climatically diverse locations in the world. The highest air pollutant emissions from forest fires in the United States, including prescribed fire, occur in the Pacific coastal states which includes California (Liu 2004, p. 3489). Lenihan et al. (2008, p. 215) studied the response of vegetation distribution, ecosystem productivity, and fire to climate change scenarios for California. In terms of fire, the findings indicated that the area burned in California would increase (9–15%) compared to historical period (1961–1990) under all three scenarios. Under the more productive less dry and cooler scenario, annual biomass consumption by fire was 18% greater than historical norm. The warmer and drier scenarios predicted that biomass consumption would be initially greater, than below or at historical norm by 2100.

Westerling and Bryant (2008, p. 231) modeled wildfire risks for California under four climatic change scenarios. The model exhibited divergent findings in terms of fire regimes. One of the findings suggested that increases in temperature would promote greater fire frequency in wetter forested areas via increased temperatures on fuel flammability. Another predicted that reduced moisture availability due to lower precipitation and higher temperatures would lead to reduced fire risks in some locations where fuel flammability may be less important than the availability of fine fuels. Property damages were also modeled. The largest damages were predicted to occur in the wildland/urban interfaces in Southern California, the Bay Area, and the Sierra foothills northeast of Sacramento.

Westerling et al. (2011, p. 459) examined climate change and growth scenarios of wildfire in California. The majority of the modeled scenarios predicted significant increases in large burned area and wildfire occurrence to occur by 2050 with substantial increases expected by 2085. The increases were attributed to the effects of projected temperature increase on evapotranspiration compounded by reduced precipitation. Under the different scenarios, the area of wildfire is expected to increase throughout the mountain forested areas of Northern California. In the

Table 8.1 Twenty largest California suppression fires since 1932

Fire name (cause)	Date	County	Acres	Structures	Deaths
1. Cedar (human)	October 2003	San Diego	273,246	2820	15
2. Rush (lightning)	August 2012	Lassen	271,911	0	0
3. Rim (human)	August 2013	Tuolumne	257,314	112	0
4. Zaca (human)	July 2007	Santa Barbara	240,207	1	0
5. Matilija (undetermined)	September 1932	Ventura	220,000	0	0
6. Witch (powerlines)	October 2007	San Diego	197,990	1650	2
7. Kamath theater complex (lightning)	June 2008	Siskiyou	192,038	0	2
8. Marble cone (lightning)	July 1977	Monterey	177,866	0	0
9. Laguna (powerlines)	September 1970	San Diego	175,425	382	5
10. Basin complex (lightning)	June 2008	Monterey	162,818	58	0
11. Day fire (human)	September 2006	Ventura	162,702	11	0
12. Station fire (human)	August 2009	Los Angeles	160,557	209	2
13. Rough (lighting)	July 2015	Fresno	151,623	4	0
14. McNally (human)	July 2002	Tulare	150,696	17	0
15. Stanislaus complex (lightning)	August 1987	Tuolumne	145,980	28	1
16. Big bar complex (lightning)	August 1999	Trinity	140,948	0	0
17. Happy Camp complex (lightning)	August 2014	Siskiyou	134,056	6	0
18. Campbell complex	August 1990	Tehama	125,892	27	0
19. Wheeler (Arson)	July 1985	Ventura	118,000	26	0
20. Simi (under investigation)	October 2003	Ventura	108,204	300	0

Sierra Nevada, the projected increase in burn area is for mid-elevation locations on the west side of range. The majority of the locations are on private land and outside the federally managed forests and parks, exposing private landowners to a substantially increased risk of wildfire.

Meanwhile, climate change has increased the length of the fire season (Flannigan et al. 2013, p. 57; Westerling et al. 2006, p. 941) and can be expected to continue or further extend the annual pattern as large wildland fire emissions in California are expected to increase with future climate scenarios (Hurteau et al. 2014, pp. 2301–2302).

Forest Fire Impacts on Health: Epidemiological Studies

Even though wildfires pose a threat to human health in the United States, only a few health studies have been conducted (Table 8.2). This is an important subject as it might impact vulnerable populations, including the old and the young and people with compromised immune systems. In a study conducted in Southern California, the strongest effect on asthma hospitalizations related to particulate matter less than 2.5 microns in aerodynamic diameter (PM2.5) during a wildfire was found for people ages 65–99 (Delfino et al. 2009, p. 192). The second strongest association was found for children ages 0–4 years of age.

Studies in the United States have found significant associations between exposure to wildfire smoke and increased self-reported respiratory symptoms (Kunzli et al. 2006, p. 1224; Mirabelli et al. 2009, p. 451) and increases in respiratory physician visits (Lee et al. 2009, p. 321), respiratory emergency department (ED) visits (Rappold et al. 2011, p. 1418), and respiratory hospitalizations (Delfino et al. 2009, p.192). Lee et al. (2009, p.321) and Mirabelli et al. (2009, p.453) reported that adults with pre-existing respiratory conditions or weakness (i.e., small airway size) were more likely to seek care or have additional symptoms after wildfire exposure than individuals without those conditions. A few studies have engaged methods to separate the effects of PM generated by fires from other sources. A recent study ran a dispersion model with and without fire emissions. The researchers found a slight but significant increase in respiratory ED visits for increases in PM2.5 from wildfires while controlling for PM2.5 from non-fire sources (Thelen et al. 2013, p. 20).

Studies have documented significantly increased ED visits (Duclos et al. 1990, p. 55; Rappold et al. 2011, p. 1418) and hospitalizations (Delfino et al. 2009, p. 192) for asthma in association with wildfire smoke exposure. Vora et al. (2011, p. 76) demonstrated no significant changes in acute lung function related to PM2.5 from wildfires among asthmatics. This may be because people with an established diagnosis of asthma are better at self-management of symptoms such as exposure avoidance and increased use of rescue medication in response to elevated levels of smoke (Vora et al. 2011, p. 76). People with asthma reported elevated levels of rescue medication usage during a wildfire in Southern California (Vora et al. 2011, p. 76; Kunzli et al. 2006, p. 1224). Kunzli et al. (2006, p. 1225) reported that children without pre-existing asthmatic conditions had a greater increase in respiratory symptoms under exposure than did other children with pre-existing asthmatic conditions. The authors suggested that children with pre-existing asthmatic conditions tended to be on medication and have better access to care, and as a result there was a smaller increase in symptoms when exposed to wildfire smoke.

Two studies, one conducted in California and the other in North Carolina, found association in ED visits for COPD related to wildfire smoke (Duclos et al. 1990, p. 56; Rappold et al. 2011). Rappold et al. (2011, p. 418) found an association with elevated risk of pneumonia and acute bronchitis in counties exposed to smoke from peat fires. Duclos et al. (1990, p. 57) found a higher number of hospitalizations for

Table 8.2 Studies on wildfire smoke and human health

Article	Location	Exposure metric	Exposure levels	Major health outcome	Effect estimate
Duclos et al. (1990)	August 1987, lightning fire in Northern California	Temporal comparison	Not used	ED visits	Observed/ expected (p value)
				Asthma	1.4(<0.001)
				COPD	1.3(0.02)
				Upper respiratory infections	1.5(<0.001)
				Pneumonia	1.0(0.4)
				Bronchitis	1.2(0.03)
Kunzli et al. (2006)	Southern California 2003	PM_{10}, Questionnaire	PM_{10} 5 day mean	Asthma attack	OR (95% CI)[a]
					1.03 (0.58, 1.80)
				Bronchitis	0.79 (0.39, 1.59)
				Medication use for symptoms	1.38 (1.03, 1.84)*
Delfino et al. (2009)	Southern California 2003	PM2.5	During fires modeled mean PM2.5 ranged from 42.1 to 76.1 μg/m^3	Hospitalizations	RR (95% CI), p value[b]
				Congestive heart failure	1.02 (0.99,1.04), 0.096
				Ischemic heart disease	1.01 (0.99,1.02), 0.313
				Dysrhythmias	99 (0.96,1.02), 0.721
				Cerebrovascular disease/stroke	1.02 (1.00,1.04), 0.971
				Respiratory	1.03 (1.01,1.04), 0.639
				Asthma	1.05 (1.02,1.08), 0.097
				COPD	1.04 (1.00,1.08), 0.320
				Acute bronchitis	1.10 (1.02,1.18), 0.223
				Pneumonia	1.03 (1.01,1.05), 0.420

(continued)

Table 8.2 (continued)

Article	Location	Exposure metric	Exposure levels	Major health outcome	Effect estimate
Lee et al. (2009)	Hoopa Valley Indian Reservation Fire of 1999	PM_{10}	PM_{10} daily mean levels ranged from 12.8 to 620 μg/m^3	Clinic visits	OR (95% CI)[c]
				Coronary artery disease (CAD)	1.48 (1.11,1.97)*
				Respiratory	1.77 (1.51,2.09)*
				Asthma	
Mirabelli et al. (2009)	2003 Southern California Fires	PM_{10}, Children Health Survey	Fire smoke 1–5 days	Respiratory symptoms	PR (95% CI)[d]
					2.07 (1.01, 4.26)
Rappold et al. (2011)	2008 peat bog fire in North Carolina, June 1–July 14, but June 10–12 were considered high exposure period	Satellite measurements of aerosol optical depth (AOD)	Lag days 0–5 after exposure	ED visits	RR (95% CI)
				Congestive heart failure or cardiac arrest	1.37 (1.01,1.85)*
				Asthma	1.65 (1.25,2.17)*
				COPD	1.73 (1.06,2.83)*
				Upper respiratory infections	1.68 (0.94,3.00)*
				Pneumonia and acute bronchitis	1.59 (1.07,2.34)*
				All respiratory diagnoses	1.66 (1.38,1.99)*
Holstius et al. (2012)	2003 Southern California Fires	Temporal comparison	Not reported	Birth weight	Effect (g), (95% CI)[e]
					−6.1, (−8.7, −3.5)

*$p < 0.05$

[a]Odds ratios for the association of smoke on all outcomes comparing communities with the highest and lowest levels of PM_{10} (~210 vs 30 μg/m^3), adjusted for baseline asthma, ethnicity, parental education, and study cohort

[b]Relative rate in relation to a 10 μg/m^3 increase in a 2-day moving average $PM_{2.5}$

[c]Adjusted odds ratios for the association PM_{10} and seeking care for the selected health outcomes

[d]Prevalence ratio of the association of smoke from fire and respiratory symptoms

[e]Estimated effect of wildfire on birth weight (g), any trimester and adjusted by fetal sex, gestational age, parity, maternal age, maternal education, maternal race/ethnicity, secular trend, and season

bronchitis and pneumonia to be associated with PM10 from wildfire. A study in southern California found that PM2.5 during a wildfire was associated with increased hospital admissions for exacerbations of COPD (Delfino et al. 2009, p. 192).

The evidence for impacts of wildfire smoke exposure to respiratory infections in general is inconsistent. Duclos et al. (1990, p. 54) found an association of ED visits for respiratory infections during major wildfires in California. This is contrary to Rappold et al. (2011, p. 1418) who found no association between ED visits for upper respiratory infections in smoke-affected counties during a peat fire in North Carolina.

Few studies have documented evidence of adverse effects for some specific cardiovascular diseases associated with exposure to wildfire smoke. One study in North Carolina showed significant increases for ED visits for congestive heart failure associated with wildfire smoke exposure (measured using satellite atmospheric optical depth measurements) during a peat fire (Rappold et al. 2011, p. 1418). However, when diseases were grouped together by age and sex, the association between cardiovascular disease and smoke exposure was not found (Rappold et al. 2011, p. 1418). Another study in Southern California found no association between hospitalizations for congestive heart failure and PM2.5 during a wildfire (Delfino et al. 2009, p. 192). Delfino et al. (2009, p. 195) also found no association between PM2.5 from wildfire and hospital admissions for cardiac dysrhythmias and no association to hospital admissions for ischemic heart disease (Delfino et al. 2009, p. 192). In a study conducted in Northern California near the Hoopa Valley Indian Reservation, particulate matter less than 10 microns in aerodynamic diameter (PM10) was a significant predictor of clinic visits for coronary artery disease (also known as heart disease) in a Native American reservation during a wildfire event (Lee et al. 2009, p. 319). More work needs to be conducted in this area, since existent studies are inconsistent and few. Thus, the association between cardiovascular outcomes and exposure to wildfire smoke is unclear at this point.

A study of a population seeking emergency relief services after a wildfire found that having difficulty breathing because of smoke or ashes was significantly associated with the probability of post-traumatic stress disorder (PTSD) or major depression 3 months after the fire occurred (Marshall et al. 2007, p.513). Duclos et al. (1990, p. 56) found no increase in mental health hospitalizations during the 1987 California fires.

Very few studies have investigated an association for exposure to smoke from wildfires and poor birth outcomes, which prevents any conclusive associations. Holstius et al. (2012, p. 1340–1345) found a small but significant decline in birth weight for babies that gestated during the 2003 southern California wildfires in comparison to babies from the same region who were born before or more than 9 months after the fires. The effects were significant for wildfire exposure during the second and third trimester of pregnancy however not during the first trimester. Since this study did not quantify air pollution exposures for the pregnant women in the study, it cannot be determined if the observed effect was due to smoke exposure to smoke from wildfires or the stress of living in an area that was experiencing a wildfire.

More epidemiological research that examines the health effects of forest fires is needed. Typical studies have only looked at short-term fire incidents, thus lack statistical power. Studies conducted for longer periods of time are required to confirm the inconsistencies and determine groups that are most affected by smoke. Additionally, the health impacts and relative risk from prescribed, managed, and wildfire (megafire) smoke must be understood for forest management to effectively produce the best health outcomes.

Forest Fire Policy and Its Impacts on the Current and Future Conditions

Fires have been widespread and frequent over a long period of history shaping the present environment (Scott 2000, pp. 335–336). Wildland fire was largely seen as an integral way to manage forested land throughout much of the west by Native American tribes (Anderson 1999, pp. 106–108, 1996, pp. 415–418). Euro-American settlers first moving west saw the importance of continuing these practices (van Wagtendonk 2007, p. 4). Losses of life in large wildfires such as the Peshtigo Fire (1871) in Wisconsin and the Santiago Canyon Fire (1889) in California began to instill the philosophy of suppression into fire management that would be important to the foundation of American firefighting policy.

Current wildland fire management and policy is a product of the 1910 fire season where 78 people died and over 8 million hectares burned and modern suppression policy originated (Silcox 1910, p. 637). This fire season also known as the "Big Burn" or "Big Blowup" was only 5 years after the US Forest Service (USFS) was established. USFS policy to put all fires out as quickly as possible was questioned even during these early years. Although light burning similar to Native American practices was used by settlers and some argued for the necessity of it being a part of sound forest policy (Koch 1935, pp. 103–104), overwhelmingly, questioning of the policy was not over burning but how to use modern techniques and management to fully suppress wildland fire (Greeley 1920, pp. 38–39). Reliance on private lumber companies and lack of USFS coordination was seen as a major obstacle to forest protection and health through suppression (Allen 1910, pp. 642–643). Cooperation was seen as what failed in the now almost exclusive held perspective of policy makers that full suppression was the only way to protect forested lands. The Weeks Act (1911) designated the USFS as the agency for federal cooperation in fire suppression and was strengthened by the Clarke-McNary Act of 1924 (Southard 2011, pp. 18–20), while the Protection of Timber Owned by the United States from Fire, Disease, or Insect Ravages (16 USC 594) was the National Park Service (NPS) equivalent.

By the mid-1930s, the policy to contain and control all fires by 10 a.m. had been adopted by the USFS, and full suppression was largely in place. In this era, wildland fire was seen as an evil that could be stopped with enough money, sound tactics, and advances in science and technology. This policy was solidified during and immediately after World War II when all fire was considered evil (Figure) and the public perception of complete suppression began despite the essential need of fire in the forest (Kauffman 2004, p. 879).

The use of wildland fire began to gather interest in the 1960s as fire management cost increased, and research began to demonstrate benefits (Kilgore 1973, pp. 498–508; Parsons and DeBenedetti 1979, pp. 29–32). In 1963, the "Leopold Report" argued that western parks should be maintained as nearly as possible to the condition when the first Euro-American settlers arrived (Leopold et al. 1963, pp. 18–21) and began to inspire policy makers to include fire management. The Wilderness Act (1964) allowed for the natural process of fire to occur and started a move to include ecologically beneficial and prescribed fire to move from fully controlled to some form of management.

The large land management organizations (typically federal, state, and tribal governments and agencies) are diverse in their missions and goals to safe and effectively manage fire at a landscape level. This creates an immediate and fundamental hurdle to a simplified one-size-fits-all policy where easy solutions for one agency are contradictory to other agencies legislative authority. The United States Forest Service (USFS) and National Park Service (NPS) frequently are located adjacent to one another spatially, but have different mandates and mission goals. The NPS, a part of the U.S. Department of Interior, is fundamentally a conservation agency with an obligation to allow natural processes to function while the USFS, a part of the U.S. Department of Agriculture, is required to incorporate sustainable harvest over much of the land they manage. Timber harvest and other anthropogenic uses are authorized in the USFS, while the NPS is required to preserve the ecological integrity of the land area they manage by eliminating to the greatest extent possible anthropogenic impacts.

The need for greater agency cooperation began to enter policy after the 1988 Yellowstone fires. The 1995 "Federal Wildland Fire Management Policy & Program Review" reflected the need to integrate fire into landscape-level management. Extensive fires in 2000 led to the "Management the Impact of Wildfires on Communities and the Environment: A Report to the President in Response to the Wildfires of 2000" to reduce risk in the Wildland Urban Interface (WUI).

The "Review and Update of the 1995 Federal Wildland Fire Management Policy" (2001) forms the basis of current wildland fire policy with the "Interagency Strategy for the Implementation of Federal Wildland Fire Management Policy" (2003) detailing the implementation.

The "Guidance for Implementation of Federal Wildland Fire Policy" (2009) is currently the primary policy direction. Current policy includes a "single cohesive federal fire policy" that directs agencies to consider long-term benefits of fire in relation to risks with the number one guiding principle being firefighter and public safety. The second guiding principle is the essential role of wildland fire as an ecological process needs to be incorporated into the planning process. Further guiding principles include requiring risk management to include the cost of either allowing or suppressing fire and the inclusion of consideration of public health and environmental quality into the decision process. All guidance is to be underpinned with fire management plans based on the "best available science."

Science-based fire management plans in the political environment faced by policy makers can be conflicting. While air quality is a rather modern concern, fire has long been understood to perform many beneficial ecosystem functions (Kilgore 1981, pp. 58–59) including helping to maximize carbon sequestration in fire-prone areas (Hurteau et al. 2008, p. 496). Recurring wildland fire additionally limits fire spread and substantially reduces fire progression under extreme weather conditions (Parks et al. 2015, pp. 1485–1486) and may provide an avenue to control emissions and the subsequent health impacts. However, past fire management policy dominated by anthropogenic factors has primarily been intended to prevent or contain wildland fire with the consequence of reducing ecological integrity in fire-adapted ecosystems (Dellasala et al. 2004, pp. 977–978).

Future policy will likely continue to be based almost exclusively around anthropogenic concerns unless the entrenched disincentives of current policy are overcome and proactive use of managed fires is supported (North et al. 2015, p. 1280). Without understanding the impacts from wildland fire smoke under a typical fire regime, it is easy to understand how suppressing all emissions for public health would appear to be sound policy. Unfortunately, this is a short-term solution where future emissions are essentially ignored and priority is given to restricting wildland fire emissions as much as possible with the assumption that future fire will not occur. Sound policy requires differences between these competing scenarios be quantified.

Fire suppression limited smoke when widespread air quality monitoring began and provided some of the first regulatory data. This time of low emissions coupled with an increase in fuel loading has created a backlog of fuels and wildland fire emissions with limited historic monitoring context. Suppression limited wildland fire emissions during the initial stages of systematic widespread air quality monitoring likely has led to inappropriate baseline estimates of air quality exposure to areas in and adjacent to the fire-adapted ecosystem of the Sierra Nevada. Smoke impacts were historically much more frequent throughout the Sierra Nevada with a more active fire regime before Euro-American settlement (van de Water and Safford 2011, pp. 29–35), but the cooler slower burning wildland fire needed to sustain the Sierra Nevada ecosystem mosaic potentially made the spatial extent of these historic wildland fire smoke events smaller.

The potential for air quality regulations to limit fire management options has long been recognized as an impediment to the use of ecologically beneficial fire (Sneeuwjagt et al. 2013, pp. 18–20). Smoke will likely be a greater concern with increased fire use and acceptability of smoke levels declines (Shindler and Toman 2003, p. 11). Numerous overarching policy considerations have been offered in the public, research, and policy sectors to help ameliorate the coming together of fire and air policy. Often air quality concerns center around current regulation with wonderfully written regulatory language that has little to no practical field applicability. Fire management needs a clear path to implementation where air quality impacts are well defined by regulators and provide a quantifiable way to manage smoke for the best health outcomes in both the short and long term.

Fire as an ecological process necessary in the Sierra Nevada has become better studied and understood in the twentieth century. Fire has been widely accepted as an important ecosystem component (Beaty and Taylor 2008, pp. 716–717; Collins et al. 2007, pp. 553–557; Kilgore 1973, p. 497; Miller et al. 2012a, pp. 10–16; Pausas and Keeley 2009, p. 593), while smoke research has largely focused on air quality and impacts to public health (Adetona et al. 2016, pp. 101–102). As population increases in California and more people move into the wildland urban interface (WUI), wildland fire policy and air regulatory policy will likely continue to conflict (Jacobson et al. 2001, p. 934).

Policy and the public may not be ready to adapt. While fire science is pointing to increased fire, a disconnection in the science and policy exists (Ayres et al. 2016, p. 80) with a smoke averse public and limited research on relative risk of wildland fire management actions (Gaither et al. 2015, p. 1418; Smith et al. 2016, pp. 137–138). Ecologically beneficial fire, or fire the size, intensity, extent, and effects historically experienced in the ecosystem, will be limited by public willingness to breathe smoke that may in part be rectified by understanding the health implications of smoke emissions and exposure under prescribed, managed, and full suppression scenarios.

Wildland fire burns have the ability to limit subsequent fire spread and lead to self-regulating landscapes (Parks et al. 2015, p. 1489). Suppression may very well be an unsustainable policy for landscape-level land management. Additionally, fire suppression policy may have created an unrealistic expectation of smoke-free air in areas which historically have seen abundant fire (van de Water and Safford 2011, pp. 32–33) and smoke. While minimal research has been conducted on risk management coupling human and natural fire-prone forest systems similar to other natural hazards (Spies et al. 2014, p. 10), smoke is almost completely ignored.

Policy decisions have a profound effect on human and ecological health. Federal land managers throughout the United States have multiple acts and policies that regulate their actions. Policies allowing natural processes that emit regulated pollutants can seemingly be in contradiction with public health. In the Sierra Nevada of California, the essential ecosystem process of wildland fire in areas set aside for conservation is one such process that is in apparent conflict with air regulations.

Forest Fire Greenhouse Gas Emissions and Its Future Impacts on Climate Change

Approximately half of forest biomass is comprised of carbon (C), which is basically created out of thin air in a thermodynamically unfavorable process driven by the energy from the sun and catalyzed by the photosynthetic machinery in green leaves of plants. When that biomass is burned, that process is reversed, and the constituent carbon in biomass rapidly recombines with oxygen in a very thermodynamically favorable process that produces three main greenhouse gases (GHGs): carbon dioxide (CO2), methane (CH3), and nitrous oxide (N2O). At the same time, this combustion process also produces many so-called "criteria" pollutants, like fine particulate matter, (PM2.5) which affect human health.

However, direct flaming and smoldering of biomass from fires is only one pathway for biomass to be converted to GHGs. In fire-prone ecosystems and at the large scale, the photosynthetic production of biomass is opposed through biomass decomposition by both combustion and respiration of that biomass back to CO2 by plants and microbes in the ecosystem. At the landscape scale, the net balance between respiration/combustion and photosynthesis, called net ecosystem productivity (NEP), determines whether a given area will absorb GHGs from the atmosphere as biomass or lose biomass, emitting GHGs back to the atmosphere.

Potter (2010, pp. 373–383) used remotely sensed data on biomass, combined with the statewide GHG emissions inventory created by the California Air Resources Board to model statewide accumulation of biomass across all natural ecosystems. He estimated that all CA ecosystems, under favorably high precipitation, accrue a maximum of between 14 and 24 million metric tonnes carbon equivalent (MMTCE), offsetting a significant fraction of annual fossil fuel emissions from the rest of California's emissions inventory (about 120 MMTCE). By contrast, under drier conditions, all ecosystems in CA showed a net loss of about 15 MMTCE. He notes that forests have been reliably sources, not sinks, for GHGs under the warmer than normal conditions in the analysis period.

These modeled, remotely sensed results generally agree with inventory-based methods that have quantified changes in stocks over time. Gonzalez et al. (2015, pp. 68–77) used inventory-based techniques to show that over the 2001–2010 period, California landscapes have lost a net 29 ± 10 million metric tonnes (Tg) of live C from biomass to the atmosphere and to the pool of dead biomass. Most of which will slowly convert to GHGs in subsequent years. If all of that C were immediately converted only to CO2 gas instantaneously (with no other more potent GHGs like CH3 and N2O produced), these losses would equate to 106 ± 37 million metric tonnes of CO2 (MMTCO2) or an average of ~11 MMTCO2 per year over the 10 year analysis period, which generally agrees with the Potter (2010) ranges described above.

Wildfires account for a large proportion of these losses. Gonzalez et al. (2015, pp. 68–77) estimate that about half of all the live, aboveground C lost from California landscapes between 2001 and 2010 was due to wildfires. Some fraction of these losses manifest as direct emissions of GHGs due to combustion, but some proportion remains on the landscape as dead material, decaying and releasing the constituent C as GHGs at a slower rate (Battles et al. 2014, pp. 44–53).

Remotely sensed methods corroborate these stock-based loss observations, showing distinct areas of mortality in the years following large fires, especially for the footprint of large, high severity fires like the recent Rim (2013), King (2014), and Rough (2015) fires (Potter 2016, pp. 1–7). This is because more intense and severe fires cause more tree mortality, with greater post-fire GHG emissions due to decay of the remaining dead material (North and Hurteau 2011, pp. 1115–1120). Fire severity and the associated mortality have been increasing significantly over the same (1987–2010) period (Miller et al. 2012b, pp. 184–203), at least partly due to a warming climate. Modeling projections show that if current trends in climate and fire severity continue, losses of C and associated wildfire emissions will also continue and possibly accelerate through the coming century (Hurteau et al. 2014, pp. 2298–2304; North and Hurteau 2011, pp. 1115–1120).

Climate, however, is only one of the drivers behind recent C losses from CA forests; the other one is related to the structure and drought resistance of CA forests brought about by aggressive fire suppression policies. Studies that have compared forest structure between measurements performed in 1911 to recent Forest Inventory Analysis (FIA) data show a doubling of canopy cover and density in mixed conifer vegetation types and a tripling of that canopy cover in lower elevation ponderosa pine forests (Stephens et al. 2015, pp. 1–16). Furthermore, that extra biomass manifests as smaller diameter trees that are not only themselves more vulnerable to fire but make the largest trees that hold the most above ground C in forest stands more vulnerable to being killed and subsequently lost to the atmosphere due to high severity fire (Lutz et al. 2012, p. e36131). The extra load of leaf area also creates additional strain on soil water resources, which in times of drought has been shown to significantly increase forest mortality (Potter 2016, pp. 1–7; van Mantgem et al. 2009, pp. 521–524).

At the large scale, this densification and ingrowth allowed by fire suppression has destabilized CA forest carbon stocks, homogenizing their structure in a way that makes them more vulnerable to large-scale, high severity fire, just as these ecosystems face unprecedented warming and drought (Collins et al. 2015, p 1174; Earles et al. 2014, pp. 732–740). While that ingrowth has temporarily created a larger carbon stock, it's also setting the stage for a large-scale reversal of that stock back into GHGs during drought periods, the beginning of which may be currently manifesting in the southern Sierra Nevada (Asner et al. 2016, pp. E249–E255; Hurteau and Brooks 2011, pp. 139–146; Potter 2016, pp. 1–7; Wiechmann et al. 2015, pp. 709–719). Research has shown that restoring low-moderate severity fire to a landscape not only confers resistance to high severity fire and a reduction in emissions (Wiedinmyer and Hurteau 2010, pp. 1926–1932), but it also reduces vulnerability to drought (van Mantgem et al. 2016, pp. 13–25). Modeling at the larger scale and into future climate scenarios has shown that under even the worst-case scenarios, large-scale application of low-moderate severity fire has the potential to reduce fire emissions by nearly half (Hurteau et al. 2014, pp. 2298–2304).

Ultimately, Sierra Nevada forests will likely continue to lose carbon back to the atmosphere—forests are simply too dense given the available water and the likely warming that will occur. The size of the carbon stock that remains, and magnitude of the air pollution and GHG emissions that result from this reversal, will depend on the pace and scale at which land managers can restore fire and drought-resistant forest structure.

Air Quality Impacts of Forest Fires

Air pollutants from a wildland fire are dependent on fuels, can be complex near the flame front, and interact with anthropogenic sources (Alves et al. 2010, pp. 3027–3031; Hosseini et al. 2013, p. 9418; Statheropoulos and Karma 2007, pp. 433–436). Smoke emission can be more toxic than urban emission during large high-intensity fires (Wegesser et al. 2009, p. 897), but there is limited understanding of the causal factors of smoke composition including fuels, fire size and intensity, and chemicals introduced when agricultural areas and houses burn. The same fire can produce large variability in smoke composition even at the same monitoring site (Wigder et al. 2013, p. 28). The variability of plume chemistry during transport along with varying dispersal conditions makes understanding individual plume toxicity challenging. It is then difficult to determine the net effects of forest fires on human health (Fowler 2003, p. 41). Wildfire smoke contains many air pollutants of concern for public health, such as carbon monoxide (CO), nitrogen dioxide (NO2), ozone (O3), particulate matter (PM), polycyclic aromatic hydrocarbons (PAHs), other hydrocarbons, volatile organic compounds (VOCs), and free radicals (Naeher et al. 2007, pp. 69–70). PM emitted from fires is most elevated compared to background levels (Naeher et al. 2007, p. 74) and is one of the best ways to assess smoke exposure (Naeher et al. 2007, p. 74; Vedal and Dutton 2006, p. 30). Thus, this section will focus on PM2.5 to consider wildland fire smoke exposure.

Particulate matter less than 2.5 microns in diameter (PM2.5) is a large portion of emissions from wildland fire (Clinton et al. 2006, p. 3692) and is easily transported over long distance (Bein et al. 2008, pp. 13–17; Dokas et al. 2007, p. 77) having a large impact on air quality (Fowler 2003, pp. 42–43; Langmann et al. 2009, pp. 112–113). Particulate matter is the most frequently studied pollutant when studying wildland fire smoke impacts in part because it can be ten times higher than non-fire background concentrations (Liu et al. 2015, pp. 128–129) and it is also a great tracer for smoke. Smoke transport can easily be detected by remote sensing (Hoff and Christopher 2009, p. 652). Quantifying ground-level concentrations of PM2.5 using remote sensing is difficult (Toth et al. 2014, pp. 6049–6056; Yao and Henderson 2013, p. 330). Remote sensing and modeling can improve remote sensing estimates of ground-level PM2.5 (Li et al. 2015, p. 4494; Reid et al. 2015, p. 3892; Yao et al. 2013, p. 1142). Remote sensing can be used to indicate exceedances from the normal of ground-level PM2.5 concentrations due to smoke in the Sierra Nevada, but ground-based monitors are necessary for accurate quantification (Preisler et al. 2015, p. 349).

Megafires

The occurrence of large wildfires (megafires) has been increasing in California (see Fig. 8.3). Thirteen of the 20 largest California wildfires in recorded history (2015 as the last year) have occurred since 2002 (Table 8.1). Smoke from these wildfires often causes the largest air quality impacts of the year in the towns and cities closer and downwind of the fire. This is particularly true for the more rural areas further away from the major anthropogenic emission sources and typically better air quality. We present some examples below.

McNally (human related), 2002

- The smoke impacts occurred in mountain communities in the eastern side of the Sierra Nevada and downwind of the fire for about 30 days. Kernville, the site closest to the fire, was the most impacted.
- Daily PM10 concentrations more than tripled over the average at some of the impacted locations.
- The California daily PM10 standard (50 ug/m^3) was exceeded 164 times during the fire and only six times before the fire at several monitor locations that were impacted by the event.
- The Federal daily PM10 standard (150 ug/m^3) was exceeded four times during the fire.
- For 4 days, PM10 hourly concentrations surpassed the 300 ug/m^3 levels reaching a maximum of 600 ug/m^3 (see Fig. 8.4).

Rim Fire (human related), 2013

- Daily 24-h average PM2.5 concentrations measured by the 22 air monitors ranged from <12 (background) to 450 μg m^{-3}.
- Locations closer and downwind of the fire experienced the highest PM2.5 impacts. The Rim Fire Camp, Tuolumne City, and Groveland sites were located closest to the fire.
- The Rim Fire Camp monitoring site reported the highest mean and maximum 24-h. average PM2.5 concentration (450 μg m^{-3}). Tuolumne City and Groveland had elevated mean 24-h average PM2.5 concentrations, 230 μg m^{-3} and 200 μg m^{-3}, respectively (see Fig. 8.5).
- PM2.5 24 h mean concentrations increase tenfold from typical background concentrations at some of the monitoring stations.

Fig. 8.3 MODIS satellite image of the Rough fire August 31, 2015

Rough Fire (lightning), 2015

- The greatest impacts were observed on locations closest to the fire (mountain communities of Fresno County) and downwind (east) of the fire with the highest hourly concentration of PM2.5 (455 μg m^{-3}) occurring on 8/28/2015.
- In Pinehurst, California, during 2015, the Rough Fire accounted for all the 2015 hourly AQI levels above moderate (unhealthy for sensitive groups, unhealthy, and very unhealthy) and additionally accounted for 137 of the moderate hourly readings for the year.
- The Rough Fire was distinctly the worse for PM2.5 air quality at Pinehurst, including the only time since 2006; when monitoring began, Pinehurst was at very unhealthy for PM2.5 (see Fig. 8.6).
- The Rough Fire accounted for the highest 10 days for PM2.5 during 2015.

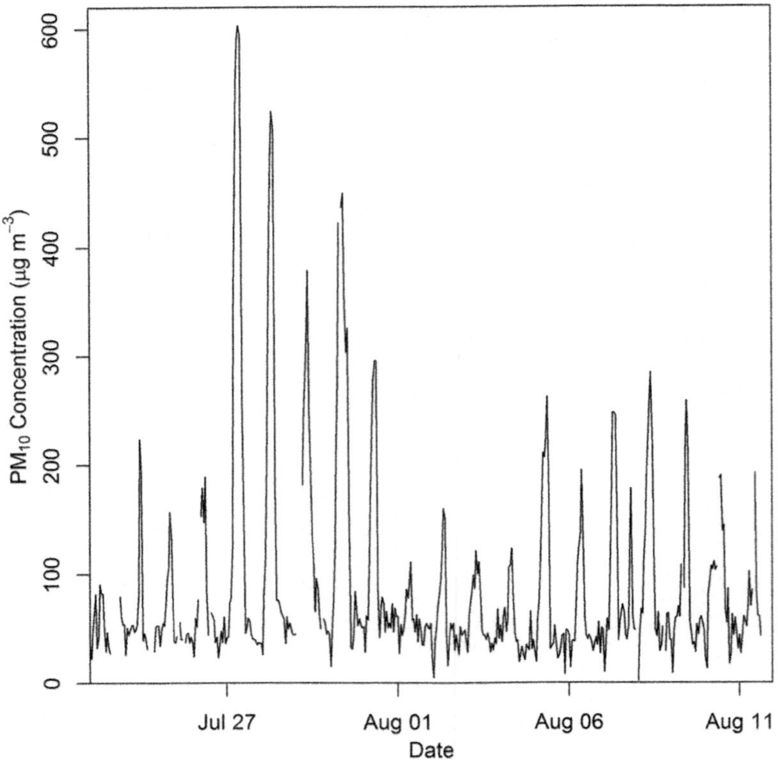

Fig. 8.4 One hour average PM10 concentrations measured in Kernville, California, during the McNally Fire in 2002

Forest Fire Impacts on Firefighter Health

In 2015, 27,000 wildland firefighters employed by the federal government were exposed to smoke while working on wildland fires that burned over 4 million hectares, the highest amount of forested land burned in the last 10 years (NIFC 2015, p. 13). Wildland firefighters suppressing wildland fires can work long hours performing physically demanding work and be potentially exposed to high levels of wood smoke and do not wear respiratory protection (Broyles 2013, p. 6; Hejl et al. 2013, p. 591). Wildland firefighters perform a variety of job tasks while working, including operating aircraft, engines, and heavy equipment, line construction, holding, staging, mop-up, and firing operations that result in exposure to a variety of air pollutants. Additionally, when working on a large wildland fire, firefighters will sleep and eat at a base camp (incident command post) that can be close to the fire to provide logistical support adding to their smoke exposure.

Past exposure assessments of wildland fires have measured levels of fine and respirable particulate matter (PM2.5–PM4), acrolein, benzene, carbon dioxide,

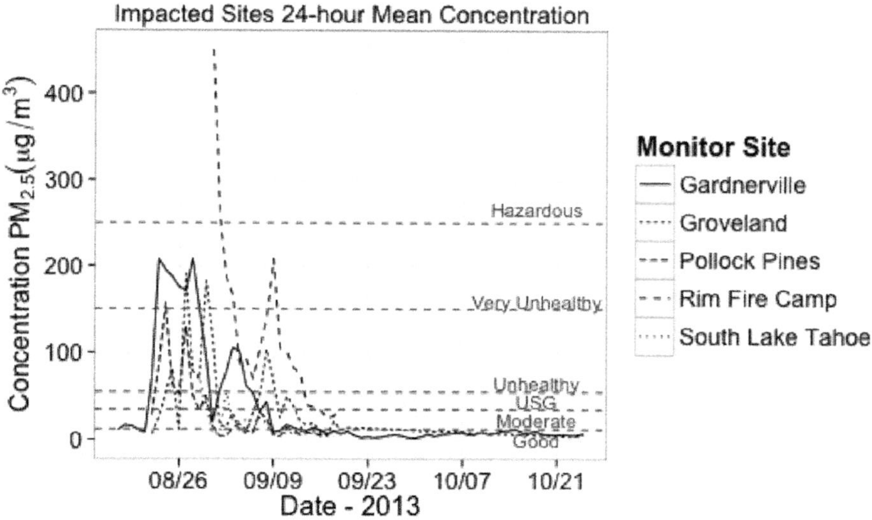

Fig. 8.5 Daily PM2.5 concentrations during the Rim Fire in 2013

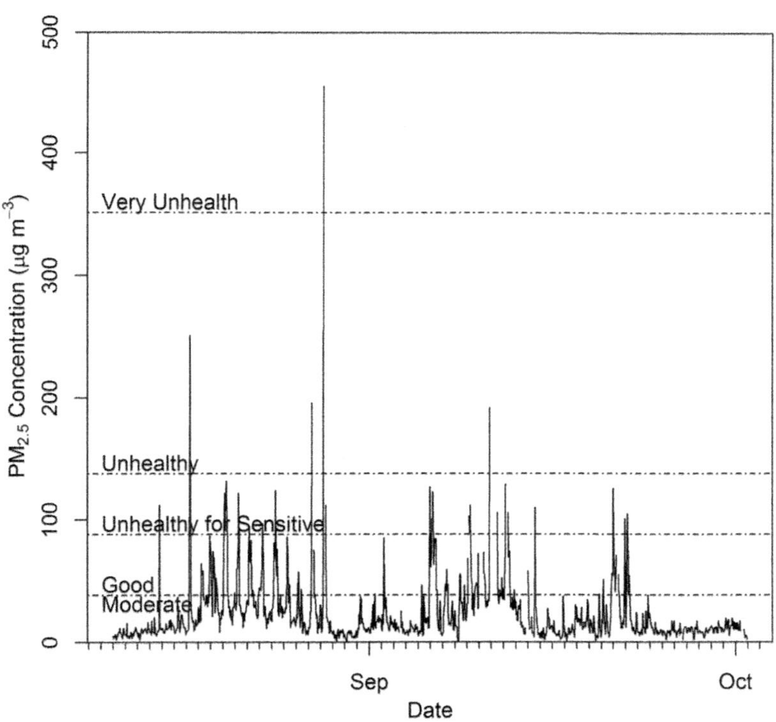

Fig. 8.6 Daily PM2.5 concentrations before, during, and after the fire monitored at Pinehurst, California, during the Rough Fire in 2015

carbon monoxide (CO), formaldehyde, crystalline silica, total particulates, and polycyclic aromatic hydrocarbons (Broyles 2013, p. 5–7). Exposure assessments of wildland firefighters have measured exposure to air pollutants while firefighters performed a variety of job tasks all over the United States. Exposure assessments for wildland firefighters have primarily focused on measuring carbon monoxide and fine and respirable particulate matter, less 4 um diameter (PM4). Studies have shown that peak exposures can surpass occupational health exposure standards.

Reinhardt and Ottmar (2000, p. 3) conducted the first large-scale exposure assessment of carbon monoxide and respirable particulate matter, benzene, acrolein, and formaldehyde at initial attack (small newly started wildfires) and project wildfires (large wildfire incidents) throughout California, Idaho, Montana, and Washington from 1991 to 1994. The study reported that the job tasks associated with higher PM4 were holding, mop-up, and fireline construction, while holding/mop-up, engine operator, and firing operation were associated with CO exposure.

Health studies have examined acute effects of smoke exposure across individual shifts and entire fire seasons. Liu et al. (1992, p. 1471) found that there were significant declines of individual lung function and an increase in airway responsive postseason compared to preseason in 63 firefighters. Swiston et al. (2008, p. 136) measured acute systemic inflammation markers and discovered that circulating levels of cytokines were higher after wildland firefighting. When examining cross-shift changes in lung function, Gaughan et al. (2014, p. 600) reported that firefighters had a significant decline in lung function and was associated with high exposure to levoglucosan.

Booze et al. (2004, p. 296–305) conducted a health risk assessment to characterize the likelihood risk of health effects from exposure smoke for firefighters. Using past studies, they examined cancer and noncancer health risks from exposure to polycyclic aromatic hydrocarbons, volatile organic compounds, respirable particulate matter, and carbon monoxide from wildfires for firefighters. The study concluded that the calculated risks of health effects were lower than expected for firefighters, but there were elevated risks of developing cancer from exposure to primarily benzene and formaldehyde and developing noncancer health effects from exposure to PM3.5 and acrolein (Booze et al. 2004, p. 303).

Recently, Semmens et al. (2016, p. 330–335) conducted the first long-term health assessment of wildland firefighters examining the association of duration of wildland firefighting career and self-reported health outcomes (Semmens et al. 2016, p. 330). The survey reported that there were significant associations between the greater number of years worked as a wildland firefighter and subclinical cardiovascular measures and having had knee surgery (Semmens et al. 2016, p. 332–333).

Wildland firefighters work under extremely arduous conditions and are exposed to potentially high levels of air contaminants in smoke that can affect their health. It is important to characterize the levels of these pollutants to be able to understand all possible health risks for wildland firefighters. Current research has not identified any long-term health consequences of firefighting associated with air pollutants from smoke. More research should be conducted to understand those long-term health risks and accurate exposures of all chemicals of concern from exposure to smoke.

Proposed Policy and Management Strategies to Deal with Air Quality Impacts in California

Until today, there have not been mitigation policies adopted in California. Federal Land Management Agencies (LMAs) in the United States have been adjusting policies after recognizing that past suppression policies are an important factor in catastrophic fires. In California, to reintroduce fire back to the forests, the National Park Service implemented prescribed burning (PB) around 1960 and the US Forest Service in the 1990s. The National Park Service and US Forest Service are the biggest federal LMAs with the most land to manage in California. It is now clear that the small-scale PB (<200 ha) will not lead to the landscape restoration sought by these agencies (Schweizer and Cisneros 2014, p. 266). Thus, the current thought is to establish a wider use of landscape-level fire, or managed fire (MF), that could burn larger areas and maximize beneficial fire effects on resources, at the same time reducing costs and increasing fire safety. MF are smaller than megafires and naturally ignited (lightning), and they burn with less intensity and have positive benefits since it is allowed to burn under favorable conditions. However, concerns with smoke exposure, economic interests, and airshed capacity (the air in the area is already heavily polluted by human activities) issues raised by air regulators hindered full implementation of MF. Fire emissions from MF are considered anthropogenic, even though they are from a natural process, making them a regulated activity under the jurisdiction of air regulatory agencies. Thus, provide conflicted policy directions where coordination with air regulators has been difficult and public opinion is heavily weighted leading to poor support for implementation of MF. In a polluted airshed with short-term air quality goals, there are no incentives for air regulators to accept additional emissions form fires. More emissions create disincentives including possible human health impacts and nuisance complaints. In summary mitigation policies, such the use of ecological beneficial landscape fires like MF, have not been fully adopted in California.

Forest and air management policy are often in conflict with regard to forest fires which in turn impacts public health. This is readily apparent in fire-prone ecosystems where smoke is routinely present. Fire has been a major natural mechanism in the Sierra Nevada Mountains of California providing evolutionary pressure which has shaped this ecosystem. As population has boomed throughout California, more people are living in and immediately adjacent to this fire-adapted ecosystem creating a conflict not only between the immediate destruction of life and property from wildland fire but additionally subjecting larger populations to the exposure of wildland fire smoke. Wildland fire smoke may be the most reviled nondestructive by-product of any natural process. Smelling smoke in the air immediately makes many people deem they are experiencing hazardous air quality even when smoke impacts are undetectable in background ambient concentrations.

Public expectation of a smoke-free environment has been instilled by suppression policy, deferring emissions to be release a later day. Thus, reliance of suppression fires in a fire-prone area is not effective in protecting human health at the population level (Schweizer and Cisneros 2016, p. 1). Consideration needs to be given to future negative health outcomes created by megafires which expose more people to extremely high levels of air pollutants. Radical change is required. One intriguing option is for beneficial wildland fire smoke to be treated as natural background and exempted from regulations (Schweizer and Cisneros 2016, p. 1).

The use of managed wildfires has the potential to restore and maintain fire-adapted ecosystems of the Sierra Nevada, and their use should be expanded particularly in remote areas to mitigate the negative consequences of suppression (Meyer 2015). While ecological benefits from fire are well established, smoke impacts are more difficult to quantify. Fuel loading, fire size, and distance from the fire are important to understanding impacts (Moeltner et al. 2013). Controlling fire size and intensity through increased use of ecologically beneficial fire may prove to be an effective tool in smoke management and reduce firefighter smoke exposure. There is potential for wildland fires of the size and intensity historically seen in the Sierra Nevada to be managed while adhering to federal health standards. Increased wildland fire has also provided data suggesting that timing and dispersal can be used to mitigate some of the health impacts (Tian et al. 2008). Modeling of smoke plume dispersal and movement has improved to provide better estimates of forest fire exposure (Yao and Henderson 2013). Further understanding of the spatial extent of smoke and public exposure levels under various fire size and intensity scenarios would help inform wildland fire policy about the role of ecologically beneficial fire. It is likely the best long-term air quality is inextricably linked to ecosystem health in California (Schweizer and Cisneros 2016, p. 3).

Managing air quality using the current PM2.5 national ambient air quality standards (NAAQS) at the most impacted areas, which happen to be mountain remote communities, will provide an opportunity to increase burning in many forests while continuing to protect public health (Schweizer and Cisneros 2014, p. 265–278; Schweizer et al. 2017, p. 345–356). Ecologically beneficial fire should be encouraged for the best possible air quality outcome in a fire-prone ecosystem within the allowable federal health standards.

Schweizer et al. (2017 p. 345–356) analyzed smoke impacts of the different types (PB, MF, and megafires) of fires on a single site's PM2.5 levels in California over multiple years (2006–2015). The study found that the area could include beneficial landscape fire or MF at levels at or above the largest size (8000 ha–20,000 acres) of MF and remain below the current NAAQS thresholds for PM2.5. These ecologically beneficial fires helped sustain this fire-adapted forest and reduced fuels with the subsequent benefits to public health via lower air pollution levels (by limiting intensity and spread of suppression fire). In contrast, the Rough Fire (2015) increased air quality impacts to PM2.5 to very unhealthy levels and led to an exceedance of the annual NAAQS 24 h. Using the NAAQS when considering air quality impacts from smoke can give land and air managers a metric for broader understanding especially when assessing ecologically beneficial fires (Schweizer et al. 2017, p. 345–356). The current practice of suppression to avoid smoke impact has proven to be an unreliable way to decrease smoke-related health impacts. Management of fires should be based on smoke impacts at monitors, making air monitoring the foundation of fire management actions giving greater flexibility for managing fires. Managed fires should be considered a natural event and exempted from current state and federal regulations.

Conclusion

Wildland fire smoke impacts will depend heavily on level of emissions, transport, and receptor distance from the fire. The economic impacts to health can be substantial when urban areas are impacted by large high-intensity fires instead of smaller fires (Rittmaster et al. 2006, p. 874–875). Megafires can result in increased asthma emergency room visits and hospital admissions and significant economic cost (Jones et al. 2016, p. 181). Protecting public health from smoke is directly dependent on controlling fire emissions. Controlling timing and quantity is essential. Timing and dispersal can be used to mitigate some of the health impacts of increased wildland fire (Tian et al. 2008, p. 2771). Complete suppression does not work. It is apparent after over 100 years of suppression in the United States that at best full suppression is a delaying tactic (Busenberg 2004, p. 148; Calkin et al. 2015, p. 1, 10; Stephens et al. 2016, p. 12–13) that can be better said to mortgage smoke exposure to subsequent generations. Climate change is only exacerbating suppression impacts by increasing season length and overall size and intensity. Forest resiliency to climate change is dependent on forest health from natural process. The natural process of fire needs widespread reintroduction to assuage long-term air quality and public health in fire-prone areas.

High concentrations of PM2.5 will be found with any fire, but reducing the spatial extent can limit exposure to populations of concern. While few associations between wildfire emissions and mortality have been observed, associations with subclinical effects have been established (Youssouf et al. 2014, p. 11773), but major and minor health outcomes due to wildland fire smoke need to be better identified (Kochi et al. 2010, p. 803).

Regional forecasting using remote sensing may eventual lead to the best understanding of fire activity impacts to human health by identifying which fires are most likely to impact a given location (Price et al. 2012, p. 1). Linking landscape ecology and epidemiological perspectives is important to reintroduction of ecologically beneficial fire into the modern world. For example, in Australia, it was noted that daily asthma presentation increase may be avoided while allowing some fire by using an airshed threshold for particulate matter (Bowman and Johnston 2005, pp. 9–80). The NAAQS as a metric where public health impacts from regional fires are used to estimate impacts from the number and size of fires over a given season may provide a way to reintroduce measured landscape-level ecologically beneficial fire in the Sierra Nevada. Regulating to present NAAQS (i.e., 3-year average concentrations for PM2.5) in the areas where the ecologically beneficial fires typically burn would provide an opportunity to increase burning in many forests while still protecting public health.

Tolerance of routine emissions from wildland fire smoke both from the public and managers is needed and must take into consideration that suppression is only deferring the risk to the future with greater impacts to more people. Public awareness of the complexity of wildland fire decisions based on air quality is necessary to provide the public support needed to allow landscape-level reintroduction of fire. Suppression of fire only mortgages the health of future generations (Schweizer and Cisneros 2016, p. 3). Understanding smoke impacts and public health advisories to protect exposure during any event is necessary and should be increased to understand impacts across the landscape.

References

Adetona O, Reinhardt TE, Domitrovich J, Broyles G, Adetona AM, Kleinman MT, Ottmar RD, Naeher LP (2016) Review of the health effects of wildland fire smoke on wildland firefighters and the public. Inhal Toxicol 28:95–139

Allen E (1910) What protective cooperation did. Am For 16:641–643

Alves CA, Goncalves C, Pio CA, Mirante F, Caseiro A, Tarelho L, Freitas MC, Viegas DX (2010) Smoke emissions from biomass burning in a Mediterranean shrubland. Atmos Environ 44:3024–3033

Anderson MK (1996) The ethnobotany of Deergrass, Muhlenbergia Rigens (Poaceae): its uses and fire management by California Indian tribes. Econ Bot 4:409–422

Anderson MK (1999) The fire, pruning, and coppice management of temperate ecosystems for basketry material by California Indian tribes. Hum Ecol 27:79–113

Asner GP, Brodrick PG, Anderson CB, Vaughn N, Knapp DE, Martin RE (2016) Progressive forest canopy water loss during the 2012–2015 California drought. Proc Natl Acad Sci 113(2): E249–E255

Ayres A, Degolia A, Fienup M, Kim Y, Sainz J, Urbisci L, Viana D, Wesolowski G, Plantinga AJ, Tague C (2016) Social science/natural science perspectives on wildfire and climate change. Geogr Compass 10:67–86

Battles JJ, Cousins SJM, Sanders JE (2014) Carbon dynamics and greenhouse emissions of standing dead trees in California mixed conifer forests. California Energy Commission, Berkeley, pp 44–53

Beaty RM, Taylor AH (2008) Fire history and the structure and dynamics of a mixed conifer forest landscape in the northern Sierra Nevada, Lake Tahoe Basin, California, USA. For Ecol Manag 255:707–719

Bein KJ, Zhao Y, Johnston MV, Wexler AS (2008) Interactions between boreal wildfire and urban emissions. J Geophys Res Atmos 113:1–17

Booze TF, Reinhardt TE, Quiring SJ, Ottmar RD (2004) A screening-level assessment of the health risks of chronic smoke exposure for wildland firefighters. J Occup Environ Hyg 1 (5):296–305

Bowman DMJS, Johnston FH (2005) Wildfire smoke, fire management and human health. EcoHealth 2:76–80

Broyles G (2013) Wildland firefighter smoke exposure, United States Department of Agriculture, Forest Service, National Technology and Development Program, 1351 1803 5100 Fire Management, 26 pages
Busenberg G (2004) Wildfire management in the United States: the evolution of a policy failure. Rev Policy Res 21(2):145–156
Calkin DE, Thompson MP, Finney MA (2015) Negative consequences of positive feedbacks in US wildfire management. For Ecosyst 2(9):1–10
Cermak RW (2005) Fire in the forest a history of forest fire control on the National Forests in California, 1898–1956. United States Department of Agriculture Forest Service, Pacific Southwest Region, R5-FR-003, 443 pages.
Clinton NE, Gong P, Scott K (2006) Quantification of pollutants emitted from very large wildland fires in Southern California, USA. Atmos Environ 40:3686–3695
Collins BM, Kelly M, van Wagtendonk JW, Stephens SL (2007) Spatial patterns of large natural fires in Sierra Nevada wilderness areas. Landsc Ecol 22:545–557
Collins BM, Lydersen JM, Everett RG, Fry DL, Stephens SL (2015) Novel characterization of landscape-level variability in historical vegetation structure. Ecol Appl 25(5):1174
Delfino RJ, Brummel S, Wu J, Stern H, Ostro B, Lipsett M, Winer A, Street DH, Zhang L, Tjoa T, Gillen DL (2009) The relationship of respiratory and cardiovascular hospital admissions to the southern California wildfires of 2003. Occup Environ Med 66:189–197
Dellasala DA, Williams JE, Williams CD, Franklin JF (2004) Beyond smoke and mirrors: a synthesis of fire policy and science. Conserv Biol 18:976–986
Dokas I, Statheropoulos M, Karma S (2007) Integration of field chemical data in initial risk assessment of forest fire smoke. Sci Total Environ 376:72–85
Duclos P, Sanderson LM, Lipsett M (1990) The 1987 forest fire disaster in California: assessment of emergency room visits. Arch Environ Health 45:53–58
Earles M, North MP, Hurteau MD (2014) Wildfire and drought dynamics destabilize carbon stores of fire-suppressed forests. Ecol Appl 24(4):732–740
Flannigan M, Cantin AS, de Groot WJ, Wotton M, Newbery A, Gowman LM (2013) Global wildland fire season severity in the 21st century. For Ecol Manag 294:54–61
Fowler CT (2003) Human health impacts of forest fires in the southern United States: a literature review. J Ecol Anthropol 7:39–63
Gaither CJ, Goodrick S, Murphy BE, Poudyal N (2015) An exploratory spatial analysis of social vulnerability and smoke plume dispersion in the U.S. south. Forests 6:1397–1421
Gaughan DM, Piacitelli CA, Chen BT, Law BF, Virji MA, Edwards NT, Enright PL, Schwegler-Berry DE, Leonard SS, Wagner GR (2014) Exposures and cross-shift lung function declines in wildland firefighters. J Occup Environ Hyg 11(9):591–603
Gonzalez P, Battles JJ, Collins BM, Robards T, Saah DS (2015) Aboveground live carbon stock changes of California wildland ecosystems, 2001–2010. For Ecol Manag 348:68–77
Greeley WB (1920) "Piute Forestry" or the fallacy of light burning. Timberman 21:38–39
Hejl AM, Adetona O, Diaz-Sanchez D, Carter JD, Commodore AA, Rathbun SL, Naeher LP (2013) Inflammatory effects of woodsmoke exposure among wildland firefighters working at prescribed burns at the Savannah River site, SC. J Occup Environ Hyg 10(4):173–180
Hoff RM, Christopher SA (2009) Remote sensing of particulate pollution from space: have we reached the promised land? J Air Waste Manag Assoc 59:645–675
Holstius DM, Reid CE, Jesdale BM, Morello-Frosch R (2012) Birth weight following pregnancy during the 2003 southern California wildfires. Environ Health Perspect 120:1340–1345
Hosseini S, Urbanski SP, Dixit P, Qi L, Burling IR, Yokelson RJ, Johnson TJ, Shrivastava M, Jung HS, Weise DR, Miller JW, Cocker DR III (2013) Laboratory characterization of PM emissions from combustion of wildland biomass fuels. J Geophys Res Atmos 118:9914–9929
Hurteau MD, Brooks ML (2011) Short- and long-term effects of fire on carbon in US dry temperate forest systems. Bioscience 61(2):139–146
Hurteau MD, Koch GW, Hungate BA (2008) Carbon protection and fire risk reduction: toward a full accounting of forest carbon offsets. Front Ecol Environ 6:493–498

Hurteau MD, Westerling AL, Wiedinmyer C, Bryant BP (2014) Projected effects of climate and development on California wildfire emissions through 2100. Environ Sci Technol 48:2298–2304

Jacobson SK, Monroe MC, Marynowski S (2001) Fire at the wildland interface: the influence of experience and mass media on public knowledge, attitudes, and behavioral intentions. Wildlife Society Bulletin 29:929–937

Jones BA, Thacher JA, Chermak JM, Berrens RP (2016) Wildfire smoke health costs: a methods case study for a southwest US 'mega-fire'. J Environ Econ Policy 5:181–199

Kauffman J (2004) Death rides the Forest: perceptions of fire, land use, and ecological restoration of western forests. Conserv Biol 18:878–882

Kilgore B (1973) The ecological role of fire in Sierran conifer forests. Its application to National Park management. Quat Res 3:496–513

Kilgore B (1981) Fire in ecosystem distribution and structure: western forests and scrublands. In: Mooney HA, Bonnicksen TM, Christensen NL (tech. cord.), Proceedings of the conference: fire regimes and ecosystem properties, USDA Forest Service General Technical Report WO-GTR-26, 58–89.

Koch E (1935) The passing of the Lolo Trail. J For 33:98–104

Kochi I, Donovan GH, Champ PA, Loomis JB (2010) The economic cost of adverse health effects from wildfire-smoke exposure: a review. Int J Wildland Fire 19:803–817

Kunzli N, Avol E, Wu J, Gauderman WJ, Rappaport E, Millstein J, Bennion J, McConnell R, Gilliland FD, Berhane K, Luhrmann F, Winer A, Peters JM (2006) Health effects of the 2003 Southern California wildfires on children. Am J Respir Crit Care Med 174:1221–1228

Langmann B, Duncan B, Textor C, Trentmann J, van der Werf G (2009) Vegetation fire emissions and their impact on air pollution. Atmos Environ 43:107–116

Lee TS, Falter K, Meyer P, Mott J, Gwynn C (2009) Risk factors associated with clinic visits during the 1999 forest fires near the Hoopa Valley Indian Reservation, California, USA. Int J Environ Health Res 19(5):315–327

Lenihan JM, Bachelet D, Neilson RP, Drapek R (2008) Response of vegetation distribution, ecosystem productivity, and fire to climate change scenarios for California. Clim Chang 87:215–230

Leopold AS, Cain SA, Cottam CM, Gabrielson IN, Kimball TL (1963) Wildlife management in the National Parks, revisiting Leopold: resource stewardship in the National Parks A Report of the National Park System Advisory Board Science Committee, 2012, 3–21

Levy S (2005) Rekindling native fires. Bioscience 55:303–308

Li X, Song W, Lian L, Wei X (2015) Forest fire smoke detection using back-propagation neural network based on MODIS data. Remote Sens 7:4473–4498

Liu YQ (2004) Variability of wildland fire emissions across the contiguous United States. Atmos Environ 38:3489–3499

Liu D, Tager IB, Balmes JR, Harrison RJ (1992) The effect of smoke inhalation on lung function and airway responsiveness in wildland fire fighters. Am Rev Respir Dis 146(6):1469–1473

Liu JC, Pereira G, Uhl SA, Bravo MA, Bell ML (2015) A systemic review of the physical health impacts from non-occupational exposure to wildfire smoke. Environ Res 136:120–132

Lutz JA, Larson AJ, Swanson ME, Freund JA (2012) Ecological importance of large-diameter trees in a temperate mixed-conifer Forest. PLoS One 7(5):e36131

van Mantgem P, Caprio AC, Stephenson NL, Das AJ (2016) Does prescribed fire promote resistance to drought in low elevation forests of the Sierra Nevada, California, USA? Fire Ecol 12(1):13–25

Marshall GN, Schell TL, Elliott MN, Rayburn NR, Jaycox LH (2007) Psychiatric disorders among adults seeking emergency disaster assistance after a wildland-urban interface fire. Psychiatr Serv 58:509–514

Martin TG, Watson EM (2016) Intact ecosystems provide best defence against climate change. Nat Clim Chang 6:122–124

Meyer MD (2015) Forest fire severity patterns of resource objective wildfire in the southern Sierra Nevada. J For 113:49–56
Miller JD, Collins BM, Lutz JA, Stephens SL, van Wagtendonk JW, Yasuda DA (2012a) Differences in wildfires among ecoregions and land management agencies in the Sierra Nevada region, California, USA. Ecosphere 3:1–20
Miller JD, Skinner CN, Safford HD, Knapp EE, Ramirez CM (2012b) Trends and causes of severity, size, and number of fires in northwestern California, USA. Ecol Appl 22(1):184–203
Mirabelli MC, Kunli N, Avol E, Gilliland FD, Gauderman JW, McConnel R, Peters JM (2009) Respiratory symptoms following wildfire smoke exposure: airway size as a susceptibility factor. Epidemiology 20:451–459
Moeltner K, Kim MK, Zhu E, Yang W (2013) Wildfire smoke and health impacts: a closer look a fire attributes and their marginal effects. J Environ Econ Manag 66:476–496
Naeher LP, Brauer M, Lipsett M, Zelikoff JT, Simpson CD, Koenig JQ, Smith KR (2007) Woodsmoke health effects: a review. Inhal Toxicol 19:67–106
NIFC (2015) National Interagency Coordinator Center-Wildland Fire Summary and Statistics Annual Report 2015
North M, Hurteau M (2011) High-severity wildfire effects on carbon stocks and emissions in fuels treated and untreated forest. For Ecol Manag 261:1115–1120
North MP, Stephens SL, Collins BM, Agee JK, Aplet G, Franklin JF, Fule PZ (2015) Reform forest fire management. Science 349:1280–1281
Parks SA, Holsinger LM, Miller C, Nelson CR (2015) Wildland fires as a self-regulating mechanism: the role of previous burns and weather in limiting fire progression. Ecol Appl 25:1478–1492
Parsons DJ, DeBennedetti SH (1979) Impact of fire suppression on a mixed-conifer forest. For Ecol Manag 2:21–33
Pausas JG, Keeley JE (2009) A burning story: the role of fire in the history of life. Bioscience 59:593–601
Potter C (2010) The carbon budget of California. Environ Sci Pol 13(5):373–383
Potter C (2016) Landsat image analysis of tree mortality in the southern Sierra Nevada region of California during the 2013–2015 drought. J Earth Sci Clim Chang 7(342):1–7
Preisler HK, Schweizer D, Cisneros R, Procter T, Ruminski M, Tarnay L (2015) A statistical model for determining impact of wildland fires on Particulate Matter (PM2.5) in Central California aided by satellite imagery of smoke. Environ Pollut 205:340–349
Price OF, Williamson GJ, Henderson SB, Johnston F, Bowman DMJS (2012) The relationship between particulate pollution levels in Australian cities, meteorology, and landscape fire activity detected from MODIS hotspots., (2012). PLoS One 7:1–10
Rappold AG, Stone SL, Cascio WE, Neas LM, Kilaru VJ, Carraway MS, Szykman JJ, Isling A, Cleve WE, Meredith JT, Vaughan-Batten H, Deyneka L, Devlin RB (2011) Peat bog wildfire smoke exposure in rural North Carolina is associated with cardiopulmonary emergency department visits assessed through syndromic surveillance. Environ Health Perspect 119:1415–1420
Reid CE, Jerrett M, Petersen ML, Pfister GG, Morefield PE, Tager IB, Raffuse SM, Balmes JR (2015) Spatiotemporal prediction of fine particulate matter during the 2008 northern California wildfires using machine learning. Environ Sci Technol 49:3887–3896
Reinhardt TE, Ottmar RD (2000) Smoke exposure at Western wildfires U.S. Department of Agriculture, Forest Service, Pacific Northwest Research Station, Research Paper PNW-RP-525
Rittmaster R, Adamowicz WL, Amiro B, Pelletier RT (2006) Economic analysis of health effects from forest fires. Can J For Res 36:868–877
Schweizer D, Cisneros R (2014) Wildland fire management and air quality in the southern Sierra Nevada: Using the Lion Fire as a case study with a multi-year perspective on PM2.5 impacts and fire policy. J Environ Manag 144:265–278

Schweizer D, Cisneros R (2016) Forest fire policy: change conventional thinking of smoke management to prioritize long-term air quality and public health. Air Qual Atmos Health. doi:10.1007/s11869-016-0405-4

Schweizer D, Cisneros R, Traina S, Ghezzhei T, Shaw G (2016) Using National Ambient Air Quality Standards for fine particulate matter to assess regional wildland fire smoke and air quality management, J Environ Manag, submitted

Schweizer D, Cisneros R, Traina S, Ghezzehei TA, Shaw G (2017) Using national ambient air quality standards for fine particulate matter to assess regional wildland fire smoke and air quality management. J Environ Manag 201:345–356

Scott AC (2000) The pre-quaternary history of fire. Palaeogeogr Palaeoclimatol Palaeoecol 164:297–345

Semmens EO, Domitrovich J, Conway K, Noonan CW (2016) A cross-sectional survey of occupational history as a wildland firefighter and health. Am J Ind Med 59:330–335

Shindler B, Toman E (2003) Fuel reduction strategies in forest communities: a longitudinal analysis of public support, J For, September, 8–15.

Silcox FA (1910) How the fires were fought. American Forestry 16:631–639

Smith AMS, Kolden CA, Paveglio TB, Cochrane MA, Bowman DMJS, Moritz MA, Kliskey AD, Alessa L, Hudak AT, Hoffman CM, Lutz JA, Queen LP, Goetz SJ, Higuera PE, Boschetti L, Flannigan M, Yedinak KM, Watts AC, Strand EK, van Wagtendonk JW, Anderson JW, Stocks BJ, Abatzoglou JT (2016) The science of firescapes: achieving fire resilient communities. Bioscience 66:130–146

Sneeuwjagt RJ, Kline TS, Stephens SL (2013) Opportunities for improved fire use and management in California: lessons from Western Australia. Fire Ecol 9:14–25

Southard L (2011) The history of cooperative forest fire control and the weeks act. For Hist Today, Spring/Fall, 17–20

Spies TA, White EM, Kline JD, Fischer AP, Ager A, Bailey J, Bolte J, Koch J, Platt E, Olsen CS, Jacobs D, Shindler B, Steen-Adams MM, Hammer R (2014) Examining fire-prone forest landscapes as coupled human and natural systems. Ecol Soc 19:1–14

Statheropoulos M, Karma S (2007) Complexity and origin of the smoke components as measured near the flame-front of a real forest fire incident: a case study. J Anal Appl Pyrolysis 78:430–437

Steel Z, Safford H, Viers J (2015) The fire frequency-severity relationship and the legacy of fire suppression in California forests. Ecosphere 6:1–23

Stephens SL, Lydersen JM, Collins BM, Fry DL, Meyer MD (2015) Historical and current landscape-scale ponderosa pine and mixed conifer forest structure in the southern Sierra Nevada. Ecosphere 6(5):1–16

Stephens SL, Collins BM, Biber E, Fule PZ (2016) U.S. federal fire and forest policy: emphasizing resilience in dry forests. Ecosphere 7(11):1–19

Swetnam T (1993) Fire history and climate change in Giant sequoia groves. Science 262:885–889

Swetnam T (2009) Multi-millennial fire history of the Giant Forest, sequoia National Park, California, USA. Fire Ecol 5:120–150

Swiston J, Davidson W, Attridge S, Li G, Brauer M, van Eeden S (2008) Wood smoke exposure induces a pulmonary and systemic inflammatory response in firefighters. Eur Respir J 32 (1):129–138

Tham R, Erbas B, Akram M, Dennekamp M, Abramson MJ (2009) The impact of smoke on respiratory hospital outcomes during the 2002–2003 bushfire season, Victoria, Australia. Respirology 14:69–75

Thelen B, French NH, Koziol BW, Billmire M, Owen RC, Johnson J, Gingsberg M, Loboda T, Wu S (2013) Modeling acute respiratory illness during the 2007 San Diego wildland fires using a coupled emissions-transport system and generalized additive modeling. Environ Health 12:94

Tian D, Wang Y, Bergin M, Hu Y, Liu Y, Russell AG (2008) Air quality impacts from prescribed forest fires under different management practices. Environ Sci Technol 42:2767–2772

Toth TD, Zhang J, Campbell JR, Hyer EJ, Reid JS, Shi Y, Westphal DL (2014) Impact of data quality and surface-to-column representativeness on the PM2.5/satellite AOD relationship for the contiguous United States. Atmos Chem Phys 14:6049–6062
Van de Water KM, Safford HD (2011) A summary of fire frequency estimates for California vegetation before Euro-American settlement. Fire Ecol 7:26–58
Van Mantgem PJ, Stepheson NL, Byrne JC, Daniels LD, Franklin JF, Fule PZ, Harmon ME, Larson AJ, Smith JM, Taylor AH, Veblen (2009) Widespread increase of tree mortality rates in the western United States. Science 323:521–524
Vedal S, Dutton SJ (2006) Wildfire air pollution and daily mortality in a large urban area. Environ Res 102:29–35
Vora C, Renvall MJ, Chao P, Ferguson P, Ramsdell JW (2011) 2007 San Diego wildfires and asthmatics. J Asthma 48:75–78
van Wagtendonk JW (2007) The history and evolution of wildland fire use. Fire Ecol 2:3–17
Watson JEM, Shanahan DF, DiMarco M, Allan J, Laurance WF, Sanderson EW, Mackey B, Venter O (2016) Catastrophic declines in wilderness areas undermine global environment targets. Curr Biol 26:1–6
Wegesser TC, Pinkerton KE, Last JA (2009) California wildfires of 2008: coarse and fine particulate matter toxicity. Environ Health Perspect 117:893–897
Westerling AL, Bryant BP (2008) Climate change and wildfire in California. Clim Chang 87:231–249
Westerling AL, Hidalgo HG, Cayan DR, Swetnam TW (2006) Warming and earlier spring increase western US forest wildfire activity. Science 313:940–943
Westerling AL, Bryant BP, Preisler HK, Holmes TP, Hidalgo HG, Das T, Shrestha SR (2011) Climate change and growth scenarios for California wildfire. Clim Chang 109:445–463
Wiechmann ML, Hurteau MD, North MP, Koch GW, Jerabkova L (2015) The carbon balance of reducing wildfire risk and restoring process: an analysis of 10-year post-treatment carbon dynamics in a mixed-conifer forest. Clim Chang 132(4):709–719
Wiedinmyer C, Hurteau MD (2010) Prescribed fire as a means of reducing forest carbon emissions in the western United States. Environ Sci Technol 44(6):1926–1932
Wigder NL, Jaffe DA, Saketa FA (2013) Ozone and particulate matter enhancements from regional wildfires observed at Mount Bachelor during 2004–2011. Atmos Environ 75:24–31
Yao J, Henderson SB (2013) An empirical model to estimate daily forest fire smoke exposure over a large geographic area using air quality, meteorological, and remote sensing data. J Expo Sci Environ Epidemiol 24:328–335
Yao J, Brauer M, Henderson SB (2013) Evaluation of a wildfire smoke forecasting system as a tool for public health protection. Environ Health Perspect 121:1142–1147
Youssouf H, Liousse C, Roblou L, Assamoi EM, Salonen RO, Maesano C, Banerjee S, Annesi-Maesano I (2014) Non-accidental health impacts of wildfire smoke. Int J Environ Res Public Health 11:11772–11804
Potter C (2010) The carbon budget of California. Environ Science & Policy 13 (5): 373-383

Ricardo Cisneros, PhD, MPH, is an assistant professor of environmental public health at the University of California, Merced. He received a PhD in environmental systems in 2008. He also completed an MPH and BS in environmental health. As an environmental scientist with specialization in environmental health and exposure science, he conducts research that recognizes the interdependence of ecological and human health with special interests in air pollution research and exposure assessment.

Donald Schweizer, PhD, is an air resource specialist with the US Forest Service. His research with the University of California, Merced, emphasizes understanding the benefits of forest health and the role of natural environmental system function in protecting human health. His concentration is on wildland fire smoke and fire management policy in the wilderness and other protected lands of the Sierra Nevada of California.

Leland (Lee) Tarnay, PhD, is an ecologist working across agencies out of the US Forest Service Region 5 Remote Sensing Lab. Lee received his BS from the University of California, Davis, in biological sciences (1995) and his PhD from the University of Nevada, Reno (2001). His core expertise is in wildland fire smoke and emission monitoring, modelling and management.

Kathleen Navarro, PhD, MPH, recently completed her PhD in environmental health sciences at the University of California, Berkeley, School of Public Health, where she completed her MPH degree in 2011. Her dissertation combined traditional exposure assessment methods with new approaches to evaluate exposures in ambient community and occupational settings to air contaminants commonly emitted from wildland fires and found in the ambient environment. She holds a BS in environmental toxicology from University of California, Davis.

David Veloz received his BS in management from the University of California, Merced, in 2013. He is currently a PhD student at the University of California, Merced, and is motivated by the desire to understand environmental justice issues created by air pollution in the California Central Valley. His primary focus is to understand the public's perception on air pollution and analyse local air quality trends.

C. Trent Procter, BS, is an air quality program manager with the US Forest Service in the Pacific Southwest Region. He completed his BS in natural resources management from the California Polytechnic State University, San Luis Obispo, in 1978. He has over 25 years of experience working for the Forest Service in Air Resources Management.

Chapter 9
Air Pollution and Climate Change in Australia: A Triple Burden

Colin D. Butler and James Whelan

Abstract This chapter mainly focuses on air pollution, with less stress on the health problems of climate change, which, conceptually, is also a form of air pollution, due to the changing composition of atmospheric trace gases. Air quality in Australia is comparatively good, by global standards, due to its large area, low population, and widespread development. However, there are areas of Australia which have significant health problems from dirty air, particularly in association with coal-burning power stations, from the combustion of wood for heating during winter and from vehicles in the large cities. Australia is also a major exporter of greenhouse gases, both as fossil fuels (coal and gas), and of beef and sheep. Much can be done to reduce this triple burden of impaired air quality, domestic climate change and exported climate change, but this requires major changes to consciousness in Australia, and greater willingness to oppose vested interests which profit from ageing paradigms of progress which discount health and environmental costs. The falling cost of renewable energy, especially, gives hope that such challenges will be increasingly successful, but additional solutions are needed to reduce the burning of wood for heat.

Keywords Air pollution • Australia • Coal mines • Climate change • Social licence • Health

Introduction: Air Pollution and Health in Australia

When the British, in 1788, began their drawn-out process of invading and occupying the southern continent now called Australia, the indigenous people they displaced from most areas had a long and rich tradition of astronomical knowledge

---

C.D. Butler (✉)
University of Canberra, Canberra, Australia

Australian National University, Canberra, Australia
e-mail: colin.butler1955@gmail.com

J. Whelan
Environmental Justice Australia, Callaghan, Australia

© Springer International Publishing AG 2018
R. Akhtar, C. Palagiano (eds.), *Climate Change and Air Pollution*,
Springer Climate, DOI 10.1007/978-3-319-61346-8_9

(Fuller et al. 2014). This tradition must have been helped, perhaps even inspired, by the brilliance of the heavens, whose glory was little impeded by significant light on the ground. However, a degree of particulate air pollution in Australia before colonisation is likely to have been frequent, due to the widespread indigenous practice of deliberately lighting fires to manage their landscape, a process today called 'firestick farming' (Gammage 2011; Jones 2012).

These traditional burning practices may have reduced the megafires which have occurred more recently in Australia (Attiwill and Adams 2013) and which have well-documented adverse health effects (Johnston et al. 2011). Today, the brilliance and inspiration of the night sky are invisible to many people globally, but the stars seen from rural Australia, on the whole, are countless and comparatively bright. Air pollution, on a continental scale in Australia, is minor, compared to Asia, due to the continent's vast size, small population and the overwhelming reliance on electricity and gas for cooking. However, there are areas of Australia which have significant health problems from dirty air, particularly in association with coal-burning power stations, from the combustion of wood for heating during winter and from vehicular emissions in large cities. The adverse health and financial impacts of air pollution in Australia are significant and can and should be reduced.

When one of the authors of this chapter commenced medical school, in 1980 (in a city then notorious for industrial air pollution, by Australian standards), he was told that the adverse health effects of air pollution were trivial. This was misinformed, even then. London, for centuries, has been called the 'big smoke' (Brimblecombe 2011). Major smog events in the heavily industrialised but narrow (temperature inversion layer-susceptible) Meuse Valley, Belgium (1930); the steel town of Donora, Pennsylvania (1948) (also in a valley); and coal fire-dependent London (1952) had each been recognised as causing much mortality and morbidity. In London, up to 4000 extra deaths occurred in a few days (Nemery et al. 2001).

While these three spectacular increases in mortality were quickly recognised, the *chronic* health effects of air pollution have proven much harder to comprehend. Almost everyone in air-polluted London in the 1940s was exposed to air pollution, as in New Delhi today. Without a control population, relatively unexposed to air pollution, chronic diseases contributed to by regularly breathing even heavily polluted air may be regarded as 'normal' (Berridge and Taylor 2005).

Recognition of the harm of air pollution, including its interaction with smoking, was also long suppressed for political reasons (Snyder 1994; Berridge 2007). Smoke, dust, smogs, inhaled irritants and fumes have long been seen as necessary companions of development and, in some cases, of basic heating, cooking and transport. Relatedly, the adverse health effects of these exposures have been downplayed, ignored and in some places suppressed.

In the last decade, however, recognition of the harm from *visible* forms of air pollution has improved. In 2014 the World Health Organization (WHO) (2016) announced that about seven million people worldwide die prematurely from air pollution, about one in eight of total deaths, and more than double earlier estimates. Furthermore, affordable alternatives for many processes which cause air pollution are now emerging; this is likely to be a powerful contributor to lifting the taboo on the health harm of air pollution and to reducing the 'social licence' of polluters (Connor et al. 2009).

A Hierarchy of Air Pollutants

Considerable effort has been expended trying to identify the 'worst' contributors to health among the scores of candidate air pollution components. The pollution episodes in the Meuse Valley and Donora were primarily a brew of industrial toxins, including particulate matter (PM) of varying sizes, sulphur dioxide (SO2), carbon monoxide and hydrofluoric acid. In the Belgian example, 30 different substances, released by 27 factories, were identified (Nemery et al. 2001). However, no single worst cause was proven (or scapegoated); then and perhaps still, it may be more realistic (and less reductionist) to consider that the health effects of air pollution accrued from a combination of exposures, whose concentration (in those cases, as is still sometimes true today) was greatly magnified by unusual weather conditions. In Donora, a zinc smelter was especially criticised, but, again, causation was eventually determined to be multifactorial, worsened (as in the Meuse Valley) by unfavourable weather and topography (Snyder 1994).

But this does not mean that all components of air pollution are either equally toxic or even that some are benign. Particulate matter is a complex mixture of solid and liquid particles, suspended in air as a result of the burning of coal, gasoline, diesel fuels and biomass such as wood (Sierra-Vargas and Teran 2012). The finest particulate matter, less than 1 micrometre (μm) in diameter (PM1), has been especially implicated in cardiovascular disease, as these particles are sufficiently tiny to not only penetrate deep into the respiratory tract but cross into the bloodstream in the alveoli, where gas exchange occurs (Martinelli et al. 2013). Larger particulate matter (PM10) has been identified as a cause of lung cancer (Raaschou-Nielsen et al. 2013) while ozone, carbon monoxide, nitrogen dioxide and sulphur dioxide all worsen asthma (Ierodiakonou et al. 2016). Diesel exhausts are much more harmful than car exhausts, containing 10–100 times the mass of particulate matter from cars, much of which has adsorbed (adherent) organic compounds derived from heavy carbon (Ristovski et al. 2012). In addition, some forms of air pollution bear heavy metals, including lead, which has been conclusively shown to impair childhood learning, above very low thresholds of exposure (McMichael et al. 1988).

In some (or many) cases, it is likely that synergisms occur between the various components of polluted air. Thresholds of exposure clearly exist, beyond which additional exposure is disproportionately harmful. Further complicating the challenge to identify the most toxic elements of air pollution is the varying susceptibility of populations. Even exposure to asbestos does not guarantee pathology (Terra-Filho et al. 2015).

A holy grail for researchers could be to determine the effects of lifelong population exposure to the various elements and combinations of air pollution, e.g. x years of exposure to a certain level of PM10, y years of exposure to ozone and z years of exposure to sulphur dioxide (average and peak). Added to this difficulty would be an estimate of the harm, acute and chronic, from numerous combinations of pollutants. But such levels of understanding are likely to take decades to evolve and may not be worth the effort. Meanwhile it is prudent to reduce exposure as much as is economically and socially possible, at the same time enhancing the resistance of exposed populations, through means such as reduced tobacco smoking and better nutrition.

Indoor and Outdoor Air Pollution

Although the burden of disease of air pollution, including in the global burden of disease studies (Lim et al. 2012), has long been divided into indoor (domestic or household) and outdoor (ambient) sources, this dichotomy has been recently been convincingly challenged. There are several reasons for this revision, particularly that solid cooking fuel such as straw, dung and wood, used indoors, with inadequate ventilation, is often sufficiently polluting and widespread to appreciably affect widespread ambient air pollution levels (Smith et al. 2014).

The Triple Burden of Air Pollution in and from Australia

The most recent estimates of the burden of disease of air pollution in Australia is low, compared to nations such as China and India (Lim et al. 2012), even on a per capita basis. However, it is far from trivial, as several case studies will illustrate.

Air pollution in Australia (and some other countries) has a triple burden. Other than tobacco, which is not further discussed in this chapter, the main forms of air pollution in Australia occur via the inhalation of airborne pollutants including particulate matter from coal dust, coal smoke and gaseous products of coal burning such as sulphur dioxide. Also important are combustion products of biomass burning including of wood (especially particulates); industrial emissions from manufacturing; refineries and chemical production; motor vehicle exhausts, including diesel fumes; and pollen. These cause direct and sometimes prolonged harm, especially to vulnerable groups, particularly people with pre-existing disease and the elderly. Health conditions known to be contributed to by air pollution include respiratory diseases (e.g. asthma, chronic bronchitis and lung cancer), some cardiovascular diseases (e.g. heart attacks and strokes), some infectious diseases and some forms of cancer, including lung cancer and, possibly, leukaemia and others (Colagiuri et al. 2012; Filippini et al. 2015).

The prolific per capita combustion of fossil fuels (mainly for transport and electricity generation) and the ingestion of meat and meat products in Australia (especially from sheep and cattle, each of which produces the greenhouse gas methane) mean Australians make a disproportionate contribution to human-made climate change, which in turn is having increasingly profound adverse health effects (Butler et al. 2016). The effects of climate change are inexorably growing and will be far higher in the future (Butler and Harley 2010).

Climate change is a form of air pollution for several reasons. Disguising this recognition, the main greenhouse gases (carbon dioxide (CO2), methane and nitrous oxide) are completely invisible and odourless at atmospheric concentrations. CO2 is essential for plant life and harmless to humans when inhaled, even at levels far higher than 400 parts per million and its present level, an increase of 45% from the pre-industrial period. Further, the harm that greenhouse gases impose on human health is different to other forms of air pollution.

However, by altering the heat-trapping characteristics of the global atmosphere, greenhouse gases contribute to extreme weather events, sea level rise, altered dynamics of some infectious diseases and other events with adverse health consequences. Some extreme weather events, such as drought, can contribute to migration and conflict, where significant other precursors for conflict exist (Bowles et al. 2015; Schleussner et al. 2016). A leaked copy of the fifth Intergovernmental Panel on Climate Change (IPCC) assessment was reported as warning of hundreds of millions people being displaced by 2100 (McCoy et al. 2014). Of interest, and consistent with the increasingly recognised way in which authorities have long downplayed the risk of air pollution, this warning was changed in the final report to the much less disturbing, unquantified statement 'climate change is projected to increase displacement of people (medium evidence, high agreement)' (IPCC 2014).

Recognising the potential health harm from greenhouse gas accumulation, the US Environmental Protection Agency (2009) identified the main greenhouse gases as air pollutants. Time will tell if this strong position survives the administration of US President Trump (Mathiesen 2016).

The third way air pollution from Australia harms health is via its exports of fossil fuels and of digastric (ruminant) sheep and cattle, which also make important contributions to climate change (McMichael et al. 2007). Australians thus not only make substantial contributions to climate change and its harm to health from their culture but also profit from it. It is a disturbing paradox but plausible that the burden of disease from climate change, due to these exports of greenhouse gases, will continue to rise, even as the burden of disease from other forms of air pollution in Australia continues to fall.

Australian Case Studies of Air Pollution

Air pollution in Australia may have a comparatively low burden of disease by global standards, but there is increasing recognition that it imposes heavy economic and social costs, including a national health bill of up to A$24.3B each year (National Environment Protection Council 2014). Recent studies point to coal mining and coal-fired power generation as major contributors to these large and growing costs.

Reducing air pollution concentrations has a significant health benefit. A study in the USA found that a reduction of 10 micrograms per cubic metre (μ/m3) in the concentration of fine particulate matter (PM2.5) explained as much as 15% of the overall increase in life expectancy in the study areas which occurred between the late 1970s and the early 2000s (Pope et al. 2009). This improvement followed determined efforts in the USA to improve air quality. Similarly, legislation in Australia has resulted in cleaner air but probably from a less polluted starting point. In lieu of comparable national-scale studies, we discuss several categories and case studies. Collectively, these examples illustrate that the health effects of air pollution in Australia are far from trivial and can and should be reduced.

Industry

Australia has been free of dramatic episodes of mortality from industrial air pollution, similar to the Meuse Valley and Donora. Pockets of industrialised air pollution exist, some of it little contaminated by pollution from traffic or domestic sources, due to small populations and isolation. Examples include Port Pirie, South Australia (the world's third largest lead-zinc smelter); Broken Hill, New South Wales (NSW); and Mount Isa, Queensland. Contamination of surfaces with dust containing lead and other heavy metals in these towns is still problematic, with exposures in children likely to reduce school performance (Taylor et al. 2013, 2014). In fact, the studies which conclusively showed that lead exposure reduced children's abilities (with, presumably, lifelong consequences) were undertaken at Port Pirie (McMichael et al. 1988). Despite attempts to reduce lead pollution in these smelting towns, problems persist. While levels are lower than at their peak, in some places they may again be worsening (Taylor et al. 2014).

Other sources of industrial air pollution include cement works, steel mills and coal-burning thermal power stations. In response to long-standing concerns about the health effects of air pollution near heavy industry, a cross-sectional study was conducted in the two steel-making cities in NSW (Newcastle and Wollongong) using data from 1993 to 1994. It found a dose-response relationship between PM10 levels and chest colds in primary school children but no relationship with SO2 exposure (Lewis et al. 1998). Each of these cities is large enough to also experience significant traffic pollution, and in fact control groups in these studies were still exposed to a significant level of PM10, of about 15 μg/m3. The authors commented that the results they found provided evidence of health effects at lower levels of outdoor air pollution in the Australian setting than was then expected. Note however, even in 2016, that the 'standard' level for PM10 exposure in Australia is 50 μg/m3 averaged over 24 h and 25 μg/m3 averaged over 1 year (NSW Environment Protection Authority and Office of Environment and Heritage 2016).

Traffic

Motor vehicles enable the movement of millions of people but have obvious drawbacks, including congestion, noise, cost, accidents and greenhouse gas emissions. In many locations, motor vehicle emissions merge with industrial and other sources of air pollution. A widely cited study from Europe (albeit using data now quite dated) concluded that about half of all mortality caused by air pollution was from motorised traffic (Künzli et al. 2000). Motor vehicles have been described as the dominant cause of air pollution in Australia (Barnett 2013); however, this is disputed by the National Environment Protection Council (2014). Certainly, in some regions and seasons, sources other than traffic, particularly wood heaters ($PM_{2.5}$ in urban areas), coal-fired power stations (SO_x, NO_x and $PM_{<2.5}$ in non-metro environments) and coal mines (PM_{10}, in non-metropolitan regions), are more important.

Air pollution from motor vehicles has been linked with the general range of respiratory and cardiac conditions, including atopy (Bowatte et al. 2015), and, possibly, congenital birth defects (Hansen et al. 2009; Padula et al. 2013). One study, based in Adelaide, South Australia, with an estimated population of 1.4 million in 2030, concluded that shifting 40% of vehicle kilometres travelled away from fossil fuel powered passenger vehicles to walking, cycling and public transport would lower annual average urban $PM_{2.5}$ concentrations by approximately 0.4 $\mu g/m^3$, saving about 13 deaths per year and preventing 118 disability-adjusted life years (DALYs) per year, due to improved air quality. It pointed out that additional health benefits may be obtained from improved physical fitness through active transport and fewer traffic injuries (Padula et al. 2013). Electric vehicles, if fuelled by renewable energy, will also improve air quality.

Diesel fumes

The carcinogenic effect of diesel exhaust products has long been suspected, and diesel was raised to Level-1 (most carcinogenic) by the International Agency for Research on Cancer in 2012 (Swanton et al. 2015). In recognition, the mayors of four major global cities have promised to ban the use of all diesel-powered cars and trucks from their streets, by 2025 (McGrath 2016). To date, no leader of an Australian city has indicated that they will match this.

Biomass and Dust

Woodsmoke

Although deliberate biofuel combustion for cooking and heating is modest in Australia compared to many low-income countries, fine particle pollution from

Fig. 9.1 In June (winter) 2016, a layer of woodsmoke settles over Armidale, a city in rural NSW of approximately 25,000 people, located at an elevation of almost 1000 m on the New England Tableland (Credit: Nathan Smith, Armidale Regional Council)

wood heaters is also a problem in some of Australia's larger cities. In Sydney, for instance, wood smoke accounts for 47% of annual PM2.5 emissions and up to 75% of particle emissions during winter (NSW Environment Protection Authority and Office of Environment and Heritage 2016). Without decisive government action to ban, replace and improve domestic wood heaters, health costs of A$8.1B are projected over 20 years in New South Wales alone (AECOM 2011).

Several urban areas in Australia experience particularly high ambient air pollution not only as a result of household use of firewood for heating but also because they are prone to inversion layers, in which a layer of warmer air above the smoke traps a cooler, polluted layer below. Three such places are the Tuggeranong valley (population c90,000) in southern Canberra (Australian Capital Territory); the smaller, regional cities of Launceston (Tasmania); and the Armidale (NSW) (see Fig. 9.1). In all these cases, winters are cold and wood fuel is comparatively cheap, abundant, and available.

Recognising the extent of air pollution in Launceston, coordinated strategies were undertaken in 2001 to reduce emissions from wood smoke, involving community education, enforcement of environmental regulations and wood heater replacement programme. A study in this city, then with a population 67,000, examined changes in daily all-cause, cardiovascular and respiratory mortality during the 6.5-year periods before and after June 2001. Mean daily wintertime concentration of PM10 fell markedly, from 44 μg/m3 (1994–2000) to 27 μg/m3 (2001–2007). This was associated with a statistically significant reduction in annual mortality among males and with lower cardiovascular and respiratory mortality during the winter months, for both males and females (Johnston et al. 2013).

Forest Fires

Smoke from bushfires in Australia is modest compared to South East Asia but is increasingly recognised to have adverse public health effects (Johnston et al. 2011; Price et al. 2012). A study of air pollution from savanna fires in Darwin, Northern Territory, examined the association between PM10 and daily emergency hospital admissions for cardiorespiratory diseases during each fire season from 1996 to 2005. It also investigated whether the relationship differed in indigenous Australians. Using modelled (rather than recorded) data, this study found an association between higher PM10 levels and daily hospital admissions that was greater in indigenous people (Hanigan et al. 2008).

Dust

Some cities in Australia experience periodic dust storms, worsened by drought and land clearing. Though fairly transient, these also impair air quality and have been found to be associated with increased mortality (Johnston et al. 2011).

Mining

Many forms of mining are associated with ill health, including from occupational exposure to toxic substances in poorly ventilated spaces including radiation daughter products, dust and fumes. Population exposure from the smelting of heavy metals (such as lead) is well documented, with exposure via inhalation and from contact with contaminated dust, including from children playing. Coal is hazardous to health not only from its mining but also its deliberate combustion (Castleden et al. 2011), which in Australia is mostly for electricity production and for steel production.

Solastalgia, Noise and Health Complaints in the Hunter Valley

The Hunter valley is a rural region of NSW, once best known for its vineyards and horse studs. However, in recent years the number of open cut coal mines has greatly increased, leading to great distress by some of its inhabitants. The term 'solastalgia' (loss of solace, formerly experienced in the same geographical setting, but gone, due to changes such as noise, industrialisation and air pollution) was coined in part to describe this distress (Albrecht et al. 2007). Additionally, in this location, many residents, civil society and local government groups have struggled to be heard by corporations and state governments, altering the region's social fabric and adding to their distress, depression, anxiety and ill health (Higginbotham et al. 2010). In limited support of these concerns, a study using general practitioner data from 1998 to 2010 found that the rate of respiratory problems in the Hunter Valley region did not fall significantly over time, in contrast to other rural areas of NSW (Merritt et al. 2013).

Coal Mining

A range of health impacts associated with power stations and coal mines has been studied. In Australia's coal mining regions, including the Hunter Valley, Latrobe Valley and Central Queensland, the vast majority of coarse particle (PM10) pollution is generated by open-cut coal mines. Adults living near coal-fired power stations have been reported as experiencing a higher risk of death from lung, laryngeal and bladder cancer, skin cancer (other than melanoma) and asthma rates and respiratory symptoms (Colagiuri et al. 2012). Children and infants are especially impacted, experiencing higher rates of oxidative deoxyribonucleic acid (DNA) damage, asthma and respiratory symptoms, preterm birth, low birth weight, miscarriages and stillbirths, impaired foetal and child growth and neurological development.

The adverse health impacts of Australia's fleet of coal-fired power stations have been estimated at A$2.6B per annum (Beigler 2009). In the Hunter Valley alone, the adverse health impacts of coal-fired power stations have been estimated at A $600M per annum (Armstrong 2015) (Fig. 9.2).

Fig. 9.2 Uncovered coal wagons in Newcastle, NSW, releasing an obvious stream of particles credit John Nella

The Morwell Coal Mine Fire

In early 2014, a fire burned for 45 days in the Hazelwood open-cut coal mine in the industrialised Latrobe Valley of Victoria started by an adjacent bushfire. This triggered one of the worst short-term episodes of air pollution in Australian history. Several communities were affected by smoke, particularly the township of Morwell, with a population of about 15,000, located less than a kilometre from the fire. The concentration of smoke contaminants was regularly monitored in several locations, by the Environment Protection Authority of Victoria, including in South Morwell (Reisen et al. 2016). The level of PM2.5 briefly peaked at over 700 μg/m3, 32 times the reporting standard of 25 μg/m3 averaged over 24 h (Fisher et al. 2015). Despite this, no one was compulsorily evacuated from Morwell nor even strongly advised to leave. Limited monitoring of the affected population is now being undertaken (Fisher et al. 2015). A Victorian Government inquiry into the mine fire concluded that there was a high probability that air pollution contributed to an increase in mortality during the fire and that the fire harmed the health of many in this community.

Black Lung in Australian Miners

Pneumoconiosis ('black lung') is a well-known occupational hazard for coal miners, occurring from overexposure to coal dust, first described in the seventeenth century. In Australia, however, which requires compulsory participation by miners in X-ray screening programs, no cases were reported for over 30 years, until recently (Cohen 2016). This was though due to better dust control in mines, a study from 2002 reported that, in 6.9% of measurements, dust exposure in 33 longwall coalmines in NSW exceeded the Australian National Standard (Castleden et al. 2011). A more recent audit of underground coal mines in Queensland found that an increasing number of workers are exposed to harmful concentrations of respirable dust, well above regulatory limits (Commissioner for Mine Safety and Health 2015). The reappearance of pneumoconiosis is thus perhaps not surprising, but what was surprising was that precautionary X-rays in miners were misread over a long period, thus contributing to complacency (Cohen 2016).

Air Pollution, Urban Forests and Pollens

Increasing the number of trees in urban areas has long been suggested as a means to reduce air pollution and lower the heat island effect (Benjamin et al. 1996). Trees reduce the quantity of particulate matter, by making available a large surface area of bark and leaves (especially of evergreens or in spring to autumn) on which gases and particles can be deposited. They can also help decompose some air pollutants, including ozone, by releasing gases (Grote et al. 2016).

However, some trees have a significant 'ozone-forming potential' (Grote et al. 2016), with some species reported to have up to four orders of magnitude more capacity to release photochemically reactive hydrocarbons than others (Benjamin et al. 1996). Eucalyptus trees, which are well known for producing a blue haze in some settings (hence the 'Blue Mountains', near Sydney, NSW), may have a significant effect in Australian settings on air pollution, by their release of hydrocarbons that may contribute to smog, but the net effect of this appears understudied. An increased urban forest, planted to improve air quality, might also elevate the risk of urban bushfires.

Some tree species also have significant quantities of wind-dispersed pollen, allergens, which can cause severe distress in vulnerable people, including asthma and possibly mood changes. For example, there are credible claims that exposure to allergens is a factor underpinning the long observed rise in suicides in spring (Kõlves et al. 2015). Grass pollens, however, may be more problematic than from trees, including in thunderstorm asthma (D'Amato et al. 2007). A study in Darwin found an association between Poaceae grass pollen and the sale of antihistamine medication (Johnston et al. 2009).

Climate Change and Health in Australia

The health effects of climate change in Australia include primary (direct, comparatively obvious) effects such as from climate change-exacerbated heatwaves, droughts, fires and floods; secondary (less obvious, indirect) including changes in allergens and atopic diseases and infectious diseases and rising food prices and impaired nutrition; and tertiary (highly indirect, catastrophic), including regional war and mass migration (Butler and Harley 2010).

Primary Health Effects

Already, extreme events, contributed to by climate change, are increasing in Australia. Although the death toll of rural suicide from droughts in Australia has recently declined (probably due to better intervention) (Hanigan et al. 2012), this improvement may not last; living with chronic depression due to loss of livelihood and other trauma (e.g. being forced to shoot suffering stock) is still likely to be high, as is the health toll from exposure to floods and, sometimes, resultant displacement. Prolonged, extreme heat in Australia is also documented to cause excessive deaths and morbidity, particularly in vulnerable sub-groups (Nitschke et al. 2011).

Secondary Health Effects

As this chapter was being finalised, the population of the Victorian state capital, Melbourne, experienced the worst episode of 'thunderstorm asthma' to ever occur in Australia. This caused the premature death of at least eight people, most or all of whom were comparatively young (Calligeros et al. 2016). Thousands were hospitalised and overwhelmed emergency services, including by generating ambulance calls every 4.5 s. This was contributed to by a wet spring, humidity and a hot day in late spring (Calligeros et al. 2016). It is plausible that climate change may make such episodes more frequent. The major source of the allergens involved in this appears to be rye grass, rather than tree pollen.

The pattern of some infectious diseases in Australia, including Ross River virus and dengue fever, is also likely to be subtly altered by climate change (Williams et al. 2016). There are many other examples, such as melioidosis and leptospirosis (Currie 2001). However, an increase in mortality from altered infectious diseases epidemiology is unlikely to be marked.

Tertiary Health Effects

Australia is a very wealthy country, though the distribution of health and other forms of security is increasingly unequal. The most dire health effects of climate change are likely to be long avoided in Australia; however, the country is already subtly affected by conflict in the Middle East, Afghanistan and parts of sub-Saharan Africa. Some of this turmoil (which also has led to the current global refugee crisis) can be attributed to climate change, interacting with social factors, including poverty, poor governance, discrimination and limits to growth (Bowles et al. 2015; Butler 2016; Schleussner et al. 2016).

The Australian government, with wide public support, has practised human rights abuses of asylum seekers for well over a decade (Newman et al. 2013). A possible explanation for this behaviour is fear, rather than overt cruelty. That is, most Australians may support a strong 'fend' (deterrence) signal to asylum seekers because they wish to prevent additional refugees seeking protection in Australia, a rich country widely perceived as underpopulated. Unfortunately, however, Australia, by cutting its foreign aid, and by aggressively exporting products that contribute to climate change, is continually seeding conditions likely to increase refugee numbers, including in countries in its region. As sea level rise and other manifestations of climate change worsen in poor, 'developing' countries in South Asia (Singh et al. 2016) and the Pacific, the number of people seeking refuge in Australia is likely to climb steeply.

Towards Solutions

Industry and Weak Legislation

In Australia, state and national air pollution laws provide few opportunities for impacted communities to seek a legal remedy. National air pollution standards are determined by Australia's nine[1] environment ministers, meeting as the National Environment Protection Council, yet are governed by state and territory laws. The Council's decision-making has been described as taking a 'lowest common denominator' approach, resulting in standards that reflect the position of the state or territory least inclined to regulate polluters. But even these low standards are not always met; each jurisdiction adopts a different approach, drawing from a regulatory toolbox that includes consent conditions for major polluters, environmental pollution licences, pollution monitoring, auditing, annual reports and various compliance mechanisms. In sharp contrast, in the USA, the US Environment Protection Authority has the power to impose sanctions on states that fail to comply with air pollution standards, which are set centrally.

In Australia, prosecutions for breaching licences or causing environmental harm from air pollution are infrequent, fall far short of the real costs of the harm caused and are generally inadequate to compel companies to invest in pollution control. Consequently, air pollution-impacted communities in Australia look to the regulatory systems in other countries for models that may be effective here.

Community Action and Organising

Air pollution consistently ranks highly among environmental concerns, particularly in communities that experience elevated levels of pollution due to specific local or regional sources. The weak legal and regulatory framework for air pollution control in Australia described above, coupled with increasing air pollution levels, leads citizens to initiate and participate in various forms of community action.

The starting point for many people is a desire to know as much as possible about what they're breathing. Residents in polluted communities assert their 'right to know' by phoning pollution hotlines, approaching polluters directly and accessing government websites and reports for monitoring data. Although state and territory governments conduct air pollution monitoring in many locations, few provide ready access to the data they collect, and there are significant 'black spots': regions that experience high levels of pollution but where governments permit and tolerate self-monitoring by industry but with no public access to these data. In the vast coalfields of Central Queensland, for instance, there is no government or independent

[1] Six state, two territorial and one federal.

monitoring for more than a million square kilometres, and community members have no legal right to access industry monitoring data. The power generators in the Latrobe Valley have, for years, monitored local pollution, free of any obligation to share their results.

In response to this suppression of information, community members have sometimes turned to citizen science. In the Hunter Valley, North West New South Wales and South East Queensland, community members have documented an increase in air pollution concentrations as coal train pass, confirming their long-held concerns.

Community members value and participate actively in dialogue with industry and regulators. In the Hunter Valley and other industrialised regions, there are community consultative committees for most major polluting facilities. These 'CCCs' create a forum for community members to air concerns, seek information and articulate their expectations. Alas, in the authors' experience, they to date rarely achieve tangible pollution reduction outcomes. Information flow is primarily one-way, that is, neither industry nor government is very responsive.

The right to know, access to reliable data and dialogue are important but not substitute for demonstrable pollution control and reduction. Too frequently, government regulators are seen to be 'captured' by polluting industries and unwilling to exercise their full statutory powers to protect polluted communities. When 'polite' mechanisms fail, as they often do, citizens need to reply on a more 'activist' suite of tools that include media commentary, parliamentary politics, legal action and protest.

Conclusion: Low-Hanging Fruit: Immediate Co-benefits for Health and Climate Change

Enough is known about the sources and impacts of air pollution to enable the development of air pollution control plans for our major cities and other polluted regions. Pollution hotspots including the Newcastle, Gladstone, coalfields of New South Wales and Queensland and Hunter and Latrobe Valleys should have action plans that incorporate 'best practice' air pollution reduction strategies that have worked elsewhere, monitoring and evaluating arrangements to facilitate adaptive management and active community involvement.

The catalogue of 'no regrets' pollution control action that have worked in other countries includes introducing strict emission standards for power stations and motor vehicles, implementing a rapid and just transition from coal-fired power generation to renewable energy, banning new wood heaters and replacing existing ones, covering and washing coal trains, enclosing coal stockpiles and facilitating the uptake of electric vehicles.

Polluters and regulators need to be much more transparent and more accountable. This requires a change in political will and almost certainly necessitates a strong national approach to air poll. Leaving states to adopt diverse approaches to air pollution, management and regulation has failed to curb air pollution in Australia. The health benefits of controlling air pollution in Australia warrant a much stronger approach. There also needs to be a much greater appreciation of the health and economic costs of air pollution and climate change. It is enormously misleading to claim that coal-fired electricity is 'cheap'. Coal mining, coal combustion and coal export cause significant health costs, in the past, present and future. Furthermore, the price of alternatives such as wind and solar continues to fall. Reducing emissions from the burning of wood and the combustion of vehicular fuel is more challenging, but much can also be accomplished in these spheres too, including electric vehicles, public transport and, in the foreseeable future, domestic production and consumption of solar energy, incorporating batteries.

References

AECOM (2011) Wood smoke control measures: cost benefit analysis Prepared for the NSW Environment Protection Authority, Sydney. http://www.epa.nsw.gov.au/resources/air/WoodsmokeControlReport.pdf

Albrecht G, Sartore G-M, Connor L et al (2007) Solastalgia: the distress caused by environmental change. Australas Psychiatry 15(1):S95–S98

Armstrong F (2015) Coal and health in the Hunter: lessons from one valley for the world, Climate and Health Alliance caha.org.au/wp-content/.../Climate-and-Health-Alliance_Report_Layout_PRINTv2.pdf. 16 Dec 2016

Attiwill PM, Adams MA (2013) Mega-fires, inquiries and politics in the eucalypt forests of Victoria, south-eastern Australia. For Ecol Manag 294:45–53

Barnett AG (2013) Time for action on air pollution. Med J Aust. https://www.mja.com.au/careers/199/199/time-action-air-pollution

Beigler T (2009) The hidden costs of electricity: externalities of power generation in Australia, report for Australian Academy of Technological Sciences and Engineering, Parkville, available at http://www.atse.org.au/Documents/Publications/Reports/Energy/ATSE%20Hidden%20Costs%20Electricity%202009.pdf

Benjamin MT, Sudol M, Bloch L et al (1996) Low-emitting urban forests: a taxonomic methodology for assigning isoprene and monoterpene emission rates. Atmos Environ 30 (9):1437–1452

Berridge V (2007) Medicine and the public: the 1962 report of the Royal College of Physicians and the new public health. Bull Hist Med 81(1):286–311

Berridge V, Taylor S (2005) The big smoke: fifty years after the 1952 London Smog. Centre for History in Public Health London School of Hygiene & Tropical Medicine, London, p 42

Bowatte G, Lodge C, Lowe AJ et al (2015) The influence of childhood traffic-related air pollution exposure on asthma, allergy and sensitization: a systematic review and a meta-analysis of birth cohort studies. Allergy 70(3):245–256

Bowles DC, Butler CD, Morisetti N (2015) Climate change, conflict, and health. J R Soc Med 108 (10):390–395

Brimblecombe P (2011) The big smoke: a history of air pollution in London since Medieval Times (Routledge Revival Edition). Routledge, Abingdon

Butler CD (2016) Planetary overload, limits to growth and health. Curr Environ Health Rep 3 (4):360–369

Butler CD, Harley D (2010) Primary, secondary and tertiary effects of the eco-climate crisis: the medical response. Postgrad Med J 86:230–234
Butler CD, Hales S, Beggs PJ et al (2016) From Paris towards 1.5 degrees C. In: Butler CD (ed) Climate change and global health. CABI, Wallingford, pp 293–305
Calligeros M, Dow A, Donelly B (2016) Thunderstorm asthma: three people critical, 8500 treated in Melbourne hospitals. The Age. http://www.theage.com.au/victoria/thunderstorm-asthma-three-people-critical-8500-treated-in-melbournehospitals-20161124-gswkdb.html. Accessed 22 July 2017
Castleden WM, Shearman D, Crisp G et al (2011) The mining and burning of coal: effects on health and the environment. Med J Aust 195(6):333–335
Cohen RA (2016) Resurgent coal mine dust lung disease: wave of the future or a relic of the past? Occup Environ Med 73:715–716
Colagiuri R, Cochrane J, Girgis S (2012) Health and social harms of coal mining in local communities. Report for Beyond Zero Emissions (Australia). https://sydney.edu.au/medicine/.../PDF_Mining_Report_FINAL_October_2012.pdf. Accessed 22 July 2017
Commissioner for Mine Safety and Health, Q. G (2015) Queensland mines inspectorate annual performance report 2014–15
Connor L, Freeman S, Higginbotham N (2009) Not just a calamine: shifting grounds of community opposition to coal mining in Southeastern Australia. Ethnos 74(4):490–513
Currie BJ (2001) Environmental change, global warming and infectious diseases in Northern Australia. Environ Health 1(4):35–44
D'Amato G, Liccardi G, Frenguelli G (2007) Thunderstorm-asthma and pollen allergy. Allergy 62 (1):11–16
Environmental Protection Agency (2009). Greenhouse gases threaten public health and the environment. https://yosemite.epa.gov/opa/admpress.nsf/0/08D11A451131BCA585257685005BF252. Accessed 23rd Nov 2016
Filippini T, Heck JE, Malagoli C et al (2015) A review and meta-analysis of outdoor air pollution and risk of childhood leukemia. J Environ Sci Health C 33(1):36–66
Fisher G, Torre P, Marshall A (2015) Hazelwood open-cut coal mine fire. Air Qual Clim Chang 49(1)
Fuller RS, Trudgett M, Norris RP et al (2014) Star maps and travelling to ceremonies: the Euahlayi people and their use of the night sky. J Astron Hist Herit 17(2):149–160
Gammage B (2011) The biggest estate on earth: how aborigines made Australia. Sydney, Allen and Unwin
Grote R, Samson R, Alonso R et al (2016) Functional traits of urban trees: air pollution mitigation potential. Front Ecol Environ 14(10):543–550
Hanigan IC, Johnston FH, Morgan GG (2008) Vegetation fire smoke, indigenous status and cardio respiratory hospital admissions in Darwin, Australia, 1996–2005: a time-series study. Environ Health 7:42
Hanigan IC, Butler CD, Kokic PN et al (2012) Suicide and drought in new South Wales, Australia, 1970–2007. Proc Natl Acad Sci USA 109(35):13950–13955
Hansen CA, Barnett AG, Jalaludin BB et al (2009) Ambient air pollution and birth defects in Brisbane, Australia. PLoS One 4(4):e5408
Higginbotham N, Freeman S, Connor L et al (2010) Environmental injustice and air pollution in coal affected communities, Hunter Valley, Australia. Health Place 16(2):259–266
Ierodiakonou D, Zanobetti A, Coull BA et al (2016) Ambient air pollution, lung function, and airway responsiveness in asthmatic children. J Allergy Clin Immunol 137(2):390–399
IPCC (2014) Summary for policymakers. Climate change 2013: the physical science basis: working group I contribution to the fifth assessment report of the intergovernmental panel on climate change. In: Stocker TF, Qin D, Plattner GK et al. (eds). Cambridge University Press, Cambridge, pp 3–32
Johnston F, Hanigan I, Bowman D (2009) Pollen loads and allergic rhinitis in Darwin, Australia: a potential health outcome of the grass-fire cycle. EcoHealth 6(1):99–108

Johnston F, Hanigan I, Henderson S et al (2011) Extreme air pollution events from bushfires and dust storms and their association with mortality in Sydney, Australia 1994–2007. Environ Res 111:811–816

Johnston FH, Hanigan IC, Henderson SB et al (2013) Evaluation of interventions to reduce air pollution from biomass smoke on mortality in Launceston, Australia: retrospective analysis of daily mortality, 1994–2007. BMJ 346:e8446

Jones R (2012) Fire-stick farming. Fire Ecol 8(3):3–8. [Reprinted from Australian Natural History 16(17): 224-228 (1969)].

Kõlves K, Barker E, De Leo D (2015) Allergies and suicidal behaviors: a systematic literature review. Allergy Asthma Proc 36(6):433–438

Künzli N, Kaiser R, Medina S et al (2000) Public-health impact of outdoor and traffic-related air pollution: a European assessment. Lancet 356(9232):795–801

Lewis PR, Hensley MJ, Wlodarczyk J et al (1998) Outdoor air pollution and children's respiratory symptoms in the steel cities of New South Wales. Med J Aust 169:459–463

Lim SS, Vos T, Flaxman AD et al (2012) A comparative risk assessment of burden of disease and injury attributable to 67 risk factors and risk factor clusters in 21 regions, 1990–2013: a systematic analysis for the Global Burden of Disease Study 2010. Lancet 380 (9859):2224–2260

Martinelli N, Olivieri O, Girelli D (2013) Air particulate matter and cardiovascular disease: a narrative review. Eur J Intern Med 24(4):295–302

Mathiesen K (2016) Polar opposites in US election. Nat Clim Chang 6(11):979–981

McCoy D, Montgomery H, Arulkumaran S et al (2014) Climate change and human survival. BMJ 348:g2351. doi:2310.1136/bmj.g2351

McGrath M (2016) Four major cities move to ban diesel vehicles by 2025. BBC news. http://www.bbc.com/news/science-environment-38170794

McMichael AJ, Baghurst PA, Wigg NR et al (1988) Port Pirie cohort study: environmental exposure to lead and children's abilities at the age of four years. N Engl J Med 319(8):468–475

McMichael AJ, Powles J, Butler CD et al (2007) Food, livestock production, energy, climate change and health. Lancet 370:1253–1263

Merritt TD, Cretikos MA, Smith W et al (2013) The health of Hunter Valley communities in proximity to coal mining and power generation, general practice data, 1998–2010. N S W Public Health Bull 24(2):57–64

National Environment Protection Council (2014) Draft variation to the National Environment Protection (Ambient Air Quality) measure: impact statement. https://www.environment.gov.au/protection/nepc/nepms/ambient-air-quality/variation-2014/impact-statement. Accessed 16 Dec 2016

Nemery B, Hoet PHM, Nemmar A (2001) The Meuse Valley fog of 1930: an air pollution disaster. Lancet 357(9257):704–708

Newman L, Proctor N, Dudley M (2013) Seeking asylum in Australia: immigration detention, human rights and mental health care. Australas Psychiatry 21(4):315–320

Nitschke M, Tucker GR, Hansen AL et al (2011) Impact of two recent extreme heat episodes on morbidity and mortality in Adelaide, South Australia: a case-series analysis. Environ Health 10 (1):42

NSW Environment Protection Authority and Office of Environment and Heritage (2016). Clean air for NSW http://www.epa.nsw.gov.au/air/clean-air-nsw.htm

Padula AM, Tager IB, Carmichael SL et al (2013) Ambient air pollution and traffic exposures and congenital heart defects in the San Joaquin valley of California. Paediatr Perinat Epidemiol 27 (4):329–339

Pope CA, Ezzati M, Dockery DW (2009) Fine-particulate air pollution and life expectancy in the United States. N Engl J Med 360:376–386

Price OF, Williamson GJ, Henderson SB et al (2012) The relationship between particulate pollution levels in Australian cities, meteorology, and landscape fire activity detected from MODIS hotspots. PLoS One 7(10):e47327

Raaschou-Nielsen O, Andersen ZJ, Beelen R et al (2013) Air pollution and lung cancer incidence in 17 European cohorts: prospective analyses from the European Study of Cohorts for Air Pollution Effects (ESCAPE). Lancet Oncol 14(9):813–822

Reisen F, Fisher G, Torre P et al (2016) Analysis of air quality during the Hazelwood mine fire. CSIRO, Canberra. CSIRO (hazelwoodhealthstudy.org.au)

Ristovski ZD, Miljevic B, Surawski NC et al (2012) Respiratory health effects of diesel particulate matter. Respirology 17(2):201–212

Schleussner C-F, Donges JF, Donner RV et al (2016) Armed-conflict risks enhanced by climate-related disasters in ethnically fractionalized countries. Proc Natl Acad Sci 113(33):9216–9221

Sierra-Vargas MP, Teran LM (2012) Air pollution: impact and prevention. Respirology 17 (7):1031–1038

Singh M, Rao M, B. C.D. (2016) Climate change, health and future well-being in South Asia. In: Akhtar R (ed) Climate change and human health scenario in south and Southeast Asia. Springer, The Netherlands, pp 11–27

Smith KR, Bruce N, Balakrishnan K et al (2014) Millions dead: how do we know and what does it mean? Methods used in the comparative risk assessment of household air pollution. Annu Rev Public Health 35(1):185–206

Snyder LP (1994) "The death-dealing smog over Donora, Pennsylvania": industrial air pollution, public health policy, and the politics of expertise, 1948–1949. Environ Hist Rev 18(1):117–139

Swanton C, Boffetta P, Peston R et al (2015) Environmental emissions, public health and lung cancer risk. Ann Oncol 27:211–212

Taylor MP, Camenzuli D, Kristensen LJ et al (2013) Environmental lead exposure risks associated with children's outdoor playgrounds. Environ Pollut 178:447–454

Taylor MP, Mould SA, Kristensen LJ et al (2014) Environmental arsenic, cadmium and lead dust emissions from metal mine operations: implications for environmental management, monitoring and human health. Environ Res 135:296–303

Terra-Filho M, Bagatin E, Nery LE et al (2015) Screening of miners and millers at decreasing levels of asbestos exposure: comparison of chest radiography and thin-section computed tomography. PLoS One 10(3):e0118585

Williams CR, Mincham G, Faddy H et al (2016) Projections of increased and decreased dengue incidence under climate change. Epidemiol Infect 144(14):3091–3100

World Health Organization (2016) 7 million premature deaths annually linked to air pollution. http://www.who.int/mediacentre/news/releases/2014/air-pollution/en/. Accessed 23 Nov 2016

Colin D. Butler is an adjunct professor of public health at the University of Canberra, Australia, and co-founder of the NGO BODHI Australia. He is founding co-chair of Health Earth and lead author for the section on health in the forthcoming flagship Global Environmental Outlook report of the United Nations Environment Programme, called Healthy Planet, Healthy People.

Dr. James Whelan is a researcher and community organiser with Environmental Justice Australia. James has been a non-government advocate on air quality issues since the mid-1990s. He represented environment groups in the development of the National Environment Protection Measures for Ambient Air Quality and the National Pollutant Inventory and has been a member of several advisory groups overseeing air quality management strategies and regulations.

Chapter 10
Epidemiological Consequences of Climate Change (with Special Reference to Malaria in Russia)

Svetlana M. Malkhazova, Natalia V. Shartova, and Varvara A. Mironova

Abstract Climatic conditions play a major role among natural factors determining human's existence. The factor of climate change is considered among other known risk factors to population health. In particular, climate leads to the changes in borders and structure of the areas of infectious and parasitic illnesses. The most serious climate changes are expected in mid- to high latitudes, especially in cities, where anthropogenic activity and air pollution cause exacerbating effect. Within the framework of this study, we try to elaborate a prognostic model of epidemiological conditions of the vivax malaria for the territory of the European part of Russia and Western Siberia. Forecasting was based on the results of climate modeling CMIP3 project under the "A2" IPCC scenario. As a result of forecasting, it is revealed that in the future (2046–2065), favorability of climatic conditions for malaria transmission will increase. The most remarkable changes are expected in the areas situated near southern limits of the considered territories.

Keywords Climate change • Malaria • Modeling • Prognosis • Epidemiological consequences • Russia

Introduction

Nowadays, climate change is considered along with other risk factors jeopardizing public health – environment pollution (including air and water pollution due to the presence in these body pollutants reducing air and water quality enough to threaten the health of people, soil pollution, residential solid waste, etc.), decrease of soil

S.M. Malkhazova (✉) • V.A. Mironova
Department of Biogeography, Faculty of Geography, Lomonosov Moscow State University, Moscow, Russia
e-mail: sveta_geo@mail.ru; mironova.va@gmail.com

N.V. Shartova
Department of Landscape Geochemistry and Soil Geography, Faculty of Geography, Lomonosov Moscow State University, Moscow, Russia
e-mail: shartova@yandex.ru

© Springer International Publishing AG 2018
R. Akhtar, C. Palagiano (eds.), *Climate Change and Air Pollution*,
Springer Climate, DOI 10.1007/978-3-319-61346-8_10

fertility, food shortage in low-income countries, deterioration in drinking water quality, etc.

Over the last few decades, the warmest years on record have been registered. The increase in global temperatures of the earth and the northern and the southern hemispheres in relation to the period 1891–1900 reached 1.08, 1.30, and 0.87 °C, respectively, in 2015, exceeding the level of 1 °C for the first time (Rosgidromet/RAN 2015). The most serious climate changes are expected in mid- to high latitudes, especially in cities, where anthropogenic activity and environment contamination (including air pollution) cause exacerbating effect. As a result, the increase of migration flows is possible due to an influx of people to areas with more favorable climate conditions. Overpopulation and excessive urbanization and insufficient public medical care may result in the resurgence of some infectious diseases and epidemic outbreaks (Malkhazova 2006).

Climate Change and Infectious Diseases

Since ancient times, it has been well known that climatic conditions play a major role among natural factors in determining human existence. It is widely acknowledged that rapid climate change is one of the most pressing environmental issues of the twenty-first century (*Atlas of health and climate* 2012) and that it may have a considerable effect on human health (Epstein 1999; Zell 2004; *Recent global changes of the natural environment* 2006; Filho et al. 2016; Wu et al. 2016; etc.). This impact may manifest itself in different ways. It may contribute to increased frequency and intensity of heat waves, growing number of floods and droughts, changes in distribution patterns of vector-borne diseases, and increased risk of disasters and malnutrition (Haines et al. 2006; Malkhazova 2006).

Infectious diseases represent a major concern because of their dependence on environmental conditions that interact with the biological agents of diseases. Alterations in climate variables (temperature, precipitation, wind, sunshine, length of the seasons, etc.) may affect survival, reproduction, or distribution of disease pathogens, hosts, and vectors. Their health effects tend to manifest as shifts in the geographic and seasonal patterns of human infectious diseases as well as changes in the frequency and severity of outbreaks. Whereas climate limits the geographical range of infectious diseases, weather affects the timing and intensity of outbreaks, especially those associated with weather extremes, such as flooding and droughts. Thus global climate change will most likely influence transmission trends of infectious disease, although the exact direction and extent of this influence remain uncertain (Zell 2004; Wu et al. 2016).

For example, the health effects of flooding may include an increased risk of symptoms associated with diarrhea and accelerating incidence of cholera, cryptosporidium infection, and other waterborne diseases (MacKenzie et al. 1994; Epstein 1999; Ahern et al. 2005; Wu et al. 2016; Aparicio-Effen et al. 2016). Unusual rainfall may cause an increase in fecal pathogens as heavy rain may stir up sediments in water leading to the accumulation of fecal microorganisms (Jofre et al. 2010). Floods can also facilitate transmission of some natural waterborne fecal diseases. Thus increased cases of leptospirosis and campylobacter enteritis have been reported after flooding in the Czech Republic (Zitek and Benes 2005). There have also been reports of flood-associated outbreaks of leptospirosis in Central and South America, South Asia, and Europe (Ahern et al. 2005; Wu et al. 2016). Globally, waterborne epidemics have shown an upward trend from 1980 through 2006, which coincides with the increased number of floods (Brown and Murray 2013).

On the other hand, concentration of effluent waterborne pathogens may be caused by droughts or low rainfall as waterbodies become shallow (Semenza and Menne 2009). Diarrheal diseases are also often associated with droughts and consequently a lack of safe fresh water in low-income countries, refugee camps, etc. There is strong evidence that climate changes influence outbreaks of *Vibrio*-related infections (e.g., human infections associated with recreational bathing and foodborne infections) worldwide, especially in the North Atlantic, which often coincide with heat waves (Vezzulli et al. 2016).

In addition to direct impact, climate change may exert indirect influence on some contagious diseases. For example, the 2014 Ebola outbreak in Western Africa is believed to be connected not only with socioeconomic conditions of affected countries but also with some alterations in the environment due to global changes. Some studies (Harris et al. 2016) suggest that climate variability forced fruit bats, which are believed to be the main natural reservoir of the Ebola virus to migrate long distances and reside near human settlements. Extreme weather events also force farmers who practice mostly subsistence farming to abandon their habitual places and venture deeper into the forests to search for land and new livelihoods. This brings them closer to infected animals and thus at higher risk of infection.

The most serious health impacts of climate change worldwide seem to occur from vector-borne infectious diseases. The greatest concern is that global climate change will result in an expansion of such diseases throughout temperate areas (Epstein 1999). Within the past decades, an increase in tick-borne and mosquito-borne disease outbreaks has been reported in different areas of the world including the mid-latitudes. This process is often connected with a warming climate (Malkhazova and Shartova 2014; Andersen and Davis 2016).

It is important to note that factors that influence transmission of vector-borne diseases are complex, so it is difficult to clearly determine the contribution of each factor. Climate is only one of many interacting determinants of vector-borne disease, but its role seems to be critical for their spread. It is obvious that ecology, development, survival, and behavior of mosquitoes and other arthropods are dependent on climatic conditions. The same factors play an important role in the life cycle of the pathogens that are transmitted by them. The development time of infectious agents such as *Plasmodium* or arboviruses is strongly determined by temperature and humidity. Moreover, these agents are tied to their vectors and depend on their life span and the conditions of their habitats. The influence of precipitation may be illustrated by diseases that are transmitted by vectors that have aquatic developmental stages (such as mosquitoes). Diseases transmitted by vectors without such stages such as ticks or sandflies are also influenced by humidity. Climate and weather conditions may also exert a range of indirect effects on environmental and human systems. The complex interplay of all these factors accounts for the overall effect of climate on the prevalence of vector-borne diseases (Reiter 2001; Campbell-Lendrum et al. 2015). Meanwhile, the nature and extent of interaction with non-climate factors vary markedly by diseases and by location.

The alterations in natural ranges of vector-borne diseases are often connected with the spread to new territories of some vectors that may get involved in the transmission of dangerous infections. Currently one of the most challenging issues is the rapid expansion of two *Aedes* mosquito species, *A. albopictus* and *A. aegypti*, which are responsible for the transmission of yellow fever, chikungunya fever, and dengue. It was well proved that *A. albopictus* induced an outbreak of chikungunya fever in Italy in 2007 (Liumbruno et al. 2008). Some prognoses suggest that the land area with environmental conditions suitable for both species' populations is expected to increase (Rochlin et al. 2013; Kraemer et al. 2015).

The greatest effect of climate change on transmission of vector-borne diseases is likely to be observed at the extremes of the range of temperatures at which transmission occurs. Extreme temperatures near the limit of physiological tolerance for the pathogen prevent its survival and impending transmission of disease. Furthermore, when a vector lives in an environment of low mean temperature, even a small increase in temperature may result in more intensive development of the parasite (Githeko et al. 2000; Patz et al. 2003). That is why the most prominent changes in the spread of vector-borne diseases are expected to occur near the margins of their geographical ranges. There is evidence of northward expansion of geographical ranges of hosts and vectors resulting in the emergence of some diseases in new areas. This process may be mutual when the increase in the range of a reservoir host leads to rising occurrence of the parasite. A study in Canada showed the expansion of the white-footed mouse which is known to be the most competent reservoir for *Borrelia burgdorferi* (the agent of Lyme disease), after changes in winter duration and winter average maximum temperature. As a result, the ticks rapidly increased so that the encounter rate between vectors and hosts increased as well. This provides enhanced conditions for the emergence and maintenance of the *B. burgdorferi* transmission cycle (Roy-Dufresne et al. 2013).

The review of possible effects of climate warming on the health of the population of Russia (Danilov-Danilian 2003) shows that permanent occurrence of such mosquito-dependent diseases such as West Nile fever (WNF), dengue hemorrhagic fever, or yellow fever is probable beyond the tropical zone. It is believed that the agents of these diseases become more active because of recent climate warming as the higher temperature hastens the reproduction of insects. This is confirmed to a degree by the spread of WNF in Russia. Since 1999, WNF cases have been recorded every year in several Russian regions, and in 2013, the disease was registered in 16 constituent territories of the Russian Federation (*Medico-geographical Atlas of Russia "Natural Focal Diseases"* 2015; Adisheva et al. 2016). Cases of Crimean-Congo hemorrhagic fever in the southern regions of Russia have also increased since the end of the 1990s due to warm winters with favorable conditions for ticks wintering in soil. Unlike previous epidemics, current outbreaks have a longer seasonal interval that is probably related to climate change and warmer winters when the ticks survive and the virus remains in their organisms for a longer time (*Medico-geographical Atlas of Russia "Natural Focal Diseases"* 2015).

Several reliable models using climate variables as drivers to predict the current and future distribution of vectors of infections such as Lyme disease, TBE, Crimean-Congo fever, dengue, and malaria clearly showed dependence on climatic characteristics (Kislov et al. 2008; Estrada-Peña et al. 2012; Caminade et al. 2014; Malkhazova and Shartova 2014; Messina et al. 2015; Nazareth et al. 2016). The results of these and other similar studies demonstrate that climate changes may often play a trigger role in the alterations of geographical ranges of vector-borne diseases.

Malaria is among the vector-borne diseases most sensitive to climate change. The global changes and their effect on malaria's geographical range have drawn the attention of many researchers (Martens et al. 1995; Githeko et al. 2000; Caminade et al. 2014; Ojeh and Aworinde 2016). Different models describing relationship between climate and disease distribution on global, regional, and local levels have been developed (Craig et al. 1999; Rogers and Randolph 2000; Lieshouta et al. 2004; Kislov et al. 2008; Parham and Michael 2010; Arab et al. 2014; Malkhazova and Shartova 2014).

In the pre-elimination era, malaria was endemic in most of Europe, including Russia. In the middle of the twentieth century, all species of malaria were eliminated, and vivax malaria was the last to disappear. Since then, short-lived episodes of autochthonous transmission following importation of *P. vivax* have been documented in a number of European countries, with Russia being the most affected. From 1997 to 2010, more than 500 autochthonous cases were recorded in European Russia. During the last quarter of the twentieth century, the favorability of weather conditions considerably improved, and receptivity of areas to malaria increased due to a more favorable combination of temperatures during summers. Since 2010, the malaria situation in Russia has improved, mostly due to the dramatic decrease in importation of the infection from Central Asian countries. However, the problem of possible reintroduction of vivax malaria in Russia is still addressed by sanitary authorities and scholars (Mironova and Beljaev 2011).

Prognosis of the Effect of Climate Change on Vivax Malaria Distribution in Russia

Considering the international experience, the present study attempts to develop a prognostic model of the epidemiology of vivax malaria within the territory of the European territory of Russia (ETR) and Western Siberia (WS) in the twenty-first century. Forecasting is based on climate modeling data within the framework of the CMIP3 (Coupled Model Intercomparison Project, phase 3).

The Method of Prognosis of Vivax Malaria Potential Spread Using Climate Modeling Data

The spread of malaria is determined by different biological and socioeconomic factors. Nevertheless, the primary factor limiting the potential geographical range of vivax malaria and its specific epidemiological features is ambient temperature (Bruce-Chwatt 1980; Lysenko and Kondrashin 1999). Thus, this model is based primarily on the analysis of temperature characteristics.

The agent of vivax malaria (*Plasmodium vivax*) was taken as an object of this research because it has the lowest temperature threshold for its development in a mosquito compared to the agents of other forms of malaria and is therefore of greatest importance for Russia.

The prognostic model was based on a postulate that there are necessary preconditions for malaria transmission both within ETR and WS. *Anopheles* mosquitoes are present in the bulk of the territories studied. *P. vivax* cases may be imported from endemic areas. The aim of the modeling was to evaluate the feasibility of vivax malaria transmission under present and forecast climatic conditions and its possible variations.

The formal territorial approach (Malkhazova and Shartova 2014) is applied to the study. The analyzed territorial unit (ATU) is a square of the degree grid on the map. The climatic data (daily average air temperatures) for ETR and WS were reanalyzed and linked to the nodes of the grid cells of 2 × 2° of latitude and longitude (Fig. 10.1).

The data on daily average air temperature were obtained for the following time series:

- Observed values for 1961–1989 interpolated for a grid cell. The period from 1961 through 1989 is characterized by modern climate and corresponds to the least time interval of 30 years that is necessary for the evaluation of climate and related changes of biotic components.
- Prognostic values for 2046–2065 on degree grid squares were derived from the results of climate modeling upon calendar-day basis for every year of this period.

To create a prognostic model (using expert evaluation) of the epidemiological features of malaria infection (Lysenko and Kondrashin 1999; Beljaev 2002; WHO 2010), the following parameters characterizing the malaria situation were selected:

- *The period of effective temperatures* – a period of a year when daily average air temperatures are permanently above +16 °C; otherwise, the development of a parasite is impossible. The term "permanently above" means that there are no

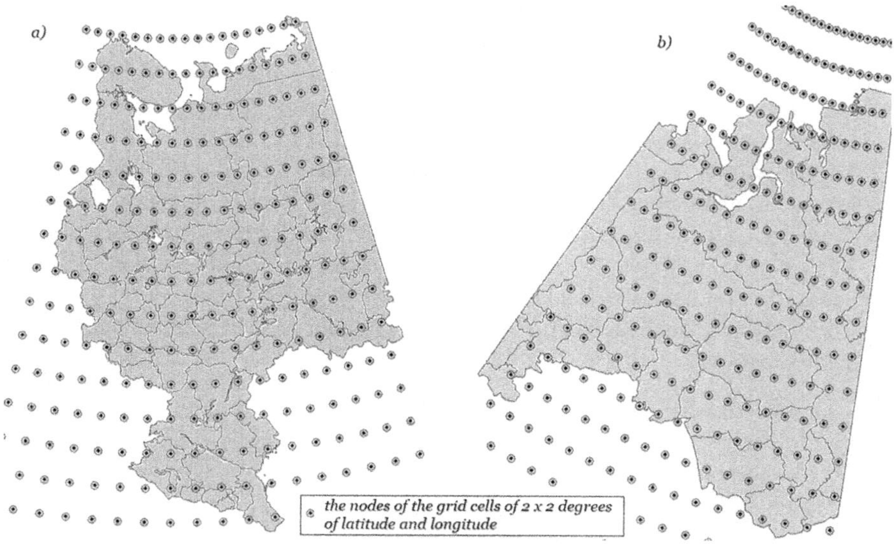

Fig. 10.1 Initial data location on the European territory of Russia (**a**) and Western Siberia (**b**)

breaks longer than 7 days when the daily average temperature falls below +16 °C.

- *The period of mosquitoes' effective infectivity* – the period during which the parasite development within a mosquito infected on a human will result in the maturation of forms capable to infect other persons.
- *Malaria transmission season* – the period during which mosquitoes with mature forms of the parasite are capable of infecting humans. The transmission season begins from the moment of the first maturing of the parasite in a mosquito, i.e., when a first infection of a human becomes possible and comes to an end with mass transition of mosquito females in the stage of diapause when they cease to consume blood and remain wintering. It is not possible to determine the exact start of wintering of mosquitoes during the whole period; therefore, for modeling purposes, the end of malaria transmission season was conditionally correlated with the end of the period of effective temperatures.
- *The number of full cycles of parasite development* characterizes the number of completed phases of development of the malaria parasite in mosquitoes and humans and indicates the degree of epidemiological risk of a territory.

The total annual sum of effective temperatures and duration of the period of effective temperatures, the beginning and the end of malaria transmission season, its duration, number of infection cycles, and other epidemiological characteristics were calculated (Malkhazova and Shartova 2014) using the S.D. Moshkovsky's method (Moshkovsky and Rashina 1951).

To determine the potential risk of a territory, the indexes of probability and intensity of infection transmission were developed.

The probability of malaria transmission of a territory exists if the minimum sum of effective temperatures (105°) could be accumulated during a year within the territory if daily average air temperatures are permanently above +16°. Parasite development in a mosquito, transmission of infection to a human, and a case of malaria become possible under such circumstances.

The intensity of malaria transmission is determined by the number of full cycles of parasite development. If more cycles become possible, the intensity of malaria transmission and consequently the epidemic risk of the territory increase.

The calculation of the abovementioned parameters was conducted separately by calendar days for each year and each ATU.

The analysis of the results allowed us to single out the following characteristics of malaria season:

- The total annual sum of effective temperatures, which makes the parasite development possible
- The duration of malaria transmission season
- The probability of malaria transmission
- The intensity of malaria transmission

Furthermore, a set of schematic maps was created using a geographic information system. They represent surface maps derived by interpolation of point data from the nodes of the data layer. This information became the basis for the analysis of possible changes in characteristics of epidemiological parameters, the probability, and intensity of malaria transmission for each of two periods under study. The results of the analysis are discussed below.

Possible Changes of Potential Spread of Vivax Malaria in Russia in the Twenty-First Century

The current climatic conditions provide quite a favorable environment for malaria parasites on the ETR. The most favorable conditions are developing in the southern part of the ETR, southward of 48°N, where the annual sum of effective temperatures equals more than 840 °C.

Within the analyzed period of 2046–2065, the northward expansion of territory with the necessary total annual sum of effective temperatures may take place. The area with unfavorable conditions for parasite development will decrease substantially. Territory with favorable conditions conversely grows considerably: up to 52°N; the sum of effective temperatures being accumulated during a year will exceed 840 °C.

The comparative cartographic analysis for WS and ETR shows that temperatures in the WS in both the modern and prognostic periods are less favorable for the development of malaria parasites. ATU with more than 840 °C are present only in the extreme southwest of this area.

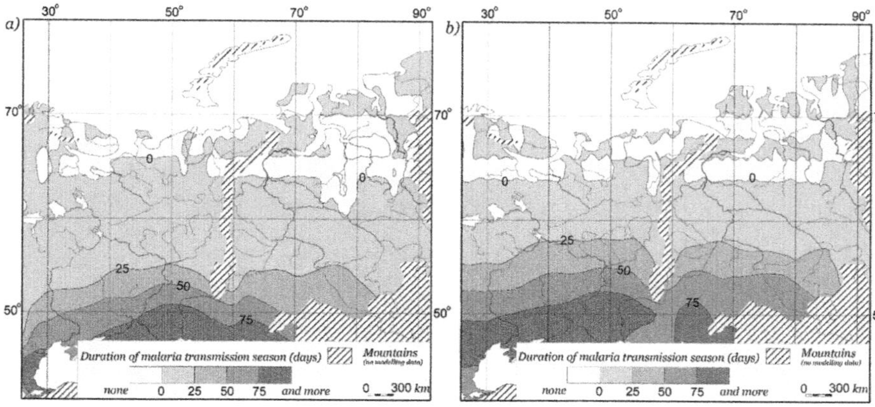

Fig. 10.2 Duration of the malaria transmission season in 1961–1989 (**a**) and 2046–2065 (**b**)

Under current climatic conditions, the duration of the malaria transmission season in the bulk of the ETR does not exceed 25 days; between 55 and 48°N, it ranges from 26 to 50 days; and south of 48°N – it exceeds 75 days (Fig. 10.2).

During the prognostic period, a northward expansion of the territory where the malaria transmission season lasts more than 75 days is taking place up to 52°N. It is important to note that the increase in the duration of the malaria transmission season is more prominent in the southern part of ETR. In the north, the territory where the malaria season does not manifest due to lack of heat will not change its margins.

In WS under current climatic conditions, the malaria transmission season does not exceed 75 days. North of 65°N, there is no malaria season due to lack of heat. In the remaining part of WS, the malaria transmission season ranges from 1 to 50 days, but as a rule, it does not exceed 25 days.

During the prognostic period, the territory without a malaria season will decrease. The borders of the regions where malaria transmission season ranges from 1 to 25 days will move northward. The territory with a malaria season of 51–75 days will shift in the southeast direction.

The pattern of changes in epidemiological parameters points to an increased risk of human infection in the future. It is most clearly shown in the changing probability of malaria transmission. The scale of the probability of malaria transmission was developed in relation to the percentage of years within the period being considered when transmission is possible. The scale is as follows: absent, transmission is impossible during the entire period; very low, during 10–30% of years; low, during 40–60% of years; medium, during 70–90% of years; and high, during all years.

Under the current climatic conditions, the high probability of malaria transmission is observed in a considerable part of the ETR, approximately up to 56°N (Fig. 10.3).

The territory with medium probability is represented by a narrow belt. The area up to 60°N ETR is characterized by a low probability of malaria transmission.

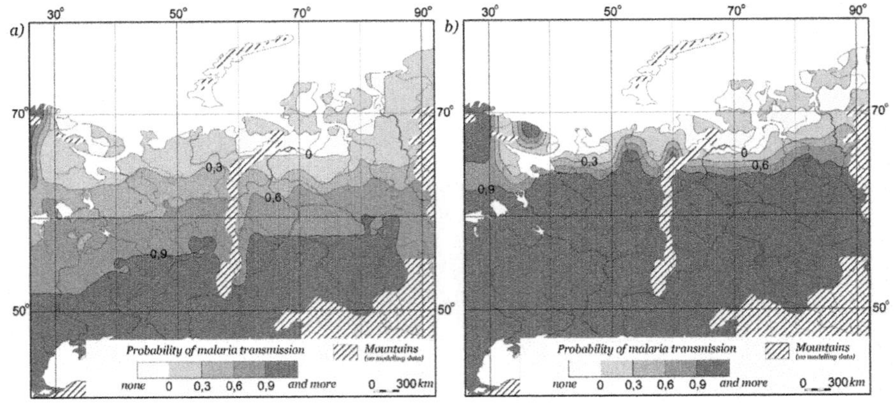

Fig. 10.3 Probability of malaria transmission in 1961–1989 (**a**) and 2046–2065 (**b**)

Further to the north, transmission is impossible in the bulk of the territory, although very small areas with very low transmission probability do exist.

During the prognostic period, almost the whole ETR up to 64°N will be characterized by high probability of malaria transmission. The area where transmission is impossible will be represented by small localities.

Under current climatic conditions, the territory of WS with a probability of malaria transmission varies considerably. In the WS area south of 60°N, it is estimated as high. When moving northward, the probability of malaria transmission decreases and is estimated as low. North of 66°N, malaria transmission is impossible.

During the prognostic period, the territory with a high probability of malaria transmission will expand northward, and the territory where malaria transmission is impossible will decrease somewhat. The area with low probability decreases by several times compared to the current climate conditions.

The annual risk of transmission and degree of disease manifestation reflects the index of intensity of infection transmission. As our analysis shows, this index demonstrates similar trends for ETR and WS (Fig. 10.4).

In general, future conditions for malaria transmission both in WS and ETR will be more favorable, and therefore the potential geographical range of vivax malaria will increase. For regions of Russia that are sensitive to environmental and climatic changes (the densely populated areas of European Russia, as well as the submontane regions of the Caucasus, the Ciscaucasia, and the Caspian Sea region), improved climatic conditions for malaria parasite development, and therefore increased malaria transmission, are forecast. However, some regions in the extreme south may become unfavorable due to excessively high temperatures and lack of breeding places for mosquitoes.

Finally, it should be noted that this work evaluates only one factor influencing malaria transmission. Malaria, as a typical anthroponosis, may be transmitted only in the presence of an infection source, e.g., a person with parasites in the blood, so

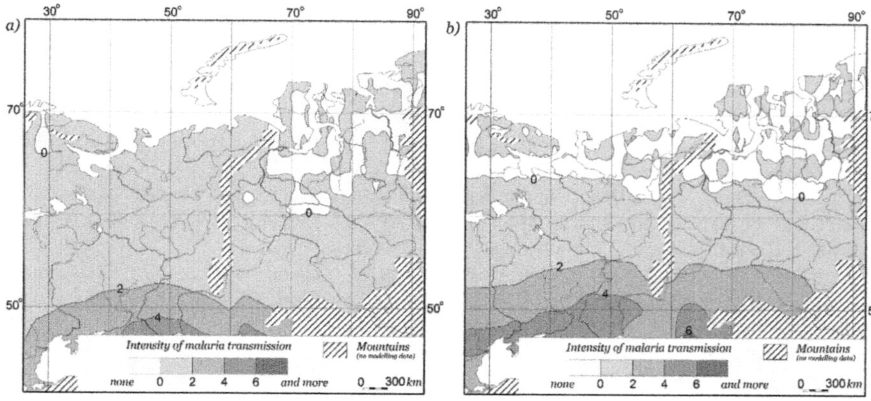

Fig. 10.4 Intensity of malaria transmission in 1961–1989 (**a**) and 2046–2065 (**b**)

while favorable climatic conditions are very important, they are not the sole precondition for malaria emergence.

Conclusions

- Global climate change may cause significant epidemiological consequences, especially in relation to vector-borne diseases.
- The analysis of the current and prognostic climatic conditions favoring vivax malaria transmission has allowed us to evaluate possible changes in potential geographical range of the infection in the ETR and WS due to climate change.
- Under current climate conditions, the ETR exhibits a better environment for malaria transmission than WS. The scenario for the mid-twenty-first century (2046–2065) suggests a similar situation.
- In the future, favorable climatic conditions for malaria transmission will increase. This will be evident in the increased sum of effective temperatures, a longer malaria transmission season, and northward extension of territory with high probability of malaria transmission. The most remarkable changes are expected in areas situated near the southern limits of the considered territories.
- Drawing prognostic medico-geographical maps facilitates spatially differentiated preventive activities focused on mitigation of the negative effects of climate change on public health. The method proposed in this work may be used as a basis for forecasting the influence of climate on the spread of other naturally determined diseases.

References

Adisheva OS, Malkhazova SM, Orlov DS (2016) Distribution of West Nile Fever in Russia//MSU Vestnik. Series 5. Geography 4:48–55

Ahern M, Kovats RS, Wilkinson P, Few R, Matthies F (2005) Global health impacts of floods: epidemiologic evidence. Epidemiol Rev 27:36–46

Andersen LK, Davis MDP (2016) Climate change and the epidemiology of selected tick-borne and mosquito-borne diseases: update from the International Society of Dermatology Climate Change Task Force. Int J Dermatol. Epub ahead of print

Aparicio-Effen M, Arana I, Aparicio J, Ramallo C, Bernal N, Ocampo M, Nagy GJ (2016) Climate change and health vulnerability in Bolivian Chaco ecosystems. Climate Change and Health, Springer International Publishing, Cham, pp 231–259

Arab A, Jackson MC, Kongoli C (2014) Modelling the effects of weather and climate on malaria distributions in West Africa. Malar J 13:12

Atlas of Health and Climate (2012) Geneva, WHO/WMO, p 26

Beljaev AE (2002) Determinants of malaria in the Middle East and North Africa. In: The contextual determinants of Malaria. RFF Press, Washington, DC, pp 137–166

Brown L, Murray V (2013) Examining the relationship between infectious diseases and flooding in Europe. Disaster Health 1(2):1–11

Bruce-Chwatt LJ (1980) Essential malariology. William Heinemann Medical Books Ltd., London. 354 pp

Caminade C, Kovats S, Rocklov J, Tompkins AM, Morse AP, Colón-González FJ, Stenlund H, Martens P, Lloyd SJ (2014) Impact of climate change on global malaria distribution. Proc Natl Acad Sci USA 111(9):3286–3291

Campbell-Lendrum D, Manga L, Bagayoko M, Sommerfeld J (2015) Climate change and vector-borne diseases: what are the implications for public health research and policy? Phil Trans R Soc B: Biol Sci 370:1665

Craig MH, Snow RW, Sueur D (1999) A climate based distribution model of malaria transmission in Sub-Saharan Africa. Parasitol Today 15:104–105

Danilov-Danilian VI (2003) Climatic changes: viewed from Russia. TEIS, Moscow. 416 p (In Russian)

Epstein PR (1999) Climate and health. Science 285:347–348

Estrada-Peña A, Ayllón N, de la Fuente J (2012) Impact of climate trends on tick-borne pathogen transmission. Front Physiol 3:64

Filho WL, Azeiteiro UM, Alves F (2016) Climate change and health. Springer International Publishing, Cham, p 532

Githeko AK, Lindsay SW, Confalonieri UE, Patz JA (2000) Climate change and vector-borne diseases: a regional analysis. Bull WHO 78(9):1136–1147

Haines A, Kovats RS, Campbell-Lendrum D, Corvalan C (2006) Climate change and human health: impacts, vulnerability and public health. Public Health 120:585–596

Harris A, Dumbuya B, Hynie M, Idahosa P, Keil R, Perkins P (2016) The social and political dimensions of the Ebola response: global inequality, climate change and infectious diseases. Climate Change Management, Springer International Publishing, Cham, pp 151–169

Jofre J, Blanch AR, Lucena F (2010) Water-borne infectious disease outbreaks associated with water scarcity and rainfall events. In: Water scarcity in the Mediterranean: perspectives under global change. Springer International Publishing, Heidelberg, p 234

Kislov AV, Yevstigneyev VM, Malkhazova SM, Sokolikhina NN, Surkova GV, Toropov TA, Chernyshev AV, Chumachenko AN (2008) Forecast of climatic resources of the East European plain under the 20st century climate warming. MAX Press, Moscow. 290 p (in Russian)

Kraemer MU, Sinka ME, Duda KA, Mylne AQ, Shearer FM, Barker CM, Moore CG, Carvalho RG, Coelho GE, Van Bortel W, Hendrickx G, Schaffner F, Elyazar IR, Teng HJ, Brady OJ, Messina JP, Pigott DM, Scott TW, Smith DL, Wint GR, Golding N, Hay SI (2015) The global distribution of the arbovirus vectors Aedes aegypti and Ae. Albopictus. elife 4:e08347

Lieshouta M, Kovats RS, Livermorec MTJ, Martens P (2004) Climate change and malaria: analysis of the SRES climate and socio-economic scenarios. Glob Environ Chang 14:87–99
Liumbruno GM, Calteri D, Petropulacos K, Mattivi A, Po C, Macini P, Tomasini I, Zucchelli P, Silvestri AR, Sambri V, Pupella S, Catalano L, Piccinini V, Calizzani G, Grazzini G (2008) The Chikungunya epidemic in Italy and its repercussion on the blood system. Blood Transfus 6 (4):199–210
Lysenko AJ, Kondrashin AV (1999) Malarialogy. Otkrytye systemy, Moscow. 248 p (in Russian)
MacKenzie WR, Hoxie NJ, Proctor ME, Gradus MS, Blair KA, Peterson DE, Kazmierczak JJ, Addiss DG, Fox KR, Rose JB (1994) A massive outbreak in Milwaukee of cryptosporidium infection transmitted through the public water supply. N Engl J Med 331:161–167
Malkhazova SM (2006) Change of the medical-geographical picture of the world. In: Modern global changes of environment, vol 2. Nauchny Mir, Moscow, pp 558–576 (in Russian)
Malkhazova SM, Shartova NV (2014) Malaria in Russia: past, present and future. In: Health and environment change in urban areas. China Environment Press, Beijing, pp 247–265
Martens WJ, Niessen LW, Rotmans J, Jetten TH, McMichael AJ (1995) Potential impact of global climate change on malaria risk. Environ Health Perspect 103(5):458–464
Medico-geographical Atlas of Russia "Natural Focal Diseases" (2015) In: Malkhazova SM (ed) Faculty of Geography, Lomonosov Moscow State University, Moscow, 208 p (in Russian)
Messina JP, Pigott DM, Golding N, Duda KA, Brownstein JS, Weiss DJ, Gibson H, Robinson TP, Gilbert M, William Wint GR, Nuttall PA, Gething PW, Myers MF, George DB, Hay SI (2015) The global distribution of Crimean-Congo hemorrhagic fever. Trans R Soc Trop Med Hyg 109 (8):503–513
Mironova VA, Beljaev AE (2011) Migrations and malaria in Russia. In: Actual questions of epidemiology of infectious diseases. Collection of scientific works, nr. 10. ZAO MP Gigiena, Moscow, p 680–690
Moshkovsky SD, Rashina MG (1951) Epidemiology and medical parasitology for entomologists. Medgiz, Moscow. 455 p (in Russian)
Nazareth T, Seixas G, Sousa CA (2016) Climate change and mosquito-borne diseases. In: Climate change and health: improving resilience and reducing risks. Springer International Publishing, Cham, pp 187–199
Ojeh VN, Aworinde SA (2016) Climate variation and challenges of human health in Nigeria: malaria in perspective. In: Climate change and health. Springer International Publishing, Cham, pp 171–185
Parham PE, Michael E (2010) Modelling climate change and malaria transmission. Adv Exp Med Biol 673:184–199
Patz JA, Githeko AK, McCarty JP, Hussein S, Confalonieri U, De Wet N (2003) Climate change and infectious diseases. In: Climate change and human health: risks and responses. World Health Organization, Geneva, pp 103–132
Recent global changes of the natural environment (2006) vol 1, 696 p., vol 2, 776 p. M.:Nauchny Mir (in Russian)
Reiter P (2001) Climate change and mosquito-borne disease. Environ Health Perspect 109:141–161
Rochlin I, Ninivaggi DV, Hutchinson ML, Farajollahi A (2013) Climate change and range expansion of the Asian Tiger Mosquito (Aedes albopictus) in Northeastern USA: implications for public health practitioners. PLoS One 8(4):e60874
Rogers DJ, Randolph SE (2000) The global spread of malaria in a future, warmer world. Science 289:1763–1765
Rosgidromet RAN, Earth Climate Change Monitoring Bulletin (2015) Moscow, 30 p (in Russian)
Roy-Dufresne E, Logan T, Simon JA, Chmur GL, Millien V (2013) Poleward expansion of the White-Footed Mouse (Peromyscus leucopus) under climate change: implications for the spread of Lyme disease. PLoS One 8(11):e80724
Semenza JC, Menne B (2009) Climate change and infectious diseases in Europe. Lancet Infect Dis 9:365–375

Vezzulli L, Grande C, Reid PC, Hélaouët P, Edwards M, Höfle MG, Brettar I, Colwell RR, Pruzzo C (2016) Climate influence on Vibrio and associated human diseases during the past half-century in the coastal North Atlantic. Proc Natl Acad Sci USA 113(34):E5062–E5071
WHO (2010) Practical guidelines on malaria elimination in countries of the WHO European Region. WHO, Geneva. 109 p (in Russian)
Wu X, Lu Y, Zhou S, Chen L, Xu B (2016) Impact of climate change on human infectious diseases: empirical evidence and human adaptation. Environ Int 86:14–23
Zell R (2004) Global climate change and the emergence/re-emergence of infectious diseases. Int J Med Microbiol 293(37):16–26
Zitek K, Benes C (2005) Longitudinal epidemiology of leptospirosis in the Czech Republic. Epidemiol Mikrobiol Imunol 54(1):21–26

Svetlana M. Malkhazova has a doctoral degree in geographical sciences. She is professor and head of the Department of Biogeography, Faculty of Geography, Lomonosov Moscow State University. Her main research interests relate to the problems of human ecology and medical geography. She is the author of more than 250 scientific publications, including ten books, several textbooks and medical and environmental atlases.

Natalia V. Shartova is PhD in geography; she is senior research fellow at the Department of Landscape Geochemistry and Soil Geography, Faculty of Geography, Lomonosov Moscow State University. Her main research interests are in medical geography and human ecology, particularly in relation to the problems of urban ecology and urban population health. She is the author of 35 scientific publications.

Varvara A. Mironova has a PhD in geography. She is senior research fellow at the Department of Biogeography, Faculty of Geography, Lomonosov Moscow State University. Her main research interests include medical geography, especially problems related to nosogeography of natural focal and natural endemic diseases, medico-geographical mapping and ecological and evolutional parasitology. She is the author of 35 scientific publications.

Chapter 11
Climate Change and Projections of Temperature-Related Mortality

Dmitry Shaposhnikov and Boris Revich

Abstract The impacts of increasing year-round temperatures on mortality from all non-accidental, all cardiovascular, and all non-cardiovascular causes were examined in the city of Akchangelsk in Russian North, where the climate change signal is expected to be stronger than the global average. Projections of future daily temperatures were made for IPCC B2, A1B, and A2 greenhouse gas emission scenarios using regional downscaling of the selected ensemble of 16 general circulation models. The distributed lag nonlinear models were used to estimate 30-day cumulative risks of the exposure to heat and cold. The projected changes in annual fractions of deaths attributed to nonoptimal temperatures are negative and not significant at 95% confidence level for all categories of mortality and emission pathways included in the study. The benefits of reduced cold-related mortality will most likely outweigh the negative impacts of higher heat-related mortality during the projection period 2045–2056. However, this situation may be reversed in the longer run.

Keywords Climate projections • Population health • Distributed lag nonlinear models • Nonoptimal temperatures • Attributable fractions and attributable numbers of deaths

Introduction

IPCC's Fifth Assessment report concluded that anthropogenic greenhouse gas emissions and other anthropogenic drivers "are *extremely likely* to have been the dominant cause of the observed warming since the mid-20th century" and warming will continue into the future (IPCC 2013). Dynamic downscaling of global atmosphere-ocean coupled general circulation models (AOGCM) to the regional level showed that circumpolar regions would experience greater climatic changes

---

D. Shaposhnikov (✉) • B. Revich
Laboratory of Forecasting of Environmental Quality and Human Health, Institute of Economic Forecasting of Russian Academy of Sciences, Nakhimovsky Prospect 47, Moscow 117418, Russia
e-mail: dshap2014@gmail.com

© Springer International Publishing AG 2018

R. Akhtar, C. Palagiano (eds.), *Climate Change and Air Pollution*, Springer Climate, DOI 10.1007/978-3-319-61346-8_11

than global averages. Due to several positive feedback mechanisms (most importantly, to changes in surface reflectivity caused by melting of such perfect reflectants as snow and ice), it appears that climate change in the Arctic is more rapid than elsewhere (ACIA 2005). Even under the most optimistic new Representative Concentration Pathway emissions scenario (RCP2.6), average surface temperatures in the Russian Arctic will increase by 3–4 °C by the 2080s compared to the 1990s (see Figure SPM.8a in IPCC 2013). An increasing interest of global warming and public health researchers in the Russian Arctic guided our choice of Arkhangelsk, Russian Federation, as the pilot region for this study. This region is characterized by very fragile ecosystems, which, coupled with local social and economic problems, leads to its particular vulnerability to both direct and indirect impacts of global warming, such as infectious diseases (Grjibovski et al. 2012; Tokarevich et al. 2011).

One of the most direct health impacts of climate change relates to changes in annual mortality rates caused by exposure to ambient temperatures. In this context, Arkhangelsk, being one of the largest cities in Russian North, is particularly interesting because of very large seasonal variations of daily mean temperatures: from −36 °C in January to 26 °C in July. Hence, the specific aim of this chapter is to answer the question: will climate change induce any significant changes in population attributable fractions (AFs or PAFs) of deaths experienced due to annual exposition to nonoptimal temperatures? To provide an informed answer, the authors used current and future distributions of daily temperatures and estimated associated changes in cold- and heat-related attributable fractions (AF_{cold} and AF_{heat}).

Making projections inevitably involves a lot of explicit and implicit assumptions about possible futures. Our intention was to minimize the number of such assumptions, especially when it comes to adaptation to future public health hazards. Making projections, we relied on the three IPCC Special Report Emission Scenarios and tried to follow "all other things being equal" principle in all subsequent calculations. To reduce uncertainty in climate projections, we used the 2050s instead of the 2080s in our projections, attributable fractions instead of attributed numbers of deaths, and chose "no acclimatization no adaptation" scenario. In other words, we assumed that the "historic" temperature-mortality relationship estimated for 1999–2010 would not change until the 2050s, which implied that the minimum mortality temperature (MMT) would not change. This may be questionable, as other authors speculated that the adaptation would likely lead to gradual increase in the MMT with time.

The physiologic mechanisms of cold-related deaths mainly involve cardiorespiratory pathways. Cardiovascular deaths also make a major part of all excess non-accidental deaths caused by exposition to heat. However, it should be noted that many other than cardiovascular causes also showed significant increases in death rates during extreme cold and heat (Shaposhnikov and Revich 2016; Shaposhnikov et al. 2014). Along with all non-accidental (natural) deaths, we studied all cardiovascular deaths as a more sensitive subgroup, where we expected to observe the greatest impacts of changing temperatures, and all non-cardiovascular deaths, as the complimentary subgroup.

Methods

Data and Climate Projections

Climate simulations for this project were performed in 2011, based on methodology of IPCC AR4, using 1980–1999 as the baseline period for climate simulations, and the Special Report on Emissions Scenarios (SRES). Daily mortality data for the city of Arkhangelsk was available from Russian Federal Statistical Service (*Rosstat*) for the period 1999–2010, and the basic temperature-mortality relationship was derived for this period. This relationship reflects adaptation of population to local climate, and cannot change noticeably over few years, unless there are massive migrations. Therefore, we considered the period 1980–1999 as the baseline for subsequent projections of changes in temperature-dependent mortality. Daily mean temperatures for daily mortality modeling were calculated from 3-h temperatures recorded in Arkhangelsk and available from the website of Russian Institute of Hydrometeorology Information http://aisori.meteo.ru/ClimateR.

Dynamic downscaling of the ensemble of 16 comparable global AOGCMs to obtain monthly average temperature anomalies (scenario-based departures from the baseline values) in Arkhangelsk Region for the projection period 2041–2060 was performed in the Voeikov Main Geophysical Observatory, St. Petersburg, Russian Federation, by the workgroup formed by the WHO project "Climate change health impact and adaptation assessment for the north of the Russian Federation" (see Acknowledgments). Such global models simultaneously simulate the Earth's atmosphere and oceans, land, and sea ice.

A medium-range climate projection period was preferred because climate models already showed significant climate change signal, while health projections avoided unwarranted assumptions about far more distant futures. The baseline period 1980–1999 was compared with 2040–2059 projection period, and the monthly temperature anomalies were calculated as the respective 20-year averages. In climate simulations, "temperature anomaly" means the estimated difference between the future and the baseline temperature values. It is averaged across model runs for each model and across the outputs of different models included in the ensemble. The confidence intervals around temperature anomalies are partly attributed to intra-model and partly to inter-model differences.

Although the latest IPCC Assessment Report AR5 introduced new emission scenarios called Representative Concentration Pathways, they have mostly inherited the assumptions built in the "old" AR4 SRES scenarios. This project made use of SRES scenario "families" B1, A1B, and A2 (Nakicenovic et al. 2000). Although B1 is often regarded as "low emission," A1B as "medium emission," and A2 as "high emission" scenarios, the difference between A1B and A2 scenarios in terms of the associated increments of global surface temperatures will remain negligible until the 2050s, as our climate simulations confirmed. The IPCC workgroup did not attach probabilities to each particular SRES scenario, preferring to treat them as equally sound "possible futures."

Calculation of Fractions of Deaths Attributable to Temperatures

The attributable fraction AF_x and attributable number AF_x for a given exposure x can be provided by

$$AF_x = 1 - \exp(-\beta_x); AN_x = n \times AF_x \tag{11.1}$$

where β_x represents the risk associated with the exposure and n is total number of exposed cases. The coefficient β_x usually corresponds to the logarithm of a ratio measure such as relative risk (so-called log-relative risk), relative rate, or odds ratio. It is generally obtained from Poisson regression models which explain exponential relationship (11.1) while adjusting for potential confounders. Poisson models, in turn, are used in time series analysis of the dependent variables – outcomes per unit of time which follow an (overdispersed) Poisson distribution.

With temperature as an exposure, additional complexity rises from *lagged* effects of the exposure, when the effect is distributed over certain period of time after the exposure occurs. Another important phenomenon associated with acute exposure to temperature is short-term harvesting, when the additional deaths caused by the exposure deplete the pool of susceptible subpopulation, so that noticeable reduction in deaths follows the exposure a few days later. Although medium- and long-term harvesting were observed after unusually strong and long-lasting heat waves, such events are very rare and are not likely to happen within any given 20-year projection period (Shaposhnikov et al. 2015). In most studies of short-term harvesting, 30-day or even 21-day follow-up period was considered enough to capture the cumulative effect of an acute exposure to both heat and cold (Gasparrini et al. 2010, p. 2229; 2015, p. 370). To estimate the total burden of additional deaths, associated with the exposure, we had to account for the lagged effects and for the short-term harvesting – i.e., exclude the deaths which were forward-displaced by only a few days. With this purpose, we defined the overall relative risk, accumulated within L days after the exposure to temperature T_i on day i as

$$RR_{overall} = \frac{\sum_{l=0}^{L} M_{l+i}}{(L+1)MM} \tag{11.2}$$

where M_i is mortality actually observed on day i and MM denotes minimum mortality, the reference value against which the relative risk is calculated. The minimum mortality corresponds to optimal temperature (minimum mortality temperature MMT), at which the estimated value of overall relative risk $\widehat{RR}$ reaches zero. Both values MMT and $\widehat{RR}$ were estimated from a distributed lag nonlinear model (*dlnm*) as described below.

Note that the attributable number of deaths $AN(T_i)$ is defined similarly to (11.2) as average daily mortality within L days from the exposure, in excess of minimum mortality:

$$AN(T_i) = \frac{\sum_{l=0}^{L} (M_{l+i} - MM)}{(L+1)} = (RR_{\text{overall}} - 1)MM \tag{11.3}$$

It is estimated from the regression model for each day i of the time series and then summed up across all days in the study period to arrive at $\widehat{AN}_{tot}$, which can be further subdivided into the partial sums across the subsets of all days with temperatures T below the optimal temperature and above the optimal temperature. These partial sums are interpreted as the numbers of additional deaths attributed to cold and heat. Then, the total attributable fraction $\widehat{AF}_{tot}$ is defined as the ratio of total attributable number of deaths $\widehat{AN}_{tot}$ with total mortality M_{tot} during the study period, and attributable to cold and heat fractions are defined similarly:

$$\widehat{AF}_{cold} = \frac{\widehat{AN}_{cold}}{M_{tot}}; \widehat{AF}_{heat} = \frac{\widehat{AN}_{heat}}{M_{tot}} \tag{11.4}$$

Thus, estimation of $\widehat{RR}(T)$ becomes the essential first step in all calculations. Note that the averaging of $AF(T_i)$ across all cold days will not produce $\widehat{AF}_{cold}$ and the averaging of $AF(T_i)$ across all hot days will not produce $\widehat{AF}_{heat}$ defined by (11.4). We calculated attributable fractions instead of attributable numbers to avoid making assumptions about future population growth and changing proportions among the cause-specific mortality rates. The R function attrdl.R was written by A. Gasparrini to calculate the attributable risks after a *dlnm* model. This function is now available in electronic supplement to (Gasparrini and Leone 2014) and works with the R package *dlnm* 2.2.0 or higher. We used this function to calculate "forward" attributable fractions and empirical confidence intervals around them, using Monte Carlo simulation and assuming normal distributions of *dlnm* model coefficients. Forward perspective interprets $AF(T_i)$ as the future burden associated with the current exposure to temperature T_i on day i and accumulated during the following L days after the exposure. Contrariwise, backward interpretation of AF calculates the contributions of L past exposures to the current risk observed on day i.

Estimation of Overall Relative Risks

Overall (cumulative) relative risk $\widehat{RR}(T)$ is calculated by summing up the contributions of the effects (log-relative risks) of temperature for lags 0, 1, ... L up to the maximum lag considered in the model. Due to nonlinear nature of temperature-mortality relationships at each lag, there was a need to develop a family of models which can simultaneously describe the effects that smoothly change along the dimension of temperature and the dimension of lags, measured in days after the exposure event. One solution was proposed by Gasparrini et al. (2010), who constructed two-dimensional "cross basis" functions within a standard generalized linear model (*glm*) framework. The algorithm, which resembles smoothing on a multidimensional grid, was implemented in an R package *dlnm* (Gasparrini 2011) which is now publicly available on the R comprehensive archive network (CRAN). We performed all calculations in R 3.3.2 statistical package (R Core Team 2016). To describe the relationship in the space of temperature, we used natural cubic spline with three internal knots placed at the 10th, 75th, and 90th centiles of year-round distribution of daily mean temperatures, which corresponded to −11 °C, 12 °C, and 16 °C in Arkhangelsk. The asymmetric choice of the middle knot (the 75th percentile) was dictated by asymmetrical shape of the underlying temperature-mortality relationship. The shape of lag-mortality relationship varied smoothly within the lag period between 0 and 30 days and described the dynamics of mortality response after exposition to a given temperature. The lagged log-relative risks $\beta_0(T)$, $\beta_1(T)$, ..., $\beta_{30}(T)$ at each temperature were fitted with a natural cubical spline of lag variable l with an intercept, with three internal knots spaced equally in the log scale: at days 1, 4, and 12. The Poisson model of daily mortality was adjusted for seasonal and long-term trend, with natural cubic splines of time with seven degrees of freedom (df) per year, and for day of week, using seven categorical variables. We did not adjust the temperature-mortality relationship for relative humidity or dew point temperature, because we had no reason to assume that the distribution of these variables would remain unchanged until the 2050s.

In the result, the complex nonlinear lagged temperature-mortality dependency was decomposed with the two-dimensional basis along the dimensions of temperature and lags. Estimated model coefficients were used to calculate overall relative risk of mortality $\widehat{RR}(T)$ accumulated over 30 days after the exposure.

Assessing the Risks of Climate Change

The arguments of attrdl.R function include the vector of exposures, the cross basis used for fitting a *dlnm* model; the vector of outcomes, the *dlnm* model (with a log link function) used for calculation of lagged risks of the exposure; and other parameters. In our study setting, the exposures were daily temperatures $\vec{T}$, and the outcomes were corresponding daily mortality counts $\vec{M}$. Here we use an arrow symbol as vector notation, meaning the complete and ordered time series of daily observations T_i and M_i. Note that the model itself depends upon the vector of exposures $\vec{T}$. Using these notations, the equation for calculation of baseline attributable fraction will look like this:

$$AF_b = \text{attrdl}(\vec{T}, \text{crossbasis}, \vec{M}, \text{model}\,(\vec{T}, \ldots)) \tag{11.5}$$

where "..." denotes other parameters included in the model. Projecting future attributable fractions AF_f, we assumed that overall relative risks of exposure to a given temperature $\widehat{RR}(T)$ should not change; only the distribution of daily temperatures shifts from the baseline $\vec{T}$ to the future $\vec{T}_f$. Thus, substituting $\vec{T}_f$ for $\vec{T}$ in the exposure vector (the first argument in (11.5)), and keeping all arguments used to calculate $\widehat{RR}(T)$ unchanged, one can estimate AF_f as

$$AF_f = \text{attrdl}\left(\vec{T}_f, \text{crossbasis}, \vec{M}, \text{model}(\vec{T}, \ldots)\right) \tag{11.6}$$

This is convenient, because AF_f under alternative scenarios of future temperatures can be calculated from the same baseline *dlnm* model. The distribution of future daily temperatures $\vec{T}_f$ can be calculated using daily temperature anomalies $\Delta\vec{T}$, generated by the output of climate simulations:

$$\vec{T}_f = \vec{T} + \Delta\vec{T}$$

We approximated daily temperature anomalies by the set of 12 monthly temperature anomalies $\Delta_{\text{monthly}}(j)$, $j = 1, 2, \ldots, 12$. These values are predicted with greater precision while retaining enough seasonal differentiation of the climate change signal. The future period for mortality projections had the same length as the baseline period for calculation of $\widehat{RR}(T)$ and AF_f and was centered at year 2050 to best fit the climate projections. For each calendar date within the 12-year future period 2045–2056, we used the daily temperature observed on the respective calendar date during the baseline period 1999–2010 *plus* the monthly temperature anomaly for the respective month. Thus, for each day i of the comparable future period, the distribution of future temperatures was modeled as follows:

$$T_f(i) = T(i) + \Delta_{\text{monthly}}(j)$$

Results

The city of Arkhangelsk, population 369,000 (1999), is one of the largest in Russian North. It is situated near the coast of the White Sea, about 220 km south from the Arctic Circle. The mean temperature of January, the coldest month, was −13 °C, and the mean temperature of July, the hottest month, was 16 °C during the study period 1999–2010. The mean daily mortality from all non-accidental causes was 11.9 cases, of which about 55% were cardiovascular deaths. Table 11.1 lists monthly temperature anomalies, calculated as the differences between the average values for the future period 2041–2060 and the baseline period 1980–1999. The differences between scenarios are explained by differing assumptions about future emissions of greenhouse gases and aerosols, population and economic growth,

Table 11.1 Monthly temperature anomalies (°C) and their standard deviations, projected by the 2050s and calculated from regional downscaling of the predictions from the ensemble of 16 general circulation models

SRES scenario	B1	A1B	A2
January	3.1 ± 1.8	4.3 ± 2.0	3.9 ± 1.4
February	3.4 ± 1.8	4.2 ± 1.7	4.3 ± 1.0
March	2.4 ± 1.6	3.2 ± 1.2	2.7 ± 1.1
April	2.0 ± 1.2	2.5 ± 1.4	2.2 ± 1.5
May	2.0 ± 1.2	2.3 ± 1.3	2.2 ± 1.5
June	1.7 ± 0.6	2.3 ± 1.1	2.0 ± 1.0
July	1.5 ± 1.0	2.0 ± 1.0	2.0 ± 1.2
August	1.6 ± 1.1	2.1 ± 1.3	1.9 ± 1.3
September	1.7 ± 1.1	2.1 ± 1.0	2.1 ± 1.1
October	1.9 ± 0.9	2.3 ± 1.2	2.1 ± 1.1
November	3.2 ± 1.2	4.3 ± 1.6	3.6 ± 1.6
December	3.7 ± 1.7	4.7 ± 1.6	4.7 ± 1.7

Standard errors are mostly attributed to inter-model differences

technological development, and adaptation strategies, as described in (Nakicenovic et al. 2000).

Greater warming in winter than in summer is clearly seen from Table 11.1, while the differences between scenarios are relatively small. The projected temperature changes are the most pronounced in scenario A1B; for this reason we will call it "pessimistic scenario" in this chapter.

Figure 11.1 shows 30-day cumulative effect following an acute exposure to a given temperature within the range of temperatures observed in Arkhangelsk during the study period. The effect is measured as an increase in mortality relative to its minimum, observed at the mean daily temperature of 15.5 °C. To characterize the extent of variation in temperature-dependent mortality, we will report 30-day cumulative RRs for all non-accidental deaths, following the exposure to extreme cold and extreme heat thresholds defined as the 2.5th and the 97.5th percentiles of location-specific year-round distribution of daily mean temperatures, following (Gasparrini et al. 2015). Extreme cold: $T = -23$ °C; $\widehat{RR} = 1.24$, 95% CI [1.16, 1.33] and extreme heat: $T = 21$ °C; $\widehat{RR} = 1.18$, [1.13, 1.23]. As expected, cardiovascular deaths showed greater relative increases during cold and heat than all non-accidental deaths, while non-cardiovascular deaths showed smaller (but still statistically significant) increases. To demonstrate this result clearly, we used the same scale of the vertical (Y) axis in all graphs.

Now we are ready to calculate attributable fractions. However, before we get to actual SRES scenarios, it is instructive to consider two hypothetical and very simple simulations of future temperatures. This example will help the reader to understand the principal behavior of attributable fractions under changing climate.

- Scenario 1: all daily temperatures will rise by 2 °C; $T_f(i) = T(i) + 2$.
- Scenario 2: all daily temperatures will rise by 4 °C; $T_f(i) = T(i) + 4$ for all days i.

The resultant changes in future attributable fractions are summarized in Table 11.2.

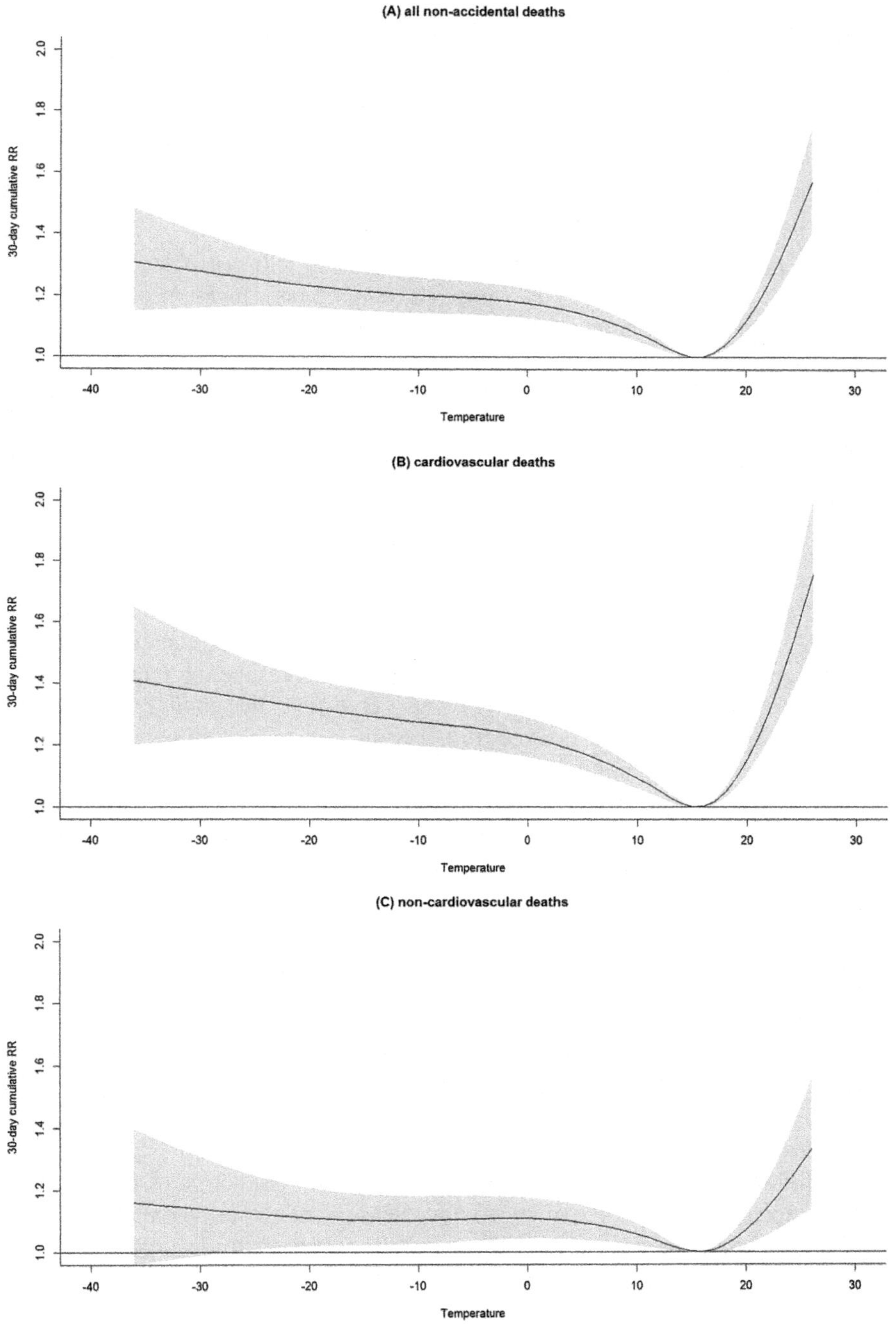

Fig. 11.1 Overall relative risks relative to MMT=15.5 °C in Archangelsk

Table 11.2 Attributable fractions of all non-accidental deaths under the baseline and the two hypothetical scenarios

AF, %	Baseline	$T_f = T + 2$	$T_f = T + 4$	$AF(\Delta T)$
Cold	10.9	9.8	8.8	(Almost) linear
Heat	0.59	1.4	2.7	Convex
Total	11.5	11.2	11.5	Non-monotonous

This table shows that AF_{cold} behaves nearly linearly with respect to temperature increments. In contrast, AF_{heat} is a pronouncedly convex function of ΔT. In the result AF_{tot} shows non-monotonous behavior with respect to ΔT. As ΔT gradually increases from zero, AF_{tot} first falls down and then goes up, reaching the baseline value again at $\Delta T = 4$ °C. One may conclude that an increase in heat-related mortality becomes greater than the reduction in cold-related mortality at $\Delta T > 4$ °C (for constant population age structure, observing "other things being equal" principle). Luckily for Arkhangelsk, the projected by the 2050s temperature increments during summer months will remain well below 4 °C (Table 11.1).

Now, let us look at our hypothetical Scenarios 1 and 2 from a different perspective. One can interpret Scenario 1 as the *mean* estimate of the projected warming, with 95% confidence interval given by the baseline scenario and Scenario 2: $\Delta T = 2$ °C, 95% CI [0°, 4°]. This seems plausible after examining standard deviations in Table 11.1, where $sd \approx 1.0$, at least for summer months. If AF_b were determined with infinite (or very high) precision, the confidence intervals around AF_f would have to be derived from the standard error of climate projections. In this case, empirical confidence interval is given by the 2.5th and 97.5th percentiles of the distribution of attributable fractions generated by Monte Carlo simulations based on a normal distribution of ΔT around the mean of 2.0 with $sd = 1.0$. We calculated these in R using 1000 simulations of

$$AF_f = \text{attrdl}\left(\vec{T} + \text{rnorm}(\text{mean} = 2.0, \text{sd} = 1.0), \text{crossbasis}, \vec{M}, \text{model}\left(\vec{T}, \ldots\right)\right)$$

where rnorm denotes a normally distributed random variable. As it turns out, the confidence interval is asymmetrical: $AF_f = 11.27$ [11.21, 11.52]. In reality, however, the AF_b estimate was not very precise. Its empirical 95% confidence interval was calculated from the simulation samples based on *dlnm* model and returned by attrdl.R function: $AF_b = 11.5$ [8.9, 14.0]. Note that the latter interval is 16 times wider than the former.

From this worked-out example, we learned that, for prediction of future health impacts, the dominant source of uncertainty stems from natural variability of daily deaths during the baseline period. Wu et al. (2014) arrived at the same conclusion after decomposition of total variance of their estimates of future heat wave mortality in Eastern US into partial variances attributed to various sources of uncertainty. Benmarhnia and coauthors (2014) also concluded that most of variability in their future mortality projections for Montreal, Canada, was related to the temperature-mortality RR, not to variability in simulations of future temperatures.

In our study setting, the relative input of uncertainty of climate projections was at least an order of magnitude smaller than the relative input of uncertainty in the baseline RR estimates.

Now, let us turn to SRES scenario-based projections. In light of the uncertainties discussed above, the inter-scenario differences are relatively small. Figure 11.2 shows the baseline attributable fractions and the projected values under the three SRES scenarios for all non-accidental deaths and cardiovascular and non-cardiovascular deaths in Arkhangelsk, with 95% confidence bands. All fractions attributed to cold and heat are statistically significant, as seen in Fig. 11.2. As expected, the deaths from cardiovascular causes are more sensitive to nonoptimal

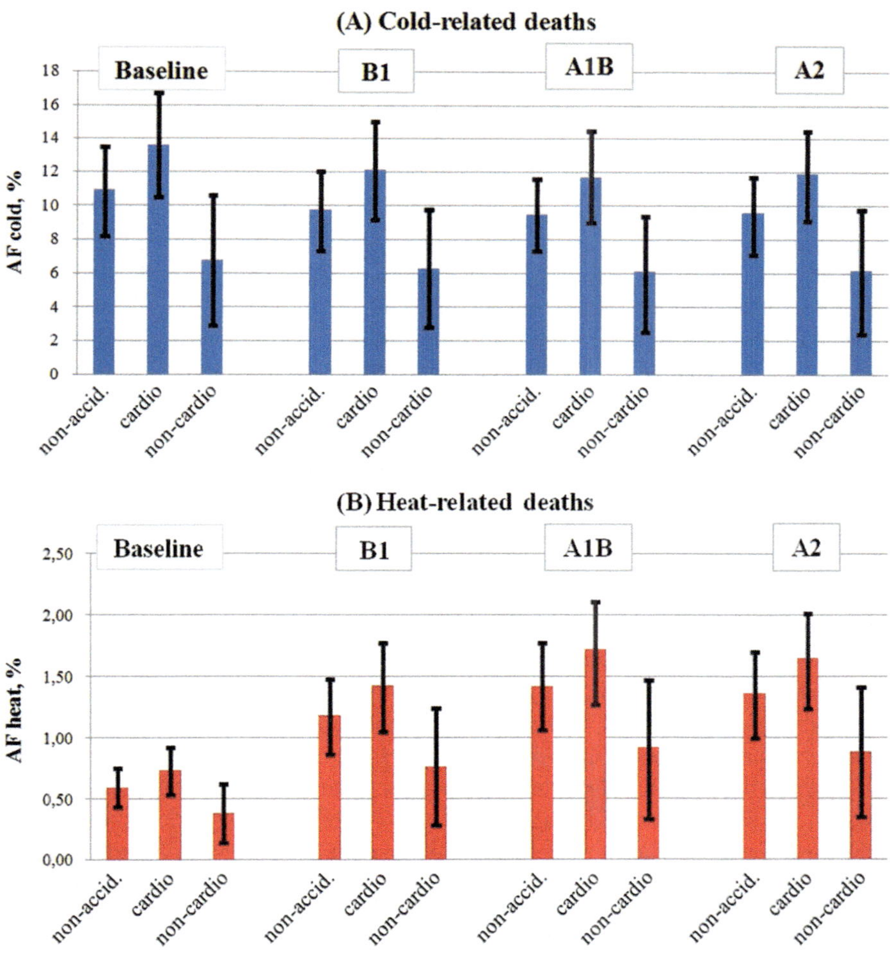

Fig. 11.2 Fractions of deaths attributed to cold (**a**) and heat (**b**) in Archangelsk under the baseline (1980–1999) scenario and the three SRES scenarios, projections for 2045–2056. Error bars show empirical 95% confidence intervals by simulating from the assumed normal distribution of the estimated *dlnm* model coefficients

Table 11.3 Projected changes by the 2050s in the fractions of deaths attributable to cold, heat, and all nonoptimal temperatures, with standard deviations

		SRES emission scenarios		
Cause of death	Temperature range	B1	A1B	A2
All non-accidental	Cold	−1.1 ± 1.8	−1.4 ± 1.7	−1.3 ± 1.8
	Heat	0.6 ± 0.2*	0.8 ± 0.2*	0.8 ± 0.2*
	Total	−0.6 ± 1.7	−0.6 ± 1.7	−0.6 ± 1.8
Cardiovascular	Cold	−1.5 ± 2.1	−1.9 ± 2.1	−1.7 ± 2.1
	Heat	0.7 ± 0.2*	1.0 ± 0.2*	0.9 ± 0.2*
	Total	−0.7 ± 2.1	−0.9 ± 2.1	−0.8 ± 2.1
Non-cardiovascular	Cold	−0.5 ± 2.6	−0.7 ± 2.6	−0.6 ± 2.7
	Heat	0.4 ± 0.3	0.5 ± 0.3	0.5 ± 0.3
	Total	−0.2 ± 2.7	−0.2 ± 2.6	−0.2 ± 2.6

Attributable fractions are measured as percentages of annual mortality from the indicated cause of death

*Statistically significant at 0.05 level

temperatures, than all non-accidental deaths, while the deaths from all non-cardiovascular causes are less sensitive. Perhaps the most important conclusion from Fig. 11.2 is the following: while the fractions AF_{cold} are greater than AF_{heat} by an order of magnitude, the future changes in AF_{cold} and AF_{heat} are oppositely directed and comparable in their absolute values. For example, the difference between AF_{cold} and AF_{heat} for all non-accidental deaths and cardiovascular deaths is almost 20-fold. The future change in AF_{cold} for non-accidental deaths under the "pessimistic" A1B scenario is −1.4%, while the difference in AF_{heat} is 0.8%. One may see that the absolute values of these changes are close so that the net change is only −0.6%.

Table 11.3 summarizes the changes in attributable fractions of deaths between the baseline period of mortality projections 1999–2010 and the comparable future period 2045–2056 under the three SRES scenarios. The differences $AF_f - AF_b$ are expressed as percentages of total mortality from the indicated cause of death during the respective period, according to Eq. 11.4.

Table 11.3 shows that the projected net changes in temperature-induced mortality rates are negative for all scenarios and all causes of death included in the analysis. The error bands around the projected changes in AF_{cold} and AF_{tot} are always much wider than their absolute values, rendering them statistically insignificant, while the projected changes in AF_{heat} can be highly significant (except for non-cardiovascular deaths). The heterogeneity among the scenario-based estimates of future changes in attributable fractions is negligibly small. Even for cardiovascular deaths, being the most temperature-sensitive subgroup, the net change in AF_{tot} is −0.7% under the "optimistic" B1 scenario and −0.9% under the "pessimistic" A1B scenario, so that the difference between the scenarios is only 0.2%. One may conclude that the divergence of estimates of attributable fractions among the alternative emission pathways will stay below the associated projection errors.

However, it is very likely that the global warming scenarios will diverge in more distant future, and by the 2080s, the heat-related increment will outweigh the cold-related decrement in deaths.

Discussion

To our best knowledge, this is the first study which implemented distributed lag nonlinear models for assessment of future impacts of global warming on mortality rates. The authors measured the impacts of climate change on mortality by the changes in attributable fractions of deaths. These fractions were calculated separately for cold and heat; the sum of these gives the fraction of deaths attributed to nonoptimal temperatures. The reference value in the applied AF measure corresponds to an imaginary situation when all days of the study period have the optimal temperature. Thus, the existence of such temperature becomes an essential prerequisite, and the effect measure is based on a "counterfactual condition" meaning that the reference state never actually occurred.

All nonoptimal temperatures will cause excess deaths. For example, all days with $T < 15.5$ °C will produce cold-related deaths in Arkhangelsk, even though these days cannot be considered "cold" in the ordinary sense of the word. As the average temperatures of June and August in Arkhangelsk are close to 13 °C, most summer days will contribute to cold-related deaths. Surely, some of cold-related and heat-related deaths can be avoided, but the extent to which temperature-related deaths can be prevented is not discussed here.

It is important to note that the attributable fractions are calculated by dividing the attributable numbers by total mortality M_{tot} (Eq. 11.4). Therefore, the projected *changes* in the attributable fractions will always have M_{tot} in the denominator. For this reason, these changes seem to be fairly small, and surely not as impressive as the results reported in many other studies of anticipated future burdens of global warming. An informative synthesis of such results can be found in a systematic review by Huang et al. (2011). The reason is that other studies used different metric: they usually reported the projections of future heat-related mortality, which could increase by several times compared to the current heat-related mortality. In other words, these studies used different denominator, i.e., the baseline AN_{heat}. Our results may be easily recalculated in this way. For example, an increase in AF_{heat} for non-accidental deaths from 0.59% in the baseline to 1.42% in A1B in Table 11.2 means more than twofold increase in heat-related deaths, which corresponds to the conclusion of Cheng et al. (2009), who projected that heat-related mortality in four Canadian cities would more than double by the 2050s. Of course, such recalculation cannot change the main conclusion of this paper: the projected net decrease in *total* temperature-related deaths.

In most international studies of future impacts of global warming on public health, heat waves have gained much focus and attention. In this study, we purposefully left heat waves and cold spells out of the equation, because the relative inputs of heat waves and cold spells in total temperature-related mortality are negligibly small. Gasparrini and Armstrong (2011) distinguished between the main effect of temperature on mortality during heat waves and the added effect. The main effect was attributed to independent effects of daily temperatures, while the added effect (the wave effect) was attributed to the duration of heat for several consecutive days. According to their estimates, added effect arises in heat waves lasting for more than 4 days and peaks at around 7 consecutive days of heat, but its contribution to total effect of heat is substantially smaller than that of the main effect. Under the widely accepted definition of heat waves as ≥4 days of continuous temperatures above the 97th percentile of year-round site-specific temperature distribution, the main effect is eight times greater than the added effect. Hajat et al. (2014, p. 643) estimated an increase in heat-related mortality in the UK by the 2050s as +257% relative to the baseline (1990s) heat-related mortality, while the change in the added (heat wave) effect in London was only 28% of the baseline heat-related mortality, which is an order of magnitude smaller. In our previous research in Arkhangelsk, we also estimated the contribution of added effect of heat waves and cold spells in total change in heat-related and cold-related deaths due to climate change and concluded that the relative contribution of added effect was several times smaller than that of the main effect (Shaposhnikov et al. 2011, p. 82).

Many literature sources emphasized that the elderly were the most susceptible subpopulation in terms of temperature-dependent mortality (Gosling et al. 2009; Kinney et al. 2008). We modeled 30-day cumulative risks of acute exposure to ambient temperatures for the subgroup over 60 years of age. The baseline estimates of fractions of non-accidental mortality attributed to cold, heat, and all nonoptimal temperatures for ≥60 years age group were $AF_{cold} = 10.2\%$, $AF_{heat} = 0.64\%$, and $AF_{tot} = 10.9\%$. The reader may compare these with the baseline AFs reported in the second column of Table 11.2. Because the modeling results did not indicate any steeper increases in relative risks for the elderly compared to all ages, we chose not to report the projections for this age group.

In conclusion, our study projected a larger reduction in cold-related deaths compared with the increase in heat-related deaths by the 2050s in Arkhangelsk. This result has been confirmed in many site-specific studies conducted elsewhere, e.g., in London (Hajat et al. 2014, p. 643). However, this proportion can be reversed in the longer run. Perhaps, the most important message from this paper could be inferred from the illustrative example in Table 11.2. It relates to the shape of the underlying temperature-mortality relationship. The *dlnm* modeling showed that the left tail of this curve was close to linear, while the right tail was not only steeper but also *convex*. For this reason, the projections based on such relationship will produce relatively faster increases in AF_{heat} and relatively slower decreases in AF_{cold} as the future temperatures rise. At some point in time between the 2050s and 2080s, the total number of deaths attributable to nonoptimal temperatures is bound to exceed its current value under all scenarios that are "worse" than RCP2.6, as follows from Figures 7a and 8a in IPCC (2013).

Acknowledgments Funding source: The research has been supported by the grant program of Russian Science Foundation, Project No. 16-18-10324. "Human in Megalopolis: Economic, Demographic and Ecological Features". The authors highlight the input of their colleagues from Voeikov Main Geophysical Observatory in Saint-Petersburg, Russian Federation, who developed regional climate projections for this study: Valentin Meleshko, Veronika Govorkova, and Tatyana Pavlova.

References

ACIA (2005) Impacts of a warming arctic: arctic climate impact assessment. Cambridge University Press, New York. http://www.acia.uaf.edu/pages/overview.html

Benmarhnia T, Sottile MF, Plante C, Brand A, Casati B, Fournier M, Smargiass A (2014) Variability in temperature-related mortality projections under climate change. Environ Health Perspect 122:1293–1298. http://dx.doi.org/10.1289/ehp.1306954

Cheng CS, Campbell M, Li Q, Li G, Auld H, Day N et al (2009) Differential and combined impacts of extreme temperatures and air pollution on human mortality in south-central Canada. Part II: future estimates. Air Qual Atmos Health 1(4):223–235

Gasparrini A (2011) Distributed lag linear and non-linear models in R: the package dlnm. J Stat Softw 43(8):1–20. http://www.jstatsoft.org/v43/i08/

Gasparrini A, Armstrong B (2011) The impact of heat waves on mortality. Epidemiology 22 (1):68–73. http://doi.org/10.1097/EDE.0b013e3181fdcd99

Gasparrini A, Leone M (2014) Attributable risk from distributed lag models. BMC Med Res Methodol 14:55. http://www.biomedcentral.com/1471-2288/14/55

Gasparrini A, Armstrong B, Kenward MG (2010) Distributed lag non-linear models. Stat Med 29 (21):2224–2234. http://doi.org/10.1002/sim.3940

Gasparrini A, Guo Y, Hashizume M, Lavigne E et al (2015) Mortality risk attributable to high and low ambient temperature: a multicountry observational study. Lancet 386:369–375. doi:10.1016/S0140-6736(14)62114-0

Gosling SN, Lowe JA, McGregor GR et al (2009) Associations between elevated atmospheric temperature and human mortality: a critical review of the literature. Clim Chang 92:299–341

Grjibovski AM, Bushueva V, Boltenkov VP et al (2012) Climate variations and salmonellosis in northwest Russia: a time-series analysis. Epidemiol Infect 141(2):1–8. doi:10.1017/S0950268812000544

Hajat S, Vardoulakis C, Sotiris E, Heaviside B (2014) Climate change effects on human health: projections of temperature-related mortality for the UK during the 2020s, 2050s and 2080s. J Epidemiol Community Health 68:641–648. doi:10.1136/jech-2013-202449

Huang C, Barnett AG, Wang X, Vaneckova P, FitzGerald G, Tong S (2011) Projecting future heat-related mortality under climate change scenarios: a systematic review. Environ Health Perspect 119(12):1681–1690. http://doi.org/10.1289/ehp.1103456

IPCC (2013) Summary for policymakers. In: Stocker TF, Qin D, Plattner G-K, Tignor M, Allen SK, Boschung J, Nauels A, Xia Y, Bex V, Midgley PM (eds) Climate change 2013: the physical science basis. Contribution of working group I to the fifth assessment report of the intergovernmental panel on climate change. Cambridge University Press, New York

Kinney PL, O'Neill MS, Bell ML et al (2008) Approaches for estimating effects of climate-change on heat-related deaths: challenges and opportunities. Environ Sci Pol 11(1):87–96. http://dx.doi.org/10.1016/j.envsci.2007.08.001

Nakicenovic N, Davidson O, Davis G, Grübler A, Kram T, Rovere ELL et al (2000) Special report on emissions scenarios: a special report of working group III of the IPCC. Cambridge University Press, New York

R Core Team (2016) R: a language and environment for statistical computing. R Foundation for Statistical Computing, Vienna. https://www.R-project.org/

Shaposhnikov D, Revich B (2016) Towards meta-analysis of impacts of heat and cold waves on mortality in Russian north. Urban Clim 15:16–24. doi:10.1016/j.uclim.2015.11.007

Shaposhnikov D, Revich B, Meleshko V, Govorkova V, Pavlova T (2011) Climate change may reduce annual temperature-dependent mortality in subarctic: a case study of Archangelsk, Russian Federation. Environ Nat Resour Res 1:75–91. doi:10.5539/ enrr.v1n1p75

Shaposhnikov D, Revich B, Bellander T, Bedada GB, Bottai M, Kharkova T, Kvasha E, Lezina E, Lind T, Semutnikova E, Pershagen G (2014) Mortality related to interactions between heat wave and wildfire air pollution during the summer of 2010 in Moscow. Epidemiology 25:359–364. doi:10.1097/EDE.0000000000000090

Shaposhnikov D, Revich B, Bellander T, Bedada GB, Bottai M, Kharkova T, Kvasha E, Lezina E, Lind T, Pershagen G (2015) Long-term impact of Moscow heat wave and wildfires on mortality. Epidemiology 26:e21–e22. doi:10.1097/EDE.0000000000000251

Tokarevich N, Tronin A, Blinova O, Buzinov R, Boltenkov V, Yurasova E, Nurse J (2011) The impact of climate change on the expansion of Ixodes persulcatus habitat and the incidence of tick borne encephalitis in the north of European Russia. Global Health Act 4:8448. doi:10. 3402/gha.v4i0.8448

Wu J, Zhou Y, Gao Y, Fu JS, Johnson BA, Huang C, Kim YM, Liu Y (2014) Estimation and uncertainty analysis of impacts of future heat waves on mortality in the eastern United States. Environ Health Perspect 122:10–16. http://dx.doi.org/10.1289/ehp.1306670

Dmitry Shaposhnikov Dmitry Shaposhnikov Laboratory of Forecasting of Environmental Quality and Human Health, Institute of Economic Forecasting of Russian Academy of Sciences, Nakhimovsky Prospect 47, Moscow 117418, Russia

Boris Revich Laboratory of Forecasting of Environmental Quality and Human Health, Institute of Economic Forecasting of Russian Academy of Sciences, Nakhimovsky Prospect 47, Moscow 117418, Russia

Chapter 12
Climate Change and Air Quality in Southeastern China: Hong Kong Study

Yun Fat Lam

Abstract As climate change continues to unfold over the next several decades in response to increasing levels of greenhouse gases (GHGs) in the atmosphere, the effects of climate change and future air quality will be more noticeable and observable. Understanding future climate and air quality has become one of the highest priorities for many countries and individual cities, where mitigation and adaptation could be planned. In Hong Kong, local government has pledged to reduce the GHG emissions by 60–65% from the 2005 level (i.e., 40 million tonnes CO_2 equivalent (CO_2e) in 2005) by 2030. The reduction focuses mainly on local energy saving, alternative transportation, and green energy generation. As Hong Kong moves into less carbon-intense technologies in both transportation and energy sectors, this much needed change will benefit the city's local air quality. Currently, no long-term carbon reduction plan for 2050 has been identified in the government.

In terms of future air quality projections, strong relationships between emissions and pollutant concentrations have been observed in Southeastern China under the IPCC AR5 scenarios, where the reduction of regional emissions (e.g., SO_2, NOx, and PM) has a great effect on future $PM_{2.5}$ air quality. Overall, $PM_{2.5}$ air quality over Pearl River Delta region has shown a clear improvement in 2050 under RCP8.5 emission scenario, with a mean concentration reduction of 5–15% (up to 12 $\mu g/m^3$). For ozone, a slight increase (i.e., 0–3%) of annual mean has been projected, which may be due to the combined effect of slow emission reduction of NMVOCs and less NOx titration in the VOCs limited regime. In addition, some studies also projected the increase of typhoons tracking near Taiwan Strait in the future climate would increase the occurrence of summer ozone episodes in Hong Kong.

Keywords Climate change • Carbon reduction • Hong Kong • Future air quality • O_3 • $PM_{2.5}$ • Tropical cyclone

---

Y.F. Lam (✉)
School of Energy and Environment, City University of Hong Kong, HKSAR, Kowloon Tong, Hong Kong

Guy Carpenter Asia-Pacific Climate Impact Centre, City University of Hong Kong, HKSAR, Kowloon Tong, Hong Kong
e-mail: yunflam@cityu.edu.hk

© Springer International Publishing AG 2018

R. Akhtar, C. Palagiano (eds.), *Climate Change and Air Pollution*, Springer Climate, DOI 10.1007/978-3-319-61346-8_12

Introduction

The earth's climate is changing dramatically as a result of human activities. Continued emission of greenhouse gases and air pollutants (i.e., short-lived climate forcers) has modified the natural balance of solar radiation on our planet. Greenhouse gases (GHGs) have caused heat to be retained in our earth system, triggering global warming and climate change, which results in rising global average air and ocean temperatures, changing distribution of precipitation, modifying regional air circulation, and even air quality. It is a global challenge faced by everyone, regardless of their size of government and its role in the national status. This chapter summarizes (1) present status of Hong Kong air pollution, (2) climate mitigation plan for Hong Kong, (3) potential co-benefits of air quality from climate mitigation, and (4) projection of future air quality in Southeastern China.

Characteristics of the Study Area

Climate and Geographical Location

Hong Kong is located at the estuary of the Pearl River Delta (PRD) in China surrounded by mountains and ocean. It lies between latitude 22° 08′ North and 22° 35′ North and longitude 113° 49′ East and 114° 31′ East, with subtropical climate that tends toward temperate climate for half of the year (HKO 2003). It has four distinct seasons, which are warm and humid spring (March and April), hot and rainy summer (May, June, July, and August), pleasant and sunny autumn (September and October), and cool and dry winter (November, December, January, and February). The daily average temperature ranges from 12 to 31 °C and can reach up to 36 °C in some areas due to the enhancement of the urban heat island (UHI), which intensifies the urban temperature as a result of poor ventilation and heat trapped by buildings. The prevailing direction of wind in Hong Kong follows the large-scale East Asia monsoon circulation. In summer, direction tends to be southwesterly, bringing humid marine air to the land. In winter, a persistent northeastern wind carries relatively cold air from the north. Occasionally, Hong Kong experiences tropical cyclones from the Western North Pacific or the South China Sea in summer and autumn. These tropical cyclones not only bring the potential issue of storm surge, mudslides, and heavy precipitation but also extreme heat and air pollution to Hong Kong.

Ambient Air Quality

Ambient air quality in Hong Kong is regulated under Air Pollution Control Ordinance (Cap. 311), which sets out several Air Quality Objectives (AQOs). These AQOs govern the major air pollutants including SO_2, PM_{10}, $PM_{2.5}$, NO_2, O_3, CO, and lead. The threshold limits and their average timings follow the standards of Air Quality Guidelines (AQGs) or Interim Target (IT-1, IT-2, or IT-3) levels from the World Health Organization (WHO), as shown in Table 12.1; it is mandated to be reviewed once every 5 years. In 2015, four air pollutants (i.e., PM_{10}, $PM_{2.5}$, O_3, and NO_2) in some ambient stations of Hong Kong exceeded short-term (1 h, 8 h, and 24 h) and long-term (annual) air quality standards. Although the government is continuously trying to reduce local emissions (e.g., speed up retirement of Pre-Euro III commercial diesel vehicles), the results to improve local air quality seemed to be slow, as it only reduced 14–29% of ambient pollutant concentrations (PM_{10}, $PM_{2.5}$, and NO_2) from 1999 to 2015 with the local emission reductions of 28% and 65% for NOx ($NOx = NO + NO_2$) and PM_{10}, respectively (HKEPD 2016). For ozone, a 24% increase of ambient concentration was observed. The worsening ozone air quality is attributed to multiple reasons including (1) the increase of regional transport of air pollutants from mainland China; (2) the rise of ambient temperature which triggers the increase of ozone photochemistry and biogenic VOC emissions; and (3) the reduction of NOx emission influencing the rate of O_3 destruction during NOx titration process in the urban environment (Huang et al. 2009; Fu et al. 2012; Lam et al. 2011; Wang et al. 2017). In 2015, the annual average ambient concentrations of PM_{10}, $PM_{2.5}$, NO_2, and O_3 were reported as 38.5 $\mu g/m^3$, 25.2 $\mu g/m^3$, 46.3 $\mu g/m^3$, and 45 $\mu g/m^3$, respectively. Table 12.1 summarizes the concentration limits on AQOs and the number of reported exceedances for each major pollutant (excluding lead).

Local and Regional Contribution to Hong Kong Air Pollution

Hong Kong's air pollution has been a serious problem since the early 1970s; the city was the central hub of a global manufacturing center. In the last three decades, it has slowly transformed into a financial and tourist center, where the majority of its factories had moved to China. With the steady increase of population and transportation networks, power generation and mobile sectors have become the major contributors for deteriorating local air quality. These two sectors alone account for 67–90% of overall local particulates (PM) and NOx emissions (Wan et al. 2016). Along with local emissions, the regional transport of air pollution from Mainland China, particularly Pearl River Delta, also plays a significant role contributing to Hong Kong air pollution. Kwok (2017) reported that the contributions of PM_{10} and NO_2 from Mainland China could be as much as 35–65% in winter, while it is only 10–15% in summer. Similar results have also been reported by Huang et al. (2009) and Lau et al. (2007), confirming the importance of background pollutant

Table 12.1 Summary of Air Quality Objectives (AQOs) and number of AQO exceedances in 2015

	SO_2		PM_{10}		$PM_{2.5}$		NO_2		O_3	CO	
Pollutant (μg/m^3)	10 min	24 h	24 h	Ann	24 h	Ann	1 h	Ann	8 h	1 h	8 h
AQO limit	500	125	100	50	75	35	200	40	160	30,000	10,000
Reference WHO standard	AQGs	IT-1	IT-2	IT-2	IT-1	IT-1	AQGs	AQGs	IT-3	AQGs	AQGs
Number of exceedances[a]	0	0	18 (2)	0	11 (2)	0	67 (3)	1 (9)	24 (6)	0	0
Annual Average concentration[a]	9.3		38.5		25.2		46.3		45	674	

[a]Based on 12 general air quality stations and number of stations involved "()"; "Ann" for annual; AQO standard for lead is not shown

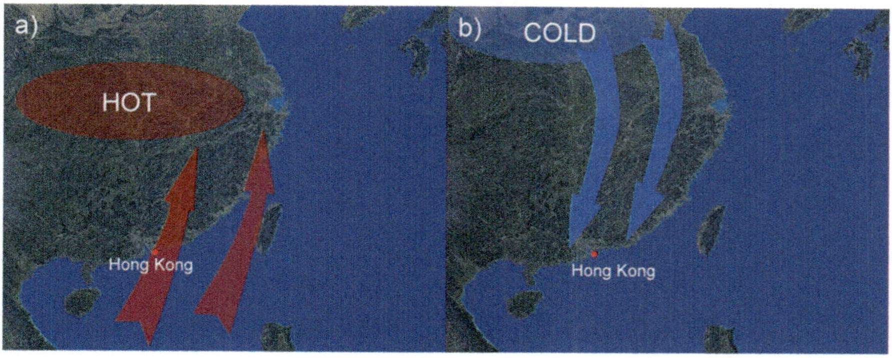

Fig. 12.1 Prevailing wind direction for (**a**) late spring and summer (AMJJA) and (**b**) late autumn and winter (ONDJF) (HKO 2010)

enhancement from Mainland China on Hong Kong air quality. Figure 12.1a, b illustrates the change of season in the context of air pollution. In early spring and summer, clean moist marine boundary air blows from the south to north, providing a pleasant condition for good air quality in South China. Frequent precipitation during these seasons promotes wet deposition, which removes air pollutants from the air. Conversely, in late autumn and winter, polluted air with high concentrations of NOx, PM and VOCs from the north arrives in Hong Kong and mixes with local air pollutants. This relatively dry and cold air promotes cloudless and sunny skies with stable atmospheric conditions, which encourages the accumulation of air pollutants and the formation of secondary pollutants (e.g., O_3). Consequently, major pollution episodes occur frequently in Hong Kong during late autumn and winter. It is clear that local air quality in Hong Kong is strongly affected by regional climate and air circulation.

Different air pollutants exhibit different seasonal patterns, depending on their sources and sinks under different environmental conditions (i.e., meteorological conditions). Pollutants such as PM derive a large portion from primary emission, while the secondary formation is also significant (Cheng et al. 2015). In some cases, primary emission of PM from certain seasonal activities appears to have a strong dependence on local meteorology (e.g., temperature and relative humidity). For example, biomass burning in South China often occurs in the dry season from late September to October contributing substantial VOCs, CO, and PM to the atmosphere (Chen et al. 2017). Residential coal burning for heating in winter (i.e., Nov, Dec, Jan, and Feb) also emits enormous amounts of PM and CO to the environment. The practice of using coal for heating is highly dependent on ambient temperature (Xiao et al. 2015). These seasonal sources add extra burdens to the existing polluted condition (from industrial emissions) in the late autumn and winter in China, which enhances the background concentration of regional pollutants, influencing Hong Kong air quality. As illustrated in Fig. 12.2, PM (e.g., 24-h $PM_{2.5}$) episodes/ exceedances (in blue) in Hong Kong are mostly clustered in October, November, December, and January, which is under the influence of transboundary pollution

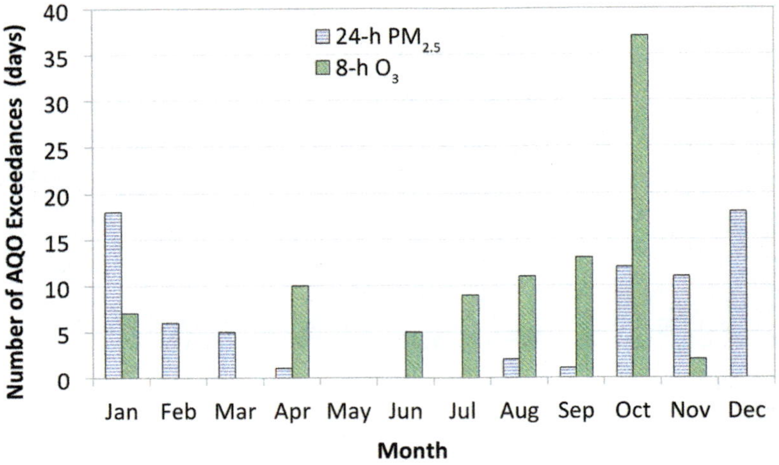

Fig. 12.2 Average monthly exceedances of 24-h $PM_{2.5}$ and 8-h ozone during 2013–2015 for Hong Kong

from the northeast winter monsoon. These 4 months have constituted more than 80% of overall PM episodes in Hong Kong. Concerning ozone, the seasonal distribution of 8-h O_3 episodes/exceedances (in green) peaks in October with some cases in summer. The high exceedances in October are contributed by the effect of transboundary pollution from the northeast winter monsoon and local emissions. Strong solar radiation and high temperatures in October (i.e., climatological average of 28.5 °C) stimulate the rapid formation of secondary O_3 on that month. For the other summer exceedances, the events are mostly associated with the presence of tropical cyclones in the vicinity of the Taiwan Strait. With the counterclockwise and sinking motion induced by the outer ring of the tropical cyclone, high pressure with stable, clear sky is produced at Hong Kong. The counterclockwise wind pattern brings air pollutants from the industrial Pearl River Delta that mix with local pollutants to form photochemical episodes (see Fig. 12.3). In general, ozone exceedances under the influence of tropical cyclones could be much stronger than from the influence of northeast winter monsoon in late autumn. The hourly and average 8-h concentrations of ozone could reach as much as 400 μg/m^3 and 337 μg/m^3, respectively.

Carbon Emission and Mitigation Plan

Carbon Emission and Emission of Air Pollutants

Hong Kong is one of the densest cities on the planet. It has a population of 7.3 million people, located on densely constructed vertical buildings on 42 km^2 of land. The heavily built-up environment reduces air circulation within the city, enhancing both the UHI and street-level air pollution. In Hong Kong, the annual production of

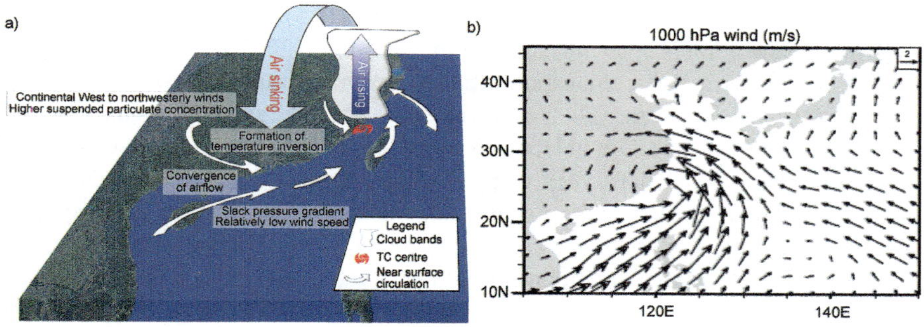

Fig. 12.3 Illustration of typhoon-induced ozone episodes in Hong Kong: (a) meteorological pattern and (b) prevailing wind direction (Leung and Wu 2015; Lam et al. 2017)

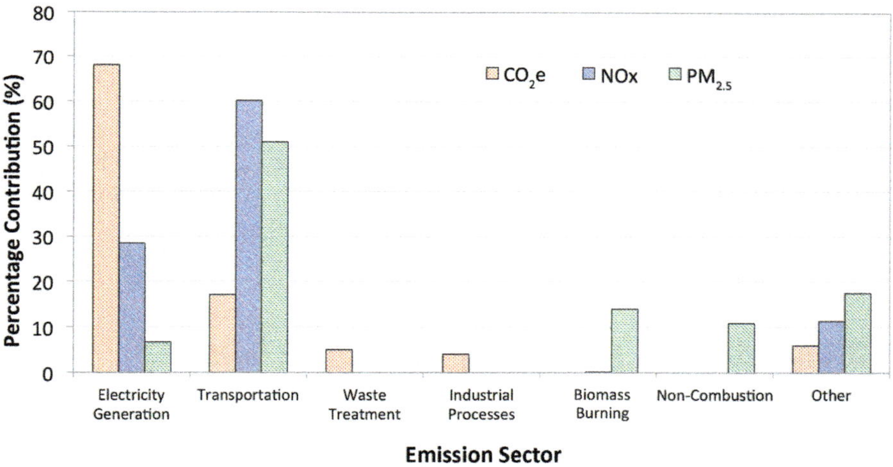

Fig. 12.4 Summary of emission contributions of carbon and air pollution emissions in 2015

greenhouse gases (CO_2, CH_4, N_2O, HFCs, PFCs, and SF_6) is reported to be about 40 million tonnes of CO_2 equivalent (CO_2e), which translates to about 6 tonnes of CO_2e on a per capita basis (dated 2005), and is ranked 69 out of 217 regions/ countries in the world (World Bank 2016). Since the listed value does not include aviation nor international marine transportation, which has been accounted for in the Chinese greenhouse gases (GHGs) inventory (to avoiding double counting), the actual CO_2e on a per capita basis in Hong Kong should even be higher, as those two sectors, in fact, are the major business areas in Hong Kong. In terms of categorical breakdown (see Fig. 12.4), the major sector of GHGs is from building-related electricity usage (68% of overall CO_2e), which supports ~42,000 buildings in Hong Kong. The second largest sector is from local transportation, which accounts for about 17% of overall CO_2e. The remaining 15% comes from industrial, agriculture process, and waste treatment (Environment Bureau 2015). The current fuel mix for electricity generation is 53% in coal, 22% in natural gas, 23% in nuclear, and 2% in others, where fossil fuel combustion accounts for more than 75% of CO_2e in the electricity generation sector. The values (i.e., coal and natural gas) translate to about 42% and 9% of overall CO_2e in Hong Kong, respectively (assuming CO_2e emission in natural gas is about half when compared with coal). Besides carbon emission, fossil fuel combustion also emits a significant amount of air pollutants. According to the Environmental Protection Department of Hong Kong (HKEPD), the electricity generation sector accounts for around 7% of $PM_{2.5}$ and 28% of NOx emissions from a total of ~4,300 and ~92,000 tonnes, respectively (HKEPD 2017). As shown in Fig. 12.4, the categorical breakdown of air pollutant emissions is slightly different from the carbon emissions, where the major source of NOx and $PM_{2.5}$ comes from the transport sector which accounts for more than 50% of overall emissions and it is about two to seven times higher than the electricity generation sector. The low contribution of $PM_{2.5}$ and NOx emissions from the energy sector is mainly attributed to the success of installing a retrofitted electrostatic precipitator (ESP) and Selective Catalytic Reduction (SCR) system in the coal-fire power plant, which reduces more than 80–90% of $PM_{2.5}$ and NOx from stack emissions, as compared with the uncontrolled carbon emission in the GHGs inventory. In terms of VOCs, the majority of VOCs (a total of 26,600 tonnes) comes from evaporative VOCs, such as paint solvent, while electricity generation and transportation only account for about 2% and 37%, respectively.

Climate Mitigation Plan and Air Quality Co-benefits in Hong Kong

In the Copenhagen Accord of 2009 and Cancun Agreement of 2010, an agreement of controlling global temperature within 2 °C has emerged. A common consent of developing a global carbon emission inventory for monitoring the growth of GHGs was also adopted. In 2012, the Hong Kong government had rigorously put forward a short-term reduction plan to adapt a 50–60% carbon emission by 2020. As in the Conference of Parties (COP21) under the United Nations Framework Convention on Climate Change (UNFCCC), also known as the World Paris Climate Conference, the Chinese government has pledged to modernize its power production from coal burning and to reduce its emissions by 60–65% from 2005 levels by 2030 (Environment Bureau 2015). In addition, the Chinese government also agreed to increase its non-fossil fuel sources in energy production to about 20% by 2030. The proposed plan built upon the fact that coal-fire combustion in China accounts for more than 70% of its electricity generation. Reducing and modernizing coal combustion facilities bring huge co-benefits to air quality in China, which aligns with the national agenda of tackling local air quality problems.

As Hong Kong is one of the special administrative regions within the People's Republic of China, a new target of carbon emission has been adopted, which is 60–65% from the 2005 level (i.e., 40 million tonnes CO_2e in 2005) by 2030. In order to achieve the carbon reduction, the government has carried out a series of studies related to potential options for reducing carbon emission (e.g., public consultation on future development of the electricity market) (Environment Bureau 2014). The major suggestion received from these studies includes reduction of carbon emission by reducing coal usage from local electricity generation, maximization of energy efficiency in buildings, and expanding the sustainable/green rail system. Specifically for electricity generation, it suggests increasing the portion of natural gas usage from 22 to 50% and importing more nuclear power (23–25%) from Mainland China. The revamping of the electricity fuel mix significantly reduces carbon emission, as natural gas produces only half the amount of carbon emission than coal. The 28% conversion from coal to natural gas would reduce 5.8 million tonnes (14.5% reduction) of overall annual carbon emission in Hong Kong, while 2% more in nuclear power could reduce 0.8 million tonnes of annual carbon emission. In terms of energy saving, the practice of energy saving in buildings is expected to reduce 5% in energy every 4–5 years, which translates to about 10–15% carbon emission by 2030. The co-benefits of revamping electricity fuel mix and building energy savings in Hong Kong are expected to reduce $PM_{2.5}$ and NOx emissions by 10.6 tonnes and 956 tonnes, respectively. Although the magnitude of local reduction seems to be large, this reduction may not show a noticeable improvement in ambient air quality in Hong Kong, as the major pollutant contributors are from the long-range transport and mobile sector. Nevertheless, as the Chinese government continues following the pledged reduction plan for carbon emission, the influence of long-range transport of air pollutants on Hong Kong air pollution would be gradually reduced. The overall co-benefits due to the action of carbon reduction would certainly be positive for primary pollutants, while it is uncertain for secondary pollutants as their ambient concentration does not solely depend on emission source.

Climate-Air Quality Interaction and Future Air Quality Projections

Impacts of Air Quality on Climate Change

Recent studies show that some air pollutants have similar absorption properties as GHGs and have comparable thermal effects on our climate (Chen et al. 2007; Ramanathan and Feng 2009). The main difference between these pollutants and GHGs is that the lifetime of these pollutants is much shorter and has distinct temporal and spatial patterns based on regional emission characteristics. They are emitted either from natural or anthropogenic processes or form in the atmosphere through the secondary chemistry. Figure 12.5 shows various primary air pollutants

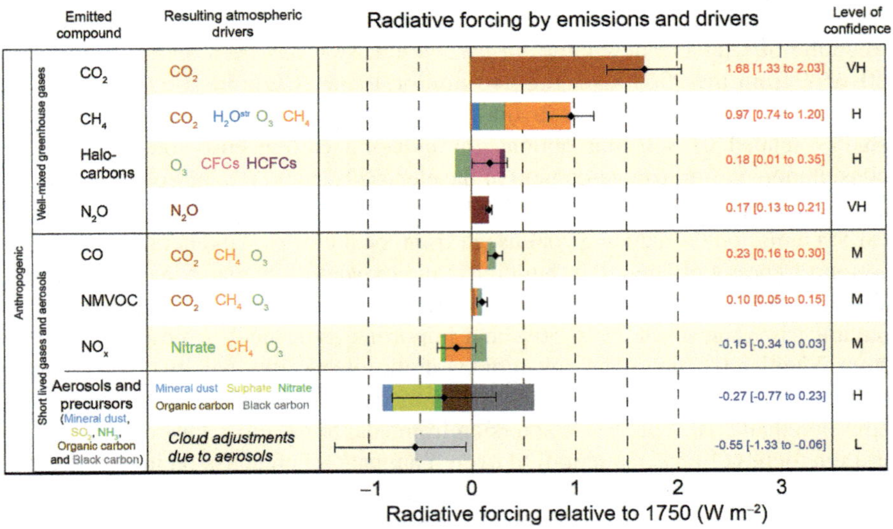

Fig. 12.5 Radiative forcing estimates on different emissions and drivers (Adapted from IPCC 2013)

with their respective ability to influence radiative forcing (IPCC 2013). These pollutants (i.e., CO, non-methane VOCs (NMVOCs), NOx, and aerosols) either contribute to atmospheric warming or cooling. For example, NOx can be oxidized to form nitrate particulates, which have a cooling effect on the atmosphere, while CO can contribute to the formation of CO_2 or O_3 which has a warming effect. The air pollutants with warming ability such as O_3 and black carbon particulates are referred to as short-lived climate forcers (SLCF). The average lifetime of a typical SLCF is from a few days to weeks, and its distribution is usually localized in megacities. These are unevenly distributed across the globe making it difficult to evaluate their effects on climate. Moreover, some SLCF are particulates, which not only affect the direct radiative balance in the atmosphere but are also involved in the formation of clouds by participating as cloud condensation nuclei (CCN) (Chen et al. 2010). The involvement in cloud formation is profound and has a great impact on radiative balance and rainfall distribution. This is currently an active research area aimed at quantifying the impacts of SLCF interaction on climate change as well as the climate co-benefits of reducing SLCF for short-term mitigation planning (e.g., 2030). According to recommendations from the United Nations Environmental Program (UNEP) and the United States Environmental Protection Agency (USEPA), climate change issues should be addressed using an integrated climate and air quality approach, which means that regional and local air quality management should be a part of the integrated platform for remedying the effects of climate change (UNEP 2011). All SLCF including black carbon and ozone and its precursors should be regulated in addition to existing GHGs.

Impacts of Climate Change on Air Quality

Historical Studies

Future climate change is known to affect the future air quality through the change of local meteorology. Factors, such as increase of local temperature, stagnation of regional circulation, intensification of urban heat island effect (UHI), and reduction of raining frequency, have been found to have direct impacts on the future air quality. Indirect factors such as increase of natural emissions (i.e., biogenic VOCs from plant and DMS from ocean) through the increase of temperature have also been found to affect the future air quality (Hu et al. 2003; Yu et al. 2004, 2009; Ramanathan and Feng 2009). Very limited studies have tied the climate change to future air quality in Hong Kong where the majority of these studies have looked at the historical data (1980–2010) on how recent climate change affected local air quality. Lee et al. (2014) investigated the linkage between temperature and local ozone air quality and found an increase of 1.0–1.6 $\mu g/m^3$ per year for ozone in recent years, while Fu and Tai (2015) studied the impact of climate and land cover changes on ozone and found a 2–10 ppbv (i.e., ~4–20 $\mu g/m^3$ for the last 20 years) of increase in summer ozone in East Asia solely from climate change. Lam et al. (2017) investigated historical typhoon data and found that more typhoons have been observed in the vicinity of Taiwan in last decade (2000–2010), which produces more frequent summer ozone episodes in Hong Kong.

Future Climate and Air Quality Studies

More recent studies have applied climate and air quality models to evaluate the effect of climate change on future air quality under the predefined conditions for East Asia. These adopted climate conditions (most updated one) are referred to as IPCC AR5 scenarios, which is suggested in the Fifth Assessment Report (AR5) of the Intergovernmental Panel on Climate Change (IPCC). A total of four scenarios in the AR5 were established, which are RCP2.6, RCP4.5, RCP6.0, and RCP8.5. The suffix value after the "Representative Concentration Pathways (RCP)" signifies the range of increase on radiative forcing values in 2100 relative to preindustrial value (i.e., 1900). For instance, RCP2.6 contains the scenario in which global radiative forcing has increased by 2.6 W/m^2 at 2100. In the future climate projections, Kwok (2017) downscaled the CESM outputs, one of the general circulation models (GCMs) in the AR5, into the WRF-CMAQ climate and air quality model to study the future air quality in Hong Kong for 2030 and found that the average ozone concentration under RCP4.5 in autumn has increased by 14% from 2002. Li et al. (2016) and Zhu and Liao (2016) have applied the nested version of the GEOS-Chem model in present climate to assess the changes of 2000–2050 in $PM_{2.5}$ and O_3 air quality in China under all 4 AR5 emission scenarios. In their studies, they have developed emission reduction plans on major air pollutants based on the RCP

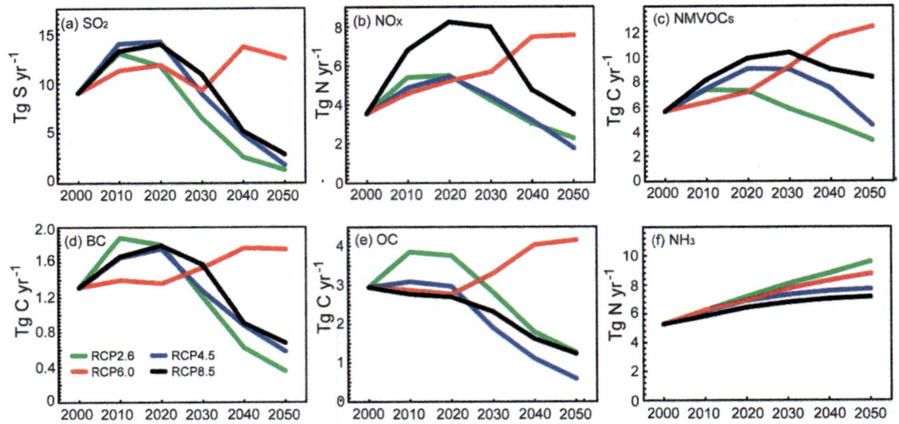

Fig. 12.6 Emission projections for RCP scenarios: (**a**) SO2, (**b**) NOx, (**c**) NMVOCs, (**d**) black carbon, (**e**) organic carbon, and (**f**) ammonia (Li et al. 2016; Zhu and Liao 2016)

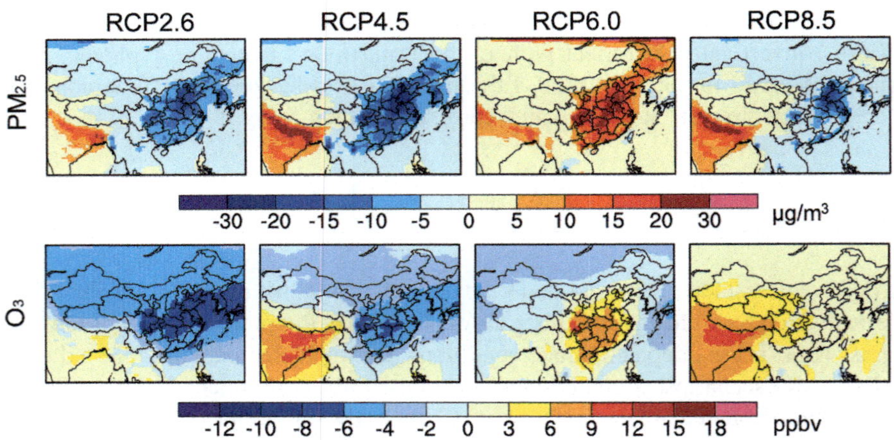

Fig. 12.7 Projected future $PM_{2.5}$ and O_3 in East Asia (Li et al. 2016; Zhu and Liao 2016)

scenarios. Figure 12.6 shows the emission changes for the RCP scenarios for NOx, NMVOCs, BC, and organic carbon (OC). In the RCP2.6, 4.5, and 8.5, the emissions increase at a different rate and reach maximum at around 2020–2030 and sharply decrease by 2050, while in RCP6.0, the emissions are continuously increased till 2050 and drastically drop after 2050. The maximum value and turnover year are slightly different for each scenario, reflecting different carbon reduction plans adopted in the AR5 scenarios. Figure 12.7 shows the summary of future $PM_{2.5}$ and ozone under the influence of emission change. In terms of annual $PM_{2.5}$, RCP2.6, 4.5, and 8.5 lead to a concentration reduction of 6.4–7.4 μg/m^3 (43–49%) in PRD area, while RCP6.0 shows a strong increase of 7.4 μg/m^3

(+50%). The large reductions of ambient $PM_{2.5}$ in those three scenarios mainly originate from the substantial reduction of primary emissions (e.g., EC/OC) and its precursors (e.g., NOx and SO_2). Eliminating EC, OC, and SO_2 in 2050 has made ammonia nitrate become the most abundant PM species. Therefore, reducing agricultural NH_3 and automobile NOx is suggested for further reducing annual $PM_{2.5}$ concentration in PRD. In terms of annual ozone, a slight decrease (i.e., ~4.0 ppbv) under RCP2.6 and 4.5 has been observed in 2050 in PRD, which comes from the significant reduction of NMVOCs and NOx in those two scenarios. The highest reduction (~6.6 ppbv) of ozone has been observed in autumn and early winter, which indicates that the influence of long-range transport from PRD would be less in the future due to the emission reduction. As a result, ozone air quality in Hong Kong would be expected to improve under those scenarios. It is observed that no exceedance (using limit of 160 $\mu g/m^3$) of maximum daily average 8-h (MDA8) ozone is found in these future emission scenarios. On the other hand, a slight increase (~0–3 ppbv) of annual ozone is observed under RCP6.0 and 8.5 in 2050 for PRD area. The increase of ozone could be attributed to the fact that the rate of NOx reduction is much faster than the rate of VOCs reduction under the VOCs limited environment, which triggers an increase of ozone (Wang et al. 2010). Therefore, careful implementation of a reduction plan for ozone precursors would be important, particularly for PRD area where biogenic sources may play a significant role in the formation of ozone as global warning continues to rise (Cheng et al. 2010). It should also be noted that the maximum increase (+6.2–6.6 ppbv) of annual ozone occurs in 2040 (not 2050) under RCP6.5, and 8.5 indicates that there will still be ozone air quality problems in the next 20 years. The exceedance of MDA8 ozone would peak in 2030 (i.e., 34 incidents) and gradually reduce to no exceedance in 2050 when sufficient anthropogenic VOCs are reduced. With respect to the influence of tropical cyclone on ozone air quality in Hong Kong, some researchers have projected the frequency of tropical cyclones (TC) in the vicinity of Taiwan would be increased (Wang et al. 2011). As a TC produces a similar transport pattern as in winter bringing pollutants from PRD to Hong Kong, it is projected that there would still be an increase of summer ozone episodes till 2030 and would gradually improve as the projected ozone after 2030 is reduced under RCP8.5, which lowers the background ozone precursors from PRD during a TC event.

Summary

As one of the densest cities on the planet, Hong Kong has adopted a stronger (60–65% carbon reduction by 2030) mitigation plan for combating climate change. Although it may not be significant from a global perspective, it shows a strong commitment as a global citizen. In Hong Kong, the major reduction of GHGs is focused on the energy sector, where changing carbon-intensity fossil fuel (i.e., coal) into less intense fuel such as natural gas, or nuclear, reducing building-related energy usage, and adopting more green transportation. These mitigation plans have moved Hong Kong toward becoming a healthier city. In terms of air pollution, these mitigation plans carry some co-benefits on local air quality, where reduction of coal/gasoline burning would reduce $PM_{2.5}$ and NOx emitted into both roadside and ambient environments. Under the future emission projections (IPCC AR5), $PM_{2.5}$ air quality for Hong Kong in 2050 would be improved under RCP2.6, 4.5 and 8.5 due to the reduction of primary PM and its precursors, while it is increased under RCP6.0. In terms of ozone, less exceedance of ozone (based on MDA8) is projected in 2050 under RCP2.6, 4.5 and 8.5 in PRD area.

Acknowledgments This work was conducted under the financial support of Guy Carpenter Asia-Pacific Climate Impact Centre project, 9360126, and CityU project, 7004692. I thank Prof. Hong Liao and Dr. Ke Li for providing important figures in this chapter.

References

Chen WT, Liao H, Seinfeld JH (2007) Future climate impacts of direct radiative forcing of anthropogenic aerosols, tropospheric ozone, and long-lived greenhouse gases. J Geophys Res-Atmos 112. doi:10.1029/2006jd008051

Chen W-T, Nenes A, Liao H, Adams PJ, Li J-LF, Seinfeld JH (2010) Global climate response to anthropogenic aerosol indirect effects: present day and year 2100. J Geophys Res 115:D12207. doi:10.1029/2008JD011619

Chen J, Li C, Ristovski Z, Milic A, Gu Y, Islam MS, Wang S, Hao J, Zhang H, He C, Guo H, Fu H, Miljevic B, Morawska L, Thai P, Lam YF, Pereira G, Ding A, Huang X, Dumka UC (2017) A review of biomass burning: emissions and impacts on air quality, health and climate in China. Sci Total Environ 579:1000. http://dx.doi.org/10.1016/j.scitotenv.2016.11.025

Cheng HR, Guo H, Saunders SM, Lam SHM, Jiang F, Wang XM, Simpson IJ, Blake DR, Louie PKK, Wang TJ (2010) Assessing photochemical ozone formation in the Pearl River Delta with a photochemical trajectory model. Atmos Environ 44(34):4199–4208

Cheng Y, Lee SC, Gu Z, Ho KF, Zhang YW, Huang Y, Chow J, Watson JG, Gao JJ, Zhang RJ (2015) PM2.5 and PM10-2.5 chemical composition and source apportionment near a Hong Kong roadway. Particuology 18:96–104. http://dx.doi.org/10.1016/j.partic.2013.10.003

Environment Bureau (2014) Future fuel mix for electricity generation consultation document. Retrieved from environment bureau website: http://www.enb.gov.hk/en/resources_publications/policy_consultation/electricity_generation.html

Environment Bureau (2015) Hong Kong climate change report 2015. Retrieved from environment bureau website: http://www.enb.gov.hk/sites/default/files/pdf/ClimateChangeEng.pdf

Fu Y, Tai APK (2015) Impact of climate and land cover changes on tropospheric ozone air quality and public health in East Asia between 1980 and 2010. Atmos Chem Phys 15:10093–10106. doi:10.5194/acp-15-10093-2015

Fu J, Dong X, Gao Y, Wong DC, Lam YF (2012) Sensitivity and linearity analysis of ozone in East Asia: The effects of domestic emission and intercontinental transport. J Air Waste Manag. Assoc., 62(9), pp 1102–1114

HKEPD (2016) Pollutant long-term trends (1990–2015). Retrieved from Hong Kong environmental protection department website: http://www.aqhi.gov.hk/api_history/english/report/files/aqt15e.pdf

HKEPD (2017) Data & Statistics: Emission Inventory for 2015. Retrieved from Hong Kong environmental protection department website: http://www.epd.gov.hk/epd/english/environmentinhk/air/data/emission_inve.html
HKO (2003) Climate of Hong Kong. Retrieved from http://www.hko.gov.hk/cis/climahk_e.htm
HKO (2010) The monsoons and climate change. Retrieved from http://www.weather.gov.hk/blog/en/archives/00000071.htm
Hu ZZ, Yang S, Wu RG (2003) Long-term climate variations in China and global warming signals. J Geophys Res-Atmos 108. doi:10.1029/2003jd003651
Huang X-F, Yu JZ, Yuan Z, Lau AKH, Louie PKK (2009) Source analysis of high particulate matter days in Hong Kong. Atmos Environ 43:1196–1203
IPCC (2013) Summary for policymakers. In: Climate change 2013: the physical science basis. Contribution of working group I to the fifth assessment report of the intergovernmental panel on climate change, Cambridge University Press, Cambridge/New York, pp 1–30, doi:10.1017/CBO9781107415324.004/
Kwok LK (2017) Development of hybrid statistical-dynamical model for air quality prediction: establishment and application. Doctoral dissertation
Lam YF, Fu JS, Wu S, Mickley LJ (2011) Impacts of future climate change and effects of biogenic emissions on surface ozone and particulate matter concentrations in the United States. Atmos Chem Phys 11(10):4789–4806
Lam YF, Cheung HM, Yan CC (2017) Impact of tropical cyclone track change on regional air quality. Sci Total Environ. Under review
Lau AKH, Lo A, Gray J, Yuan Z, Loh C (2007) Relative significance of local vs. regional sources: Hong Kong's air pollution, civic exchange
Lee Y, Shindell D, Faluvegi G, Wenig M, Lam YF, Ning Z, Hao S, Lai C (2014) Increase of ozone concentrations, its temperature sensitivity and the precursor factor in South China. Tellus B 66. doi:http://dx.doi.org/10.3402/tellusb.v66.23455
Leung YKJ, Wu MC (2015) Influence of tropical cyclones on visibility in Hong Kong. Retrieved from Hong Kong observatory website: http://www.hko.gov.hk/education/article_e.htm?title=ele_00089
Li K, Liao H, Zhu J, Moch JM (2016) Implications of RCP emissions on future $PM_{2.5}$ air quality and direct radiative forcing over China. J Geophys Res Atmos 121. doi:10.1002/ 2016JD025623
Ramanathan V, Feng Y (2009) Air pollution, greenhouse gases and climate change: global and regional perspectives. Atmos Environ 43:37–50. doi:10.1016/j.atmosenv.2008.09.063
UNEP (2011) Climate change adaptation. Retrieved from http://www.unep.org/climatechange/adaptation/InformationMaterials/News/PressRelease/tabid/6710/Default.aspx?DocumentId=2645&ArticleId=8780
Wan Z, Leung R, Huang H (2016) 2014 Hong Kong emission inventory report (Report No. EPD/TR/1/16). Hong Kong
Wang R, Wu L, Wang C (2011) Typhoon track changes associated with global warming. J Clim 24:3748–3752
Wang X, Zhang Y, Hu Y, Zhou W, Lu K, Zhong L, Zeng L, Shao M, Hu M, Russell AG (2010) Process analysis and sensitivity study of regional ozone formation over the Pearl River Delta, China, during the PRIDE-PRD2004 campaign using the Community Multiscale Air Quality modeling system. Atmos Chem Phys 10(9):4423–4437
Wang T, Xue L, Brimblecombe P, Lam YF, Li L, Zhang L (2017) Ozone pollution in China: a review of concentrations, meteorological influences, chemical precursors, and effects. Sci Total Environ 575:1582–1596. https://doi.org/10.1016/j.scitotenv.2016.10.081
World Bank (2016) World development indicators. Retrieved from http://databank.worldbank.org/data/reports.aspx?source=2&series=EN.ATM.CO2E.PC&country=#
Xiao Q, Ma Z, Li S, Liu Y (2015) The impact of winter heating on air pollution in China. PLoS One 10(1):e0117311. http://doi.org/10.1371/journal.pone.0117311
Yu RC, Wang B, Zhou TJ (2004) Tropospheric cooling and summer monsoon weakening trend over. East Asia Geophys Res Lett 31. doi:10.1029/2004gl021270

Yu Y, Niu SJ, Niu HW, Wu ZY, Wu XH (2009) The weakening relationship between the precipitation in north china during rainy season and the two high in the southern hemisphere. 2008 international workshop on education technology and training and 2008 international workshop on geoscience and remote sensing, 1, pp 261–264, doi:10.1109/Ettandgrs.2008.139
Zhu J, Liao H (2016) Future ozone air quality and radiative forcing over China owing to future changes in emissions under the Representative Concentration Pathways (RCPs). J Geophys Res Atmos 121:1978–2001. doi:10.1002/2015JD023926

Yun Fat Lam is an Assistant Professor in the School of Energy and Environment at City University of Hong Kong. He specializes in regional air quality management, emission inventory development, and future air quality modeling under climate change scenarios.

Part III
Case Studies: Developing Countries/Regions

Chapter 13
Trends and Seasonal Variations of Climate, Air Quality, and Mortality in Three Major Cities in Taiwan

Mei-Hui Li

Abstract The interactions among climate change, air pollution, and human health are multiple and complex. Many epidemiological studies in Taiwan have consistently demonstrated the effects of short-term exposures to extreme weather events, particulate matter, and traffic-related air pollutants on a variety of health effects. However, these findings might not explain or predict overall seasonal mortality patterns to provide insights into the drivers of mortality acting on society levels for public health policy and practice. There are very limited studies on seasonality of weather, air pollution, and mortality in Taiwan. The objectives of this study are to evaluate if there are any changes in trends and seasonality of mortality in three major Taiwanese cities from 1991 to 2010 and examine its association with climatic condition and air pollution. Among these major Taiwanese cities, seasonal mortality patterns are similar in two subtropical cities, Taipei and Taichung, compared to another tropical city, Kaohsiung. Taipei had significantly increased trends in most monthly temperature variables and the number of hot days examined during 1991–2010 compared to the other two cities. Winter/summer ratios of mortality only showed a decreased trend in Taipei, but not in Taichung or Kaohsiung. Mean monthly ambient temperature was also found as the most optimal temperature variable for predicting all-cause monthly mortality at all three cities in this study. Seasonal mortality patterns in three cities were with higher levels of deaths from December to March. Trends in air quality are showing mixed patterns over the past two decades. SO_2, CO, and NO_x concentrations have decreased significantly and steadily, while O_3 has significantly increased in recent years. In three major Taiwanese cities, O_3 and PM_{10} are major air pollutants of current concerns. The results of this study showed that monthly mean O_3, PM_{10}, and NO_x levels and monthly mortality were not closely related, but temperature-related variables were positively associated with monthly mortality among three major Taiwanese cities. Moreover, changes in other socioeconomic and demographic factors may also play a key role in determining seasonality mortality and morbidity and need to be considered in future studies.

M.-H. Li (✉)
Department of Geography, National Taiwan University, Taipei, Taiwan
e-mail: meihuili@ntu.edu.tw

© Springer International Publishing AG 2018

R. Akhtar, C. Palagiano (eds.), *Climate Change and Air Pollution*,
Springer Climate, DOI 10.1007/978-3-319-61346-8_13

Keywords Seasonal mortality patterns • Urbanization • Air pollution • Climate change

Introduction

The global urban population has exceeded rural population since 2014 (United Nations 2014). Urbanization is a process of intensive human activities in land use and economic development. Urbanization has numerous negative effects on air pollution worldwide, and urban areas are the significant emission sources of greenhouse gases due to concentrate industries, transportation, and households. The urban areas are also at great risk affected by climate change with increases in the frequency and intensity of heavy rainfalls, heat waves, and other extreme weather events (Lankao 2008; Romero-Lankao et al. 2012). Furthermore, air quality is strongly dependent on weather and is sensitive to climate change. Both climate change and air pollution are the most challenging global issues we face today. Many processes of urbanization contribute to climate change and air pollution such as combustion of fossil fuels and land use changes; therefore, cities have become research hotspots to understand the link between climate change and air pollution on human health.

Seasonal variations of mortality and disease in human society are well known. Proper assessment of seasonal mortality in a population is with important scientific and public health implications. While climate change may lead to alter seasonality of atmospheric condition, seasonal mortality patterns can be also influenced by these changes. Especially, air pollution and climate change can influence each other through complex interactions in the atmosphere and affect human health in different regions. There are many short-term effects or epidemiological studies on the relationships between air pollution and health or temperature and mortality in Taiwan. Several recent studies have already reported significant associations between daily temperature and daily mortality or cardiopulmonary diseases in Taiwanese cities (Liang et al. 2008, 2009; Lin et al. 2011, 2012, 2013a, b; Wang et al. 2012; Sung et al. 2013; Wang and Lin 2014). Moreover, there is growing evidence that particulate matter is responsible for mortality and cardiorespiratory diseases in Taiwanese cities (Tsai et al. 2010, 2014a, b, 2015; Chang et al. 2015a, b; Cheng et al. 2015; Wang and Lin 2015). However, these recent findings might not explain or predict overall seasonal mortality patterns. In fact, there are very limited studies on seasonality of weather, air pollution, and mortality in Taiwan.

The objectives of this study are to evaluate if there are any changes in trends and seasonality of weather, air pollution, and mortality in three Taiwanese cities from 1991 to 2010. First, the seasonal patterns of mortality, climate, and air quality are described in three major Taiwanese cities. Second, any changes in trends of mortality, climate, and air quality are examined in these three cities. Third, relationships between climate, air pollution, and mortality are investigated.

Methods

Study Area

Three metropolitans, Taipei, Taichung, and Kaohsiung, were selected for this study. Taipei is the largest and capital city of Taiwan at northern Taiwan. Kaohsiung is the second largest city and an industrial city located on the southwestern coast of Taiwan. Taichung is the third largest metropolitan area located in the west-central part of Taiwan. Table 13.1 shows some basic characteristics of these three cities. At the end of 2010, both Taichung and Kaohsiung cities were merged with Taichung and Kaohsiung counties to form large special municipalities, respectively. Therefore, monthly all-cause mortality, weather, and air quality data were analyzed from 1991 to 2010 for these three cities in this study.

Mortality Data

Monthly all-cause mortality data were retrieved online from the Ministry of Health and Welfare website during the period from 1991 to 2010 for Taipei, Taichung, and Kaohsiung. The seasonality index (100-Index) and winter/summer ratio were applied to assess seasonal mortality. A 100-Index was estimated by each month death relative to the average month death for each year and multiply it by 100. A winter/summer ratio was calculated as the number of winter deaths (December to March) divided by the number of summer deaths (June–September) for each year. On 21 September 1999, the Jiji earthquake occurred in central Taiwan, causing 87 and 112 deaths in Taipei and Taichung, respectively. Such deaths were excluded from calculating winter/summer ratio in Taipei and Taichung for 1999.

Climatological Data

Taipei (station no 466920), Taichung (station no 467490), and Kaohsiung (station no 467440) weather stations of the Central Weather Bureau (CWB) are located at urban centers with the most representative of the population's exposure in

Table 13.1 Characteristics of three Taiwanese cities

City	Area (km^2)[a]	Density in 2009 (persons/km^2)[a]	Topography	Climate
Taipei	271.8	9653	Taipei Basin	Subtropical monsoon
Taichung	161.9	6631	Taichung Basin	Subtropical monsoon
Kaohsiung	146.6	9948	Jianan Plain	Tropical monsoon

[a]Urban and Regional Planning Statistics, 2010, from Department of National Spatial Planning and Development for National Development Council, R.O.C. (Taiwan)

these three cities (Fig. 13.1). The climatological data were extracted from these three CWB weather stations from 1991 to 2010, with the monthly data including mean daily ambient temperature, relative humidity, atmospheric pressure, rainfall, hours of sunshine, diurnal temperature range, maximum and minimum temperatures, etc.

Air Quality Data

Air quality monitoring stations were fully automated and provided daily readings of SO_2 (by ultraviolet fluorescence), PM_{10} (by beta-ray absorption), NO_2 (by ultraviolet fluorescence), carbon monoxide (CO) (by nondispersive infrared photometry), and ozone (O_3) (by ultraviolet photometry) by the Taiwanese Environmental Protection Administration (EPA). Five, two, and six air quality monitoring stations in Taipei, Taichung, and Kaohsiung were selected to analyze average monthly data for SO_2, CO, PM_{10}, O_3, and NO_x from July of 1993 to December of 2010, respectively (Fig. 13.1). During the period of January 1991–June 1993, air pollution data only existed from one and three air quality monitoring stations in Taipei and Kaohsiung, respectively. There was no air quality data available for Taichung from January 1991 to June 1993. Therefore, air quality records between 1994 and 2010 were used for trend analysis in three cities.

Statistical Analysis

Because climate, air quality, and mortality data do not follow a normal distribution and can show seasonal changes within a year, nonparametric statistic methods are applied in all data analysis. Seasonal Mann-Kendall (MK) trend tests which defined each month as a "season" were used to assess monthly data change over 20 years. Classic MK trend test was also performed to assess and determine the presence of a trend on winter/summer ratios and annual mean metrological variables or air qualities. In this study, the magnitude of changes in metrological variables during the study period was determined by Sen's estimator method (Sen 1968), while the statistical significance was analyzed through MK test by using the NIWA's Time Trends and Equivalence software version 3.31 (Jowett 2012). Comparison of air qualities among three cities was determined by nonparametric Kruskal-Wallis test followed by Mann-Whitney test as post hoc test. The associations between mean monthly mortality and monthly temperature-related variables or air pollutant concentrations were evaluated by quadratic regression analysis. Pearson correlation coefficient was also used to estimate the correlation of monthly temperature-related variables or air pollutant concentrations with monthly mortality 100-Index.

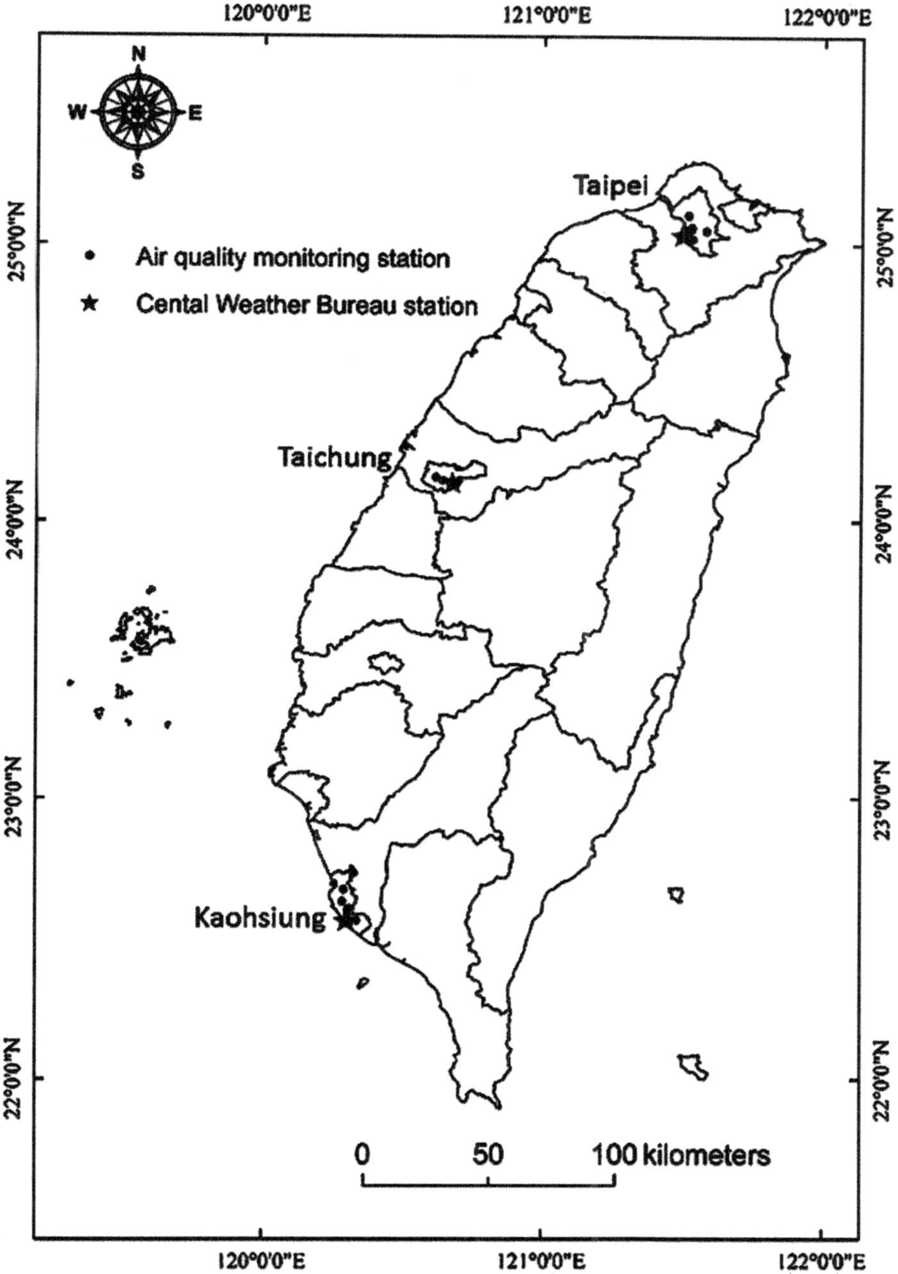

Fig. 13.1 The locations of weather and air quality monitoring stations in three major cities in Taiwan

Results

Seasonal Variations of All-Cause Mortality

Except winter/summer ratio of 2002 in Kaohsiung which was less than 1, all-cause mortality was higher in the winter (December to March) than in the other seasons at three cities during 1991–2010 (Fig. 13.2). Winter/summer ratios of mortality in Taipei showed a decreased trend ($P = 0.041$) from 1991 to 2010 as examined by MK test. No significant trend was observed for Taichung ($P = 0.256$) or Kaohsiung ($P = 0.230$) during the same period. The mean winter/summer ratio of 1.08 in Kaohsiung was the lowest among three cities with a range of 0.962–1.166. On the other hand, the mean winter/summer ratios in Taipei and Taichung were 1.13, but the mean winter/summer ratio in Taichung was with the highest variation ranging from 1.006 to 1.312.

Overall seasonal mortality (100-Index) patterns in three cities were with generally higher levels of deaths from December to March (Fig. 13.3). Mortality in July was also slightly higher than monthly average mortality in Kaohsiung, but not in Taipei or Taichung (Fig. 13.3). The 100-Index of Taipei ($P = 0.048$) and Taichung ($P = 0.015$) in March exhibited a decreased trend during a 20-year period as determined by MK trend tests. Furthermore, the 100-Index of Kaohsiung in August ($P = 0.041$), September ($P = 0.025$), and December ($P = 0.01$) all showed increased trends during a 20-year period.

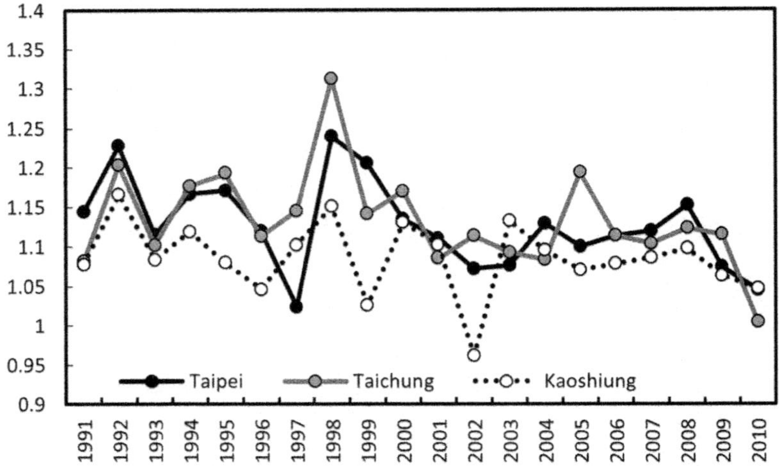

Fig. 13.2 Winter/summer ratios of mortality in three Taiwanese cities from 1991 to 2010 (Mortality data were retrieved from the Ministry of Health and Welfare of Taiwan)

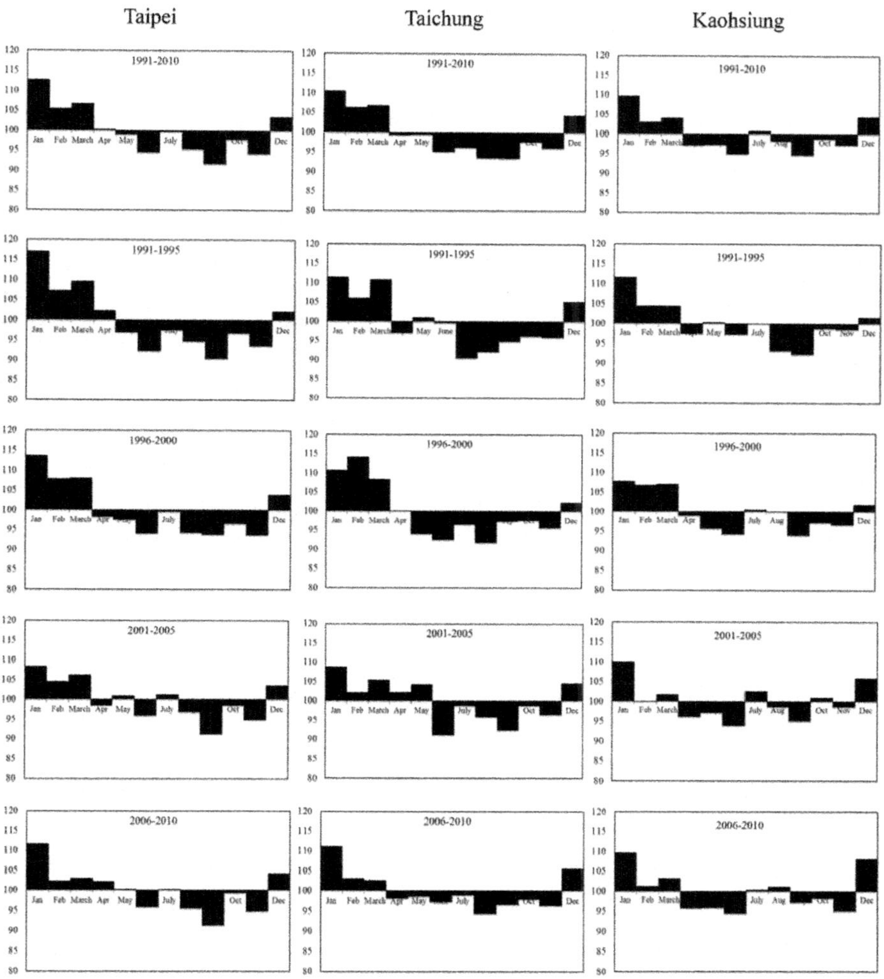

Fig. 13.3 Seasonality in mortality in three Taiwanese cities during the period 1991–2010 and every 5-year period (Mortality data were retrieved from the Ministry of Health and Welfare of Taiwan)

Trend and Seasonal Changes of Climatic Conditions

Taipei, Taichung, and Kaohsiung weather stations demonstrated significantly positive trends with a Sen slope value averaging 0.044, 0.020, and 0.020 °C/year in mean monthly temperature over 20 years, respectively (Table 13.2). Three weather stations also showed increased trends in monthly maximum relative humidity and mean minimum temperature (Table 13.2). Maximum temperature-related variables in Taipei and Kaohsiung displayed increased trends, but showed no changes in Taichung between 1991 and 2010 (Table 13.2). Monthly mean diurnal temperature

Table 13.2 Summary of different climatic trends determined by using the seasonal Mann-Kendall test and Sen's slope methods during the period 1991–2010 in three major Taiwanese cities

		Taipei			Taichung			Kaohsiung	
Variable	*N*	Monthly M-K test P	Sen slope	*N*	Monthly M-K test P	Sen slope	*N*	Monthly M-K test P	Sen slope
PP01	240	0.667	0.412	237	0.099	0.600	237	0.800	0.015
PS01	240	0.245	0.017	240	**0.029**	**−0.036**	240	**0.001**	**−0.050**
RH01	240	0.902	0.000	240	0.061	0.000	240	0.783	0.000
RH02	240	0.300	0.071	240	0.792	0.000	240	0.125	−0.111
RH04	137	**0.000**	**5.500**	137	**0.000**	**5.588**	137	**0.000**	**5.500**
SS01	240	0.085	0.659	240	0.772	−0.104	240	**0.000**	**1.801**
SS02	240	0.075	0.200	240	0.888	−0.017	240	**0.000**	**0.490**
TX01	240	**0.000**	**0.044**	240	**0.029**	**0.020**	240	**0.021**	**0.020**
TX02	240	**0.000**	**0.044**	240	0.342	0.007	240	0.077	0.020
TX04	240	**0.000**	**0.055**	240	0.077	0.020	240	**0.000**	**0.006**
TX06	240	**0.007**	**0.050**	240	**0.036**	**0.033**	240	0.352	0.012
TX08	240	**0.000**	**0.060**	240	0.632	0.000	240	**0.000**	**0.050**
TX09	240	**0.000**	**0.043**	240	**0.000**	**0.033**	240	**0.030**	**0.018**
TX10	240	**0.029**	**0.014**	240	**0.005**	**−0.019**	240	**0.001**	**0.018**
TX11	240	0.319	0.014	240	**0.006**	**−0.036**	240	**0.017**	**0.025**
DY03	240	**0.018**	**0.000**	240	0.879	0.000	240	**0.001**	**0.000**
DY04	240	**0.007**	**0.000**	240	0.519	0.000	240	0.523	0.000
DY05	240	0.972	0.000	240	0.945	0.000	240	0.924	0.000

The bold values represent the significant trend at the 5% level

PP01 precipitation (mm), *PS01* mean station pressure (hPa), *RH01* mean relative humidity (%), *RH02* minimum relative humidity (%), *RH04* maximum relative humidity (%), *SS01* sunshine duration (hour), *SS02* rate of sunshine (%), *TX01* mean ambient temperature (°C), *TX02* dew point temperature (°C), *TX04* absolute maximum temperature (°C), *TX06* absolute minimum temperature (°C), *TX08* mean maximum temperature (°C), *TX09* mean minimum temperature (°C), *TX10* mean diurnal temperature range (°C), *TX11* maximum diurnal temperature range (°C), *DY03* number of days with maximum temperature ≧30 °C, *DY04* number of days with maximum temperature ≧35 °C, *DY05* number of days with minimum temperature ≦10 °C

range showed significantly increased trends in Taipei and Kaohsiung, but a significantly decreased trend in Taichung. Taipei had significantly increased trends in most monthly temperature variables and the number of hot days examined during 1991–2010 compared to the other two cities (Table 13.2).

The average diurnal temperature range from Taipei weather station observations is larger during summer (May–August) than during other months (Fig. 13.4). In contrast, the average diurnal temperature ranges from Taichung and Kaohsiung weather stations are larger during winter (December–March) than during other months (Fig. 13.4). The sunshine duration and rate of sunshine in Kaohsiung displayed positive trends during the 20-year period, but not in Taipei or Taichung.

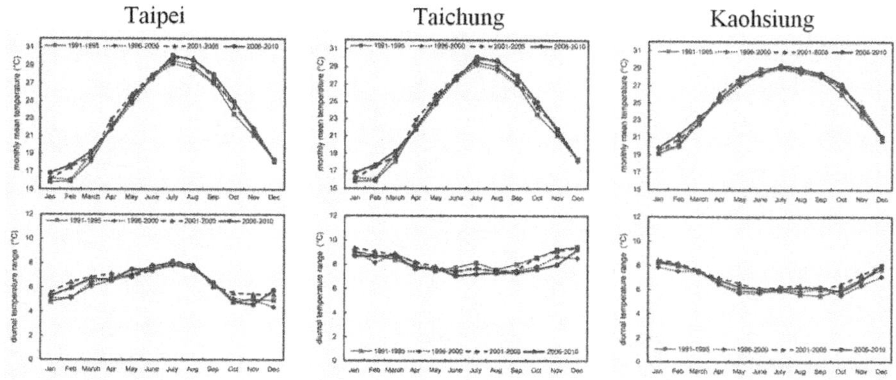

Fig. 13.4 Seasonality in mean temperature and diurnal temperature range in three weather stations at three cities during each 5-year period from 1991 to 2010 (The climatological data were obtained from Central Weather Bureau)

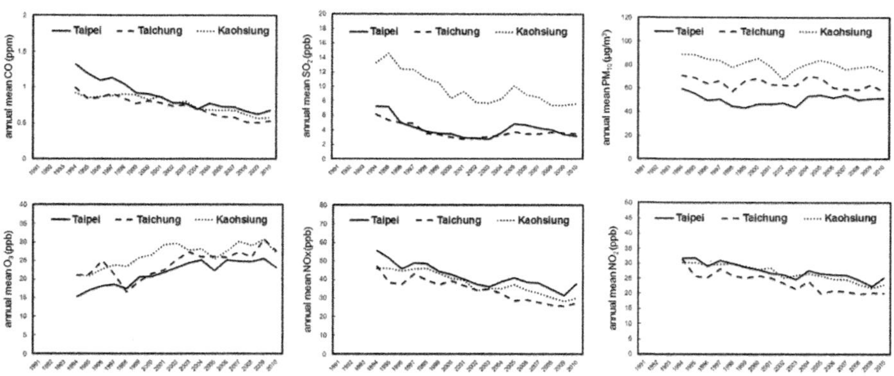

Fig. 13.5 Annual mean concentrations of air pollutants in three major Taiwanese cities during 1994–2010 (Air quality data were obtained from Taiwanese EPA)

Interestingly, the mean station pressure in Taichung and Kaohsiung showed negative trends during 1991 to 2010 (Table 13.2).

Trend and Seasonal Changes of Air Qualities

Based on the results of seasonal M-K trend tests, trends of all air quality parameters were significantly changed in all three cities with p values less than 0.01 during the study period 1994–2010. Trends in air quality are showing mixed patterns over the past two decades. SO_2, CO, and NO_x concentrations have decreased significantly and steadily, while O_3 has significantly increased in recent years (Fig. 13.5). On the

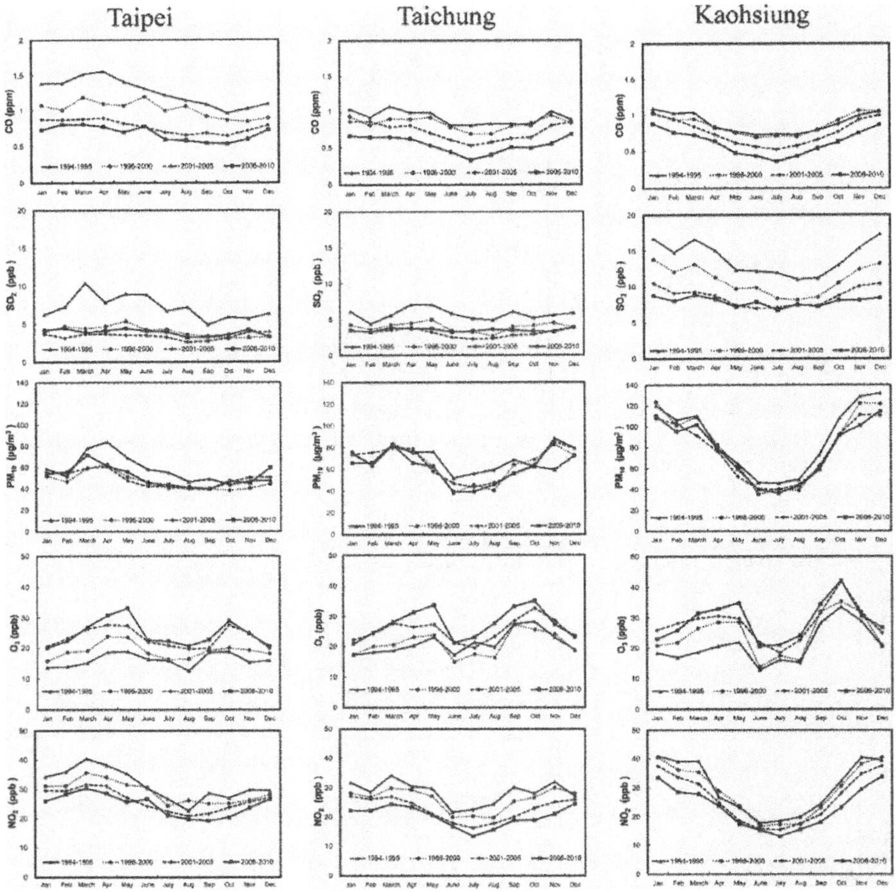

Fig. 13.6 Monthly variation of air qualities in three major Taiwanese cities, 1994–2010 (Air quality data were obtained from Taiwanese EPA)

other hand, traffic-related air pollutants, such as NO_2 and PM_{10}, have been kept constant over the past decade (Fig. 13.5). Overall, O_3 and PM_{10} are major air pollutants of current concerns in three major Taiwanese cities. Among three cities, the concentrations of SO_2, O_3, and PM_{10} in Kaohsiung were higher than those in Taipei and Taichung ($P < 0.001$). The levels of CO and NO_x in Taipei were higher than those in Taichung and Kaohsiung. The levels of NO_2 in Taichung were lower than those in Taipei and Kaohsiung (Fig. 13.5). The O_3 levels showed two peaks in May and October in all three cities, respectively (Fig. 13.6). The concentrations of CO, PM_{10}, and NO_2 showed a seasonal pattern with a peak in winter (January and December) in Kaohsiung, but not in Taipei or Taichung (Fig. 13.6).

Associations Between Climate, Air Pollution, and Mortality

Figures 13.7 and 13.8 present the monthly mortality 100-Index in relation to the monthly temperature-related variables and air pollutant concentrations in these three cities during the study period. Mean ambient temperature was found to be the most effective temperature variable among the temperature-related variables for predicting all-cause mortality 100-Index in all three cities (Fig. 13.7). Quadratic regression analysis in association with air pollutant concentrations and monthly mortality was not statistically significant in all three cities, and regression equations were not shown in Fig. 13.8. By calculating Pearson correlation coefficients, mean monthly O_3 concentrations showed no significant correlation with the monthly mortality at three cities (Fig. 13.8). In contrast, mean monthly PM_{10} and NO_x concentrations showed significant correlation with the monthly mortality 100-Index at three cities ($P < 0.01$). Interestingly, monthly mean diurnal temperature range was negatively correlated with the monthly mortality 100-Index at Taipei ($r = -0.266$; $P < 0.001$), but was positively correlated with 100-Index at both Taichung ($r = 0.318$; $P < 0.001$) and Kaohsiung ($r = 0.538$; $P < 0.001$).

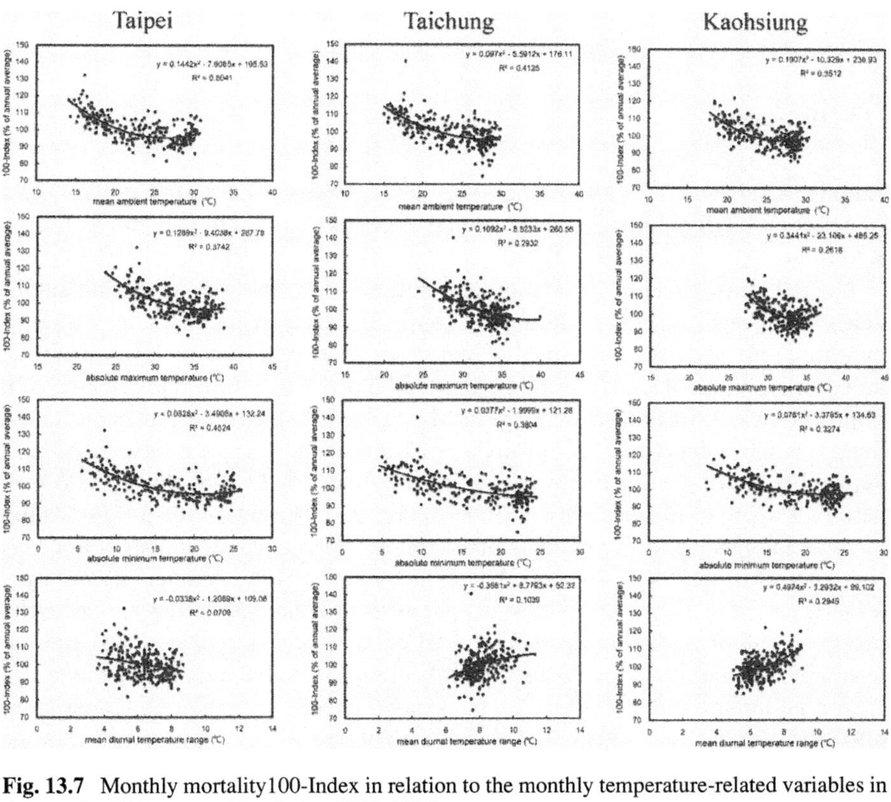

Fig. 13.7 Monthly mortality100-Index in relation to the monthly temperature-related variables in three major Taiwanese cities during 1991–2010

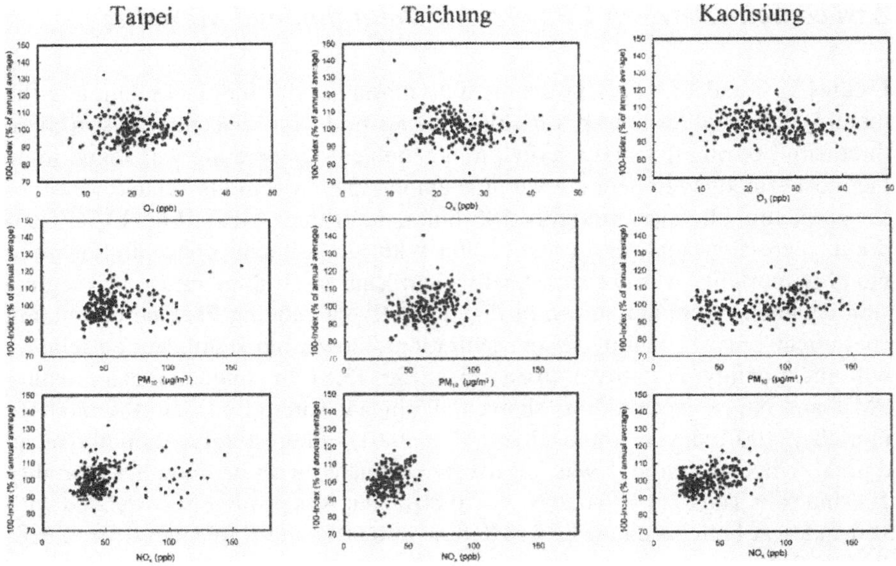

Fig. 13.8 Monthly mortality100-Index in relation to the monthly mean O_3, PM_{10}, and NO_x in three major Taiwanese cities during 1994–2010

Discussion

Among these major Taiwanese cities, seasonal mortality patterns are similar in two subtropical cities, Taipei and Taichung, compared to another tropical city, Kaohsiung.

Overall, seasonality index of mortality in three cities showed decreasing amplitude of seasonal variations during the past 20 years. Winter/summer ratios of mortality only showed a statistically significant decreased trend in Taipei, but not in Taichung or Kaohsiung. Monthly analyses showed that 100-Index of two subtropical cities, Taipei and Taichung, in March exhibited a significantly decreased trend. On the other hand, the 100-Index of Kaohsiung, a tropical city, in August, September, and December showed significantly increased trends during a 20-year period. Taipei is the most densely populated city in Taiwan and had significantly increased trends in most monthly temperature variables and the number of hot days examined during 1991–2010 compared to the other two cities. Ambient temperature was suggested as the most optimal temperature variable among high-temperature indices for predicting all-cause daily mortality in Taiwan (Lin et al. 2012). Similar results were also found for all-cause monthly mortality at all three cities in this study.

Air pollutants did not show to be a good predictor for monthly mortality 100-Index for all three cities. In Taiwan, ambient air quality has improved in the last two decades. However, there is a large body of evidence suggesting that exposure to air pollution, even at the current levels, leads to adverse health effects. In Kaohsiung, higher levels of ambient air pollutants increase the risk of hospital admissions for cardiovascular diseases (Chang et al. 2015a), respiratory diseases (Tsai et al. 2014b; Cheng et al. 2015), and daily mortality for all causes (Tsai and Yang 2014; Tsai et al. 2015). In Taipei, particulate matter and traffic-related air pollutants, CO, O_3, and NO_x, were positively associated with increased risk of hospital admissions for cardiovascular diseases (Yang 2008; Chiu et al. 2013), asthma (Chan et al. 2009), respiratory diseases (Yu and Chien 2016), emergency room visits for stroke in the warm seasons (Chen et al. 2014), and daily mortality for all causes (Tsai et al. 2014a). On the other hand, many epidemiological studies showed that air pollution level and daily mortality lack a strong association either in Taipei or Kaohsiung (Tsai et al. 2003; Yang et al. 2004; Tseng et al. 2015). The result of this study also showed that monthly mean O_3, PM_{10}, and NO_x levels and monthly mortality were not closely related at these three cities.

In conclusion, monthly mean temperature-related variables, but not monthly mean air qualities, are positively associated with monthly mortality among three major Taiwanese cities. Moreover, the changes in other socioeconomic and demographic factors may also play a key role in determining seasonality mortality and morbidity and shall be considered in future studies.

References

Chan TC, Chen ML, Lin IF, Lee CH, Chiang PH, Wang DW, Chuang JH (2009) Spatiotemporal analysis of air pollution and asthma patient visits in Taipei, Taiwan. Int J Health Geogr 8(1):1

Chang CC, Chen PS, Yang CY (2015a) Short-term effects of fine particulate air pollution on hospital admissions for cardiovascular diseases: a case-crossover study in a tropical city. J Toxicol Environ Health A 78(4):267–277

Chang CC, Chiu HF, Yang CY (2015b) Fine particulate air pollution and outpatient department visits for headache in Taipei, Taiwan. J Toxicol Environ Health A 78(8):506–515

Chen SY, Lin YL, Chang WT, Lee CT, Chan CC (2014) Increasing emergency room visits for stroke by elevated levels of fine particulate constituents. Sci Total Environ 473:446–450

Cheng MH, Chiu HF, Yang CY (2015) Coarse particulate air pollution associated with increased risk of hospital admissions for respiratory diseases in a tropical city, Kaohsiung, Taiwan. Int J Environ Res Public Health 12(10):13053–13068

Chiu HF, Peng CY, Wu TN, Yang CY (2013) Short-term effects of fine particulate air pollution on ischemic heart disease hospitalizations in Taipei: a case-crossover study. Aerosol Air Qual Res 13:1563–1569

Jowett I (2012) Time trends and equivalence software, version 3.31

Lankao PR (2008) Urban areas and climate change: review of current issues and trends – issues paper for the 2011 global report on human settlements, 101p. http://www.Ral.Ucar.Edu/staff/prlankao/GRHS_2011_IssuesPaperfinal.Pdf. Accessed 18 August 2016

Liang WM, Liu WP, Chou SY, Kuo HW (2008) Ambient temperature and emergency room admissions for acute coronary syndrome in Taiwan. Int J Biometeorol 52(3):223–229

Liang WM, Liu WP, Kuo HW (2009) Diurnal temperature range and emergency room admissions for chronic obstructive pulmonary disease in Taiwan. Int J Biometeorol 53(1):17–23

Lin YK, Ho TJ, Wang YC (2011) Mortality risk associated with temperature and prolonged temperature extremes in elderly populations in Taiwan. Environ Res 111(8):1156–1163
Lin YK, Chang CK, Li MH, Wu YC, Wang YC (2012) High-temperature indices associated with mortality and outpatient visits: characterizing the association with elevated temperature. Sci Total Environ 427:41–49
Lin YK, Chang CK, Wang YC, Ho TJ (2013a) Acute and prolonged adverse effects of temperature on mortality from cardiovascular diseases. PloS One 8(12):e82678
Lin YK, Chang CK, Chang SC, Chen PS, Lin C, Wang YC (2013b) Temperature, nitrogen dioxide, circulating respiratory viruses and acute upper respiratory infections among children in Taipei, Taiwan: a population-based study. Environ Res 120:109–118
Romero-Lankao P, Qin H, Dickinson K (2012) Urban vulnerability to temperature-related hazards: a meta-analysis and meta-knowledge approach. Glob Environ Chang 22(3):670–683
Sen PK (1968) Estimates of the regression coefficient based on Kendall's tau. J Am Stat Assoc 63:1379–1389
Sung TI, Wu PC, Lung SC, Lin CY, Chen MJ, Su HJ (2013) Relationship between heat index and mortality of 6 major cities in Taiwan. Sci Total Environ 442:275–281
Tsai S, Yang CY (2014) Fine particulate air pollution and hospital admissions for pneumonia in a subtropical city: Taipei, Taiwan. J Toxicol Environ Health A 77(4):192–201
Tsai SS, Huang CH, Goggins W, Wu TN, Yang CY (2003) Relationship between air pollution and daily mortality in a tropical city: Kaohsiung, Taiwan. J Toxicol Environ Health A 66 (17):1341–1349
Tsai DH, Wang JL, Chuang KJ, Chan CC (2010) Traffic-related air pollution and cardiovascular mortality in central Taiwan. Sci Total Environ 408(8):1818–1823
Tsai SS, Chang CC, Liou SH, Yang CY (2014a) The effects of fine particulate air pollution on daily mortality: a case-crossover study in a subtropical city, Taipei, Taiwan. Int J Environ Res Public Health 11(5):5081–5093
Tsai SS, Chiu HF, Liou SH, Yang CY (2014b) Short-term effects of fine particulate air pollution on hospital admissions for respiratory diseases: a case-crossover study in a tropical city. J Toxicol Environ Health A 77(18):1091–1101
Tsai SS, Weng YH, Chiu YW, Yang CY (2015) Short-term effect of coarse particles on daily mortality rate in a tropical city, Kaohsiung, Taiwan. J Toxicol Environ Health A 78 (23–24):1409–1420
Tseng E, Ho WC, Lin MH, Cheng TJ, Chen PC, Lin HH (2015) Chronic exposure to particulate matter and risk of cardiovascular mortality: cohort study from Taiwan. BMC Public Health 15(1):1
United Nations (2014) World urbanization prospects: the 2014 revision, highlights, department of economic and social affairs, population division, New York, (ST/ESA/SER.A/352)
Wang YC, Lin YK (2014) Association between temperature and emergency room visits for cardiorespiratory diseases, metabolic syndrome-related diseases, and accidents in metropolitan Taipei. PloS One 9(6):e99599
Wang YC, Lin YK (2015) Mortality associated with particulate concentration and Asian dust storms in Metropolitan Taipei. Atmos Environ 117:32–40
Wang YC, Lin YK, Chuang CY, Li MH, Chou CH, Liao CH, Sung FC (2012) Associating emergency room visits with first and prolonged extreme temperature event in Taiwan: a population-based cohort study. Sci Total Environ 416:97–104
Yang CY (2008) Air pollution and hospital admissions for congestive heart failure in a subtropical city: Taipei, Taiwan. J Toxicol Environ Health A 71(16):1085–1090
Yang CY, Chang CC, Chuang HY, Tsai SS, Wu TN, Ho CK (2004) Relationship between air pollution and daily mortality in a subtropical city: Taipei, Taiwan. Environ Int 30:519–523
Yu HL, Chien LC (2016) Short-term population-based non-linear concentration–response associations between fine particulate matter and respiratory diseases in Taipei (Taiwan): a spatiotemporal analysis. J Expo Sci Environ Epidemiol 26(2):197–206

Mei-Hui Li is a professor in the department of Geography at National Taiwan University. She received her Ph.D. degree from the University of Illinois at Urbana-Champaign. Her research interests extend across a wide range of environmental problems, with a particular focus on how anthropogenic activities affect the environment and human health. Her current work includes the development of biomarkers and bioindicators to monitor aquatic ecosystems, as well as how environmental change impacts ecosystem services and human health.

Chapter 14
Climate Change and Urban Air Pollution Health Impacts in Indonesia

Budi Haryanto

Abstract Climate change in Indonesia greatly affects economy, poor population, human health, and the environment. It influences air pollutant emissions as higher emissions of carbon dioxide (CO_2) have caused rapidly worsening air pollution. Urban areas being most affected by air pollution. The transportation sector contributes the most (80%) to the air pollution followed by emissions from industry, forest fires, and domestic activities. The large number of vehicles together with lack of infrastructure results in major traffic congestions resulting in high levels of air polluting substances, which have a significant negative effect on public health. Current air pollution problems are greatest in Indonesia as it caused 50% of morbidity across the country. Diseases stemming from vehicular emissions and air pollution include acute respiratory infection, bronchial asthma, bronchitis, and eye, skin irritations, lung cancer, and cardiovascular diseases. The prevalence and incidence rate of diseases related to air pollution is predicted to become worse in the near future since the range growth of energy consumption is about 6–8% per year. It is impacted to the increasing of NOx up to 51% (from 814 kt/year in 2015 to 1,225 kt/year in 2030), $PM_{2.5}$ up to 26% (from 87.7 kt/year in 2015 to 110.5 kt/year in 2030), as well as other pollutants such as SO_2, PM_{10}, VOC, and O_3. Most recently, some studies on developing scenarios for reducing emission have been conducted. These include analysis of fuel economy and the time effective for Euro 4 standard implementation as compliment to transportation improvement policy in Indonesia, in which it suggested that the government of Indonesia must enhance energy security and mitigate CO_2 emissions, improve efficiency in energy production and use, increase reliance on non-fossil fuels, and sustain the domestic supply of oil and gas through decreased fossil fuel consumption, support the use of proposed breakthrough technologies, and protect human health from air pollution by conducting more research on health vulnerability and implementing more effective adaptation of human health.

B. Haryanto (✉)
Department of Environmental Health, Faculty of Public Health, Research Center for Climate Change, University of Indonesia, Depok, Indonesia
e-mail: bharyant@cbn.net.id; bharyanto@ui.ac.id

© Springer International Publishing AG 2018

R. Akhtar, C. Palagiano (eds.), *Climate Change and Air Pollution*,
Springer Climate, DOI 10.1007/978-3-319-61346-8_14

Keywords Climate change • Air pollution • Health impacts • Reduction emission scenarios

Indonesia and Climate Change

Indonesia is the world's largest archipelagic state encompassing over 17,000 islands and home to over 260 million inhabitants (WPR 2016), which makes it the fourth most populated country in the world and a significant emitter of greenhouse gases due to deforestation and land-use change (WRI 2005). The Indonesian archipelago lies between Asia and Australia. It is bounded by the South China Sea in the north, the Pacific Ocean in the north and east, and the Indian Ocean in the south and west. More than 80% of Indonesia's territory is covered with water; the land area is about 1.9 million square kilometers. The urbanization rate is very high (4.4%). Two-thirds of the total population and more than half of the poor (57%) reside on Java (BPS 2016). Nearly 60% of Indonesia's terrestrial area is forested. However, deforestation and land-use change is estimated at two million hectares (ha) per year and accounts for 85% of the Indonesia's annual greenhouse gas emissions (WRI 2002). Indonesia's forested land also supports extremely high levels of biodiversity, which, in turn, support a diverse array of livelihoods and ecosystem services. Vulnerability for food security is high due to the country's dependence on the production of rice, the primary staple food, which is projected to decrease as a result of climate change. Poverty of a large part of the population (110–140 million live on less than USD 2 per day) decreases their adaptive capacity to the effects of climate change (World Bank 2014). The combination of high population density and high levels of biodiversity makes Indonesia one of the most vulnerable countries to the impacts of climate change.

Trends and Future Climate

Indonesia is extremely vulnerable to climatic hazards, including a sea-level rise, and the frequency of natural hazards appears to be increasing (Suryanti 2006). Since 1990, the temperature in Indonesia has increased by 0.3 °C, and it is expected to increase in the range of 1.5–3.7 °C by 2100, with a mean increase across models of 2.5 °C (IPCC 2007a). The increase in GHG emissions will also continue to affect the "natural" climate variability, thus leading to more intense weather events (Case et al. 2007).

Indonesia also experienced intense rainfall due to the impact of climate change which is predicted to result in about 2–3% more rainfall in Indonesia each year (Sari et al. 2007). The amplified rainfall is expected to persist and result in a shorter rainy season, with a substantial increase in the risk of floods. For example, the Jakarta flood in February 2007 affected 80 districts and caused traffic chaos paralyzing the affected cities. In the flood more than 70,000 houses had water levels ranging from 5 to 10 cm, and an estimated 420,000 to 440,000 people were displaced from their homes (Case et al. 2007).

Climate change will also increase the average sea level as a result of the increased volume of warmer water and the melting of polar ice caps. The mean sea level in the Jakarta Bay will rise as much as 0.57 centimeters (cm) annually, and the land surface will decline as high as 0.8 cm per year. In Indonesia, the combination of rising sea levels and land subsidence will move the coastline inland, which will cause an increasing risk of flooding (ADB 2009).

Impacts of Climate Change

Climate change in Indonesia greatly affects many aspects of the country, including economy, poor population, human health, and the environment. Vulnerability studies have illustrated that the economically productive areas of Bali, Java, Sumatra, and Papua are particularly vulnerable to the effects of climate change (World Bank 2009). The poor communities that live on the coast and those dependent on agriculture will greatly be affected by droughts, sea-level rises, floods, and landslides (World Bank 2010). Despite these hazards, the annual benefit of adopting measures to combat climate change is likely to exceed its expected costs by the year 2050 (World Bank 2010). Thus, adopting methods and policies to mitigate climate change now will promote the potential development of Indonesia and help to preserve the country's rich biodiversity.

Changing climate is already affecting the timing of seasons in Indonesia, with the onset of the wet season delayed by up to 20 days in the period 1991–2003 compared to 1960–1990 in parts of Sumatra and Java, and it is expected that climate change will cause a longer dry season and more intense wet season over much of Indonesia. El Niño has a large impact on Indonesian climate. Its effect includes decreased rainfall and water storage and an increased area affected by drought and fire among others, whereas La Niña increases precipitation and is linked to flooding.

Approximately 60% of Indonesians live in low-lying coastal cities and extremely vulnerable to sea-level rise, with the 42 million people who live less than 10 meter (m) above sea level. A 1 m rise in sea level could inundate 405,000 ha of land and reduce Indonesia's territory by inundating low-lying islands which mark its borders, and a 50 cm rise in sea level, combined with land subsidence in Jakarta Bay, could permanently inundate densely populated areas of Jakarta and Bekasi with a population of 270,000 (PEACE 2007). The sea-level rise, along with the observed sinking in the Jakarta Bay region, will have massive influences on infrastructure and businesses (Case et al. 2007). The rise will also reduce coastal livelihoods and farming. The sea-level rise will most likely affect the production of both fish and prawn, with an estimated loss of over 7000 tons, worth over 0.5 million US dollars, in the Krawang and Subang districts. The Citarum Basin is also expected to experience a loss of 15,000 tons of fish, shrimp, and prawn yield. The overall effect of this sea-level rise will result in the reduction of potential average income. For example, it is predicted that in the Subang region alone, 43,000 farm laborers will lose their jobs. Also, more than 81,000 farmers will have to seek other sources of income due to the flooding of farms from rising sea levels (Sari et al. 2007).

Climate change will also pose a threat to food security in Indonesia. One of the major concerns for Indonesia is the risk of a reduced food security due to climate change. Climate change will affect evaporation, precipitation, and run-off soil moisture and water, hence affecting agriculture and food security. For example, the 1997 El Nino droughts affected approximately 426,000 hectares of rice. A model that simulated the impacts of climate change on crops at the Goddard Institute of Space Studies in the United Kingdom depicted a decrease of crop harvest in East and West Java. Along with these effects, climate change will also lower soil fertility by 2–8%, which will result in the estimated decrease of rice yields by 4% per year and maize by 50% per year (Sari et al. 2007).

Human health in Indonesia will be both directly affected by climate change, through deaths from floods and other disasters, and also indirectly affected due to increased infections and diseases. The more frequent prolonged heat waves, extreme weather, floods, and droughts caused by climate change will also lead to increased injury, sickness, and mortality (Case et al. 2007). The direct effects – higher temperatures, sea-level rising, and frequent floods and heat waves – will lead to more injury and deaths. Extreme occurrences influenced by climate change in Indonesia, such as floods, hurricanes, tidal waves, landslides, droughts, and forest fires, are happening more often than before. There are 300 events of extreme occurrences from January to August 2008, resulting in 263 deaths, 1927 critically injured, 66,988 with mild injuries, 7 missing, and 92,210 refugees. Those refugees are susceptible to easily spreading communicable diseases, even worsened by unpredictable climate. A rise in seawater temperature has contributed to the spread of diseases such as malaria, dengue fever, diarrhea, cholera, and other vector-related diseases (La Niña years). A change in temperature, humidity, and wind speed is also contributing to the increase in vector population, increasing their life-span and also widening their spread. This in turn may intensify the occurrence of vector-related communicable diseases such as leptospirosis, malaria, dengue fever, yellow fever, schistosomiasis, filariasis, and plague (Haryanto 2016). Many people in Indonesia will also experience enlarged respiratory effects as a result of increased burning and air pollution. Numerous studies have also observed the association between climate-related factors – severe floods, droughts, and warming temperatures – with diarrheal diseases such as malaria, hepatitis, cholera, and dengue fever (Case et al. 2007). The rise in sea levels, precipitation changes, and increased flooding may also degrade the quality of freshwater and potentially contaminate drinking water. Thus, water-borne disease will become more common in the region. Once again, the poor in Indonesia are going to be the most impacted by the threat to human health posed by climate change. Many of the region's poor live in coastal areas, and most of the small farmers and fisherman are too poor to acquire access to sufficient health services. Thus, the poor lack a safety net to protect them against the threats that climate change causes.

Climate Change and Air Pollution

Weather conditions influence air quality via the transport and/or formation of pollutants (or pollutant precursors). Weather conditions can also influence air pollutant emissions, both biogenic emissions (such as pollen production) and anthropogenic emissions (such as those caused by increased energy demand) (Haryanto 2016). Higher emissions of carbon dioxide (CO_2) have caused rapidly worsening air pollution that wreaks havoc on the environment and people's health, a problem that Indonesia knows far too well. Air pollution from fuel burning and forest fire is well known as the main contributor driver for climate change in Indonesia.

Urban areas are being most affected by air pollution. The transportation sector contributes the most (80%) to the air pollution followed by emissions from industry, forest fires, and domestic activities. The large number of vehicles together with lack of infrastructure results in major traffic congestions (mainly in urban centers) resulting in high levels of air polluting substances, which have a significant negative effect on public health, quality and quantity of crops, forests buildings, and surface water quality.

The average of sulfur content used for diesel fuel in Indonesia is 2156 ppm (between 400 and 4600 ppm) in 2007 (Bappenas and Swisscontact 2006). The sulfur concentration is higher than 2006 (1494 ppm). In 29 cities, sulfur concentration is found above 1000 ppm. Index of PM_{10} concentration in Jakarta in 2001–2015 (Air Quality Monitoring System) shows the yearly average about three times higher than WHO standard (20 μg/m^3) (Pusarpedal KLHK 2015). The source of PM_{10} in Jakarta is from fuel burning and soil. The excess of PM10 and SO3 concentration also occurred in the cities of Surabaya and Bandung (BPLHD DKI 2009). Air quality monitoring using non-AQMS in 30 cities shows high concentration for NO_2 (0–30 ppm) and SO_2 (0–50 ppm). A number of vehicles used on the road increase annually with the average of 12% (motorcycle 30%) which are linear with the increasing of fuel consumption. Emission test in Jakarta 2005 found that 57% vehicles did not pass the test. Meanwhile the traffic jams among cities continue and worsen. Kerosene is the cooking fuel used by 45% of the households sampled (BPS SUSENAS 2005). Fuelwood is used by 42% of the households sampled (in 12 provinces, >50% households used fuelwood for cooking).

Indonesia holds the world's third largest tropical forests, covering almost two-thirds of the country's land area, and globally significant biodiversity. Over the past 50 years, Indonesia has lost over 40% of its total forest cover. Currently the deforestation rate is very high (1.8% annually). Between 2000 and 2005, forest loss rate per year is 1.1 million ha (MoE 2009) and 0.4 million ha from 2009 to 2011

(MoF 2013). This is alarming as the forest sector provides important ecosystem services, significantly supports the country's economic development, and contributes to livelihoods, particularly for the rural poor. The Indonesian forests are threatened by logging and agricultural clearance that results in deforestation.

Fires associated with agricultural and plantation development in Indonesia impact ecosystem services and release emissions into the atmosphere that degrade regional air quality and contribute to greenhouse gas concentrations. Primary forest clearance in Indonesia totaled 6.02 Mha from 2000 to 2012 (Margono et al. 2014), with some of the highest deforestation rates observed in carbon-rich peatland forests in Sumatra and Kalimantan (Miettinen et al. 2011; Margono et al. 2014). Forty-five percent of Indonesia's deforestation from 2000 to 2010 was observed on oil palm, timber, logging, and coal mining concessions (Abood et al. 2015), and by 2010, industrial plantations covered 2.3 Mha of peatlands in Sumatra and Kalimantan, with approximately 70% developed since 2000 (Miettinen et al. 2012a). Fires are considered to be a cheap and effective method to clear and maintain land for agricultural and plantation development (Simorangkir 2007), but also damage biodiversity, reduce carbon storage potential, and can severely degrade regional air quality.

In 2015, within June to October, it is estimated that more than 2.6 million ha of Indonesia's forest and peatland are burned, bumping the country's annual emission from sixth largest emitter in the world to fourth largest. Various data also recorded an increase of at least 55% more hotspots in 2015 compared to 2014, where Sipongi (Ministry of Environment and Forestry's database) shows that in 11 prioritized provinces, there are more than 108,622 hotspots, with Central Kalimantan and South Sumatra ranked number one and two, respectively, with 30,204 and 28,327 hotspots (MoEF 2015). A more urgent and devastating consequence of wildfires is its effect on people's health, directly and immediately affecting people who live in haze-affected areas. Pollutant standard index (PSI) reached the highest level in September and October 2015, far above the very dangerous level of 400. In Central Kalimantan, the PSI reached the highest level of 3300, ten times the dangerous level of 300.

Diseases Related to Air Pollution: Research Evidence

Ambient air pollution, which is mainly caused by the combustion of nonrenewable fossil fuels for electricity generation, transport, and industry, has been worsening over the past five decades (Rowshand et al. 2009; Ying et al. 2015). Many epidemiological studies have indicated that air pollutants such as particulate matter (PM), nitrogen dioxide (NO_2), sulfur dioxide (SO_2), and ozone (O_3) are responsible for increasing mortality and morbidity in different populations around the world, especially from respiratory and cardiovascular diseases (CVD) (Rowshand et al. 2009; Samet and Krewski 2007; Tsai et al. 2014; Tsangari et al. 2016). A global study of the burden of diseases in the year 2000 suggested that nearly two-thirds of the estimated 800,000 deaths and 4.6 million lost years of healthy life worldwide caused by exposure to air pollution in that year were in the developing countries of Asia (World Health Organization 2002), and this phenomenon has continued until very recently (World Health Organization 2014). Air pollution in major cities, especially in developing countries, has reached a crisis point. The bad air quality is responsible for the death of three million people each year and presents a dilemma for millions worldwide that suffer asthma, acute respiratory diseases, cardiovascular diseases, and lung cancer (MOE and KPBB 2006). In Indonesia, exposure to air pollutants can have many serious health effects, especially following severe pollution episodes. Long-term exposure to elevated levels of air pollution may have greater health effects than acute exposure. Current air pollution problems are greatest in Indonesia as it caused 50% of morbidity across the country (Haryanto and Franklin 2011).

Air pollution is proven as a major environmental hazard to residents in Jakarta, regardless of their socioeconomic status. Transportation comprises 27% of Indonesia's GHG emissions, and traffic congestion is a huge problem in Jakarta. Diseases stemming from vehicular emissions and air pollution include acute respiratory infection, bronchial asthma, bronchitis, and eye and skin irritations, and it has been recorded that the most common disease in northern Jakarta communities is acute upper respiratory tract infection – at 63% of total visits to health-care centers (Haryanto 2008). The prevalence of acute respiratory infection exceeds the national prevalence (25.5%) in 16 provinces, whereas the top 10 highest rank of the prevalence are in the city/district Kaimana (63.8%), Manggarai Barat (63.7%), Lembata (62%), Manggarai (61.1%), Pegunungan Bintang (59.5%), Ngada (58.6%), Sorong Selatan (56.5%), Sikka (55.8%), Raja Ampat (55.8%), and Puncak Jaya (56.7%). The prevalence of cough in 2007 is 45% and flu 44% without any significant different between urban and rural.

National Basic Health Research 2007 reported that the prevalence of acute respiratory infection exceeds the national prevalence (25.5%) in 16 provinces. The prevalence of cough in 2007 is 45% and flu 44% without any significant difference between urban and rural. The prevalence of pneumonia exceeds the national prevalence (2.18%) in 14 provinces. The prevalence of tuberculosis (TB) exceeds the national prevalence (0.99%) in 17 provinces. In 2007, a number of 232,358 cases found out of 268,042 TB cases (86.7%). The prevalence of asthma exceeds the national prevalence (4%) in nine provinces (Ministry of Health 2008). Pneumonia is overall the number one killer disease for infants (22.3%) and children under 5 years of age (23.6%) and is among the top 10 diseases that result in deaths among the adult population. The WHO in 2002 estimates acute lower respiratory infection (ALRI) deaths attributable to solid fuel use (for children under 5 years) in Indonesia at 3130 population, while chronic obstructive pulmonary disease (COPD) deaths attributable to solid fuel use (for 30 years old and more) were estimated at 12,160 population (Haryanto 2016).

Air pollution of leaded gasoline exposure impact studies found that blood lead levels (BLLs) of elementary school children in Bandung was 66% above the CDC (Centers for Disease Control and Prevention, USA) level of 10 ug/dl in 2005 and

53% in 2006 (Haryanto et al. 2015). The Committee of Leaded-Gasoline Phasing Out found 90% of children under 5 years old living near road and children's road have BLLs above 10 ug/dl in Makassar in 2005. An indoor air study found about 50% of Jakarta's professionals reported their symptoms of sick building syndrome in the average of five times during 3 months observations due to indoor air quality at their workplace (Haryanto and Sartika 2009).

Pneumonia is overall the number one killer disease for infant (22.3%) and children under 5 years of age (23.6%) and among the top 10 diseases that result in deaths among the adult population. The WHO, in 2002, estimates acute lower respiratory infection (ALRI) deaths attributable to solid fuel use (for children under 5 years old) in Indonesia at 3130, while chronic obstructive pulmonary disease (COPD) deaths attributable to solid fuel use (for 30 years old and more population) were estimated at 12,160 population.

Air quality impacts are not limited to source regions (primarily in Central and Southern Sumatra and Southern Kalimantan) but can be transported in the atmosphere to affect transboundary locations such as Singapore (Hyer and Chew 2010; Atwood et al. 2013; Reddington et al. 2014; Kim et al. 2015). Air pollution from forest fire source in Sumatra and Kalimantan has had affected millions of human health in Sumatra, Kalimantan, and neighborhood countries, Singapore and Malaysia (Haryanto 2016).

The human cost of air pollution in Indonesia is shocking: the 2015 haze caused more than 28 million people are exposed, at least ten deaths from haze-related illness, and 560,000 people suffer from haze-related respiratory problem – the real number is likely to be higher as people living in remote areas and villages did not go to hospital or local health center (MoH 2015) and firefighting costs are pushing $50 million per week. In 2010, 57.8% of the population in Jakarta was reported to have suffered from different air pollution-related illnesses (e.g., asthma, bronchopneumonia, and chronic obstructive pulmonary disease or COPD, among others). Associated costs were estimated at IDR 38.5 trillion – with the effect of decline in productive days on economic growth (Safrudin 2015). Moreover, the national 35,000-megawatt development project is expected to increase the number of premature deaths from 6500 to 28,300 people per year due to impending air pollution from coal-fired power plants (Greenpeace 2015). Thus, emissions and consequent air pollution from mobile sources (i.e., motor vehicles) affect a range of sectors and contribute to or affect the economy as a whole. Increased health costs and reduced activity days (lower productivity) from air pollution-related illnesses directly cause a lower quality of life and indirectly reduce GDP for a specific city or country. Total economic cost, including direct damage (crops, forests, infrastructures), cost of responding to the wildfires, and losses in other economic trades, is estimated to exceed US $16 billion (IDR 221 trillion), more than double the costs of 2004 tsunami and three times the national health budget in 2015 (World Bank 2015). This number is higher than the estimates of economic losses from 1997 forest fires.

Emission Status and Prediction in Indonesia

As mentioned earlier, the climate change drives air quality in Indonesia to become worse with its impacts to the huge number of severe diseases and mortality as the consequences. A lot of money had been spent for the treatments. To make more matters, let's see the current air pollution status and its near future. The main sources of emission in Indonesia are from fossil fuel burning (coal, oil, and natural gas) and tropical deforestation. As accounting for 37.5% of the region's total primary energy demand in 2011 (IEA 2013), Indonesia is the largest energy consumer in ASEAN and the world currently. The range growth of energy consumption is about 6–8% per year. This condition does not balance yet with the energy supply (ESDM 2014). Total energy consumption is the quality of energy consumed in industrial (growth 2–8% per year), households (growth 2–4% per year), commercial (growth 1–2% per year), transportation sectors (growth 3–11% per year), and nonenergy consumption (growth 1–4% per year).

Figure 14.1 shows the trend of energy consumption by sector in Indonesia (included biomass) from 2000 to 2013. Overall, it can be seen that the energy consumption by sector fluctuated over the period. To begin, in 2000, the most energy was used on household sector, at approximately 1700 PJ and then fluctuated level through the following decade. Industrial sector appeared to follow the opposite pattern to household using. It started lower than household at about 1400 PJ per year, fluctuated in the following year, and then increased significantly to finish at just under 2500 PJ in 2013. Transportation, which at just over 500 PJ, accounted for the lower than industrial sector at the beginning of the period, fluctuated

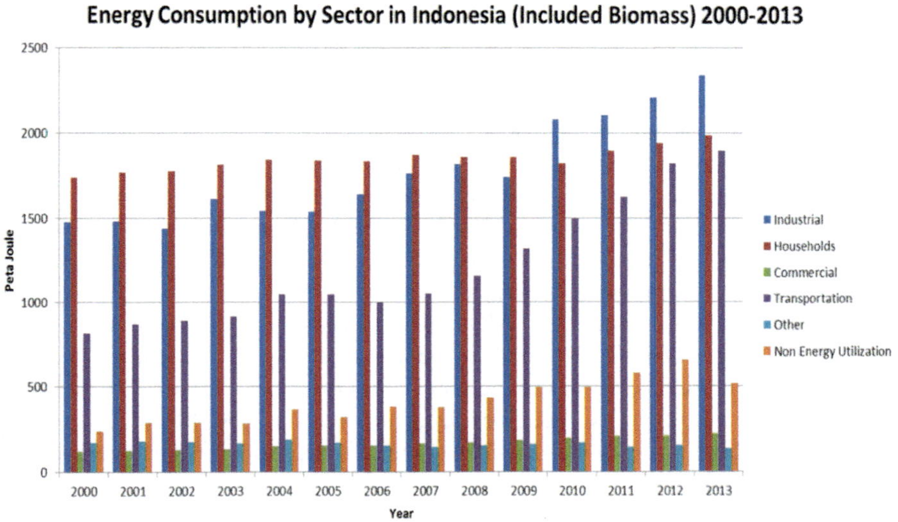

Fig. 14.1 Energy consumption by sector in Indonesia (included biomass) 2000–2013 (ESDM 2014)

dramatically over the time frame, and then jumped to just under 2000 PJ in the final year. Energy consumption in transportation sector is projected to increase at an average rate of 5.9% per year in 2012–2035, driven by the rising demand for mobility and subsidies. The lack quality of public transport is expected to continue to underpin a major expansion of vehicle ownership. Indonesia's fleet of passenger light-duty vehicles (PLDVs) rises from 10.4 million in 2012 to 21.3 million in 2020 and then 37.5 million in 2035. The further development of Indonesia's mass public transportation system could significantly alter these projected trends. The use of energy in transport remains dominated by oil. The household sector has the lowest average growth rate of all the end use sectors, at 0.8% per year, in line with ongoing switching from the inefficient use of traditional biomass energy to more efficient energy sources by households. Growth of energy consumption on commercial and other sectors is 6.6% and 5.5%, respectively. Increasing market for electrical appliances and electrification ratio improvement increased electricity consumption on household and commercial sector by 5.7% between 2012 and 2035.

Based on the current time-series data related to emission in Indonesia from 1990 to 2010 reported by the Ministry of Environment; National Agency for Meteorological, Climatology, and Geophysics; National Bureau for Statistics; Ministry of Industry; Ministry of Agriculture; Ministry of Health; Ministry of Energy and Natural Resources; Indonesia Institute of Science; universities; and other potential environmental monitoring stations, the Research Center for Climate Change University of Indonesia (RCCC-UI) from 2013 up to 2016 analyzed the prediction of general air pollutants and greenhouse gases using the GAINS model (greenhouse gases – air pollution interaction and synergies) which was developed by the International Institute for Applied Systems Analysis (IIASA) Austria (http://gains.iiasa.ac.at). GAINS describes the pathways of atmospheric pollution from anthropogenic driving forces to the most relevant environmental impacts (Amann et al. 2004). It brings together information on future economic, energy, and agricultural development, emission control potentials and costs, atmospheric dispersion, and environmental sensitivities toward air pollution. The model addresses threats to human health posed by fine particulates and ground-level ozone, risk of ecosystems damage from acidification, excess nitrogen deposition (eutrophication), exposure to elevated levels of ozone, and long-term radiative forcing. These impacts are considered in a multi-pollutant context, quantifying the contributions of sulfur dioxide (SO_2), nitrogen oxides (NO_x), ammonia (NH_3), non-methane volatile organic compounds (VOC), and primary emissions of fine ($PM_{2.5}$) and coarse ($PM_{2.5}$-PM_{10}) particles. GAINS also accounts for emissions of the six greenhouse gases that are included in the Kyoto Protocol, i.e., carbon dioxide (CO_2,), methane (CH_4), nitrous oxide (N_2O), and the three F-gases. The scenario emission reduction had also been analyzed utilizing the GAINS model. Among others, the following are the current status of major pollutants and some component of greenhouse gases and its prediction up to 2030 in Indonesia:

Figure 14.2 shows the current status of NO_x and $PM_{2.5}$ concentration from 1990 to 2010 and its prediction with the scenario "business as usual" from 2015 to 2030 which are slightly increased over time. Total percentage increase of NO_x is

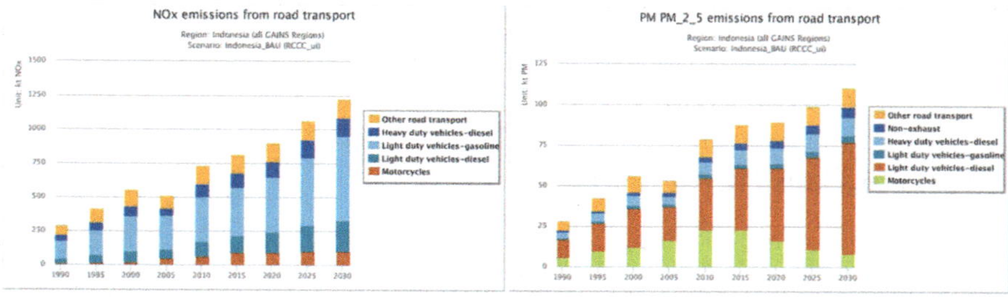

Fig. 14.2 Current status and prediction of NO_x and $PM_{2.5}$ 1990–2030 by the source of exposures

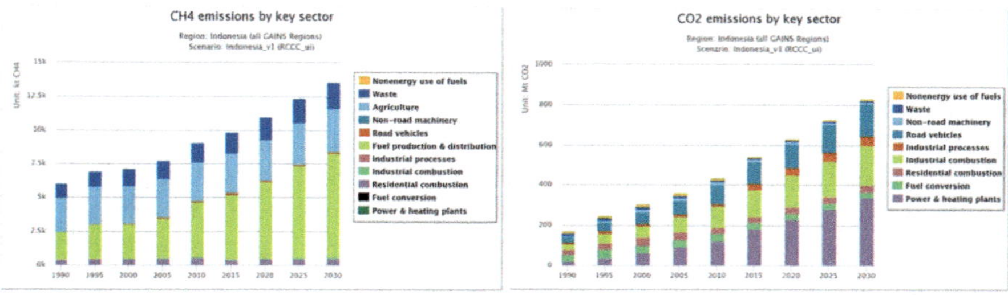

Fig. 14.3 Current status and prediction of CH_4 and CO_2 1990–2030 by key sectors

predicted up to 51% (from 814 kt per year in 2015 to 1225 kt/year in 2030) with the proportion of emission source dominated by light-duty vehicles-gasoline (from 44% in 2015 to 63% in 2030) and followed by light-duty vehicles-diesel, other road transport, heavy-duty vehicles-diesel, and motorcycles, respectively. For $PM_{2.5}$, total percentage increase is predicted up to 26% (from 87.7 kt per year in 2015 to 110.5 kt/year in 2030) with the proportion of emission source dominated by light-duty vehicles-diesel (from 43% in 2015 to 50% in 2030) and followed by other road transport, heavy-duty vehicles-diesel, motorcycles, non-exhaust, and light-duty vehicles-gasoline, respectively.

The other pollutants such as SO_2, PM_{10}, VOC, and O_3 are also found to increase over time from the year 2015 to 2030.

Figure 14.3 shows the current status of two components of the greenhouse gases, CH_4 and CO_2, emission from 1990 to 2010 and its prediction with the scenario "business as usual" from 2015 to 2030 which are slightly increased over time. Total percentage increase of CH_4 is predicted up to 38% (from 9842 kt/year in 2015 to 13,570 kt/year in 2030) with the proportion of emission sector dominated by fuel production and distribution (from 48% in 2015 to 57% in 2030) and followed by agriculture, waste, residential combustion, road vehicles, industrial combustion, and others, respectively. For CO_2, total percentage increase is predicted up to 53% (from 542 million tons per year in 2015 to 831 Mt/year in 2030) with the proportion of emission sector dominated by power and heating plants (from 34% in 2015 to 41% in 2030) and followed by industrial combustion, road vehicles, industrial processes, fuel conversion, residential combustion, and others, respectively.

The increasing trend over time is also found among the other components of greenhouse gases such as N_2O, CO, NH_3, and non-CO_2 GHG.

Efforts to Combat Air Pollution

Several activities have been developing, implementing, and improving to prevent and control climate change and air pollution in Indonesia. In 1999, the Indonesian Ministry of Environment suggested several steps to the Indonesian government in order to combat the effects and impacts of climate change. Among others, it is to prevent forest fires in areas that are prone to such fires due to forestry is the main cause of Indonesia's high greenhouse gas emission, and thus it must be the country's primary concern. A decade later, in May 2011, Indonesia announced a moratorium on granting new concession licenses in primary forests and peatlands while working toward land-use planning reforms that would help Indonesia achieve its greenhouse gas reduction targets (Austin et al. 2012). However, recent work analyzing the effect of this moratorium indicates that it would have offered only slight reductions (~5%) in national greenhouse gas emissions from deforestation if the policy had been in place over the prior decade (Busch et al. 2015). In addition, it remains unclear how much fire activity was associated with deforestation and management within different concession types during this time period. Logging concessions tend to have much lower deforestation rates than oil palm or timber concessions (Abood et al. 2015; Busch et al. 2015). However, given the tendency of logging concessions to be reclassified into other types of plantations (Gaveau et al. 2013) and with 35% of Indonesia's remaining forest area located within industrial-scale (not smallholder) concessions (Abood et al. 2015), it is crucial to understand differences among various industries regarding both deforestation and fire activity, along with the subsequent impacts on air quality and public health.

To make more focus, the efforts to combat the increasing trend of air pollution in Indonesia should be differentiated into two general efforts, reducing the sources of pollutants (mitigation) and preventing wider and more severe impacts to the population (adaptation). These efforts of mitigation and adaptation to combat air pollution in Indonesia are the following.

Mitigation

Mitigation refers to anthropogenic actions to reduce the emissions of greenhouse gases to the atmosphere, and thus reduce the magnitude of future climate change (IPCC 2007b), and is important in Indonesia due to its status as the third largest emitter of greenhouse gases, principally from large emissions from deforestation. Policies exist to reduce deforestation and thus emissions but are currently not well implemented. Indonesia's energy policy is to increase the use of fossil fuels, in particular coal, with the result that emissions from the energy sector are expected to triple by 2030 (PEACE 2007; MOE 2007). Policies are in place to support the use of renewables, but there is a lack of financial incentives to support these policies and encourage uptake. The government is also expanding the production of biofuel, for both domestic use and export. This is largely produced from palm oil and will require an extra 200,000 ha of plantations in 2009, driving deforestation (PEACE 2007). Biofuel produced from Jatropha has the potential to rehabilitate degraded land and provide a source of rural livelihoods, but issues around deforestation and conflict over land remain to be resolved.

It is estimated that Indonesia has the potential for 235 million tons of CO2 equivalent ($mtCO_2e$) in emissions reductions through the Clean Development Mechanism (CDM); however there are currently only eight projects registered with the Executive Board of the CDM, accounting for $13mtCO_2e$ of reductions. GTZ and the Asian Development Bank have been building the capacity for CDM in Indonesia; however, compared to neighboring countries in Asia, CDM is underdeveloped in Indonesia (PEACE 2007). Indonesia is currently lobbying the UNFCCC to include the proposal on avoided deforestation (REDD), whereby developing countries would receive compensation for preventing deforestation, as part of the international agreement on climate change.

In 2009 at the G20 Summit, Susilo Bambang Yudhoyono, the previous president, called for the emissions target that become the basis for Indonesia's Intended Nationally Determined Contributions (INDC) in 2015, a 26% reduction in greenhouse gas (GHG) emissions below business as usual by 2020 and up to 41% reduction by 2020 with international assistance. The current INDC stands at 29% reduction by 2030 and the same 41% conditional target. In 2011, Yudhoyono declared Presidential Regulation Number 61 which included the National Action Plan for Greenhouse Gas Reduction (*Rencana Nasional Penurunan Emisi Gas Rumah Kaca, RAN-GRK*). Presidential Regulation No. 61 was the outcome of the G20 summit and the Conference of the Parties (COP) meetings in Cancun and Copenhagen. The decree intended to use *RAN-GRK* as a reference document for GHG emissions in any government development planning. *RAN-GRK* has been expanded since the decree. It identifies the actions that Indonesia will take to reduce its GHG emissions. In 2012, *Bappenas* (Board of National Development Planning) established a secretariat for *RAN-GRK*. The executive branch has largely developed and implemented *RAN-GRK*.

RAN-GRK is the "plan of action" for Indonesia's emissions reductions targets. It requires the participation of government ministries and institutions to reduce GHG emissions. *RAN-GRK* identifies five major sectors that will be essential to achieve local action plan (*RAD*)-*GRK*'s emission reduction target. These are forestry and peatlands, agriculture, energy, industry, transportation, and waste. The responsible government ministries are *Bappenas*, the ministries of environment, forestry, agriculture, public works, industry, transportation, energy, and finance. Although *RAN-GRK* is a national action plan, it also lays the foundation for the actions of provinces, localities, and private enterprises to implement GHG reductions. *RAN-GRK* mandates that Indonesia's provinces develop and submit a local action plan

(*RAD-GRK*). *RAN-GRK* provides capacity building, budgets, and potential participation in domestic and international markets to local governments to incentivize them to contribute to *RAN-GRK*'s goals. *RAD-GRK*s are tailored to the development plans of each of the provinces. The Ministry of Home Affairs with the support of *Bappenas* and the Ministry of the Environment oversees and coordinates the preparation of *RAD-GRK*s. *Bappenas* creates the guidelines for each of the local action plans. Local action plans are planned with these expectations:

- Calculation of GHG inventory and of a provincial multi-sectoral business-as-usual (BAU) baseline
- Identification and selection of mitigation actions
- Development of mitigation scenarios according to selected and prioritized GHG mitigation actions in line with their local development priorities and plans
- Identification of the key stakeholders/institutions and financial resources

Local governments can also encourage the involvement of public and private companies by raising awareness of the climate change impacts and facilitating public private partnerships (among other options).

In early January 2016, *BRG* (Peatland Restoration Body, *Badan Restorasi Gambut*) was established with the target of restoring two million ha of peatland within 5 years. *BRG*, which works directly under President Joko Widodo, aims to achieve 30% or 600,000 ha by the end of 2016 in four regencies, Pulang Pisau in Central Kalimantan, Ogan Komering Ilir and Musi Banyuasin in South Sumatra, and Meranti in Riau (BRG 2016). Another effort backed by the government is One Map Policy (MSP 2016), a project aimed to create a single map for land tenure, land use, and other spatial planning in Indonesia. The unclear land ownership system, overlapping interest, and claims on land between community and plantation companies hamper the legal enforcement of wildfires and social conflict results from these complications. One Map is expected to provide a single reference map that will help the investigation of wildfires cases.

These efforts will not provide immediate results. The government of Indonesia had stated that it will take at least 3 years before any significant results can be seen. This means that the program to help society in mitigating the impacts of haze and wildfires should be the government's top priority. It has become more urgent as several fires have started in Sumatra and Kalimantan on June 2016; Riau government even declared emergency state as the province saw 45 hotspots in March 27. There are ongoing debates surrounding the practice of slash and burn in land clearing including the debate on the provision which allows smallholders to legally burn up to 2 ha of land and debate around opposing claims regarding source and causes of wildfires between palm oil plantations and small-scale farmers. There is also a lack of understanding in regard to farmers' decision-making, motivation, and aspiration in this area, considering complexity of their life, including their aspiration for children's education and well-being.

Adaptation

Adaptation can be seen as adjustments in human or physical systems in response to current or expected climate changes in order to cope with the impacts of climate change and take advantage of any new opportunities (IPCC 2007b). To achieve its goals for economic development and poverty reduction, in particular among the poorest and most marginalized sectors of population, Indonesia will need to adapt to climate change. It is also clear that many Indonesians are already adapting to climate change, for example, by building houses on stilts to respond to increased flooding, or responding to decreased reliability of fish catches by diversifying livelihoods, and that indigenous adaptation strategies should form the base for building adaptation to future change (UNDP 2007). Adaptation and mitigation in Indonesia are strongly coupled, as continued rapid deforestation will not only exacerbate the impacts of climate change but also constrain the adaptation options that are available to vulnerable communities. The priority sectors for adaptation are seen as agriculture, water, coastal, and urban areas.

There are adaptation options that are specific for each of these sectors, for example, faster-growing crops varieties in the agricultural sector; however there are also general needs to be addressed which will build capacity for adaptation across sectors. These include the development of a system to provide climate information to actors at different scales, for example, seasonal forecasts, and training in how to use this information effectively to manage climate risks. Training in vulnerability analysis and assessment of adaptation options would help to identify priorities for adaptation. Initiatives such as the development of community action plans to cope with flooding are being pursued in the field of disaster risk reduction (DRR) but are equally relevant in building community resilience to future climate change. Adaptation to climate change will be a long-term process, and as such will require long-term partnerships and cooperation between different actors at different scales. Encouraging dialogue between these different actors, in a similar way to the workshop convened to discuss the Climate Change Adaptation Program (ICCAP), will help to foster the relationships needed to enable adaptation to take place.

In order to speed innovation, the Indonesian government must focus on improving the technology and transfer of information to the farmer. It is also important to strengthen the research that is done on the development of more sustainable agricultural practices. The government must also promote innovative and improved agricultural practices that release the least amount of greenhouse gases into the atmosphere. However, such program intervention will not be as effective and relevant as it should be without understanding the experience of people who are directly affected by haze. Various reports indicate that some people are not using masks in haze-affected areas, while there is very limited or no reports that provide a better understanding toward people's willingness, awareness, and access (or lack of) to use proper mask during haze or how family manage and cope with the haze-related risks, especially in relation to their and their children's health.

In 2007, Indonesia's Ministry of Environment established National Action Plan on Climate Change Adaptation (*RAN-API*) which is coordinated by the National Council on Climate Change and is composed of 17 ministers and is chaired by the president. *RAN-API* brings together many different mitigation strategies. These include Indonesian Adaptation Strategy (Bappenas 2011), National Action Plan for Adaptation to Climate Change (DNPI 2011), Indonesia Climate Change Sectoral Roadmap (Bappenas 2010), the National Action Plan for Climate Change Mitigation and Adaptation (Ministry of the Environment 2007), and the sectoral adaptation plans by line ministries/government agencies. *RAN-API* strengthens *RAN-GRK*'s seven mitigation actions through these ways and helps achieve the 2020 target of 26% GHG emissions reductions. These mitigation actions are (1) sustainable peatland management, (2) reduction in rate of deforestation and land degradation, (3) development of carbon sequestration projects in forestry and agriculture, (4) promotion of energy efficiency, (5) development of alternative and renewable energy sources, (6) reduction in solid and liquid waste, and (7) shifting to low-emission transportation modes.

The climate change adaptation to air pollution in Indonesia is not clearly stated both in RAN-GRK and RAN-API as well as on the line ministries decrees or regulations. It resulted to inappropriate and readiness responses when the outbreak of air pollution occurs with high number of respiratory disorders and more severe impacts among population as consequences. However, environmental and public health experts have been intensively suggesting government to implement some actions related to air pollution adaptation, for example, environmental health capacity building, healthy public policy development, public education and awareness, early-alert systems for heatwaves and weather extremes, climate-proofed housing design and "cooler" urban layout, and greenhousing design standards.

Emission Reduction Scenarios

Indonesia Energy Outlook (IEO) 2013 provides an update of energy demand and supply projections based on recent macroeconomic conditions, population growth, and government policies. ALT (Alternative Policy) Scenario is based on government policies that are recently announced, including those not implemented yet and plan to implement in the next coming year. In ALT Scenario, Indonesia's total primary energy demand is projected to grow at an average of 5% per year between 2011 and 2035, rising from nearly 214.5 million tons of oil equivalent (Mtoe) to around 672 Mtoe (IEA 2013). As the largest and most populous archipelago in the world, providing modern energy access is a particular challenge, which partly explains its comparatively low levels of per capita energy consumption. Energy use per capita has been rising at a rapid pace over the last several decades, fueling strong economic growth. In the New Policies Scenario, it rises to 2.25 toe per capita in 2035. Total final energy consumption rises at a projected 5.5% per year on average through to 2035. Final energy in industry grows faster than other sectors,

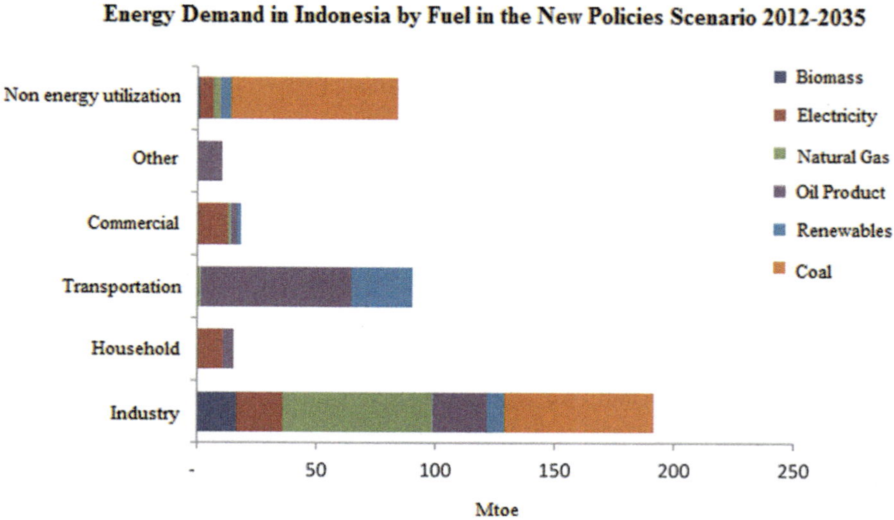

Fig. 14.4 Energy demand in Indonesia by fuel in the new policies scenario 2012–2035

rising an overall 7.3% in 2012–2035. The replacement of inefficient technology is one of the key challenges in Indonesia. The share of gas in the industrial fuel mix rises significantly from 31% in 2012 to 39% in 2035, on improving gas supply infrastructure, increasing fertilizer production capacity, and growth in the ceramic industries (Fig. 14.4 and Tables 14.1).

For the international climate negotiations in Paris 2015, Indonesia has pledged to increase its emissions over the next 25 years by 29% less than it would have under a "business-as-usual" scenario. That won't be possible without curbing forest fires and deforestation. So for Indonesia, getting a grip on palm oil producers will be even more important than going after power plants.

Emission Scenarios for Road Transport in Indonesia

Several reduction emission scenarios had been developed in Indonesia for several cities and national level by universities, NGOs, local government as well as line ministries. Some of the reports were used as the compliment for sectoral government planning and actions. Most recently, the UNEP funding supported the expert team in Indonesia to study Cost Benefit Analysis Fuel Economy in Indonesia in 2010 (MOE 2010a). The project justification was while (some) policies to reduce emissions by improving fuel efficiency have been enacted in Indonesia, implementation has been unsystematic and, often, ineffective at best. Thus, an evaluation of existing policies is warranted to determine the more appropriate course(s) of action that can and should be undertaken to raise current air quality levels in Indonesia. Nine (9) policy options were examined and assessed based on a comparison of its

Table 14.1 Energy demand in Indonesia by fuel in the new policies scenario (Mtoe)

Type	2012	2020	2025	2030	2035	2012–2035
Coal	38	90	114	127	145	6.0%
Oil	78	96	124	158	180	3.7%
Gas	43	85	131	153	172	6.2%
Hydro	2	2	2	4	7	7.2%
Bioenergy	8	16	24	28	34	6.6%
Other renewables	1	29	41	66	100	20.3%
Total	170	318	437	537	639	5.9%

estimated costs and projected benefits, and the policy alternative which yields the highest advantage per unit cost was determined. Calculations and corresponding recommendations made take into account the local, national, and regional socio-political conditions to arrive at scenarios to address air pollution levels in Indonesia. The nine (9) policy options proposed and evaluated in the study include:

- Option 1. Implementing Euro 2 in 2005, Euro 3 in 2015, and Euro 4 in 2020.
- Option 2. Enhance fuel efficiency by 10% in 2009.
- Option 3. Convert at least 1% of passenger cars and buses to compressed natural gas (CNG) in 2009, 2% in 2011, and 5% in 2021.
- Option 4. Use catalytic converters on 25% of vehicles that run on diesel: passenger cars, buses, and trucks.
- Option 5. Beginning in 2009, scrap 50% of vehicles that are more than 20 (20) years old.
- Option 6. Promote and use hybrid technology for at least 0.05% of passenger cars and buses in 2009, 0.1% in 2011, 0.5 in 2016, and 1% in 2021.
- Option 7. Convert at least 1% of passenger cars to biofuels in 2009, 2% in 2011, and 5% in 2021.
- Option 8. Owners of passenger cars and motorcycles shift to public transport by at least 5% and 1% in 2011, 10% and 5% in 2014, 20% and 10% in 2018, and 40% and 20% in 2025.
- Option 9. Implement Euro 2 in 2005 and adopt Euro 3 in 2013 and Euro 4 in 2016.

For all the proposed scenarios, it is assumed that Option 1 or the improvement of fuel quality by meeting Euro 2 standards has been implemented.

Implementing the baseline (i.e., improvement of fuel quality) alone will result in the reduction of sulfur levels below 500 ppm, leading to reduced health costs and productivity losses of IDR 38,963 billion (net present value, NPV) for the period 2005–2030 and approximately IDR 71,395 billion (NPV) in fuel savings. These are the baseline figures by which all the other policy options were measured against. Alternatively, the policy's expected economic gains and savings are that which the other eight options aim to enhance or build on. In terms of gaining the highest economic benefits and generating savings from fuel subsidies, adopting hybrid vehicle technology (Option 5) would be considered the best option with IDR

1,563,678 billion (NPV) for reduced health cost and production losses for 2005–2030 and IDR 1,098,827 billion (NPV) in fuel saving for 2009–2030. Even though, it needs the most investment cost both in the refinery and auto manufacturer. The retiring or scrapping of old vehicles (Option 6) would be considered the most cost-effective. However, because it raises social equity issues and requires high compensation costs, political and social challenges may hinder its implementation and/or effectiveness. Moreover, it assumes a reliable public transport system that can and will absorb the increase in the number of commuters who will stand to lose their motor vehicles to comply with the policy. Also, the political implications of implementing the policy make it unpopular to incumbent politicians and/or officials.

The option to enhance fuel efficiency, which builds on the baseline, yields the second highest economic gain and savings. Risks for implementing it are low, thus, the likelihood of the government promoting and undertaking it is high. Setting up incentives for the auto industry to produce more fuel-efficient vehicles should accompany policy implementation to ensure its effectiveness. Promotion and adoption of the use of CNG, hybrid technology in vehicles, and biofuels all yield positive net economic benefits and fuel savings, with CNG showing the highest economic gains and the use of biofuels providing the largest savings. However, all three options entail high costs: a catalytic converter to shift from conventional gas to CNG, acquiring or providing incentives for investments in hybrid technology, and the unsubsidized prices of biofuels. Option 9, i.e., implementation of Euro 4 standards in 2016, is consistent with the positive, upward trend of the expected economic gains of and fuel savings from the other policy options. The success of implementing Option 8, or encouraging the shift from private to public transport, is largely contingent on the public's behavior. However, it could also be argued that improving the current state of the country's public transport can help influence the public's attitude toward and usage of it. Nevertheless, the benefits of improved public transport in terms of reduced air pollution, fuel consumption, and traffic congestion and overall improved quality of life are underscored. These are more than enough justifications to adopt and pursue implementation of this policy option. The adoption and use of CNG, hybrid technology, and improvement of public transport appear to be the most cost-effective among all the nine options. Thus, given the projected economic gains, expected fuel savings, and least cost to reduce emissions per ton, provision and improvement of public transport seems to be the most promising, in terms of both short- and long-term effects.

Research Center for Climate Change – University of Indonesia (RCCC-UI), by support funding from Toyota Clean Air Project Japan (TCAP) and technical assistant of International Institute for Applied Systems Analysis (IIASA) Austria, has been conducting 4 years study on Reduction Emission Scenarios Development for Indonesia based on energy transportation since 2014. In this study, emission scenarios define as the combination of activity projections and control strategies. The activities data are used and input to the GAINS model for calculating emissions. Prior to the development of dataset, some calculations and mathematical conversions were conducted to meet the format of GAINS' datasets. There are four

Table 14.2 Scenarios for emission reduction of road transport in Indonesia

No.	Title	Scenario	Objective
1	BAU (business as usual)	No control strategy for road transport in Indonesia	To know the value of emission from road transport without control
2	Indonesia 2017	The implementation of EURO 4 in 2017	Implemented to new gasoline vehicles in 2018 and existing vehicles in 2020 and all diesel vehicles in 2020
3	Indonesia 2020	The implementation of EURO 4 in 2020	Implemented to all existing and new gasoline and diesel vehicles in 2020
4	Indonesia 2023	The implementation of EURO 4 in 2023	Implemented to all existing and new gasoline and diesel vehicles in 2023

scenarios of the implementation of EURO 4 fuel standard as the following (Table 14.2):

GAINS calculates current and future emissions based on activity data specified in the activity pathway, the "uncontrolled" emission factors, the application rates of emission control measures, and their emission removal efficiencies. Among others, some examples of analysis model comparison between BAU and PM2.5 and NOx for EURO 4 implementation scenario in 2023 and between BAU 6 °C and CH_4 and CO_2 for 2° scenario implementation 2015 are the following:

The scenario to implement the national fuel energy started in 2023 (Fig. 14.5), with all of its preparations, will reduce the emission of $PM_{2.5}$ and NO_x gradually up to 304% in 2050 (304.5% reduction of $PM_{2.5}$ or about 119 kt and 131.6% reduction of NO_x or about 972 kt). The most reduction emission proportion of $PM_{2.5}$ is dominated by the sector of light-duty vehicles-diesel. Meanwhile, the most reduction emission proportion of NO_x is dominated by the sector of light-duty vehicles-gasoline.

The scenario to implement all efforts to reduce temperature up to 2 °C started in 2015 in Indonesia (Fig. 14.6), with all of its preparations, will reduce the emission of CH_4 and CO_2 gradually up to 191% in 2050 (55% reduction of CH_4 or 6739 kt and 191% reduction of CO_2 804 kt) compared with the BAU of 6 °C. The most reduction emission proportion of CH_4 is dominated by the sector of fuel production and distribution. Meanwhile, the most reduction emission proportion of NOx is dominated by the sector of transportation losses.

Conclusion

Higher emissions of carbon dioxide (CO_2) have caused rapidly worsening air pollution in Indonesia with fuel burning and forest fire as its main contributor drivers. Climate change in Indonesia greatly affects many aspects of the country, including economy, poor population, human health, and the environment. Air pollution affects mostly urban areas since the transportation sector contributes the most (80%) followed by emissions from industry, forest fires, and domestic activities. The large number of vehicles together with lack of infrastructure results in

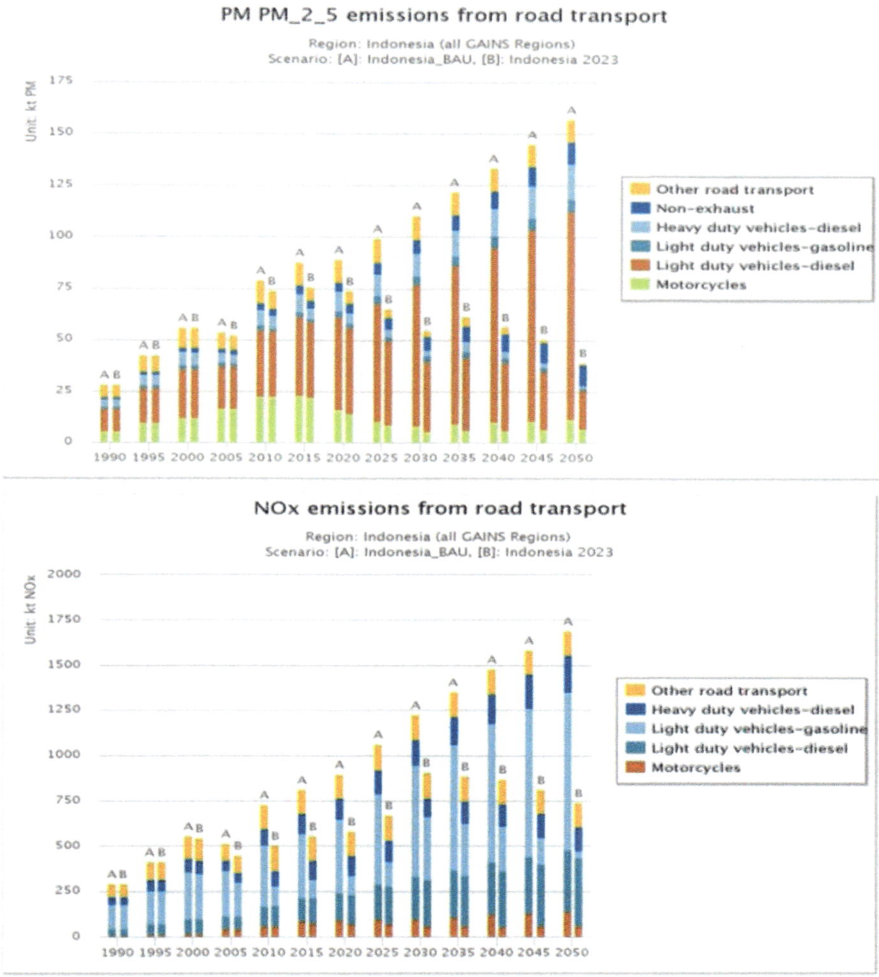

Fig. 14.5 $PM_{2.5}$ and NO_x road transport emission scenarios BAU vs EURO 4 2023

major traffic congestions resulting in high levels of air polluting substances, which have a significant negative effect on public health.

Current air pollution problems are greatest in Indonesia as it caused 50% of morbidity across the country. Air pollution is proven as a major environmental hazard to residents in Jakarta. Diseases stemming from vehicular emissions and air pollution include acute respiratory infection, bronchial asthma, bronchitis, and eye and skin irritations, and it has been recorded that the most common disease in northern Jakarta communities is acute upper respiratory tract infection – at 63% of total visits to health-care centers. The number of diseases related to air pollution cases had been predicted to be higher and more severe as the source of air pollution,

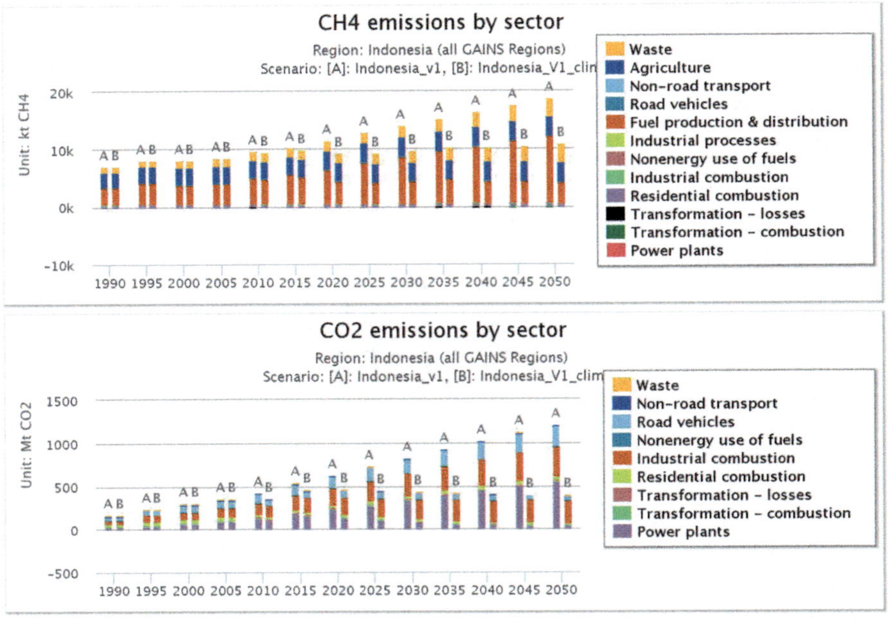

Fig. 14.6 CH_4 and CO_2 road transport emission scenarios BAU (6 °C) vs. 2 °C

energy demand, projected to be sharply increased up to 2050 and indeed, will directly affect to the increasing of air pollutant parameters.

Some efforts had been conducted to combat air pollution problems, but the effective implemented actions have reported almost no significant results found on the reduction of both the emission and health impacts. However, in order to support government efforts, most currently some studies on developing scenarios for reducing emission have been conducted. These include analysis of fuel economy and the time effective for EURO 4 standard implementation as compliment to transportation improvement policy in Indonesia.

This paper suggests Indonesia, in energy sector, must enhance energy security and mitigate CO_2 emissions in order to protect strategic reserves, improve efficiency in energy production and use, increase reliance on non-fossil fuels, and sustain the domestic supply of oil and gas through decreased fossil fuel consumption. In addition, the government must support the use of proposed breakthrough technologies, including the diffusion and deployment of clean-energy technologies. In the transportation sector, Indonesia should adopt European emission standards (Euro 4 and Euro 6 standards), switching the basic mode of transportation and attempting to mitigate current emissions by enforcing a low-sulfur fuel and low-emission vehicle policy. In health sector, Indonesia must protect human health from air pollution by conducting more research on health vulnerability and implementing more effective adaptation of human health.

References

Abood SA, Lee JSH, Burivalova Z, Garcia-Ulloa J, Koh LP (2015) Relative contributions of the logging, fiber, oil palm, and mining industries to forest loss in Indonesia Conserv. Lett 8:58–67

Amann M, Cofala J, Heyes C, Klimont Z, Mechler R, Posch M, Schopp W (2004) Rains review 2004. Interim Rep IIASA Austria (February 2004)

Asian Development Bank (2009) The economics of climate change in Southeast Asia: a regional review (Manila, The Philippines: Asian Development Bank, April); available at: http://www.adb.org/Documents/Books/Economics-Climate-Change-SEA/PDF/Economics-Climate-Change.pdf

Atwood SA, Reid JS, Kreidenweis SM, Yu LE, Salinas SV, Chew BN, Balasubramanian R (2013) Analysis of source regions for smoke events in Singapore for the 2009 El Niño burning season Atmos. Environment 78:219–230

Austin K, Sheppard S, Stolle F (2012) Indonesia's moratorium on new forest concessions: key findings and next steps WRI working paper. World Resources Institute, Washington, DC. www.wri.org/publication/indonesia-moratorium-on-new-forest-concessions

Badan Restorasi Gambut (2016) Peraturan Presiden No. 1 Tahun 2016 tentang Badan Restorasi Gambut. Sekretariat Kabinet Republik Indonesia, Januari 2016, Jakarta

Bappenas & Swisscontact (2006) National strategy and action plan of urban air quality improvement. Pelangi, 1999 on Haryanto, B. (2008). Climate change and public health adaptation in Indonesia. Working paper for the workshop on mapping causal complexity in climate change impacts and adaptation. The Nautilus Institute – the Royal Melbourne Institute of Technology University, Melbourne 21–22 October 2008

Biro Pusat Statistik (2005) Survey Sosial Ekonomi (Susenas) – National Survey on Social and Economic Report 2004. Jakarta

Biro Pusat Statistik (2016) at https://www.bps.go.id/linkTableDinamis/view/id/1119

BPLHD DKI (2009) http://bplhd.jakarta.go.id/filing/materi%202_Bahan%20FGD%20pencegahan%20Dampak%20LLAJ%20pemprov%20DKI%20LH.pdf

Busch J et al (2015) Reductions in emissions from deforestation from Indonesia's moratorium on new oil palm, timber, and logging concessions. Proc Natl Acad Sci U S A 112:1328–1333

Case M, Ardiansyah F, Spector E (2007) Climate change in Indonesia. Implications for humans and nature. World Wide Fund for Nature (WWF), Gland, Switzerland. Available at: http://assets.panda.org/downloads/indoesian_climate_change_impacts_report_14nov07.pdf

Gaveau DLA et al (2013) Reconciling forest conservation and logging in Indonesian Borneo. PLoS One 8:e69887

Greenpeace (2015) Human cost of coal power: how coal-fired power plants threaten the health of Indonesians. August 2015, http://www.greenpeace.org/seasia/id/PageFiles/695938/full-report-human-cost-of-coal-power.pdf

Haryanto B (2016) Climate change and human health scenario in South and Southeast Asia. Akhtar R, editor. Springer ISSN 1879-7180, 2016

Haryanto B, Franklin P (2011) Air pollution: a tale of two countries. Rev Environ Health 26 (1):75–82

Haryanto B, Sartika RD (2009) The effect of anti-oxidant supplementation in reducing the frequency of sick building syndrome among professionals in Jakarta 2008–2009. Indonesian Public Health Association & Faculty of Public Health University of Indonesia.

Haryanto B, Djaja IM, Sutrisna B (2015) Effect of calcium supplementation on school children's blood lead levels in Indonesia. Int J Sci Res 4(5):1620–1625

Hyer EJ, Chew BN (2010) Aerosol transport model evaluation of an extreme smoke episode in Southeast Asia Atmos. Environment 44:1422–1427

International Energy Agency (IEA) (2013) World energy outlook 2013. OECD/IEA, 2013

IPCC (2007a) Working group I, chapter 11: regional climate projections. In: Solomon S, Qin D, Manning M, Chen Z, Marquis M, Averyt KB, Tignor M, Miller HL (eds) Climate change 2007: the physical science basis. Contribution of working group I to the fourth assessment report of the intergovernmental panel on climate change. Cambridge University Press, Cambridge

IPCC (2007b) Appendix 1: glossary. In: Parry ML, Canziani OF, Palutikof JP, van der Linden PJ, Hanson CE (eds) Climate change 2007: impacts, adaptation and vulnerability. Contribution of working group II to the fourth assessment report of the intergovernmental panel on climate change. Cambridge University Press, Cambridge, pp 273–313

Kementerian Enerji dan Sumber Daya Mineral (ESDM) (2014) Statistik Minyak Bumi 2012. Pusat Data dan Teknologi Informasi ESDM

Kim PS, Jacob DJ, Mickley LJ, Koplitz SN, Marlier ME, DeFries RS, Myers SS, Chew BN, Mao YH (2015) Sensitivity of population smoke exposure to fire locations in equatorial Asia. Atmos Environ 102:11–17

Margono BA, Potapov PV, Turubanova S, Stolle F, Hansen MC (2014) Primary forest cover loss in Indonesia over 2000–2012. Nat Clim Chang 4:730–735

Menuju Satu Peta (MSP), One Map Policy (2016) Peraturan Presiden Nomor 9 Tahun 2016 tentang Percepatan Pelaksanaan Kebijakan Satu Peta. Sekretariat Kabinet Republik Indonesia, February 2016, Jakarta

Miettinen J, Shi C, Liew SC (2011) Deforestation rates in insular Southeast Asia between 2000 and 2010. Glob Chang Biol 17:2261–2270

Miettinen J, Hooijer A, Shi C, Tollenaar D, Vernimmen R, Liew SC, Malins C, Page SE (2012a) Extent of industrial plantations on Southeast Asian peatlands in 2010 with analysis of historical expansion and future projections. GCB Bioenergy 4:908–918

Ministry of Environment (2007) Climate variability and climate change, and their implications. MOE Republic of Indonesia, Jakarta

Ministry of Environment (2009) Indonesian environmental status 2008. Jakarta

Ministry of Environment and Forestry (2015) http://sipongi.menlhk.go.id/home/karhutla_monitoring_system

Ministry of Forestry of Indonesia (2013) Statistics of forest planology 2012. Directorate general of forest planology, ministry of forestry of Indonesia 2013

Ministry of Health (2008) National basic health research report 2007. Institute for Health Research and Development, Jakarta

Ministry of Health (2015) Laporan Masalah Kesehatan akibat Kabut Asap Kebakaran Hutan dan Lahan Tahun 2015

MoE & KPBB (2006) Indonesian fuel quality report 2006. Clean fuel: a requirement for air quality improvement

PEACE (2007) Indonesia and climate change: current status and policies. PEACE, Jakarta

Pusarpedal KLHK (2015) http://pusarpedal.menlh.go.id/wp-content/uploads/2014/07/5.Pemantauan-Kualitas-Udara-Ambien-dengan-Peralatan-AQMS.pdf

Reddington CL, Yoshioka M, Balasubramanian R, Ridley D, Toh YY, Arnold SR, Spracklen DV (2014) Contribution of vegetation and peat fires to particulate air pollution in Southeast Asia. Environ Res Lett 9:094006

Rowshand GR, Khoshakhlagh F, Negahban S et al (2009) Impact of air pollution on climate fluctuations in Tehran city. Environ Sci 7:173–191

Safrudin A (2015) Low sulfur fuel, vehicle emission and fuel economy standard (presentation at the Conclave of Champion Cities of Asia and Africa in Clean Air and Sustainable Mobility, New Delhi, April 9–10, 2015), http://www.cseindia.org/userfiles/ahmad-safrudin-delhi-presentation.pdf

Samet J, Krewski D (2007) Health effects associated with exposure to ambient air pollution. J Toxic Environ Health A 70:227–242

Sari AP, Maulidya M, Butarbutar RN, Sari RE, Rusmantoro W (2007) Executive summary: Indonesia and climate change – working paper on current status and policies. The World Bank, Washington, DC and Department for International Development (DFID), London, UK, March. Available at: http://www.conflictrecovery.org/bin/PEACEClimateChange-ExecSum.pdf

Simorangkir D (2007) Fire use: is it really the cheaper land preparation method for large-scale plantation? Mitig Adapt Strateg Glob Chang 12(1):147–164

Suryanti Y (2006) Summary on awareness raising and capacity building to address vulnerability and adaptation to climate change. Ministry of Environment, Republic of Indonesia, Jakarta

Tsai SS, Chiu HF, Liou SH, Yang CY (2014) Short-term effects of fin particulate air pollution on hospital admissions for respiratory diseases: a case-crossover study in a tropical city. J Toxicol Environ Health Part A 77:1091–1101

Tsangari H, Paschalidou AK, Kassomenos AP, Vardoulakis S, Heaviside C, Georgiou KE et al (2016) Extreme weather and air pollution effects on cardiovascular and respiratory hospital admissions in Cyprus. Sci Total Environ 542:247–253

World Bank (2009) Indonesia rising: mainstreaming climate change for sustainability. World Bank Office Jakarta, Jakarta. available at: http://go.worldbank.org/JIGX6UTVJ0

World Bank (2010) Indonesia: climate Change. The World Bank, Washington, DC. online resource; available at: http://go.worldbank.org/HQQFW5MV70

World Bank (2014) World development report 2014: risk and opportunity. At http://siteresources.worldbank.org/resources

World Bank (2015) http://www.worldbank.org/en/news/feature/2015/12/01/indonesias-fire-and-haze-crisis

World Health Organization (2002) World health report 2002: reducing risk, promoting healthy life. WHO, Geneva

World Health Organization (2014) Burden of disease from ambient air pollution for 2012. http://www.who.int/phe/health_topics/outdoorair/databases/AAP_BoD_results_March2014.pdf. Last accessed February 2016

World Population Review (2002) A guide to world resources 2002–2004. At http://pdf.wri.org/wr2002_execsumm

World Population Review (2005) The wealth of the poor: Managing ecosystems to fight poverty. At http://www.wri.org

World Population Review (2016) at http://worldpopulationreview.com/countries/indonesia-population/

Ying Z, Gong WS, Xia MY, Zheng SK, Fan CY, Xu L et al (2015) Association between ambient air pollution and hospital emergency admissions for respiratory and cardiovascular diseases in Beijing: a time series study. Biomed Environ Sci 28:352–363

Dr. Budi Haryanto is a Lecturer in the Department of Environmental Health and the Deputy Director of Research Center for Climate Change at Universitas Indonesia, BOD of the PBC Environment and Health and Co-chair of Environmental Health for APRU. He has published numerous air pollution and climate change health impacts studies and is actively involved in scientific climate talks at IPCC's COP 15 Copenhagen, COP 21 Paris and COP 22 Marrakech.

Chapter 15
Climate Change and Air Pollution in Malaysia

Nasrin Aghamohammadi and Marzuki Isahak

Abstract Air pollution due to anthropological activities and natural disasters are the major challenges for environmental issues for last few decades. Human activities and population growth aggregate the atmospheric composition and damaged Earth's atmosphere. Southeast Asia (SEA) is facing with natural disasters such as flood and tsunami that are challenging international attempts to address these issues for climate change. Transboundary haze is one of the significant environmental issues in SEA since 1983. The transboundary haze pollution has adverse impacts on environment due to greenhouse gases (GHGs) emissions as well as ecosystem and biodiversity which caused climate changes in recent decades.

Keywords Haze • Tropical country • Malaysia • Open burning • Health impact • Forest fire

Introduction

Air pollution due to anthropological activities and natural disasters are the major challenges for environmental issues for last few decades. Human activities and population growth aggregate the atmospheric composition and damaged Earth's atmosphere. Southeast Asia (SEA) is facing with natural disasters such as flood and tsunami that are challenging international attempts to address these issues for climate change. Transboundary haze is one of the significant environmental issues in SEA since 1983. The transboundary haze pollution has adverse impacts on environment due to greenhouse gases (GHGs) emissions as well as ecosystem and biodiversity which caused climate changes in recent decades.

Land use changes and land clearing using open burning in SEA caused the haze with significant density level that considered as transboundary haze pollution. The

N. Aghamohammadi (✉) • M. Isahak
Centre for Occupational and Environmental Health, Department of Social and Preventive Medicine, Faculty of Medicine, University of Malaya, 50603 Kuala Lumpur, Malaysia
e-mail: Nasrin@ummc.edu.my

© Springer International Publishing AG 2018
R. Akhtar, C. Palagiano (eds.), *Climate Change and Air Pollution*,
Springer Climate, DOI 10.1007/978-3-319-61346-8_15

wind direction and the El Niño phenomenon caused drier condition which deteriorates transboundary haze pollution and prolonged duration of haze episode in SEA. Urbanization, industrialization and population growth are the major factors that trigger air pollution due to local emissions. Many studies during these decades confirmed the different levels of air pollutant during the haze episode that triggered transboundary pollution in Malaysia and neighbouring countries.

Air pollution due to transboundary haze pollution causes climatic changes which have significant impact on human health and lifestyle as the pollution has adverse health impact along with natural disaster.

Some of epidemiological data correlated between air pollution, morbidity and premature mortality. The number of cases for morbidity and/or premature mortality associated with air pollution was determined. Studies by Aouizerats et al. (2015) and Behera et al. (2015) found that the visibility was reduced to 0.5 km during the haze due to significant concentration of particulated matter with aerodynamic size below 10 μm (PM_{10}). This proved biomass burning which is the most contributor to the haze episode reduces visibility as well as affects human health as the reduction in visibility may cause accidents during the haze episode. Particulated matter can penetrate into human respiratory system with aerodynamic size below 10 as they may be trapped in upper respiratory system, while it will be more harmful when the size reduces below 2.5 ($PM_{2.5}$) due to deeper penetration in lower respiratory system and reach into the bloodstream.

As the transboundary haze is the critical issue in Malaysian air quality during dry season annually, therefore this chapter will discuss on haze pollution disaster which has significant impact on climate change of Malaysia and may cause natural disasters.

Haze Scenario in Malaysia

Malaysia has the first record of disturbing haze in 1983; the forest fires in Sumatra caused haze in 1991 for the second time that occurred during the month of September. The main cause of the problem was identified as forest fires in Kalimantan and Southern Sumatra. Subsequently, haze polluted Malaysia in 1997 with the dry weather and stable atmospheric conditions coupled with emissions from local pollution sources such as from motor vehicles, industries and open burning of wastes also aggravated the situation (Keywood et al. 2003). This haze episode was considered one of the worst situations due to co-occurrence of El Niño, which prolonged the dry season in that year. In 2005, haze emergency was declared in the month of August as the Air Pollution Index (API) announced unhealthy, and few flights were suspended; few years after 2013, a short period of haze with highest API in the month of June occurred due to transboundary pollution where forest fires were happening in Sumatra. At this time many schools closed due to haze emergency, and API exceeded to hazardous point. 2015 has the longest duration of haze episode in Malaysia due to massive forest fires and biomass burning in Sumatra and Kalimantan. In this time, many schools and universities closed in Kuala Lumpur, Selangor, Sarawak and Melaka. Haze came back to Malaysia in September 2016 for a very short time, and API reading was lesser than 100 only for a day.

Haze Monitoring and Air Pollution Index in Malaysia

These monitoring stations of air quality in Malaysia are located in residential, industrial and business areas to detect any changes in the air quality which may cause adverse health impact on human and the environment. The Department of Environment (DOE) of Malaysia monitors the country's ambient air quality through a network of 52 stations. Malaysian air quality reported the Air Pollution Index (API) five pollutants: carbon monoxide (CO), sulphur dioxide (SO_2), nitrogen dioxide (NO_2), particulated matters with 10 μm (PM_{10}) and ground-level ozone (O_3). These five criteria of air pollutant are measured in the stations, and the API calculated based on the measurements. The API is an indicator of the air quality including the haze and was developed based on scientific assessment to indicate, in an easily understood manner, the presence of pollutants in air and its impact on human health. Table 15.1 shows API in Malaysia based on Environmental Protect Agency (EPA).

Table 15.1 Value of Air Pollution Index (API) and its relation with health effect

API	Status	Health effect	Health advice
0–50	Good	Low pollution without any bad effect on health	No restriction for outdoor activities to the public. Maintain healthy lifestyle
51–100	Moderate	Moderate pollution that does not pose any bad effect on health	No restriction for outdoor activities to the public. Maintain healthy lifestyle
101–200	Unhealthy	Worsen the health condition of high-risk people with heart and lung complications	Limited outdoor activities for the high risk people. Public need to reduce the extreme outdoor activities
201–300	Very unhealthy	Worsen the health condition and low tolerance of physical exercises to people with heart and lung complications. Affect public health	Old and high-risk people are advised to stay indoor and reduce physical activities. People with health complications are advised to see a doctor
301–500	Hazardous	Hazardous to high risk people and public health	Old and high-risk people are prohibited for outdoor activities. Public are advised to prevent from outdoor activities

Source: DOE Malaysia

Sources of Haze in Malaysia and Its Effect on Climate Change

Fire is commonly used in Indonesia as well as in Southeast Asia to clear land and to get rid of the agricultural waste, crops and debris for the establishment of plantations as it is the cheapest and cost-effective method of clearance. Most of the time, the fires flash out of control during the dry seasons, and the flames engulf vast areas. The combustion is not completed due to lack of oxygen during the burning, and acres of peatlands are covering in the region, thus causing thick smoke and brownish haze to cover the region.

Wild land fires and wildfires have been a characteristic of Southeast Asia ecology for centuries. It may happen by reducing the period of rainfalls especially during dry season; the past fires were smaller in area and more spread out over time. Forest fires and biomass burning in Borneo, Sabah, Sarawak and Sumatra have been reported a number of times over the last century. The sources of fires in forest are by human activities such as agricultural activities, ecotourism, camping in the forest and making fires as well as lightening caused fires that have insignificant impact on forest fires.

Forest clearing and peatland drainage associated with one of these projects, the Mega Rice Project, contributed substantially to the emissions observed during the 1997 El Niño (Page et al. 2002; Field et al. 2009).

The argument of forest conversion by showing that the native forests of Borneo have been impacted by selective logging, burning and land use conversion to extraordinary scales since industrial-scale extractive industries began in the early 1970s supported by Gaveau et al. (2014a). This study estimated that the reduction of Borneo's forested area was about 737,188 km^2 (30.2%) until 1973. Gaveau et al. (2014b) assessed the pollution levels generated, estimated climatic conditions prior to the fires and calculated the area burned prior vegetation cover and land ownership preceding the fires in Sumatra using satellite imageries. This study shown that 84,717 ha which is 52% of the total burned area was within concessions, i.e. land allocated to stakeholders and companies for plantation development. However, 60% of burned areas in concessions (50,248 ha or 31% of total burned area) were also occupied by communities. This scenario made attribution of fires problematic. The remaining 48% of the total burned land (79,012 ha) was owned by Indonesia's Ministry of Forestry (under central government). Another source of the haze is slash and burn of the remains of agricultural activities. There are three groups responsible for the fires: traditional cultivators, small-scale investors and large-scale investors. The traditional cultivators are the inactive farmers who burn their small plots of land after harvest to rejuvenate the soil and to keep their land free of weeds (Wosten et al. 2008). Others include the shifting cultivators who practice the slash-and-burn technique to clear a stretch of the forest for cultivation. Slash and burn is a cheap land clearing technique usually done for agricultural development especially in Western Africa, South America and Southeast Asia (Nganje et al. 2001; Varma 2003). Slash and burn is also part of traditional livelihood where small farmer practiced this system from one generation to another specifically in developing country. The slash-and-burn practice also has negative economic impact. A study by Varma (2003) estimated loss of USD20.1 billion in the economic impact of slash and burn that caused the forest fire in 1997/1998 in SEA. Slash-and-burn practice was discussed widely for its contribution to the forest alteration and large-scale forest burning. Slash and burn was criticized as the factor that causes the biodiversity loss and reduction of carbon sink and increases GHGs which is the main causes of global warming and climate change (Varma 2003).

The wide range usage of oil palm from food industries to household cleaning production and as biofuel or biodiesel triggers its unprecedented plantation expansion and unfortunately is responsible for large-scale forest conversions. This extensive tropical land conversion contributes to significant carbon emissions and global warming. Therefore oil palm plantation is another contributor source of haze in Malaysia. Ansari (2011) estimated that there will be higher demand of oil palm product in the next two decades because of the European countries' target for the use of biofuel for transport by year 2020. Based on reports by Sulaiman et al. (2011), about 85% of world's crude oil palm is supplied by Malaysia and Indonesia. Mosarof et al. (2015) investigated that to date about 19.667 million tonnes of palm oil has been produced from about 5.392 million hectares of land with the largest plantation area located in Sabah, Malaysia. Oil palm plantation area is expanded intensively in Malaysia for the last few decades as shown in Fig. 15.1. The oil palm trees are not considered typical tropical vegetation, mostly originated from Africa with scientific species *Elaeis guineensis*. The species grows well in tropical and rainforest climate which requires paramount of sunshine and hot and humid tropic conditions with high level of rainfall (Awalludin et al. 2015). Moreover, another factor that contributed to the high oil palm production was the low production cost and high productivity among major oil crops (Murdiyarso et al. 2010). The rapid

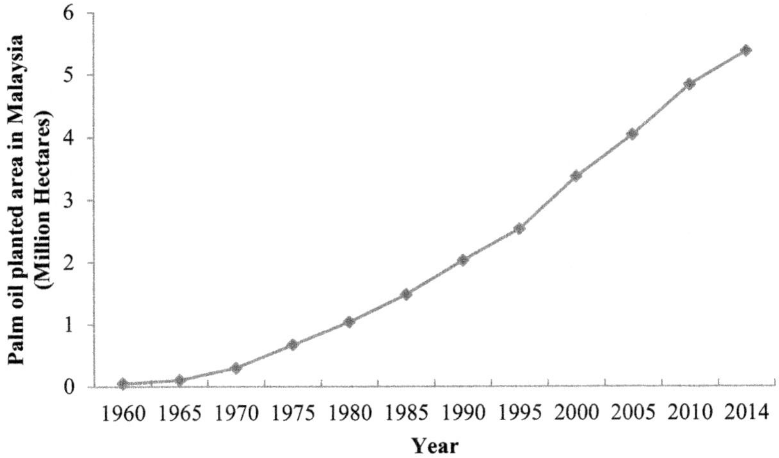

Fig. 15.1 Oil palm plantation area in Malaysia (Source: Awalludin et al. 2015)

expansion of oil palm plantation in Malaysia increases demand for large land areas which include not only natural tropical forest but also peatland forest. Figure 15.1 shows the oil palm plantation area in Malaysia since 1960 till year 2014. The plantation areas dramatically are increasing every 5 years.

The main threat to peatland is fires which are the main cause of haze pollution during forest fires in Malaysia. The exploitation of peatland includes all activities that change the pristine ecosystem of peatland such as logging, agriculture and water drainage. The gas fluxes between peatland areas and atmosphere were also affected by these destructive activities on peatland ecosystem (Miettinen and Liew 2010). According to Usup et al. (2000), fire that occurred in peatland area is due to the organic matter either already decomposed or still continue to decompose which are susceptible to fire. Dried peat is very susceptible to fire with the aid of dry season that usually lasts from May to October (Jaenicke et al. 2010). Organic peat soil combusted steadily and slowly without flame into the soil (Rein et al. 2008). This stage of burning which is considered as incomplete combustion is usually known as smouldering process. Smouldering can be described as slow, low temperature, flameless form of combustion and the most persistent type of combustion (Zaccone et al. 2014) which produce significant amount of CO_2, CO and particulate matter with harmful effect on human health. Peatland area is difficult to extinguish where it can smoulder deep underground and burn again during the next dry period (See et al. 2007; Blake et al. 2009). The fire in peat soil can persist for long period and can have enough time to spread deep underground with high production of particulate matter (Zaccone et al. 2014).

According to Keywood et al. (2003), the emission from combustion process such as vehicle emission, industrial emission and biomass burning produced high amount of particle that influences the formation of haze. Other air pollutant emissions from motor vehicle and other burning processes are NOx, CO, SO_2 aerosol which is the most important haze-producing species and carbon dioxide (CO_2). Atmospheric conversion of SO_2 to SO_4^{2-} produced sulphur in airborne particulate matter (Hopke et al. 2008). The emission of SO_2 came from motor vehicle, fossil fuel and high sulphur fuel dependency for industrial production and electric power generation (Abdullah et al. 2012).

Motor vehicle produced significant emission of air pollutant. As reported by KeTTHA (2011), there are increasing numbers of vehicles where in year 2009, more than one million units of new vehicle were registered, and there were approximately 20 million registered vehicles on the road. Increasing number of vehicle contributes to high amount of pollutant due to petrol combustion. Referring to Afroz et al. (2003), the major air pollution in Malaysia came from motor vehicle that contributing to at least 70–75% of total air pollution. Motor vehicle emissions consequently impacted the spatial and temporal distribution of ambient concentration that also determined by meteorological factors (Kim and Guldmann 2011).

Other sources of air pollution that can contribute to haze problem are stationary sources such as industrial emission and urbanization. According to Abdullah et al. (2012), in year 1998–2008, the industrial and urban areas contributed high concentration of PM_{10} which exceeded Malaysia Ambient Air Quality Guideline permissible level. Moreover, the concentration of PM_{10} in urban area is usually higher than rural area. Industrial areas in Malaysia are highly concentrated in Selangor, Sarawak, Johor, Sabah, Perak and Pahang which also producing high demand of fossil fuel and energy (Afroz et al. 2003). Heavy metals are one of the elements in air pollutant that related to industrial emission. According to Lopez et al. (2005), air pollutants with high concentration of lead, copper and nickel are relatively related to industrial sources.

Open burning source is one of the main contributors for high air pollutant concentration that enhance the haze episode. According to Lemieux et al. (2004), any combustion of materials in ambient environment described the open burning activity that include unintentional forest fires, burning of grain field for the preparation of next growing season and also fireworks at public celebration. Referring to Yu et al. (2012), open burning is the major source of global air pollutant that is responsible for 40% of all emitted CO, 32% of emitted CO_2, 20% of emitted aerosol and 50% of emitted poly aroma hydrocarbons (PAHs). In Malaysia, the penalty for open burning has been raised from RM100,000 to RM500,000 to show the seriousness of this action towards environmental pollution (Afroz et al. 2003). A study by Latif et al. (2011) found that a concentration of 31.8 μg/m3 for suspended particulate with particle size <1.5 μm was closely related to open burning in agricultural area in Sekincan, Selangor. The study by Amil N. (2016) showed that PM2.5 mass averaged at 28 ± 18 μgm3, 2.8-fold higher than the World Health Organization (WHO) annual guideline. The $PM_{2.5}$ mass ranged between 6 and 118 μgm^3 with the daily basis of WHO guideline exceeded 43%. High concentration of particulate matter with small particle size during open burning can worsen the air quality and cause severe haze pollution and higher carbon footprint which contributes to climate change. Climatic change consequences are natural disasters such as flood and tsunami. This phenomenon can contribute to adverse health impact due to waterborne diseases, food-borne diseases, poverty, loss of shelters and communicable diseases.

Impact of Haze on Human Health

Haze is not a new issue in Malaysia. Together with other countries in Southeast Asia region, Malaysia had been affected several times by haze episode due to the open forest burning in Indonesia. First haze episode was recorded in 1983 followed by 1990, 1991 and 1994. The worst episode occurred in 1997 when the whole country was covered with thick smoke haze from Kalimantan and Sumatra. During this time, Malaysian government declared emergency in some of the states such as Sarawak and Johor due to the hazardous Air Pollution Index (API) reading which was greater than 300 ($PM_{10} > 420$ $\mu g/m^3$) that leads to the closure of schools in the affected area (Othman et al. 2014; Mohd Shahwahid H.O et al.2016). Since then, several minor haze episodes were also recorded in 2005, 2006 and 2010. It was followed by severe episodes in 2013 with Muar of Johor, which recorded the highest API reading of 641 (Mohd Shahwahid et al. 2016).

Majority of the health impact were associated with respiratory condition such as asthma, acute bronchitis, allergic rhinitis and acute upper respiratory tract illness (URTI). It was also associated with conjunctivitis and eczema which include contact dermatitis (Emmanuel 2000). In addition, short-term exposures to haze can also be associated with cardiac arrhythmias, worsening heart failure and increased risk of developing acute myocardial infarction among high-risk patients (Brook et al. 2004). During the severe haze episode in 1997, casualty visit in Kuching and Kuala Lumpur showed 100% increases with majority of the cases were due to asthma or acute respiratory infection. In Singapore, similar pattern was also observed in that year with an increase of 30% outpatient cases due to haze-related illness by Emmanuel (2000).

Apart from health effects, haze can also give significant economic impacts to the affected country. The total economic impact due to haze can be in a form of cost of illness from both patient's and provider's perspective. These include medical treatment and hospitalization, medical-related leave taken due to the haze, cost of buying personal protective equipment, cost due to reduced-activity days or loss of productivity and foregone income opportunities (Mohd Shahwahid et al. 2016). A study done in Malaysia by Jamal et al. found that 'the average annual economic loss due to the inpatient health impact of haze was valued at MYR273,000' (Othman et al. 2014). Another study done on the economic impact of haze episode in Malaysia in 2013 stated that the total cost of illness due to haze was about MYR410,587,779 (Mohd Shahwahid et al. 2016).

Climatic and environmental factors play an important role in the breeding and dispersion of the *Aedes* mosquito, a primary vector of dengue fever. By monitoring these factors, it is possible to predict the emergence of a dengue endemic and subsequently reduce its spread. Study by Aghamohammadi et al. (2015) investigated the correlation between the Air Pollution Index (API) and the reported number of dengue cases in five districts of Malaysia. Data of the API and the number of dengue cases from five districts in the state of Selangor in the years 2013 and 2014 were obtained from the Malaysian Department of Environment website and the Malaysian Ministry of Health website, respectively. Average API readings for each week were assigned to either good (<50), moderate (50–100) or unhealthy (>100), and the total number of cases in each district that fell into either one of these API categories was summed up. Cumulatively, in 2013 and 2014, 66.5% of dengue cases were recorded when the API reading was within 'good' levels, while 31.8% and 1.7% of cases were recorded while the API reading were within 'moderate' and 'unhealthy' levels, respectively. Spearman's correlation, ρ, test and significance testing were carried out between the API categories and the number of recorded dengue cases in the five districts. The results were $R = -0.532$ with a p-value $(0.002) < \alpha = 0.01$ ($n = 30$). These results show that there is a statistically significant negative correlation between the dengue cases and the API value. In conclusion, the significant relationship between the API values

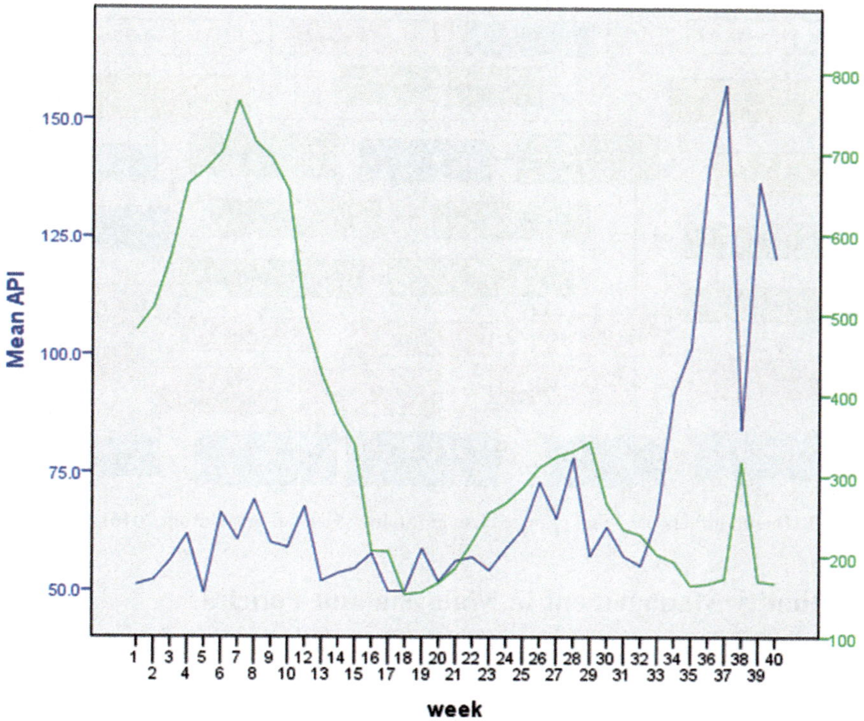

Fig. 15.2 API and dengue cases for 2015 (Source: Aghamohammadi et al. 2015)

and the recorded dengue cases suggests that an increase in the API levels causes a decrease in the number of dengue cases. This could be due to the presence of smog, dust particles and other particles that disrupt either the breeding or feeding pattern of the dengue vector (Aghamohammadi et al. 2015). The study on correlation between Air Pollution Index and dengue cases in Malaysian districts found a significant negative correlation between the number of reported dengue cases and the air quality in Malaysia as shown in Fig. 15.2 (Aghamohammadi et al. 2015).

Another study by Hashim and Hashim (2016) shows the health effects of global climate change and presented the association between climate change with environmental impact and health impact shown in Fig. 15.3.

Malaysia had faced the periodic intense exposures to particulate matter of haze from both domestic sources such as increased traffic and constructions and also international sources such as open forest fires from the neighbour country. Despite all the precautions and discussions made, the issue still persists with the latest episode recorded in 2015.

The impact of the exposures can be seen from both health and also economical perspective. The monetary burden due to economic loss and increase in healthcare expenditure was very significant and might affect the development of Malaysia.

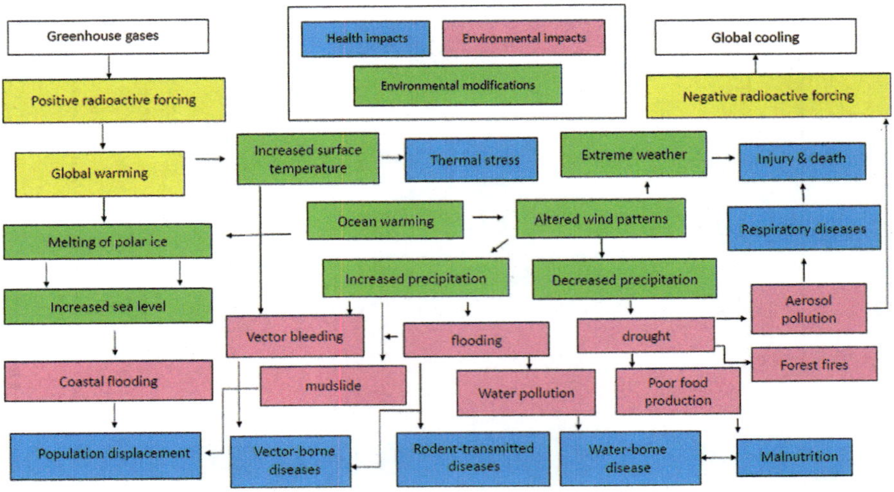

Fig. 15.3 Health effects of global climate change (Source: Hashim and Hashin 2016)

Air Quality Management in Malaysia and Policies

The first Malaysia Ambient Air Quality Guideline has been used since 1989. The New Ambient Air Quality Standard adopts six air pollutant criteria that include five existing air pollutants which are particulate matter with the size of less than 10 μm (PM_{10}), sulphur dioxide (SO_2), carbon monoxide (CO), nitrogen dioxide (NO_2), and ground-level ozone (O_3) as well as one additional parameter which is particulate matter with the size of less than 2.5 μm ($PM_{2.5}$).

The air pollutant concentration limit will be strengthened in stages until 2020. There are three interim targets set which include interim target 1 (IT-1) in 2015, interim target 2 (IT-2) in 2018 and the full implementation of the standard in 2020 shown in Table 15.2.

The Environmental Quality Act 1974 was amended in 1998 to provide a more stringent penalty for open burning offences. According to the Act, any person who contravenes shall be guilty of an offence and shall, on conviction, be liable to a fine not exceeding RM500,000 or to imprisonment for a term not exceeding 5 years or both. The Environmental Quality (Declared Activities) (Opening Burning) 2003 Order came into force on 1 January 2004. It prohibits open burning of certain activities under specified conditions and in certain designated areas.

To enhance the enforcement capacity, the Department of Environment, the agency entrusted to enforce the law against open burning, has delegated powers to officers of the fire and rescue department, the Royal Malaysia Police, the Ministry of Health and the local authorities to assist in the investigation of open burning activities.

At the operational level, ground and air surveillance to curb and prevent open burning activities in the fire-prone areas will be intensified especially during the dry seasons. At the state level, the State Department of Environment has developed a

Table 15.2 The New Ambient Air Quality Standard in Malaysia

Pollutants	Averaging time	Ambient Air Quality Standard (μg/m³)		
		IT-1 (2015)	IT-2 (2018)	Standard (2020)
Particulate matter with diameter size of less than 10 μm (PM_{10})	1 year	50	45	40
	24 h	150	120	100
Particulate Matter with diameter size of less than 2.5 μm ($PM_{2.5}$)	1 year	35	25	15
	24 h	75	50	35
Sulphur dioxide (SO_2)	1 h	350	300	250
	24 h	105	90	70
Nitrogen dioxide (NO_2)	1 h	320	300	280
	24 h	75	75	70
Ground-level ozone (O_3)	1 h	200	200	180
	8 h	120	120	100
Carbon monoxide (CO)	1 h	35	35	30
	8 h	10	10	10

Source: DOE (2016)

specific plan of action to prevent fires in their respective state. The components of the plan among others include (a) map of fire-prone areas, (b) enforcement and monitoring programmes, (c) implementation of the awareness programmes, (d) preparedness for firefighting and (e) communication network to coordinate complaints and investigate cases of open burning.

Under the DOE, the Clean Air Action Plan (2010–2020) was established in 2011, and it contains five main strategies in order to improve the air quality. The five strategies are described to reduce emissions from motor vehicles, prevent haze pollution from land and forest fires, reduce emissions from industries, build institutional capacity and capabilities and strengthen public awareness and participation.

In order to prevent haze from land and forest fires, two approaches were adopted – prevention and control at national as well as at the regional level. Among the actions taken at the national level include the implementation of fire prevention and peatland management programme and strengthening the enforcement on open burning.

In order to reduce emission from motor vehicles, the focus is on sharing the development of better fuel and engine technology as well as the development of a roadmap for the implementation of a more stringent emission standard. Further initiatives are also encouraged to further reduce the emissions from industrial activities such as reviewing existing emission standards, improving emission inventories, encouraging the concept of self-regulation and performance-monitoring of antipollution equipment by industries as well as promoting the best available air pollution control technology.

The CAAP is also aimed in addressing the need to strengthen institutional capacity such as the development of expertise in air quality prediction and modelling and the development of a new ambient air quality standard. Public awareness and public participation programmes are given a new push to attract the interest of the students, environmental practitioners, corporate leaders and decision-makers.

References

Abdullah AM, Samah MAA, Tham YJ (2012) An overview of the air pollution trend in Klang Valley, Malaysia. Open Environ Sci 6:13–19

Afroz R, Hassan MN, Ibrahim NA (2003) Review of air pollution and health impacts in Malaysia. Environ Res 92:71–77

Aghamohammadi N, Zaki R, Ragupathy K, Yahya A (2015) Correlation between air pollution index and dengue cases in Malaysian Districts, 47th Asia-Pacific Academic Consortium for Public Health, 21 Oct 2015 to 23 Oct 2015, University of Indonesia

Amil N, Latif MT, Khan MF, Mohamad M (2016) Seasonal variability of PM2:5 composition and sources in the Klang Valley urban-industrial environment. Atmos Chem Phys 16:5357–5381

Ansari AH (2011) Peatlands and global warming: a study with special reference to South-East Asian countries. Aust J Basic Appl Sci 5(7):596–605

Aouizerats B, Van Der Werf GR, Balasubramanian R, Betha R (2015) Importance of transboundary transport of biomass burning emissions to regional air quality in Southeast Asia during a high fire event. Atmos Chem Phys 15:363–373

Awalludin MF, Sulaiman O, Hashim R, Nadhari WNAW (2015) An overview of the oil palm industry in Malaysia and its waste utilization through thermochemical conversion, specifically via liquefaction. Renew Sust Energy Rev 50:1469–1484

Behera SN, Betha R, Huang X, Balasubramanian R (2015) Characterization and estimation of human airway deposition of size-resolved particulate-bound trace elements during a recent haze episode in Southeast Asia. Environ Sci Pollut Res 22:4265–4280

Blake D, Hinwood AL, Horwitz P (2009) Peat fires and air quality: volatile organic compounds and particulates. Chemosphere 76:419–423

Brook RD, Franklin B, Cascio W, Hong Y, Howard G, Lipsett M, et al. (2004) AHA Sci Statement 6083(71)

DOE (1997) Malaysia environmental quality report. Department of Environment, Kuala Lumpur

DOE (2016) Air quality standards. http://www.doe.gov.my/portalv1/wp-content/uploads/2013/01/Air-Quality-Standard-BI.pdf

Emmanuel SC (2000) Impact to lung health of haze from forest fires: the Singapore experience. Respirology 5(2):175–182

Field RD, van der Werf GR, Shen SP (2009) Human amplification of drought-induced biomass burning in Indonesia since 1960s. Nat Geosci 2:185–188

Gaveau DLA, Sloan S, Molidena E, Yaen H, Sheil D, Abram NK (2014a) Four decades of forest persistence, clearance and logging on Borneo. PLoS One 9(7):e101654

Gaveau DLA, Salim MA, Hergoualc'h K, Locatelli B, Sloan S, Wooster M, Marlier ME, Molidena E, Yaen H, DeFries R, Verchot L, Murdiyarso D, Nasi R, Holmgren P, Sheil D (2014b) Major atmospheric emissions from peat fires in Southeast Asia during non-drought years: evidence from the 2013 Sumatran fires. Sci Rep 4:1–7

Hashim JH, Hashim Z (2016) Climate change, extreme weather events, and human health implications in the Asia Pacific Region Asia Pacific. J Public Health 28(2S):8S–14S

Hopke PK, Cohen DD, Begum BA, Biswas SK, Ni B, Pandit GG, Santoso M, Chung Y-S, Davy P, Markwitz A, Waheed S, Siddique N, Santos FL, Pabroa PCB, Seneviratne MCS, Wimolwattanapun W, Bunprapob S, Vuong TB, Hien PD, Markowicz A (2008) Urban air quality in the Asian region. Sci Total Environ 404:103–112

Jaenicke J, Englhart S, Siegert F (2010) Monitoring the effect of restoration measures in Indonesian peatlands by radar satellite imagery. J Environ Manag 92(3):630–638

KeTTHA (2011) Low carbon cities: framework and assessment system. Kementerian Tenaga, Teknologi Hijau dan Air, Putrajaya

Keywood MD, Ayers GP, Gras JL, Boers JL, Leong CP (2003) Haze in the Klang Valley of Malaysia. Atmos Chem Phys Discuss 3:615–653

Kim Y, Guldmann J-M (2011) Impact of traffic flows and wind directions on air pollution concentrations in Seoul, Korea. Atmos Environ 45:2803–2810

Latif MT, Anuwar NY, Srithawirat T, Razak IS, Ramli NA (2011) Composition of Levoglucosan and surfactants in atmospheric aerosols from biomass burning. Aerosol Air Qual Res 11 (7):837–845

Lemieux PM, Lutes CC, Santoianni DA (2004) Emissions of organic air toxics from open burning: a comprehensive review. Prog Energy Combust 30:1–32

Lopez JM, Callén MS, Murillo R, Garcia T, Navarro MV, de la Cruz MT, Mastral AM (2005) Levels of selected metals in ambient air PM10 in an urban site of Zaragoza (Spain). Environ Res 99:58–67

Miettinen J, Liew SC (2010) Status of peatland degradation and development in Sumatra and Kalimantan. Ambio 39:394–401

Mohd Shahwahid HO (2016) The economic value of the June 2013 Haze impacts on Peninsular Malaysia. EEPSEA Research Report No. 2016-RR3. Economy and Environment Program for Southeast Asia, Laguna

Mosarof MH, Kalam MA, Masjuki HH, Ashraful AM, Rashed MM, Imdadul HK, Monirul IM (2015) Implementation of palm biodiesel based on economic aspects, performance, emission, and wear characteristics. Energy Convers Manag 105:617–629

Murdiyarso D, Hergoualc'h K, Verchot LV (2010) Opportunities for reducing greenhouse gas emissions in tropical peatlands. Proc Natl Acad Sci U S A 107(46):19655–19660

Nganje W, Schuck EC, Yantio D, Aquach E (2001) Farmer education and adoption of slash and burn agriculture, Agribusiness & Applied Economic Miscellaneous Report No. 190. North Dakota State University

Othman J, Sahani M, Mahmud M, Sheikh Ahmad MK (2014) Transboundary smoke haze pollution in Malaysia: inpatient health impacts and economic valuation. Environ Pollut Elsevier Ltd 189:194–201

Page SE, Siegert F, Rieley JO, Boehm H-DV, Jayak A, Limink S (2002) The amount of carbon released from peat and forest fires in Indonesia during 1997. Nature 420:61–65

Rein G, Cleaver N, Ashton C, Pironi P (2008) The severity of smouldering peat fires and damages to the forest soil. Catena 74(3):304–309

See SW, Balasubramanian R, Rianawati E, Karthikeyan S, Streets DG (2007) Characterization and source apportionment of particulate matter 2.5 μm in Sumatra, Indonesia, during a recent peat fire episode. Environ Sci Technol 41(10):3488–3494

Sulaiman F, Abdullah N, Gerhauser H, Shariff A (2011) An outlook of Malaysian energy, oil palm industry and its utilization of wastes as useful resources. Biomass Bioenergy 35:3775–3786

Usup A, Takahashi H, Limin SH (2000) Aspect and mechanism of peat fire in tropical peat land: a case study in Central Kalimantan 1997, Proceedings of the international symposium on tropical peatlands, pp 79–88

Varma A (2003) The economics of slash and burn: a case study of the 1997–1998 Indonesian forest fires. Ecol Econ 46:159–171

Wosten JHM, Clymans E, Page SE, Rieley JO, Limin SH (2008) Peat–water interrelationships in a tropical peatland ecosystem in Southeast Asia. Catena 73:212–224

Yu T-Y, Lin C-Y, Chang L-FW (2012) Estimating air pollutant emission factors from open burning of rice straw by the residual mass method. Atmos Environ 54:428–438

Zaccone C, Rein G, D'Orazio V, Hadden RM, Belcher CM, Miano TM (2014) Smouldering fire signatures in peat and their implications for palaeoenvironmental reconstructions. Geochim Cosmochim Acta 137:134–146

Dr Nasrin Aghamohammadi (Env. Health Eng. BSc, Chem. Eng. PhD) is an Environmental Health Engineer (PhD in Chemical Engineering, MSc in Civil Engineering, and BSc in Environmental Health Engineering) and joined as a senior lecturer at the Department of Social and Preventive Medicine, Faculty of Medicine, University of Malaya. Her core expertise is in environmental engineering and Health. Her ongoing project is a grand challenge project focusing on Urban Heat Island and thermal comfort for urban residents in Kuala Lumpur as world class city by 2030.

Dr Marzuki Isahak (MBBS, MPH, DrPH) is a senior medical lecturer and public health physician at the Department of Social and Preventive Medicine, Faculty of Medicine, University of Malaya (UM). He is currently the head of Occupational Safety, Health and Environment Unit in UM Medical Centre. He is also a council member in the Academy of Occupational and Environmental Medicine, Malaysia.

Chapter 16
Climate Change, Air Pollution, and Human Health in Bangkok

Uma Langkulsen and Desire Rwodzi

Abstract Background While a number of studies have published the health effects of climate change and air pollution, little has been studied in Thailand on the health effects following interactions between air pollution and climate change.
Objectives The aim of the study was to explore the interplays between climate change and air pollution and how these in turn impact on human health among residents of Bangkok, Thailand.
Methods We conducted a descriptive study based on existing data on air pollution from Thailand's Pollution Control Department, data on number of vehicles from the Transport Statistics Subdivision under Thailand's Department of Land Transport, data on rainfall and temperature from the Thai Meteorological Department, data on health outcomes from Thailand Ministry of Public Health, and demographic data from the Department of Provincial Administration.
Results As of 2016, the Pollution Control Department of Thailand had a total of 17 air pollution monitoring stations around Bangkok, including 6 roadside and 11 general area stations. While there has been a downward trend in PM_{10} concentrations from 1992 to 2015, $PM_{2.5}$ concentrations have not only been above-recommended standards but also going up. The number of registered vehicles in Bangkok peaked at more than one million in 2013, but since then a declining trend has been observed. In Bangkok, temperatures peaked around April, while rainfall peaked during the month of September. Overall, both annual minimum and maximum temperatures have been going up since 1951. The average amount of rainfall received monthly had two peaks, first in May and later in September. From 1951 to 2015, the mean annual rainfall in Thailand went below 1400 mm only in 1977, 1979, and 1992. Mortality rates due to diseases of the circulatory and respiratory system have also been going up since 2010, with mortality rates per 100,000 population higher among males than females. While the number of outpatients

U. Langkulsen (✉)
Faculty of Public Health, Thammasat University, Pathumthani 12121, Thailand
e-mail: u.langkulsen@gmail.com

D. Rwodzi
Strategic Information Hub, UNAIDS Eastern and Southern Africa Region, Johannesburg, South Africa
e-mail: rwodzidesire@gmail.com

© Springer International Publishing AG 2018

R. Akhtar, C. Palagiano (eds.), *Climate Change and Air Pollution*,
Springer Climate, DOI 10.1007/978-3-319-61346-8_16

due to diseases of the circulatory system continues to increase, outpatients due to respiratory diseases peaked around 2010, and since then a downward trend has been observed.
Conclusion Results suggest possible correlations between air pollution-climate change interactions and mortality due to diseases of the respiratory and circulatory systems.

Keywords Climate change • Air pollution • Temperature • Human health • Bangkok

Introduction

A growing body of evidence suggests that the global climate is changing rapidly, and the planet has warmed substantially as a result of increased greenhouse gas emissions largely from human activities (D'Amato et al. 2015; Franchini and Mannucci 2015). Consequently, climate change is attributed to a global rise and variability in ambient temperature, increased air pollution, an increased frequency of heat waves, of adverse weather events such as floods and drought periods, as well as an uneven distribution of allergens and vector-borne infectious diseases (D'Amato et al. 2015; Franchini and Mannucci 2015). Changes in climatic conditions as well as air quality have measurable impact on human health (Mirsaeidi et al. 2016), in part by altering the epidemiology of climate-sensitive pathogens. Climate change may modify the incidence and severity of respiratory infections by affecting vectors, and host immune responses to, for example, infections, such as avian influenza, are being experienced in areas previously unaffected, apparently because of global warming (Mirsaeidi et al. 2016).

Variability in ambient temperature is reported to have had its toll on human health in different parts of the world. In Brisbane, both hot and cold temperatures were associated with increases in emergency department admissions for childhood asthma, and their effects both appeared to be acute (Xu et al. 2013). A recent study in China showed that a 1 °C increase in diurnal temperature range corresponded to an increase in total non-accidental mortality, cardiovascular mortality, and respiratory mortality during the cool seasons (Zhou et al. 2014).

Interrelationships between air pollution and climate change are complex, and in a reciprocal interplay, various air pollutants contribute to global warming, while global warming in turn leads to the formation of various pollutant compounds (Schulte et al. 2016). A recent study evaluating associations between air pollutants and meteorological factors reported strong correlations between and among gas pollutants due to their photochemical activity, as well as positive correlation between air temperature and pollutants (Lagidze et al. 2015). D'Amato et al. (2015) posited that an individual's response following air pollution exposure depends on the source and components of air pollution, as well as the underlying meteorological conditions. Indeed, it has been observed that some air pollution-related outcomes such as asthma do not depend only on increased air pollution levels but also on atmospheric conditions favoring the accumulation of air pollutants at ground level (D'Amato et al. 2013).

Due to climate change, air pollution patterns are changing in several urbanized areas of the world, with a significant effect on respiratory health and consequences ranging from decreases in lung function to allergic diseases, new onset of diseases, and exacerbation of chronic respiratory diseases (D'Amato et al. 2013). Associations between short-term exposure to air pollutants and mortality have been reported in several studies (Guo et al. 2014; Shang et al. 2013; Tsai et al. 2014). Long-term exposures to pollutants have also been linked to mortality (Chen et al. 2012, 2013; Deguen et al. 2015). However, a growing body of evidence suggests that long-term exposures have greater effects than short-term variation of pollutants' concentrations (Beverland et al. 2012; Deguen et al. 2015).

The objective of this study is to use existing data from Thailand's Pollution Control Department and Ministry of Public Health to explore the scenario in Bangkok in terms of the interplays between climate change and air pollution and how these in turn impact on human health.

Methods

This study is based on existing data on air pollution, rainfall and temperature, and health outcomes from relevant ministries and agencies. We obtained data on air pollution from the Pollution Control Department of Thailand. This included latest data on the number and distribution of monitoring stations in Bangkok, annual average PM_{10} concentrations from 1992 through 2015, and $PM_{2.5}$ concentrations from 2011 to 2015. Additional data on the number of registered vehicles from 2006 to 2015 was sourced from the Transport Statistics Subdivision under Thailand's Department of Land Transport.

We sourced data on average temperature and rainfall in Bangkok from the Thai Meteorological Department. Data on health outcomes, including morbidity, mortality, and low birth weight, was obtained from Thailand's Ministry of Public Health. Additional data on the demographics of Bangkok was obtained from the Department of Provincial Administration.

Demographics of Bangkok

The Department of Provincial Administration reported that as of December 2015, Bangkok had a total population of 5,696,409. This comprised 5,605,672 (98%) Thai nationals and 90,737 (2%) non-Thai nationals.

Air Pollution in Bangkok

Air pollution is one of the major environmental problems affecting Bangkok. The World Bank cites transport, industry, construction, power generation, indoor air pollutants, and refuse burning as the main causes of air pollution in Bangkok. Most of the air pollution in the city is emitted within the transport sector due to the concentration of motor vehicles. The construction industry also causes high level of dust pollution. Lack of proper planning and zoning of housing areas has aggravated the seriousness of air pollution.

Number of Vehicles

Figure 16.1 below shows the annual number of registered vehicles in Bangkok from 2006 to 2015. The least number (606,901) of vehicles was reported in 2009, after which the number increased remarkably by 79% to reach a peak of 1,084,080 in 2013. Since then, the annual number of registered vehicles has been on a declining trend, declining by 25% from 2013 to 2015. As of December 2015, there were 811,222 registered vehicles in Bangkok.

Air Pollution Monitoring Stations

As of 2016, the Pollution Control Department (PCD) of Thailand had a total of 17 air pollution monitoring stations around Bangkok, and these include 6 roadside and 11 general area stations (see Fig. 16.2 below). At these stations, concentrations for criteria pollutants including sulfur dioxide (SO_2), nitrogen dioxide (NO_2), carbon monoxide (CO-1 h, CO-8 h), ozone (O_3-1 h, O_3–8 h), PM_{10} and $PM_{2.5}$ (particulate matter with aerodynamic diameter less than 10 and 2.5 μm, respectively), total suspended particulates (TSP), and lead (Pb) were measured.

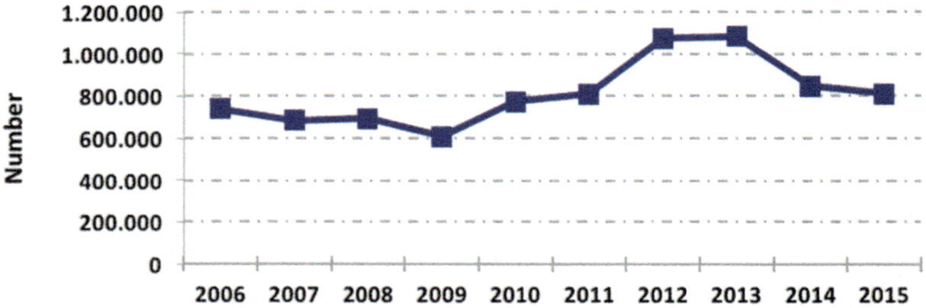

Fig. 16.1 Number of registered vehicles as of 31 December 2015 (Source: Transport Statistics Sub-Division, Planning Division, Department of Land Transport 2015)

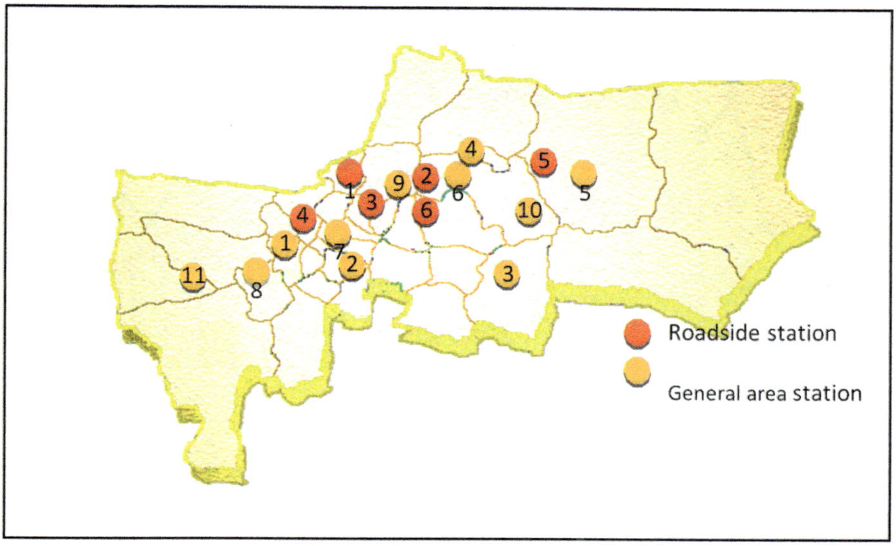

Fig. 16.2 Bangkok air quality monitoring stations, 2016

Roadside station	General area station
1. Ministry of Science and Technology	1. Bansomdejchaopraya Rajabhat University
2. Land Transport Department	2. Rat Burana Post Office
3. Chulalongkorn Hospital	3. Thai Meteorological Department, Bangna
4. Thonburi Power Substation	4. Chandrakasem Rajabhat University
5. Chokchai 4 Police Box	5. Klongjun – National Housing Authority
6. Dindaeng – National Housing Authority	6. Huaykwang – National Housing Authority Stadium
	7. Nonsi Withaya School
	8. Singharaj Pittayakom School
	9. The Government Public Relations Department
	10. Bodindecha (Sing Singhaseni) School
	11. Bang Khun Thian 2 Highway

Annual Average PM_{10} Concentrations

Figure 16.3 below shows trends for annual average PM_{10} concentrations in Bangkok from 1992 to 2015 as measured by roadside and general area monitoring stations. Overall, there has been a downward trend in PM_{10} concentrations from 1992 to 2015. In more recent years from 2013 to 2015, 19% and 7% declines in PM_{10} concentrations were observed for roadside and general area monitoring stations, respectively. The highest concentrations measured at both roadside and

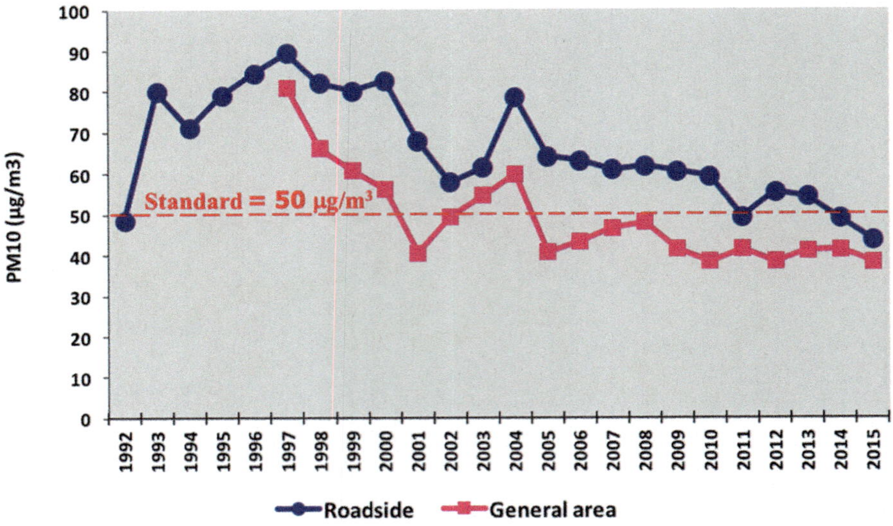

Fig. 16.3 Trends of PM_{10} in Bangkok (Source: Pollution Control Department 2016)

general area monitoring stations were reported in 1997, while the lowest concentrations were reported in 2015. Roadside PM_{10} concentrations were consistently above PM_{10} concentrations recorded at general area monitoring stations from 1992 through 2015. In addition, roadside PM_{10} concentrations were consistently above the standard of 50 μg/m^3 as recommended by the Pollution Control Department of Thailand, except for 3 years, that is, 1992, 2014, and 2015.

Annual Average $PM_{2.5}$ Concentrations

The Pollution Control Department of Thailand measured $PM_{2.5}$ at three stations comprising one roadside station situated at Dindaeng National Housing Authority and two general area stations situated at the Government Public Relations Department and Bodindecha (Sing Singhaseni) School. Figure 16.4 below shows the trend for annual average $PM_{2.5}$ concentrations from the roadside monitoring station in Bangkok from 2011 to 2015. Overall, the annual average $PM_{2.5}$ had an upward trend that was consistently above the annual average standards of 25 μg/m^3 by the PCD and 10 μg/m^3 by the World Health Organization. The annual average $PM_{2.5}$ increased by 9% from 33 μg/m^3 in 2011 to 36 μg/m^3 in 2014, after which the annual concentration leveled off at 36 μg/m^3.

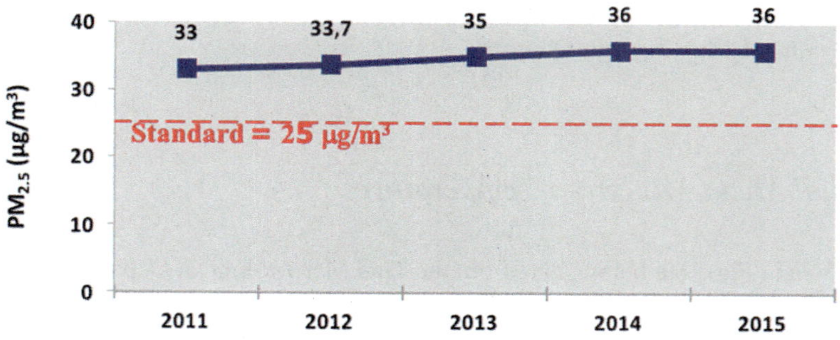

Fig. 16.4 Trends of annual average $PM_{2.5}$ concentration from a roadside monitoring station in Bangkok (Source: Pollution Control Department 2016)

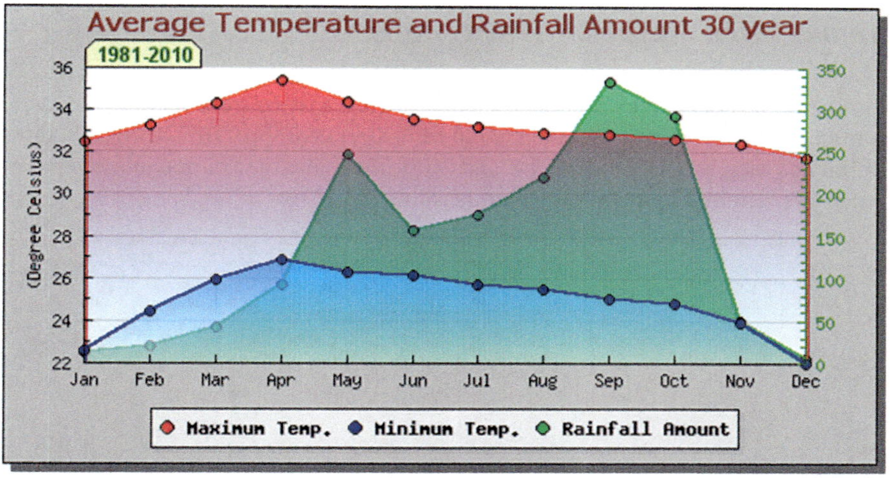

Fig. 16.5 Average temperature and rainfall in Bangkok over a 30-year period: 1981–2010 (Source: Thai Meteorological Department 2016 [Online]. Available: http://www.tmd.go.th/province_weather_stat.php?StationNumber=48455)

Climate Change in Bangkok

Rainfall and Temperature

Figure 16.5 above shows trends for average temperature and rainfall patterns in Bangkok over a 30-year period from 1981 to 2010. The average monthly temperatures peaked during April, reaching a maximum of 35.5 °C and a minimum of 26.9 °C. Average temperatures then declined steadily through the months to reach their lowest in December, coinciding with the lowest amounts of rainfall received in the same month. The average amount of rainfall received monthly had two peaks, first in May and later in September. Overall, most of the rainfall was received from the months of May through October.

Annual Mean Minimum Temperature

As shown in Fig. 16.6 below, data from the Thai Meteorological Department indicate that although fluctuating, the overall pattern for the annual mean minimum temperature is a rising trend. The lowest annual mean temperature was recorded in 1955, while minimum temperatures above 23.5 °C were recorded in 1998, 2010, and 2012.

Annual Mean Maximum Temperatures

Similar to the annual mean minimum temperatures, a rising trend has been reported for the annual mean maximum temperatures as shown in Fig. 16.7 below. On three different years, that is, 1998, 2010, and 2015, the annual mean maximum temperatures are reported to have reached at least 33.5 °C.

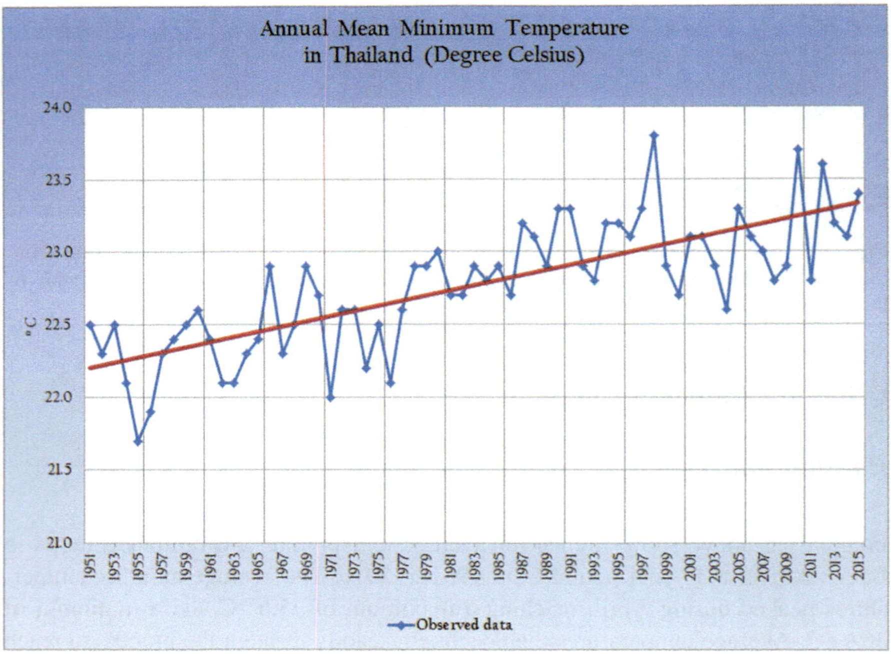

Fig. 16.6 Annual mean minimum temperature in Thailand (1951–2015) (Source: Thai Meteorological Department 2016 [Online]. Available: http://www.tmd.go.th/climate/climate.php?FileID=7)

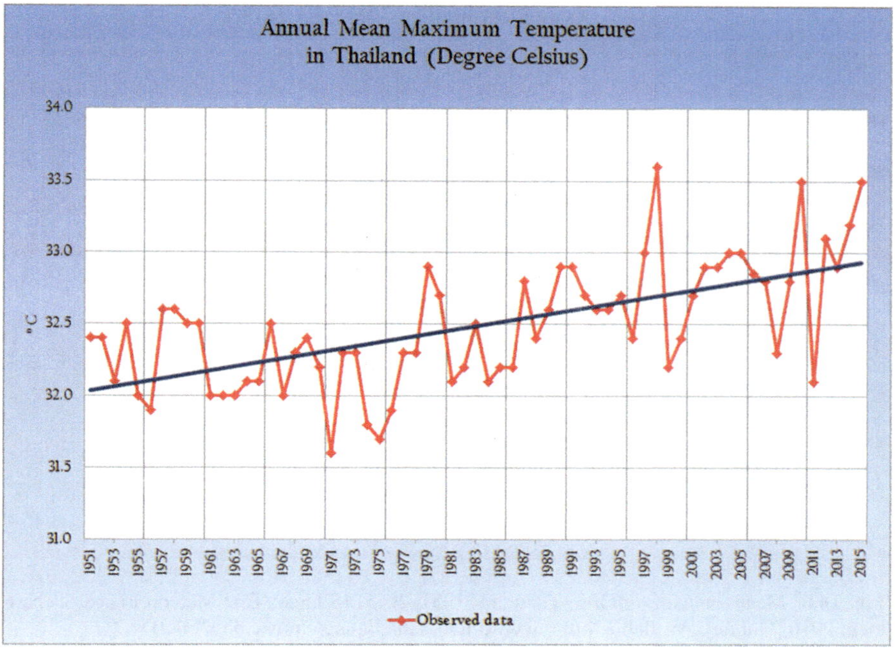

Fig. 16.7 Annual mean maximum temperature in Thailand (1951–2015) (Source: Thai Meteorological Department 2016 [Online]. Available: http://www.tmd.go.th/climate/climate.php?FileID=7)

Mean Annual Rainfall in Thailand

Figure 16.8 below shows the mean annual rainfall in Thailand from 1951 to 2015 as reported by the Thai Meteorological Department. Although the observed data shows some fluctuations, the highest mean annual rainfall was recorded in 1953 and 2011. The 3-year moving average shows an overall decline in rainfall received from 1951 to 1992, after which an upward trend is observed, but with huge fluctuations. Overall, the mean annual rainfall went below 1400 mm only in 1977, 1979, and 1992.

Air Pollution-Climate Change Interactions and Effects on Health

Results point toward a possible correlation between air pollution and climate change, in particular temperature changes in Bangkok. With increasing temperatures, $PM_{2.5}$ concentrations also showed an increasing trend. These interactions

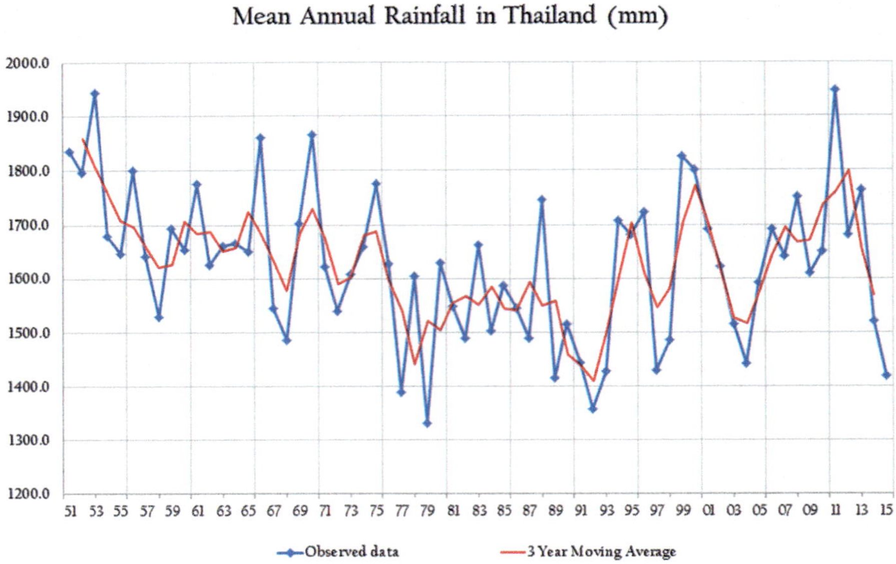

Fig. 16.8 Mean annual rainfall in Thailand (1951–2015) (Source: Thai Meteorological Department 2016 [Online]. Available: http://www.tmd.go.th/climate/climate.php?FileID=7)

between air pollution and climate change also showed some associations with human health and, in particular, morbidity and mortality due to diseases of both the circulatory and respiratory systems. However, the air pollution-climate change interactions appeared not to have any correlation with low birth weight.

Mortality Due to Diseases of the Circulatory System

According to Ministry of Public Health reports, deaths as a result of diseases of the circulatory system have been on an upward trend for both males and females from 2010 to 2014 (Bureau of Policy and Strategy; Ministry of Public Health 2015). Although deaths reported among females were lower than those for males, the percent increase in mortality rate per 100,000 population was higher for females compared to men. Mortality rate per 100,000 population due to diseases of the circulatory system increased by 50% among females from 52% in 2010 to 78% in 2014, while the mortality rate increased from 72% to 103% during the same period (Figs. 16.9 and 16.10).

Mortality Due to Diseases of the Respiratory System

Similar to mortality rates per 100,000 population as a result of diseases of the circulatory system, the Ministry of Public Health reported that the mortality rates

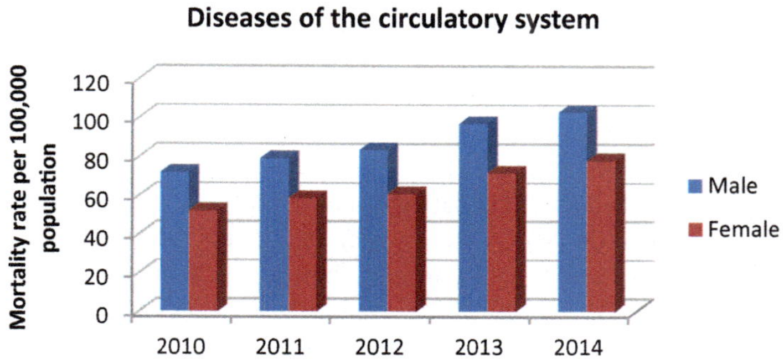

Fig. 16.9 Mortality rates per 100,000 population of disease of the circulatory (2010–2014) (Source: Bureau of Policy and Strategy; Ministry of Public Health 2015. Based on ICD mortality tabulation list 1, 10th revision)

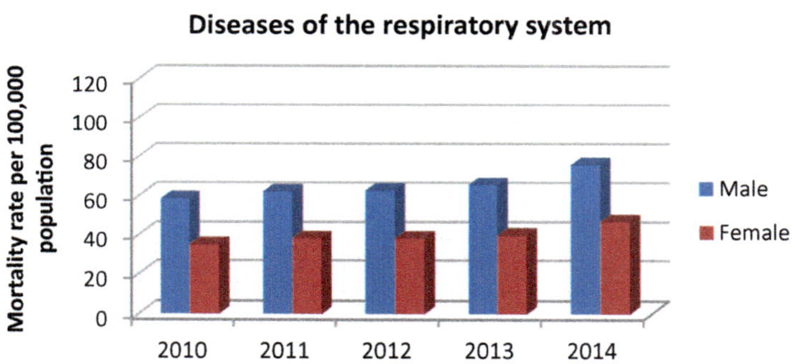

Fig. 16.10 Mortality rates per 100,000 population of disease of the respiratory system (Source: Bureau of Policy and Strategy; Ministry of Public Health (2015). Based on ICD mortality tabulation list 1, 10th revision)

due to diseases of the respiratory system were higher among males than females. Overall, the trend was increasing for both gender from 2010 to 2014. While the mortality rates per 100,000 population due to diseases of the respiratory system increased by 33% among females, rate increased by 30% among males from 2010 to 2014.

Low Birth Weight

The proportion of low birth weight babies in Thailand is fairly low as shown in Fig. 16.11 below. Overall, the proportion of low birth weight babies ranged between 8% and 10% from 1997 to 2012. With a fairly stable trend, the proportion of low birth weight babies declined to below 8% only in 2013.

Morbidity: Number of Outpatients

Figure 16.12 below shows the number of outpatients with diseases of the circulatory system compared to outpatients with diseases of the respiratory system. While the number of outpatients for diseases of the circulatory system continued to rise from 2005 to 2014, the number of outpatients due to diseases of respiratory system increased from 2005 to 2009, after which the number leveled off at below 30,000,000 and then started to decline. By the end of 2014, the number of outpatients for diseases of the circulatory system was greater than the number of outpatients reporting diseases of the respiratory system.

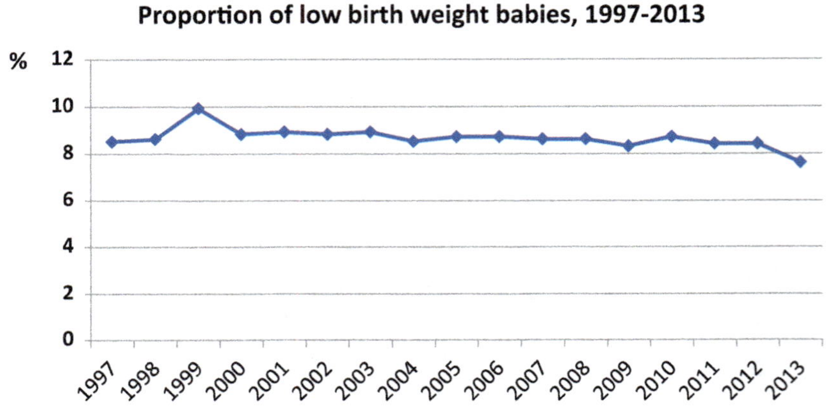

Fig. 16.11 Proportion of low birth weight (less than 2500 g), 1997–2013 (Source: Bureau of Health Promotion, Department of Health, Ministry of Public Health 2016 [Online]. Available: http://hp.anamai.moph.go.th/main.php?filename=index6)

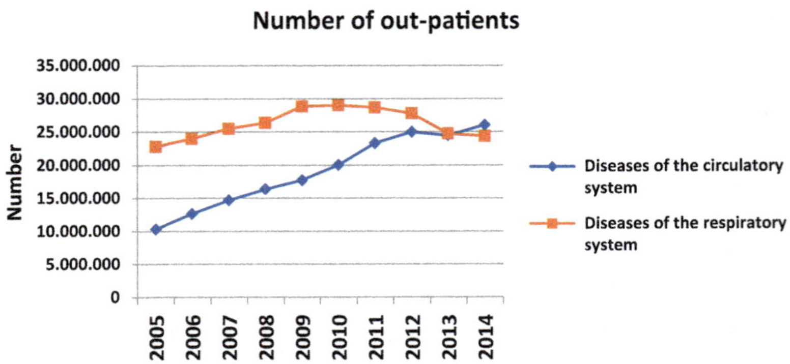

Fig. 16.12 Number of outpatients (Source: Office of the Permanent Secretary for Public Health, Ministry of Public Health 2015)

Government Response

Air Pollution Mitigation

Over the last two decades, remarkable contributions have been made by the Royal of Thai government, the Pollution Control Department (PCD), other government organizations, as well as private agencies in an attempt to resolve the air pollution challenges and preserve the environments. The main role of the local government in air quality management is in the enforcement of existing policies through inspection and public awareness raising. Bangkok Metropolitan Administration (BMA) declared 1999 as the Air Pollution Mitigation Year and implemented the following 13 measures:

- Providing free car engine tune-up service stations for the public.
- Publishing car engine maintenance manuals for public distribution.
- Setting up black-smoke inspection points in 50 districts jointly with the traffic police.
- Setting up six mobile black-smoke inspection units in six areas.
- Setting up motorcycle white-smoke and noise-level inspection units in the inner area of Bangkok.
- Reporting about air pollution in critical areas in cooperation with PCD through the display boards and air quality reports to promote pollution-free streets.
- Designating pollution-free streets, which prohibited single-occupant vehicles. Originally, there were three streets, later increased to eight streets.
- Paving road shoulders to reduce dust.
- Enforcing windscreens for buildings which were under construction.
- Enforcing dust controls for trucks by covering loads and cleaning wheels.
- Putting up campaign boards to inform the public on various measures being implemented.
- Designating car-free streets to reduce air pollution.
- Improving fuel quality by joint efforts to reduce air pollution.

Improvements in Air Quality

Bangkok's air quality has improved enormously in comparison with previous decades largely due to the city having far fewer buses, trucks, and motorcycles emitting smoke. In addition, remarkable improvements have been reported regarding Bangkok's air quality management capabilities in recent years. With stringent laws and policies on the use of unleaded fuel, as well as new vehicle emission standards, ambient lead concentrations as well as roadside concentrations of CO, NO_x, and SO_2 have been greatly reduced and put under control. The Thai government is pursuing the stringent emission standards of Europe. As such, emission controls have been progressive; however, the levels of TSP, PM, and O_3 have increased in recent years. Bangkok is still to attain a relatively "clean" urban air status; however, the integrated approach and strategies for national and local air quality management promise positive results in further improving the air quality in Bangkok.

According to projections made from the Bangkok Air Quality Management Project, it is estimated that a 10 μg/m^3 decline in the annual average of PM_{10} concentrations in Bangkok would result in the following reductions:

- 700–2000 premature deaths
- 3000–9300 new cases of chronic respiratory diseases
- 560–1570 respiratory and cardiovascular hospital admissions
- 2,900,000–9,100,000 days with respiratory symptoms severe enough to restrict a person's normal activities
- 2,200,000–74,000,000 days with minor respiratory symptoms

Discussion and Conclusions

This study investigated the correlations between climate change and air pollution and how these in turn impact on human health among residents of Bangkok, Thailand. Overall, results suggest possible correlations between increases in temperature and increases in $PM_{2.5}$ concentrations, which appeared to be correlated with increases in mortality due to diseases of the respiratory and circulatory systems.

We observed increasing trends in mortality due to cardiovascular and respiratory illnesses, and this was correlated to increases in annual temperatures as well as increases in $PM_{2.5}$ concentrations. Confirming our findings, a recent study in Thailand highlighted that increases in concentrations of major air pollutants had significant short-term impacts on non-accidental mortality, with O3 significantly associated with cardiovascular mortality, while PM_{10} was significantly related to respiratory mortality (Guo et al. 2014). High temperatures on the other hand increased the associations of PM with daily mortality in eight Chinese cities (Meng et al. 2012). Such findings do have implications on health effects of both air pollution exposure and climate change.

A number of studies have demonstrated that mortality risks following air pollution exposure to differ by weather type or season (Guo et al. 2014; Vanos et al. 2015). Guo et al. (2014) showed that the effects of all air pollutants on all mortality types were stronger during summer and winter seasons compared to the rainy season. This study, however, did not investigate the seasonality issue, the reason being that the available data on health effects was not disaggregated according to seasons.

Based on previous epidemiological investigations, associations between air pollution and mortality differ by individual characteristics (Li et al. 2016), neighborhood characteristics (Deguen et al. 2015), and underlying health conditions (Vichit-Vadakan et al. 2010). In Thailand, the associations between mortality due to all natural causes and PM_{10} exposure increased with age, with the strongest effects reported among people aged 75 years and older (Vichit-Vadakan et al. 2010). Li et al. (2016) also reported stronger air pollution effects on COPD mortality among the elderly and males. Plausible explanations suggested for gender differences in air pollution effects include confounding effects of smoking and job-related chemical exposures associated with the male gender (Li et al. 2016). Among the most susceptible groups are infants with respiratory illnesses, children less than 5 years of age with lower respiratory infections (LRIs), and people with asthma (Vichit-Vadakan et al. 2010). A recent study observed pollution-induced cardiovascular disease mortality risk both for those with and without existing cardio metabolic disorders (Pope et al. 2015). Our study, however, did not disaggregate the mortality data by characteristics related to individuals, neighborhood, or underlying health conditions.

Short-term exposure to air pollution at very high concentrations, as well as long-term exposure to relatively low concentrations of pollutants PM, ozone, and NO_2, has been linked with adverse health outcomes in previous investigations. Prolonged exposure to PM has probably a greater impact on public health in comparison with short-term exposure to peak concentrations. It has been documented in previous investigations that subjects residing in close proximity to busy roads often experience more short-term and long-term effects of traffic-related air pollution than those residing further away. In urban areas, due to the settlement patterns, up to 10% of the population may be residing in such "hot spots." The unequal distribution of air pollution exposure and subsequent health outcomes raise concerns over environmental justice and equity.

Of prime concern are the effects of long-term exposure to PM on mortality. Long-term exposure to low and moderate levels of fine PM has been associated with a reduction in life expectancy by up to some months. Some analysis has been published on the relative public health significance of short-term and long-term exposures to PM, with "disability-adjusted life years" (DALYs) estimated for both. The analysis submits that the public health significance of long-term effects of PM exposure clearly outweighs the public health significance of the short-term effects. However, this does not lessen the significance of the short-term effects of PM, which consist of attributable deaths and hospital admissions for cardiovascular and respiratory adverse outcomes.

Our study had some limitations worth mentioning. First, it is a descriptive study that shows only correlation and not causation. Secondly, the lack of data on health effects disaggregated by season made it impossible to investigate the effect of temperature on the relationships between air pollution exposure and health outcomes. In conclusion, this study highlighted that there is a possible correlation between air pollution-climate change interactions and health effects on the population of Bangkok, and this may have implications on the health impact of both air pollution exposure and climate change.

References

Beverland IJ, Cohen GR, Heal MR, Carder M, Yap C, Robertson C, Hart CL, Agius RM (2012) A comparison of short-term and long-term air pollution exposure associations with mortality in two cohorts in Scotland. Environ Health Perspect 120(9):1280–1285. doi:10.1289/ehp.1104509

Bureau of Policy and Strategy; Ministry of Public Health (2015) Public health statistics A.D. 2014. Printing Press of the War Veterans Organization of Thailand under Royal Patronage of His Majesty the King, Bangkok, 230 p. ISSN 08570-3093

Chen R, Kan H, Chen B, Huang W, Bai Z, Song G, Pan G, CAPES Collaborative Group (2012) Association of particulate air pollution with daily mortality: the China Air Pollution and Health Effects Study. Am J Epidemiol 175(11):1173–1181. doi:10.1093/aje/kwr425

Chen H, Goldberg MS, Burnett RT, Jerrett M, Wheeler AJ, Villeneuve PJ (2013) Long-term exposure to traffic-related air pollution and cardiovascular mortality. Epidemiology 24 (1):35–43. doi:10.1097/EDE.0b013e318276c005

D'Amato G, Baena-Cagnani CE, Cecchi L, Annesi-Maesano I, Nunes C, Ansotegui I, D'Amato M, Liccardi G, Sofia M, Canonica WG (2013) Climate change, air pollution and extreme events leading to increasing prevalence of allergic respiratory diseases. Multidiscip Respir Med 8 (1):12. doi:10.1186/2049-6958-8-12

D'Amato G, Holgate ST, Pawankar R, Ledford DK, Cecchi L, Al-Ahmad M, . . . Annesi-Maesano I (2015) Meteorological conditions, climate change, new emerging factors, and asthma and related allergic disorders. A statement of the World Allergy Organization. World Allergy Organ J 8(1):25. doi:10.1186/s40413-015-0073-0

Deguen S, Petit C, Delbarre A, Kihal W, Padilla C, Benmarhnia T, Lapostolle A, Chauvin P, Zmirou-Navier D (2015) Neighbourhood characteristics and long-term air pollution levels modify the association between the short-term nitrogen dioxide concentrations and all-cause mortality in Paris. PLoS One 10(7):e0131463. doi:10.1371/journal.pone.0131463

Franchini M, Mannucci PM (2015) Impact on human health of climate changes. Eur J Intern Med 26(1):1–5. doi:10.1016/j.ejim.2014.12.008

Guo Y, Li S, Tawatsupa B, Punnasiri K, Jaakkola JJ, Williams G (2014) The association between air pollution and mortality in Thailand. Sci Rep 4:5509. doi:10.1038/srep05509

Lagidze L, Matchavariani L, Tsivtsivadze N, Khidasheli N, Paichadze N, Motsonelidze N, Vakhtangishvili M (2015) Medical aspects of atmosphere pollution in Tbilisi, Georgia. J Environ Biol 36. Spec No:101–106

Li L, Yang J, Song YF, Chen PY, Ou CQ (2016) The burden of COPD mortality due to ambient air pollution in Guangzhou, China. Sci Rep 6:25900. doi:10.1038/srep25900

Meng X, Zhang Y, Zhao Z, Duan X, Xu X, Kan H (2012) Temperature modifies the acute effect of particulate air pollution on mortality in eight Chinese cities. Sci Total Environ 435–436:215–221. doi:10.1016/j.scitotenv.2012.07.008

Mirsaeidi M, Motahari H, Taghizadeh Khamesi M, Sharifi A, Campos M, Schraufnagel DE (2016) Climate change and respiratory infections. Ann Am Thorac Soc. doi:10.1513/AnnalsATS.201511-729PS

Pope CA 3rd, Turner MC, Burnett RT, Jerrett M, Gapstur SM, Diver WR, Krewski D, Brook RD (2015) Relationships between fine particulate air pollution, cardiometabolic disorders, and cardiovascular mortality. Circ Res 116(1):108–115. doi:10.1161/CIRCRESAHA.116.305060

Schulte PA, Bhattacharya A, Butler CR, Chun HK, Jacklitsch B, Jacobs T, . . . Wagner GR (2016) Advancing the framework for considering the effects of climate change on worker safety and health. J Occup Environ Hyg 1–60. doi:10.1080/15459624.2016.1179388

Shang Y, Sun Z, Cao J, Wang X, Zhong L, Bi X, Liu W, Zhu T, Huang W (2013) Systematic review of Chinese studies of short-term exposure to air pollution and daily mortality. Environ Int 54:100–111. doi:10.1016/j.envint.2013.01.010

Tsai SS, Chen CC, Yang CY (2014) Short-term effect of fine particulate air pollution on daily mortality: a case-crossover study in a tropical city, Kaohsiung, Taiwan. J Toxicol Environ Health A 77(8):467–477. doi:10.1080/15287394.2014.881247

Vanos JK, Cakmak S, Kalkstein LS, Yagouti A (2015) Association of weather and air pollution interactions on daily mortality in 12 Canadian cities. Air Qual Atmos Health 8(3):307–320. doi:10.1007/s11869-014-0266-7

Vichit-Vadakan N, Vajanapoom N, Ostro B, HEI Health Review Committee (2010) Part 3. Estimating the effects of air pollution on mortality in Bangkok, Thailand. Res Rep Health Eff Inst (154):231–268

Xu Z, Huang C, Hu W, Turner LR, Su H, Tong S (2013) Extreme temperatures and emergency department admissions for childhood asthma in Brisbane, Australia. Occup Environ Med 70 (10):730–735. doi:10.1136/oemed-2013-101538

Zhou X, Zhao A, Meng X, Chen R, Kuang X, Duan X, Kan H (2014) Acute effects of diurnal temperature range on mortality in 8 Chinese cities. Sci Total Environ 493:92–97. doi:10.1016/j.scitotenv.2014.05.116

Uma Langkulsen received a doctoral degree in environmental management from the National Research Center for Environmental and Hazardous Waste Management (NRC-EHWM), Chulalongkorn University. In 2007, she received a postdoctoral scholarship from the Commission on Higher Education of Thailand under the programme Strategic Scholarships Fellowships Frontier Research Networks. She continued to produce original research in health impact of climate change on occupational health and productivity.

Desire Rwodzi received his Bachelor of Science in Health Education and Promotion from University of Zimbabwe in 2007 and Master of Public Health in Global Health from Thammasat University, Thailand in 2012. Mr. Desire worked for two years serving as Data Analyst Consultant, UNAIDS RST East and Southern Africa, South Africa.

Chapter 17
Climate Change, Air Pollution and Human Health in Delhi, India

Hem H. Dholakia and Amit Garg

Abstract Over centuries, the Indian capital of Delhi has been the seat of power for several empires. Today, however, Delhi finds itself in the unenviable position of being among the world's most polluted cities. Mitigating air pollution as well as greenhouse gases in Delhi without adversely impacting development remains a crucial goal. Further, climate change has profound impacts that Delhi must adapt to. From a health perspective, in addition to health impacts of pollution, addressing health impacts of climate change such as heatwaves is important.

This chapter understands the transitions of key drivers of energy use such as population, vehicle use and per capita incomes that in turn drive emissions of pollutants and greenhouse gases. It provides estimates of greenhouse gas and pollutant emissions from Delhi. It estimates pollution as well as future heat-related mortality for Delhi. Finally, it argues that policies for GHG as well as pollutant mitigation require to be better aligned. This will ensure that health co-benefits are accrued for Delhi.

Keywords Delhi • Air pollution • GHG • Health impacts • Co-benefits

Introduction

Over centuries, the Indian capital of Delhi has been the seat of power for several empires. Culture, history, art and economy are complexly interwoven into the fabric of the city that has drawn people from around the world. Today, however, Delhi finds itself in the unenviable position of being among the world's most polluted cities (Fig. 17.1). Urbanisation, population growth, rising incomes, increase in vehicle ownership, growing energy demand and proximity to industrial hubs have all contributed to the steady rise in pollution levels over time. Associated with

H.H. Dholakia (✉)
Council on Energy, Environment and Water, New Delhi, India
e-mail: hem.dholakia@ceew.in

A. Garg
Indian Institute of Management, Ahmedabad, India
e-mail: amitgarg@iima.ac.in

© Springer International Publishing AG 2018

R. Akhtar, C. Palagiano (eds.), *Climate Change and Air Pollution*,
Springer Climate, DOI 10.1007/978-3-319-61346-8_17

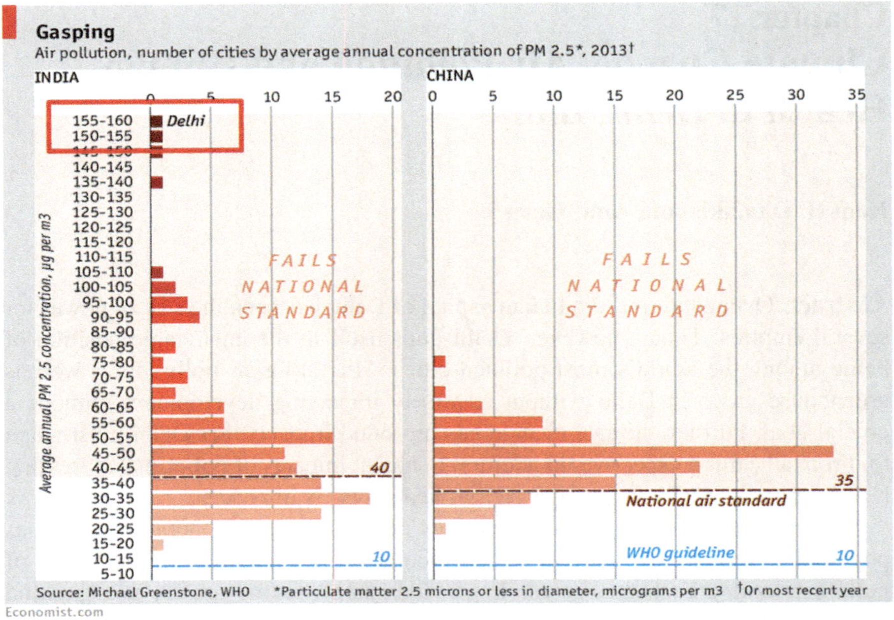

Fig. 17.1 Air pollution in Indian and Chinese cities (Source: Economist 2015. http://www.economist.com/news/asia/21642224-air-indians-breathe-dangerously-toxic-breathe-uneasy)

pollution are the increased emissions of greenhouse gases (GHG). Patterns of energy use have made urban areas the largest contributors of greenhouse gas emissions as well as most polluted geographies across the world.

Climate change and air pollution can be conceptualised as two sides of the same coin as they share several common sources (mainly fossil fuel use). However, they differ on spatial and temporal scales. Climate change is a long-term global phenomenon, whereas pollution is short term and local in nature. Further, it is important to recognise that climate change impacts the air we breathe. Weather patterns can modify the concentrations of indoor and outdoor air pollutants thereby modifying health risks (Fann et al. 2016). The impacts of climate change on pollutants such as ozone (Jacob and Winner 2009), particulate matter (Dawson et al. 2014; Pernod et al. 2014) and aeroallergens (Bielory et al. 2012) are fairly well documented. Therefore, it is critical to find policy synergies that can simultaneously mitigate pollution and GHG emissions.

A related issue is the reverse impacts of climate change on development and populations. Climate change can have profound impacts on built environment, health of people and water and energy systems (Garg et al. 2015). These impacts are related to the amount of mitigation that is achieved globally. Higher amounts of mitigation will result in lesser global warming and, consequently, lower impacts.

The current chapter aims to capture these different dimensions in the context of Delhi. We present the underlying patterns in energy use, discuss emissions, air pollution and associated health implications as well as provide estimates for climate-related risks such as heat-related mortality. Finally, we suggest policy insights to address some of these issues.

Economic and Population Trends

Since economic reforms (1992–1993), there has been a significant increase in the per capita gross domestic product (GDP) in India. This increase in GDP has also led to job creation especially in the services sector and migration of people from rural areas to metros (like Delhi).

Delhi has one of the highest per capita incomes in the country of INR ~240,800 (current prices) in 2014–2015. In relative terms, Delhi's per capita income was three times the average per capita income of India. The gross state domestic product (GSDP) of Delhi recorded a 15% growth in 2014–2015 as compared to 2013–2014, and the economy is expected to grow around 8% in the years to come (Government of Delhi 2016). These changes in GDP have enhanced the purchasing power of citizens bringing about a change in lifestyles. It has been documented that in urban areas (especially in developing countries), slight increases in income impact consumption patterns, standard of living and food habits (Schoot et al. 2011). This is reflected in the number of households that own electrical appliances such as geysers, refrigerators, air conditioners and ownership of private vehicles.

Over time, the population of Delhi has grown steadily making it the second most populous megacity in India (~16 million people as per Census 2011). Over the last century, Delhi has transformed from being 57% urban (in 1911) to being 97% urban by 2011 with an average population density of >11,000 persons per square kilometre (Government of Delhi 2016). This population increase has been a consequence of natural growth as well as in-migration from neighbouring states (estimated at roughly 16–18% each year). A rising population has been a key driver of increased demand for services such as energy and transport. This can be corroborated by the exponential increase in demand for private vehicle ownership.

The number of registered vehicles in Delhi increased (Fig. 17.2) from 31.64 lakhs (in 1999–2000) to 88.27 lakh in 2014–2015. In other words, the number of registered vehicles increased about 180% over a 15-year period. The highest increases (219%) were observed in cars and jeeps followed by increases in two-wheelers (173%). In terms of ownership, Delhi has 85 cars per 1000 population as compared to the national average of eight cars per 1000 population (SOE 2010). However, it must be noted that there is no system of deregistration of vehicles in India. As a result, the actual number of vehicles plying on the road may be lesser than those registered. Considered together, economic factors and population growth-associated patterns of urban development are intricately linked to energy

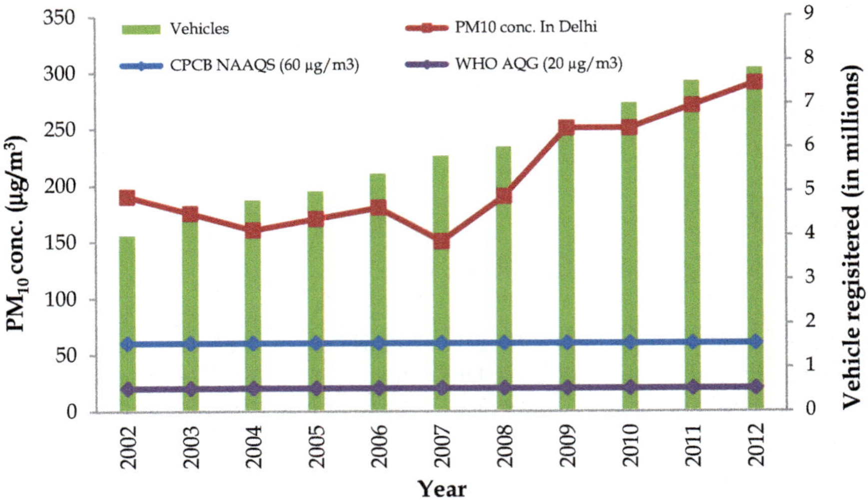

Fig. 17.2 PM_{10} levels and registered vehicles in Delhi (2002–2012) (Source: Center for Science and Environment (2015), Central Pollution Control Board (2009), World Health Organization (2005))

consumption. This energy consumption (primarily from fossil fuel sources) is a driver of greenhouse gas as well as pollutant emissions.

Greenhouse Gas (GHG) Emissions

It is well understood that increasing energy use (especially fossil fuel) results in increased emissions of greenhouse gases as well as local pollutants. Most estimates for GHG emissions are available at the national level. For instance, in 2000, India emitted 1,523,777.44 Gg CO2e across energy, industry processes, agriculture and waste management sectors (MoEF 2012). Excluding land use change and forestry sectors, the emissions of carbon dioxide (CO_2), methane (CH_4) and nitrous oxide (N_2O) at the national level were 1,024,772.84 Gg, 19,392.3 Gg and 257.42 Gg, respectively (MoEF 2012).

However, not many studies have estimated GHG emissions at the city level for India. Ramachandra and colleagues estimated emissions of three major greenhouse gases – carbon dioxide (CO_2), methane (CH_4) and nitrous oxide (N_2O) – across several sectors including electricity, households, transportation, industry, agriculture, livestock and waste for eight major cities in India (Ramachandra et al. 2014). They found that among the cities studied (Table 17.1), Delhi had the highest carbon footprint (CO_2e) of 38,633.2 Gg/year (Ramachandra et al. 2014).

Table 17.1 Carbon footprint across Indian cities (2009 as baseline)

No	City	CO2e (Gg/year)
1	Delhi	38,633.2
2	Greater Mumbai	22,783.1
3	Chennai	22,090.5
4	Greater Bangalore	19,796.6
5	Kolkata	14,812.1
6	Hyderabad	13,734.6

Source: Adapted from Ramachandra et al. (2014)

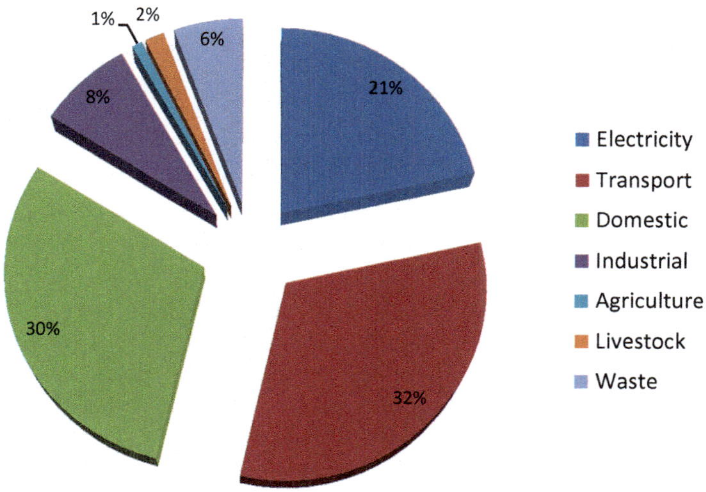

Fig. 17.3 Sectoral contribution of GHG for Delhi (Source: Adapted from Ramachandra et al. (2014))

The chief sources of greenhouse gases for Delhi are given in Fig. 17.1. The major contributor to GHG emissions in Delhi is the transport sector (32%) followed by the domestic and electricity sectors, respectively. Together, these three sectors constitute more than 80% of Delhi's GHG emissions. There exist multiple opportunities across all these sectors for GHG abatement (Fig. 17.3).

Outdoor Air Pollution

Pollution levels in Indian cities are found to be several times higher than the standards prescribed by the World Health Organisation (WHO). Outdoor air pollution is among the top ten risk factors in India, and associated health impacts are staggering. The Global Burden of Disease Study estimated that in India, 670,000 deaths (in 2010) could be attributed to outdoor air pollution alone. Other studies found that on average, Indians lose 3.2 years of life expectancy and 2.1 billion life years as a consequence of high air pollution (Greenstone et al. 2015).

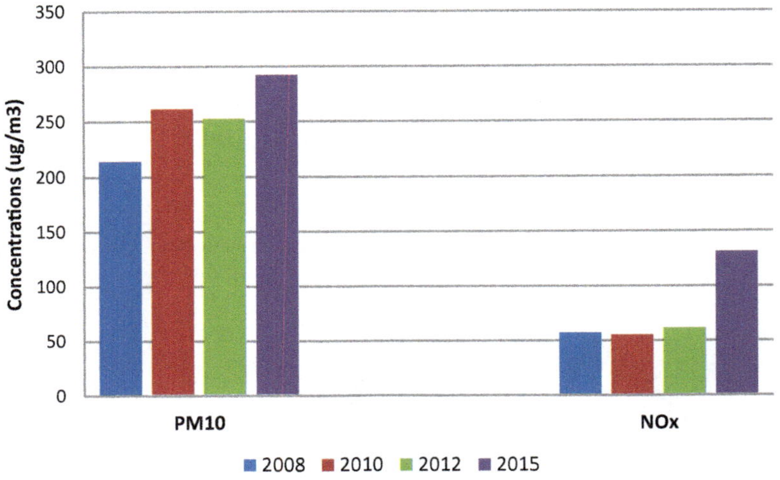

Fig. 17.4 Pollutant concentrations for Delhi over the years (annual average) (Source: Central Pollution Control Board). PM_{10} particulate matter less than 10 microns, NOx oxides of nitrogen

It is well recognised that Delhi is among the most polluted cities in India. Under the National Ambient Air Monitoring Programme (NAMP), four criteria pollutants are routinely monitored (suspended particulate matter, respirable suspended particulate matter, i.e. PM_{10}, oxides of nitrogen, oxides of sulphur). It was in 2010, during the Commonwealth Games that particulate matter less than 2.5 microns ($PM_{2.5}$) was monitored for the first time. With the increase in pollution, routine monitoring of $PM_{2.5}$ has commenced since 2015. Air quality trends have worsened over time. In addition, the air pollution challenge in Delhi has been difficult to manage despite several policy interventions. One of the reasons is that modern-day pollution in Indian cities is a complex phenomenon. It is a combination of vehicular exhaust, construction, waste burning, industrial emissions, thermal power plant emissions as well as transport of pollutants from neighbouring areas due to varied reasons such as burning of crop residue. This implies that a portfolio of stringent pollution control measures across sectors is required.

To understand the key contributors of pollution, there have been several source apportionment studies for Delhi. Each study has adopted different analytical methods and has been carried out at different points in time for different size fractions of particulate matter. In addition, different authors have interpreted source profiles differently, making direct comparisons difficult (Pant and Harrison 2012). However, across studies, several common patterns emerge. The most common sources of pollution in Delhi include crustal resuspension from road dust, vehicular sources, biomass burning, industrial emissions, waste incineration and coal burning.

The most recent source apportionment study for Delhi was undertaken in 2016 by the Indian Institute of Technology, Kanpur (Fig. 17.4). The study found that for particulate matter of size 10 microns (PM_{10}), the key sources are secondary particle

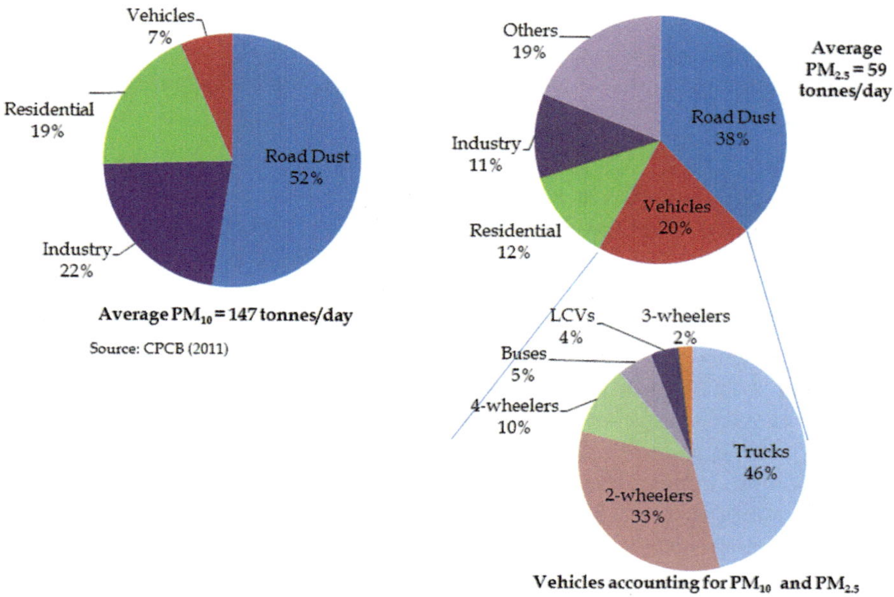

Fig. 17.5 Source apportionment of PM_{10} and $PM_{2.5}$ for Delhi (2013) (Source: IIT Kanpur Sharma and Dikshit 2016)

formation (8–32%), biomass burning (2–28%), coal and fly ash (7–50%), soil and road dust (8–34%), vehicles (4–24%), solid waste burning (2–18%) and construction material (2–5%) (Sharma and Dikshit 2016). On the other hand, the key sources for $PM_{2.5}$ include secondary particle formation (13–39%), biomass burning (3–35%), coal and fly ash (1–35%), soil and road dust (1–36%), vehicles (6–29%), solid waste burning (3–15%) and construction material (1–5%) (Sharma and Dikshit 2016). A strong seasonal variation is observed, wherein concentrations are higher in winter as compared to summer months. It is clear that reducing pollution in Delhi will require a portfolio of policies across all these sectors (Fig. 17.5).

Indoor Air Pollution

Though discussed to a lesser extent in the context of urban areas, lack of access to clean cooking energy is a major contributor to indoor air pollution. It is well established that solid fuel usage results in exposure to high amounts of indoor air pollution and remains a large cause for morbidity and mortality especially in women and children. The Census (2011) estimates that ~25% rural and ~10% urban houses in Delhi lack access to clean cooking energy (e.g. LPG, solar cookers, PNG, etc.). Most of these households use solid fuels such as dung, wood, crop residue, etc. It has been estimated that 16–25% of indoor air pollution contributes to outdoor air pollution levels (Smith et al. 2013). Provision of clean cooking energy remains important in urban areas not only to reduce pollution but protect the health of people.

Box 1 Addressing Brick Kilns
As the second largest producer of coal-fired bricks (150–200 million annually), India's brick sector is saddled with older traditional technology, making it one of the most polluting sectors. It is estimated that brick production in India consumes 25 million tonnes of coal annually. One of the barriers that have prevented adoption of cleaner technology is the production costs. Using cleaner technology (Table B1) such as vertical shaft nearly doubles the production costs relative to traditional technologies (down draft kiln). However, there is a significant reduction in PM_{10} as well as $PM_{2.5}$ with cleaner technology implementation. As India implements urban development programmes such as 'Smart Cities Mission' and 'Atal Mission on Urban Rejuvenation and Transformation (AMRUT)', addressing issues in the brick sector will prove important to minimise pollution as well as GHG emissions. This will require a policy push as well as finance.

Delhi set up an Air Ambience Fund that collected a cess of diesel sale. As of March 2015, this fund had INR 385 crore. Whereas industries cannot escape installing pollution control equipment, some of this money could be used as loan guarantees or viability gap funding to promote clean technologies for the brick kiln sector.

Health Impacts

Air Pollution

The physiological basis of air pollution impacts on human health is complex and multifaceted in nature. Most of the underlying evidence for physiological impacts comes from animal model studies, and there is general consensus that cellular injury and inflammation play a key role (USEPA 2009). The impacts of pollution not only affect the pulmonary system but also extend to cardiovascular, haematopoietic as well as central nervous system.

Air pollution may impact health in different ways. The impacts of particulate matter inhalation may be acute or chronic in nature. This depends on whether exposure to particulate matter is short term or long term in nature. Air pollution may (1) increase risk of underlying diseases, leading to frailty and higher risk of short-term deaths in frail individuals; (2) increase risk of chronic diseases leading to frailty but may not be related to timing of death; or (3) increase the risk of short-term death in frail individuals but may not be related to risk of chronic disease

Table B1 Sales and emissions data from brick kilns around Delhi

	Brick kiln A (down draft kiln)	Brick kiln B (bull trench kiln)	Brick kiln C (vertical shaft kiln)
Sales			
Annual production (million)	1.2	1.2	1.2
Weight per brick (kg)	2.95	2.95	2.95
Production cost per brick (in cents)	2.7	3.6	5.4
Price per brick (in cents)	5.4	6.3	8.1
Emissions (particulate matter) CPCB standards[a]			
SPM (g/kg of fired bricks)	0.004–0.009	0.006–0.008	0.001
PM_{10} (g/kg of fired bricks)	0.0013–0.0082	0.0018–0.0073	0.0003–0.001
$PM_{2.5}$ (g/kg of fired bricks)	0.0004–0.0024	0.0005–0.0022	0.0001–0.0003
Emissions (particulate matter) actuals from survey[b]			
SPM (g/kg of fired bricks)	1.56 ± 1.41	0.86 ± 0.74	0.1 ± 0.02
PM_{10} (g/kg of fired bricks)	0.47–2.67	0.26–1.44	0.03–0.11
$PM_{2.5}$ (g/kg of fired bricks)[b]	0.97 ± 0.47	0.19 ± 0.07	0.09 ± 0.06

[a]Source: Emission Standards for Brick Kilns (2009) [http://www.cpcb.nic.in/Industry-Specific-Standards/Effluent/472-1.pdf]
[b]Lalchandani and Maithel (2013)

(Künzli et al. 2001). Whereas several studies have been carried on health impacts of pollution globally, these are lacking in the Indian context.

For Delhi, most studies relating health and pollution are cross-sectional in nature. For example, Foster and Kumar (2011) quantified the effects of air pollution regulation – specifically closing of polluting industries and adoption of compressed natural gas by buses – in Delhi city. They surveyed 1576 households and monitored pollution at 113 sites over a 6-month period (July to December 2003). They found that stringent regulation was positively associated with improved respiratory function, though these effects varied by gender and income class (Foster and Kumar 2011). Identifying the need for more such studies, short-term effects of air pollution on daily mortality were recently studied for two Indian cities – Delhi and Chennai (Balakrishnan et al. 2011; Rajarathnam et al. 2011). Using daily all-cause mortality and pollution data from 2002 to 2004, both studies ran a series of Poisson regression models to measure the association between PM_{10} and daily deaths. Delhi showed a 0.15% (95% confidence interval = 0.07 to 0.23) increase in daily all-cause mortality with every 10 $\mu g/m^3$ increase in PM_{10} concentrations.

Table 17.2 Health benefits of pollution reduction in Delhi

Health Outcome	Cost/incident (USD)
Premature mortality	1,167–9,000
Chronic bronchitis (adults)	2.52–6.31
Chronic bronchitis (children)	0.1–0.26
Respiratory hospital admissions	121–301
Emergency room visits	3.4–8.5
Restricted activity days	2.77

Range of individual health costs for an average individual due to reduction in impacts possible if ambient $PM_{2.5}$ concentration levels are brought down 60 μg/m^3 in Delhi
Source: Garg (2011)

A key aspect that is the distribution of these health impacts is often discussed to a lesser extent. Garg (2011) attempted to study the pro-equity health benefits of pollution reduction on health as well as GHG for Delhi (Garg 2011). The study found that highest relative health benefits of pollution reduction accrued to lower-income groups, followed by middle and higher income groups, thereby showing pro-equity effects of pollution and GHG mitigation policies (Garg 2011). Further, it estimated that in addition to health effects, there remain strong economic benefits of pollution control (Table 17.2). However, cohort studies with strong design that look at specific health end points of cardiovascular disease or stroke are lacking in the Indian context. This lack of evidence is one of the reasons why stringent standard setting has been difficult.

Climate Change

Scientific evidence for warming of the climate system is unequivocal. There is high degree of confidence that climate change will adversely impact human health both directly and indirectly. Heat- and cold-related morbidity and mortality due to shifts in temperature means and extremes are some of the anticipated direct impacts. Malnutrition and diarrhoea due to food and water system degradation and an increased incidence of vector-borne diseases are some examples of indirect effects of climate change on human health. Of these different impacts, this chapter focuses on heat-related mortality for Delhi.

Multiple studies suggest that average as well minimum and maximum temperatures in India are expected to increase in the future (INCCA 2010). India has experienced a series of heatwaves in the past (De and Mukhopadhyay 1998) that reveal their significant mortality impacts. For instance, in the year 1998, the state of Orissa faced an unprecedented heatwave situation as a result of which 2042 people lost their lives (OSDMA 2007). In another instance, 1421 people were killed in

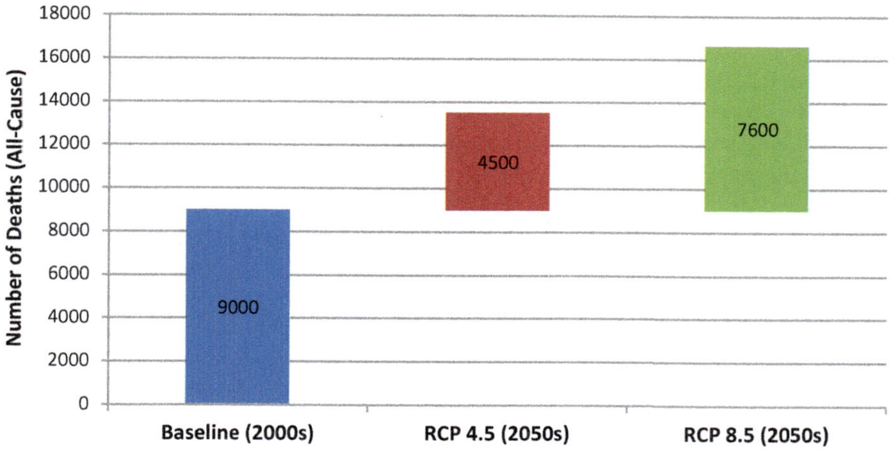

Fig. 17.6 Current and future heat-related mortality for Delhi (Source: Dholakia, Mishra and Garg (2015) estimated the additional heat-related deaths due to a changing climate for Delhi)

Andhra Pradesh from a heatwave in 2003 (Jafri 2003). Delhi, with its extreme weather, puts a large population at risk for future heat-related mortality.

Historically, the average maximum temperatures for Delhi during the summer months have been 36 °C. Data from 23 global climate models indicate that depending on the climate change scenario, these temperatures may increase by 1.6 °C (RCP 4.5 scenario) and 2.2 °C (RCP 8.5 scenario), respectively, in the 2050s. The corresponding estimated increases in all-cause mortality in the future are 5500 additional deaths (RCP 4.5 scenario) and 7600 additional (RCP 8.5 scenario) in the 2050s (2050–2059) as compared to the baseline (2000–2009) period (Dholakia et al. 2015). In addition, extremes of temperature are known to impact human productivity, implying that the expected economic losses are likely to be very high. Therefore, we need to institute heat-health warning systems. Delhi can learn from the example of Ahmedabad which instituted a heat-health warning system in 2010. Since the implementation of a heat-health system, morbidity and mortality related to heatwaves in Ahmedabad have significantly declined (Fig. 17.6).

Policy Synergies for Climate Change and Air Pollution

It is well known that Delhi has instituted several policy measures over the last few years to mitigate air pollution. In the transport sector, India's Auto Fuel Policy (2003) mandated Euro IV equivalent standards from April 1, 2010, in 20 major cities including Delhi. Further, following a supreme court order, 100,000 vehicles were retrofitted with CNG including 3000 buses (Kathuria 2002). Further, highly polluting industries and brick kilns (classified as 'red' category) were relocated outside the jurisdiction of Delhi. Further, coal-based power plants were converted to gas-based plants. Studies indicate that these measures did help to bring down the pollution levels (Reynolds and Kandlikar 2008). However, these benefits were short-lived due to increase in vehicle numbers over the subsequent years.

Modelling studies show that stringent pollution control measures across different sectors can play an instrumental role in meeting National Ambient Air Quality Standards. For instance, Dholakia et al. studied the future air quality implications of current policies for Delhi (Dholakia et al. 2013). They found that policies such as shifting to Euro VI standards for vehicles, introduction of electric vehicles, use of high-efficiency de-dusters in power plants and industries to control stack emissions, etc. could help Delhi meet its air quality standards by 2020. Of course, this would require tremendous coordination across different ministries.

Table 17.3 shows that in addition to coordination, a long-term perspective on pollution control is needed. First, articulate a clear goal for air pollution control. For instance, China aims to reduce $PM_{2.5}$ levels by 10% in the year 2017. Such goal setting is crucial in the case of Delhi. For Delhi, the goal could be to reach India's National Ambient Air Quality Standards in a 5-year time frame (i.e. reduce annual average levels $PM_{2.5}$ levels to 40 $\mu g/m^3$ by 2020). This goal will help determine the portfolio of policies (across transport, energy, waste and transboundary issues) required to meet this goal. Having achieved this goal, the next step would be to reach the World Health Organisation (WHO) Standards.

Second, enhance the capacity of Central and State Pollution Control Boards (CPCB, SPCBs). Both these institutions play a critical role in providing scientific inputs to policymakers. However, there is dearth of capacity (technical as well as manpower) in these institutions. Independent studies show that CPCB in 2010 would need to fill 308 posts immediately to meet its targets. This has implications for controlling pollution from industrial clusters in and around Delhi (e.g. Faridabad, Ghaziabad). Upskilling of existing staff knowledge and coordination between CPCB and SPCBs are essential.

Third, leverage technology for innovative solutions. Transboundary sources such as crop burning in Punjab and Haryana as well as industrial clusters in Faridabad are known to contribute 20–30% towards Delhi's pollution. There exist opportunities for innovative business models by which farmers can secure revenue from waste-to-energy projects or providing pollution control technologies to industrial clusters of small and medium enterprises.. If the respective State Pollution Control Boards are lacking in resources, some financial assistance could be provided from the Air Ambience Fund (that had INR 385 crores until 2015). This could be used as loan guarantees of viability gap funding for technology penetration. Without this long-term perspective, there is a risk of choosing populist policies at the peril of deeper reforms that are required for pollution control and protecting the health of people.

As Delhi has transformed over the years, its demand for energy has increased exponentially. Fossil fuels have been the mainstay of the energy system making Delhi one of the cities with highest GHG as well as pollutant emissions. Further, Delhi remains vulnerable to the impacts of climate change. All of this has had

Table 17.3 Policy portfolio for air pollution reduction in Delhi

Sector	Policies	Measures	Implementation strategies	Key institutions[a]
Power sector	Efficiency improvements	Emission targets, emission standards	NCR as sub-grid in northern grid; increasing generation capacity; load management through smart grid connections; reduced distribution losses; strict control on diesel generator sets	MoEF, CERC, EMC
	Fuel switch	Taxation mechanisms	Technology transfer; infrastructure development for renewable energy through PPP[b]	MoEF, MNRE
Transport	Efficiency improvements	Emission standards	Leapfrog to Euro VI standards	MoPNG
	Technology push	Subsidy mechanisms	Penetration of electric and hybrid vehicles	MoEF
	Process improvements	Awareness, education	Traffic light synchronisation, road dust management systems; better linkages between metro and outer areas of Delhi; creating unified transport authority	MoRTH
Industry	Process improvements and recycling	Standards, tax, awareness	Strict monitoring and correction; adoption of vertical shaft brick kilns	Regulatory bodies, industry associations
	Raw material improvements and switch	Industry standards	Industry leadership, supply chain management	
Industry	Shifting polluting industries	Create SEZ for polluting industries	Subsidies for shifting, provide finance, create market for their cleaner products (e.g. fly ash bricks)	Delhi govt., financial institutions, industries
Transboundary effects	Efficiency and process improvements	Emission targets, emission standards, awareness	Enhancing metro connectivity	MoEF, Delhi, Haryana and Uttar Pradesh state governments
	Agriculture crop residue burning		Creating economic opportunities for crop residue instead of open burning	
All	Monitoring PM levels	Many more stations (100 plus for Delhi), real time	Real-time data sharing through Internet/display boards/apps	SPCB
	Emergency measures	Cloud seeding	Compulsory cloud seeding above 125 μg/m3 for heavily populated areas	Delhi govt., NDMC (national disaster management)
		Shutting down schools	Above 60 $\mu g/m^3$	Delhi govt.

(continued)

Table 17.3 (continued)

Sector	Policies	Measures	Implementation strategies	Key institutions[a]
	Reducing exposures to population			
	Stop more pollution	Stop all construction and road vehicles	Above 60 μg/m^3	Delhi govt.

Source: Adapted from Garg et al. (2002)
[a]MoEF (Ministry of Environment and Forests), CERC (Central Electricity Regulatory Authority), EMC (Energy Management Centre), MNRE (Ministry of New & Renewable Energy), MoPNG (Ministry of Petroleum & Natural Gas), MoRTH (Ministry of Road Transport & Highways)
[b]*PPP* Public Private Partnership

significant impact on the health of people, in terms of cardiovascular disease, respiratory disease as well as heat-related mortality. There remains the opportunity for Delhi to leverage technology and modify policies to address climate change as well as pollution.

References

Balakrishnan K, Ganguli B, Ghosh S, Sankar S, Thanasekaran V, Rayadu VN, Caussy H (2011) Short term effects of air pollution on mortality: results of a time-series analysis in Chennai, India (Public Health and Air Pollution in Asia (PAPA): Coordinated Studies of Short Term Exposure to Air Pollution and Daily Mortality in two Indian cities No. Research Report 157). Health Effects Institute

Bielory L, Lyons K, Goldber R (2012) Climate change and allergic disease. Curr Allergy Asthma Rep 12:485–494

Central Pollution Control Board (2009) National ambient air quality status. Retrieved July 20, 2017, from http://cpcb.nic.in/upload/Publications/Publication_514_airqualitystatus2009.pdf

Center for Science and Environment (2015) Unlocking the traffic gridlock in our neighbourhoods. Retrieved July 20, 2017, from http://cseindia.org/userfiles/MOVE-FREE-Unlocking-the-traffic-gridlock-in-our-neighbourhoods.pdf

Dawson J, Bloomer B, Winner D, Weaver C (2014) Understanding the meteorological drivers of U.S. particulate matter concentrations in a changing climate. Bull Am Meteorol Soc 95:521–532

De US, Mukhopadhyay RK (1998) Severe heatwave over the Indian subcontinent in 1998, in perspective of global climate change. Curr Sci 75(12):1308–1311

Dholakia HH, Purohit P, Rao S, Garg A (2013) Impact of current policies on future air quality and health outcomes in Delhi, India. Atmos Environ 75:241–248. https://doi.org/10.1016/j.atmosenv.2013.04.052

Dholakia HH, Mishra V, Garg A (2015) Predicted Increases in Heat related Mortality under Climate Change in Urban India. IIM-A Working Paper 2015–05-02

Fann N, Brennan T, Dolwick P, Gamble J, Ilacqua V, Kolb L, . . . Ziska L (2016).Air quality impacts. In: The impacts of climate change on human health in the United States: a scientific assessment. US Global Change Research Program, Washington, DC, pp 69–98

Foster A, Kumar N (2011) Health effects of air quality regulations in Delhi, India. Atmos Environ 45(9):1675–1683. https://doi.org/10.1016/j.atmosenv.2011.01.005

Garg A, Kapshe M, Shukla PR, Ghosh D, (2002) Large point source (LPS) emissions from India: regional and sectoral analysis. Atmos Environ 36(2):213–224

Garg A (2011) Pro-equity effects of ancillary benefits of climate change policies: a case study of human health impacts of outdoor air pollution in New Delhi. World Dev 39(6):1002–1025. https://doi.org/10.1016/j.worlddev.2010.01.003

Garg A, Mishra V, Dholakia H (2015) Climate change and India: adaptation gap. New Delhi

Government of Delhi (2016) Economic survey of Delhi 2014–15. Delhi. Retrieved from http://www.indiaenvironmentportal.org.in/files/file/economic%20survey%20of%20Delhi%202014-15.pdf

Greenstone M, Nilekani J, Pande R, Ryan N, Sudarshan A, Sugathan A (2015) Lower pollution, longer lives: life expectancy gains if India reduced particulate matter pollution. Econ Polit Wkly 8:40–46

INCCA (2010) Climate change: India 4 × 4 assessment. Analysis. Indian Network on Climate Change Assessment, New Delhi

Jacob D, Winner D (2009) Effect of climate change on air quality. Atmos Environ 43:51–63

Jafri S (2003) Andhra Pradesh finally gets respite from heat wave. Retrieved 20 Oct 2013, from http://www.rediff.com/news/2003/jun/13rain.htm

Kathuria V (2002) Vehicular pollution control in Delhi. Transp Res Part D: Transp Environ 7 (5):373–387

Künzli N, Medina S, Kaiser R, Quenel P, Horak F, Studnicka M (2001) Assessment of deaths attributable to air pollution: should we use risk estimates based on time series or on cohort studies? Am J Epidemiol 153(11):1050–1055

Lalchandani D, Maithel S (2013) Towards cleaner bricks kilns in India. Retrieved July 20, 2017, from http://www.gkspl.in/reports/energy_efficiency/Towards%20Cleaner%20Brick%20Kilns%20in%20India.pdf

MoEF (2012) India: second National Communication to the United Nations framework convention on climate change. Ministry of Environment and Forests, Government of Delhi, New Delhi. Retrieved from http://unfccc.int/resource/docs/natc/indnc2.pdf

OSDMA (2007) Heat wave. Retrieved 20 Oct 2013, from http://v3.osdma.org/ViewDetails.aspx?vchglinkid=GL002&vchplinkid=PL008

Pant P, Harrison RM (2012) Critical review of receptor modelling for particulate matter: a case study of India. Atmos Environ 49:1–12. https://doi.org/10.1016/j.atmosenv.2011.11.060

Pernod A, Zhang Y, Wang K, Wu S, Leung L (2014) Impacts of future climate and emission changes on U.S. air quality. Atmos Environ 89:533–547

Rajarathnam U, Seghal M, Nairy S, Patnayak RC, Chhabra S, Kilnani KV, Ragavan S (2011) Time series study on air pollution and mortality in Delhi, Research report no. report no. 157. Health Effects Institute, Boston

Ramachandra TV, Sreejith K, Bharath HA (2014) Sector-wise assessment of carbon footprint across major cities in India. In: Muthu SS (ed) Assessment of carbon footprint in different industrial sectors, volume 2. Springer Singapore, Singapore, pp 207–267. Retrieved from http://link.springer.com/10.1007/978-981-4585-75-0_8

Reynolds CCO, Kandlikar M (2008) Climate impacts of air quality policy: switching to a natural gas-fueled public transportation system in New Delhi. Environ Sci Technol 42(16):5860–5865. https://doi.org/10.1021/es702863p

Schoot Uiterkamp BJ, Azadi H, Ho P (2011) Sustainable recycling model: a comparative analysis between India and Tanzania. Resour Conserv Recycl 55(3):344–355. https://doi.org/10.1016/j.resconrec.2010.10.009

Sharma M, Dikshit O (2016) Comprehensive study on air pollution and green house gases (GHGs) in Delhi. Indian Instittute of Technology, Kanpur, Kanpur. Retrieved from http://delhi.gov.in/DoIT/Environment/PDFs/Final_Report.pdf

Smith KR, Frumkin H, Balakrishnan K, Butler CD, Chafe ZA, Fairlie I, . . . Schneider M (2013). Energy and human health. Annu Rev Publ Health, 34(1):159–188. https://doi.org/10.1146/annurev-publhealth-031912-114404

SOE (2010) State of the environment report for Delhi. Ministry of Environment and Forests, Government of Delhi, Delhi

USEPA (2009) Integrated science assessment for particulate matter. (No. EPA/600/R-08/139F). United States Environment Protection Agency

World Health Organisation (2005) Air quality guidelines – global update 2005. Retrieved July 20, 2017, from http://www.who.int/phe/health_topics/outdoorair/outdoorair_aqg/en/

Hem H. Dholakia is a senior research associate with the Council on Energy, Environment and Water, New Delhi. His research addresses the linkages between energy, environment, human health and public policy in India. He was a recipient of the Young Scientist Summer Award (2012) at the International Institute of Applied Systems Analysis (Austria). He has published six papers in international peer-reviewed journals and three book chapters and coauthored a book. He holds a PhD from IIMA and a master's from Brighton University (UK).

Amit Garg is a professor with the Public Systems Group at the Indian Institute of Management Ahmedabad. His research interests include the water-energy-agriculture nexus, energy plantation, corporate accounting of greenhouse gases and vulnerability assessment and adaptation due to climate change. He has worked on several research and consulting assignments for international and Indian organizations. He has coauthored six books and 18 international research reports and published extensively in international peer-reviewed journals. He has been a lead author of four reports for UN's Intergovernmental Panel on Climate Change. He holds a PhD from IIMA and a master's from IIT Roorkee.

Chapter 18
Climate Change and Air Pollution in Mumbai

S. Siva Raju and Khushboo Ahire

Abstract Climate change and global warming are potential threats to the existence of living beings, and it is increasingly noticed in recent years. Consistent increase in population growth and activities undertaken for furthering socio-economic development, with the application of technologies, not only exhaust resources but also pollute environment, thereby resulting in environmental degradation. Climate change affects all sections of population and more to the vulnerable sections like elderly and children. Amongst various adverse climatic conditions, air pollution is a major one, as it affects health and wellbeing of the population. Epidemiological studies in cities like Mumbai have revealed that with raised pollution levels, there was an increased occurrence of dyspnoea, chronic and intermittent cough, frequent colds, chronic bronchitis, cardiac disorders, high blood pressure and deaths due to non-tuberculosis respiratory and ischaemic heart diseases. The city of Mumbai, which is considered as a case study for the paper, is the capital city of Maharashtra state. The Maharashtra Pollution Control Board (MPCB) is implementing various environmental legislations in the state along with various other organisations which are promoting good practices of afforestation, solid waste management and traffic diversions of road ways to curtail the pollutants in the environment. To mention a few, with the projects like Eastern Freeway, Santa Cruz-Chembur Link Road and Andheri-Ghatkopar Link Road, it is expected that the connectivity of various areas of the Mumbai city is well networked and these measures are greatly contributing to combat air pollution in the region. To tackle further the issues related to environmental degradation, it is important to act at individual level as well as collectively. Hence, the city dwellers have a major role to play, in protecting the ecosystem of the city, and to actively participate in anti-pollution measures. The paper focuses on various aspects related to climate change scenario and its impacts, with a specific reference to Mumbai by critically analysing various reports and secondary data on climate change and air pollution issues.

S. Siva Raju (✉)
Tata Institute of Social Sciences (TISS), Hyderabad Campus, Mumbai, India
e-mail: sivaraju@tiss.edu

K. Ahire
School of Development Studies, Tata Institute of Social Sciences, Mumbai, India
e-mail: khushboo.ahire5@gmail.com

© Springer International Publishing AG 2018

R. Akhtar, C. Palagiano (eds.), *Climate Change and Air Pollution*,
Springer Climate, DOI 10.1007/978-3-319-61346-8_18

Keywords Climate change • Ecological degradation • Air pollution • Environmental risk factors • Epidemiology • Anti-pollution measures

Introduction

Climate change and global warming, two mounting issues, are potential threats to the existence of living beings. Some impacts of climate change are melting water from the glaciers, flash floods, inconsistent rainfall, sudden changes in atmospheric temperature and extinction of endangered species. Through processes of rapid industrialisation and urbanisation, human activities have led to adverse effects like ‘greenhouse gas emissions’ and air pollution which, over the long term, have contributed to the problem of climate change. Industrialisation and urbanisation also have significance in contributing to the high growth rate of population, leading to issues related to overcrowding and environmental pollution, especially in the developing countries like India. Since urbanisation in most of the developing countries is limited and concentrated to a few cities, the burden of population and pressure on civic amenities is higher in such cities.

Mumbai, being the economic capital of India, has a wide range of income opportunities to offer to populations across India. Against the background of India striving to improve its economic growth rate through stimulating economic activities, the trend of migration during the last 10 years is largest in Greater Mumbai amongst urban agglomerations (UAs) (Census GoI 2011). The data related to the proportion of in-migrants to that of total population amongst all the UAs indicates that Greater Mumbai stands first, accommodating approximately 18.4 million in-migrants, followed by Delhi (16.3 millions) and Kolkata (14.1 millions). Mumbai, which had witnessed 30.47% population growth during 1991–2001, has slowed down to 12.05% during 2001–2011 (Census GoI 2011). Though majority of the industries in Mumbai have transformed from manufacture sector to the service sector, industrial pollution to a certain extent has been replaced by vehicular pollution. Continuous vehicular activities in Mumbai have contributed significantly in deteriorating the quality of air, causing various health issues. ‘Recording a slum population of 77.55 percent and a Human Development Index of 0.05 in Deonar region, 256 slum settlements and 13 large resettlement colonies in this ward are reflective of the creation of a ghetto in global city’ (TISS 2015).

Industrial Development

Industrialisation is the process by which an economy is transformed from primarily agricultural to one based on the manufacturing of goods. In this process, individual manual labour is often replaced by mechanised mass production.

Historically, industrialisation has played a crucial role in developed countries with its positive after effects of settled global trade that leads to long-term economic growth. Various innovations that occurred during the industrial revolution as with transport and manufacturing products created a global economy stimulating economic growth in these countries. As a consequence of the transmission of global trade, capital and population migration flows considerably to developing countries. The development of first-world countries is largely due to factors like speed of transmission of the industrial revolution, institutional readiness, developed capitalist institutions in factor markets and supportive economic and political institutions, whereas, for developing countries, such factors of development are still in the pipeline, and the main engine of growth is exports which varied across countries.

Natural resources are being extensively used for rapid industrialisation and to fulfil the needs of growing populations that leads to the degradation of environment.

Environmental degradation is one of the most significant issues in the world today. The United Nations International Strategy for Disaster Reduction (1999) characterises environmental degradation as the lessening of the limit of the earth to meet social and environmental destinations and needs.

Environmental degradation is the disintegration of the earth or deterioration of the environment through mal-consumption of natural resources, for example, air, water and soil, the destruction of environments and extinction of species of various wildlife. It is characterised as any such aggravation to nature's turf. Ecological effect or degradation is caused by expanding human populace and their consistent activities for socio-economic development with the application of technologies that exhaust resources and pollute environment.

Causes of Environmental Degradation

Amongst several life species in the environment, some species require specific areas to help procure food, living space and other resources. At the point when the biome is divided, vast patches of living space do not exist anymore, which makes it difficult for wildlife to get the assets they need in order to survive. As detailed out by Rinkesh (2009), the environment goes on, even though the animals and plant life are not there to help sustain it properly.

Land disturbance: Land damage is basic cause of environmental degradation. Various foreign and obtrusive plant species, for instance, garlic mustard, adversely impact due to rupture in *environmental* surroundings by growing rapidly while eliminating the local greenery. Such invasive growth of species limits the food assets and creates disturbances to other environmental life.

Pollution: Pollution, in whatever form, air, water, land or noise, is harmful. *Air pollution* causes health issues for the population. Water pollution degrades the quality of water that we use for drinking purposes. *Land pollution* results in degradation of the earth's surface. *Noise pollution* arisen due to large sounds like honking of vehicles on a busy road or machines producing large noise in a factory or a mill can cause irreparable damage to our ears when exposed continuously.

Overpopulation: Rapid population growth, due to a decrease in the mortality rate and increased lifespan, puts strain on natural resources and thereby results in degradation of environment. More population simply means more demand for basic needs – food, clothes and shelter – which all requires additional space and resources. This results in deforestation which is another factor of environmental degradation.

Landfills: Landfills pollute the environment and destroy the beauty of a city. Landfills come due to the large amount of wastes that get generated by various sources like households, industries, factories and hospitals. *It* poses a high risk to the environment and to the people. Landfills produce foul smell when burned and cause environmental degradation.

Deforestation: Rapid growth in population and *urban sprawl* are two major causes of deforestation. Apart from that, the use of forest land for agriculture, animal grazing and harvest for firewood and logging are some of the other causes of deforestation. It contributes to a great extent to global warming as decreased forest size puts carbon back into the environment.

Natural causes: Avalanches, quakes, tidal waves, storms and wildfires affect nearby animal and plant groups. This can either come to fruition through physical demolition as the result of a specific disaster or through the long-term degradation of assets by the introduction of an obtrusive foreign species to the environment.

Of course, humans are not completely to blame. Earth itself causes ecological issues. While environmental degradation is most normally connected with human activities, the environment is always changing. With or without the effect of human exercises, a few biological systems degrade to the point where they cannot support the life that is supposed to live there.

Effects of Environmental Degradation

Human health may be impacted by environmental degradation. Fine particulate of air pollution adversely affects human health and gives rise to various pulmonary diseases such as pneumonia and asthma. A recent study by Arden and Douglas (2012) shows that air pollution causes mortality, cardiovascular diseases and chronic obstructive pulmonary diseases.

Many organisms are susceptible to the pollutants. Increased air pollution and degradation of environment has implications on loss of biodiversity. Deforestation, global warming, overpopulation and pollution are a few major causes for loss of biodiversity.

Ozone layer is responsible for protecting earth from harmful UVB radiations. The greenhouse gases like carbon monoxide, carbonyl sulphide, chlorofluorocarbons and other compounds in the atmosphere cause threats to the ozone layer and living beings.

Environmental degradation can have a big economic impact also. The economic impact can also be in terms of losses to tourism and other industries. Restoration of green cover, cleaning up of landfills and protection of *endangered species* are some of the measures essential for the holistic development of the country.

Climate Change and Its Effects

Climate change is increasingly noticed in recent years, and it is adversely affecting the past climatic conditions of the earth. Just in the last 650,000 years, according to NASA research, 'there have been seven cycles of glacial advance and retreat, with the abrupt end of the last ice age about 7000 years ago marking the beginning of the modern climate era and of human civilisation' (https://climate.nasa.gov/evidence/). These changes started with the minute variations in the orbit of the earth while having impacts on the solar energy. Scientific research, such as of NASA, have inferred that the greenhouse gases and carbon dioxide have the ability to affect the transfer of infrared energy through the atmosphere and the increasing amount of such gases leads to global warming. Events of ice cores drawn from regions like Greenland, Antarctica and tropical mountain glaciers are predicted to be earth's climate response to the raising greenhouse gas levels.

Climate change affects all sections of population and more to the vulnerable sections like elderly and children. According to the UNICEF (2016), 'children in certain countries are at greater risk from the impacts of climate change; more than 600 million children live in the 10 countries that are most vulnerable to climate change'. On the contrary, children are the least responsible for causing climate change, and yet they are highly vulnerable to bear the significant impacts.

Similarly, for elderly, the adverse climatic conditions, especially effects of air pollution, are not conducive for their health and wellbeing. According to Siva Raju and Smita (2016), 'an older person's sensitivity and risk of injury or loss increases in proportion to their level of physical and/or cognitive impairment, level of social isolation and financial dependency'.

They are also of the view that given a range of environmental exposures, supporting older persons to maintain good health and be physically active is a key strategy in building resilience to and reducing vulnerability to climate change.

Air Pollution

Air pollution is 'a contamination of the indoor or outdoor environment by any chemical, physical or biological agent that modifies the natural characteristics of the atmosphere' (WHO 2016). Some of the common sources of air pollution include household combustion devices, motor vehicles, industrial facilities and forest fires. Harmful pollutants include particulate matter, carbon monoxide, ozone, nitrogen dioxide and sulphur dioxide, which can cause respiratory and other diseases.

The concept of an Air Quality Index (AQI) has been developed and used effectively in many developed countries. An AQI transforms weighted values of individual air pollution-related parameters (SO2, CO, visibility, etc.) into a single number or set of numbers (National Air Quality Index 2015). There have not been significant efforts to develop and use AQI in India, primarily due to the absence of a dedicated air quality assessing and monitoring programme until 1984 and lack of public awareness about air pollution, till recently. The challenge of communicating with the people in a comprehensible manner about air pollution has two dimensions: (i) translate the complex scientific and medical information into simple and precise knowledge and (ii) communicate with citizens in the historical, current and futuristic senses. Addressing these challenges and thus developing an efficient and comprehensible AQI scale is much needed in the present day context (Table 18.1).

Indian metropolitan cities remain exposed to high levels of air pollutants mainly due to high vehicular movements and poor roads. Although the concentration of these pollutants varies according to the traffic density, type of vehicles and time of day, some people by virtue of their occupation are more exposed to high levels of traffic-related air pollutants (TERI 2015). These people include filling station workers, traffic policemen, professional drivers and toll-booth workers because of the proximity and high emissions from vehicle idling, deceleration and acceleration. In a recently conducted study on air quality monitoring (PM2.5, CO, NOX, SO_2 and EC/OC) at highway toll plazas, municipality toll plazas and control sites, it was found that there was a high level of air pollution at almost all locations with PM2.5 values exceeding the national permissible limit (60 $\mu g/m^3$) except at a few control sites. The study found that pollutant concentrations were highest at municipality toll plazas with minimum protective work areas. The observed reduction in lung function indices was significant over years of occupational exposure even after making adjustments for age, amongst non-smoking outdoor workers (CPCB 2015).

According to a World Bank study in India as cited in TERI (2015), in 2009, about 1100 billion INR (1.7% of GDP) and more than 800 billion INR (1.3% of GDP) were estimated as the annual cost of environment damage caused by ambient air pollution and household air pollution, respectively, in India. This translates to that about 52% of the relative share of damage cost by environment category was due to ambient and household air pollution put together (World Bank 2013). Data from the country's apex environmental regulator, the Central Pollution Control Board (CPCB), reveals that 77% of Indian urban agglomerations exceeded National Ambient Air Quality Standard (NAAQS) for reparable suspended particulate matter PM10 in 2010 (CPCB 2010). Estimates from the WHO suggest that 13 of the

Table 18.1 India – AQI category and range

AQI category	AQI range
Good	0–50
Satisfactory	51–100
Moderate	101–200
Poor	201–300
Very poor	301–400
Severe	4001–500

Source: National Air Quality Index (www.cpcb.nic.in)

20 cities in the world with the worst fine particulate PM2.5 air pollution are in India, including Delhi, the worst-ranked city ranked 7th. Air pollution also leads to a reduction in life expectancy. Using a combination of ground-level in situ measurements and satellite-based remote sensing data, it has been estimated that 660 million people, over half of India's population or nearly every Indian (1204 million people or 99.5% of the population), live in areas that exceed the Indian National Ambient Air Quality Standard for fine particulate pollution. Reducing pollution in these areas to achieve the standard would increase life expectancy for these Indians by 3.2 years on an average for a total of 2.1 billion life years (Greenstone et al. 2015).

Apart from there, studies around the world conclusively showed that air pollution is a serious environmental risk factor that causes or aggravates acute and chronic diseases in living beings. A study conducted in six cities of India, viz. Chennai, Delhi, Hyderabad, Indore, Kolkata and Nagpur, by Ghosh and Mukherjee (2010), has inferred that 'an increase in ambient air pollution significantly increases child morbidity, especially respiratory problems and high prevalence of allergy in them'.

A study carried out by Awasthi et al. (1996) noticed a close relation of between ambient air pollutants and respiratory symptoms complex (RSC) in preschool children, of 1 month to 4.5 years.

A study by Sinha and Bandyopadhyay (1998) has tried to capture the metallic constituents of aerosol present in biosphere, which have been identified as potential health hazards to human beings. The study examined the concentration of Cd (cadmium), Zn (zinc), Fe (iron), Pb (lead) and Cr (chromium) in ambient air of Delhi, Mumbai, Calcutta and Chennai cities in India. The health survey conducted in 1997–1998 by All India Institute of Medical Sciences (AIIMS), on individuals residing in the residential areas of Delhi, revealed that the air pollution led to irritation of the eyes (affecting about 44.4% of the subjects surveyed), cough (28%) and respiratory problems (5.9%) (Kumar 1999).

Similarly, a study by the National Environmental Engineering Research Institute (NEERI) revealed 'open burning and landfill fires of municipal solid waste (MSW)' as being the major sources of air pollution in Mumbai (CPCB 2010). The survey results show that about 2% of total generated MSW is burnt on the streets and slum areas and 10% of the total generated MSW is burnt in landfills by management authorities or due to accidental landfill fires, thereby emitting large amounts of CO, PM, carcinogenic HC and NOX.

According to Sharma and Tiwari (2000), coastal cities like Mumbai are undergoing social, economic and political transition. They noted that 'this is an appropriate time to rejuvenate these cities and protect them from further deterioration; otherwise, they will lose their comparative advantages to newer cities which have been more environmentally oriented'. Another factor that coastal cities like Mumbai needs attention is the policy of reclaiming land. Increasing reclamation for accommodating population density is not only depleting the coastal biodiversity but also threatening the existence of the city due to rise in sea water levels and the consequential submergence and flooding.

Mukhopadhyay (2003) opines that some of the changes in population distribution are due to the Development Control Rules of Mumbai that were originally formulated under the Bombay Town Planning Act of 1955. These rules have undergone considerable modifications over time. For instance, changes in the FSI in different parts of the city have affected population distribution. For example, Chembur is an area where a cluster of sensitive installations like oil refineries, the Bhabha Atomic Research Centre (BARC), a fertiliser plant and naval ammunition depot had prompted the government to initially limit FSI to 0.5. However, this was increased to 0.75 and later in 1998 to 1.00. It led to a spurt in conversion of bungalows into high-rise apartments and consequent population growth, which has close bearing on environmental pollution.

Rode Sanjay (2000) studied rising solid wastes in Mumbai Metropolitan Region. Such rise in solid waste generation was also observed in Brihanmumbai, Thane, Mira-Bhayander, Kalyan-Dombivali, Ulhasnagar, Navi Mumbai and Bhiwandi-Nizampur Municipal Corporation. The study accentuated that due to urbanisation, population increase, over-transportation and food habits, solid waste has been increased tremendously. Inefficient solid waste management has resulted in significant rise in epidemics of the population residing in these areas. The study strongly recommended for improved solid waste management system in the city.

Several epidemiological studies in Mumbai have revealed that with moderately raised pollution levels, there was an increased occurrence of dyspnoea, chronic and intermittent cough, frequent colds, chronic bronchitis and cardiac disorders, high blood pressure and deaths due to non-tuberculosis respiratory and ischaemic heart diseases (Kamat 2000). Another study in Mumbai, Parikh and Hadkar (2003), has specifically highlighted the high health costs spent by patients on the treatment of severe attacks related to air pollution. Greater emphasis therefore is required in urban planning and infrastructure development.

Mumbai City: A Profile

Mumbai is a metropolitan city and has a population of 12,442,373 (Census 2011). The annual growth rate of Municipal Corporation of Greater Mumbai was 1.90% in census 2001, which has reduced to 0.38% by 2011.

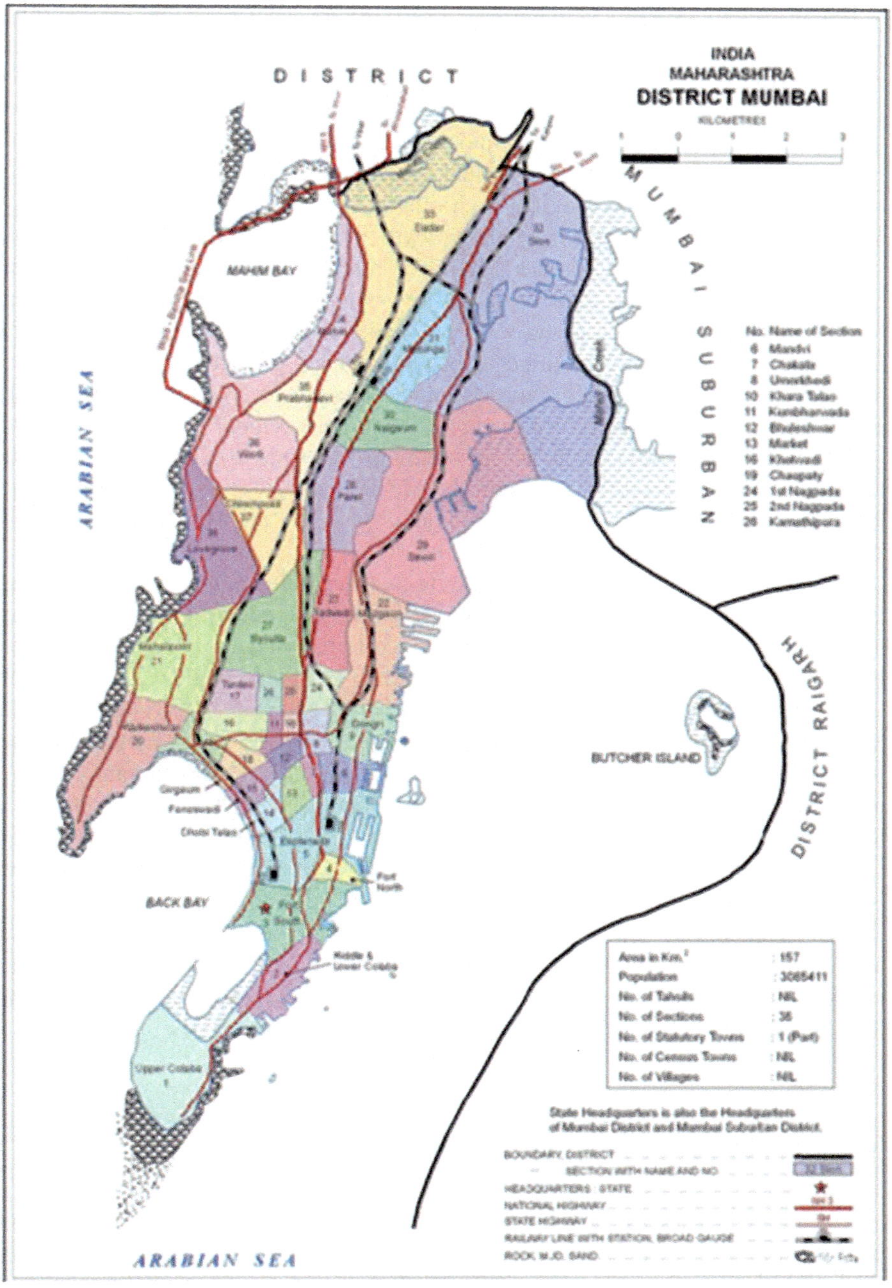

Source: District Census Handbook, Mumbai 2011

The urban dynamics in Mumbai appears to be a mix of a perpetually expanding global city, with new ideas, institutions and opportunities and persistent, emerging and complex forms of poverty and widening deficits in human development (TISS 2015). The city pays highest income tax, as a commercial capital of the nation (MCGM), and is home to 42% of population residing in slums (Census 2011).

Early development in Greater Mumbai revolved around the port and the mills to its south. As the city grew, it expanded northwards along its twin suburban railway networks, and till 1968, most of the growth in Mumbai Metropolitan Region (MMR) was confined to Greater Mumbai. Post-1968, the suburbs in Greater Mumbai grew along with areas surrounding Greater Mumbai, viz. Thane, Kalyan, Mira-Bhayander, Vasai-Virar and Navi Mumbai. The suburban rail networks have been crucial in this story of urban expansion. Since 1980, the MMR has been witnessing a higher decadal growth rate than Greater Mumbai. The MMR added 3.44 million people in the last decade. However, the annual compound decadal growth rate (2001–2011) of the MMR, which is 1.65%, is lower (2.79%) than the previous decade (1991–2001). Within the MMR, the fastest-growing cities in the last decade (2001–2011) are Navi Mumbai, Vasai-Virar, Mira-Bhayander and Thane.

The Mumbai Metropolitan Region lies to the west of the Sahyadri hill range and is part of the North Konkan region. It broadly lies between the rivers Tansa in the north and Patalganga in the south. The southwestern boundary extends beyond the Patalganga River and includes the town of Alibag and Pen. On the west, the MMR is bounded by the Arabian Sea; on the southeast, it extends to the foothills of the Sahyadris; and in the north-east, its extent is contiguous with the administrative boundaries of Bhiwandi, Kalyan and Ambernath Tehsils. The geography of the region is a significant determinant of urbanisation in MMR. The MMR is largely a low land, though not plain. A series of north-south trending hill ridges bring significant local elevational variation, though the average elevation of most areas is below 100 m above sea level.

The significant geographical features of the region include hills, rivers, lowlands and a long coastline, which in turn determine the nature of land uses prevalent in the region. The MMR is typical of the Deccan basaltic terrain with flat-topped mountains bordering a low-lying coastal region traversed by five rivers.

Mumbai City: Climate and Environmental Issues

The climate of MMR can be described as warm and humid. MMR receives ample rainfall from the southwestern monsoons during the wet monsoon season between June and September every year. The annual rainfall ranges between 180 and 248 cm. The monsoons are followed by three short cooler winter months between December and February. The rest of the months are hot.

Temperature: Typically, January is the coldest month of the year with May being the warmest, in accordance with the course of the sun. During the monsoons, the temperature is nearly uniform at around 27 °C. In October, the temperature rises before beginning to fall gradually reaching its minimum in January.

Wind: The normal seasonal prevailing wind direction during the dry season is west-north-west except during the monsoons when it is southwest. During December, the wind direction fluctuates between west-north-west and east-north-east. There is considerable diurnal and seasonal variation though there is little fluctuation in the velocity in the dry season. The winds are light and variable at 8 kmph during the dry season and reach a peak of around 13 kmph during the monsoon. Southern parts of MMR have higher wind velocities at around 10 kmph during the dry season peaking to about 25 kmph during the wet season. Squalls are common during the monsoon and accompanied by gusty winds.

Monsoons: The southwestern monsoons generally arrive in the Mumbai area during the second week of June and continue till late September. The average rainfall in the region is over 2000 mm, with the coastal areas receiving much less rain than the interior plains typically, though they receive the first onslaught of the rains. Due to local topographical conditions, Matheran which is situated at 760 m above MSL receives the highest rainfall in the region.

Climate variation: Though typically the climate of Mumbai and its surrounds are termed as equable with no large seasonal fluctuations of temperature (due to the proximity to the sea and the relatively large amount of humidity in the atmosphere), over the years, however, with increasing urbanisation, there are variations in the climate. Studies in the long-term trends of rainfall reveal a significant increase in monsoon rains for Mumbai between 1901 and 2000, significant reduction in wind speeds (59%) along with significant changes in frequencies of occurrence of warmer days (maximum temperatures above certain threshold) and colder days (minimum temperatures below certain threshold).

Private motorised transport, with an annual growth rate of 15.5%, is fast becoming the most preferred mode for intra-city travels, primarily due to intolerable crowding levels in suburban trains. Within Greater Mumbai, the proportion of cars is the highest compared to Eastern Suburbs.

The traffic survey in Greater Mumbai shows high volumes of traffic exchanges, along the connection to Vasai-Virar towards north and along connections towards Navi Mumbai.

All the main arterial roads have a high percentage of private vehicle traffic (cars) ranging 24–51% of total transport volume, whereas percentage of bus traffic is very low. This is one of the major reasons for traffic congestion along the corridors. Good vehicles ply highest on Sion Panvel Highway from and towards Navi Mumbai (42% of total traffic volume) due to the presence of JNPT and other goods that are handled in the area.

Existing Road Networks

Roads constitute 8.16% of the total area and 14% of the developed areas in Greater Mumbai (Mumbai City Development Plan 2005–2025). Street networks in most of Greater Mumbai are old and narrow, and their capacity is reduced considerably due to on-street parking, pedestrian spillover on the streets and hawkers and other encroachments. Station areas throughout the city are typically congested. With commercial establishments and informal markets nearby and high-density vehicular and pedestrian traffic, they are subject to bad traffic snarls during peak hours. Most areas in the island city, such as Navy Nagar, Marine Drive, Horniman Circle, Colaba, Mazgaon, Parel, Dadar, Matunga, Sion and Mahim, are planned developments, with gridded network of streets. However, the bazaar areas in the island city, including Null Bazar and Bhendi Bazaar areas, experience traffic conflicts due to their narrow streets, bazaar activity and high pedestrian movements. Gaothans and Koliwadas face similar issues arising out of narrow pedestrian road networks. East \west connectivity across the Western and Eastern Suburbs is limited to the Jogeshwari-Vikhroli Link Road, the Andheri-Ghatkopar Link Road and the recently opened Santa Cruz-Chembur Link Road, which is insufficient. Further, in some parts of the Western Suburbs, the east-west connectivity between road and rail lines is poor.

The natural systems of Greater Mumbai consist of hills and bays, coastal ecosystem, natural drainage system including rivers and the forest areas. Greater Mumbai has 26 km of coastline along its western edge. A third of the area of Greater Mumbai is under natural open spaces including forests, water bodies, mangroves and wetlands. It is also one of the few cities in the world to have a national park (Sanjay Gandhi National Park) within city limits. Greater Mumbai has three lakes (Powai, Vihar and Tansa), four rivers (Mithi, Oshiwara, Dahisar and Poisar) and several creeks and hills. However, large areas under marsh and mangroves have been reclaimed to accommodate an ever-growing population which creates flooding in several areas during the monsoon season. Environment Status Report Sec 63B of the Mumbai Municipal Corporation Act makes it mandatory for the municipal commissioner to place before the corporation before 31 July every year 'a report on the status of environment, from time to time', as may be specified by the state government, in the last financial year. The objective is to continue to obtain comparable data on environmental benchmarks and take necessary steps for improving the city environment. The overall status of the environment is analysed in terms of standard indicators that measure air quality, water quality and noise level.

The below table contains the summary of readings for the six pollutants vis-à-vis the CPCB standards. Three of the pollutants are within prescribed limits, while three are found in excess at some locations. Seasonal fluctuation due to wind direction, monsoon, etc., and variations in air quality could be noted (Table 18.2).

The table below shows that transport sector is the single major contributor to air pollution in MCGM (Table 18.3).

Table 18.2 Comparison with CPCB standards (annual avg.) at fixed air monitoring sites in 2010–2011

Sr. No	Unit	SO_2	NO_2	NH_3	SPM	Lead	B(a)P[a] 1
1	Range[b]	7–10	14–50	37.242	125–642	0.07–0.37	0.3–0.9
2	Maximum at	Maravli and Bhandup	Maravli	Maravli	Maravli	Maravli	Maravli and Khar
3	CPCB standards annual average	50 μg/m^3	40 μg/m^3	100 μg/m^3	140 μg/m^3	0.5 μg/m^3	1 ng/m^3
4	Comparison with standards	Not exceeded	Exceeded at Maravli and Khar	Exceeded at Maravli	Exceeded at all the sites except Borivali	Not exceeded	Not exceeded

Source: Environmental Status of Brihanmumbai 2010–2011, MCGM and benzo(a)pyrene
[a]Unit ng/m^3, benzo(a)pyrene
[b]Unit μ/m^3

Table 18.3 Emission load of Mumbai City in the year 2010–2011 (tons/day)

Sr No.	Use	SO_2	Particulate matter	NOX	CO	HC	Total
1	Domestic	4.41	9.15	29.23	93.81	34.74	171.34
2	Industrial	24.01	0.21	0.05	–	–	24.27
3	Refuse burning	0.16	1.56	0.32	5.99	2.22	10.25
4				Transport			
4.1	Transport (diesel)	5.96	2.48	34.15	18.12	7.16	67.87
4.2	Transport (petrol)	0.66	0.18	18.2	265.3	39.05	323.39
	Total	35.2	13.58	81.95	383.22	83.17	597.12

Source: EIG, MCGM

The table below shows the trend of pollution across 3 years (2008–2008 to 2010–2011) at six locations, two each in the three zones of Greater Mumbai (Table 18.4).

Environmental Vulnerability

Greater Mumbai areas are prone to three potential natural hazards of heavy rainfall, flooding, landslides and earthquake. Of these, flooding is the major threat because of its greater impact on life and property. Its estuarine setting, coupled with continuous reclamation in marsh lands and low-lying areas, has led to an obstruction in the natural flow of water bodies and drains. Most of Greater Mumbai is on reclaimed lands that are almost flat, which makes the city naturally prone to flooding. Prime city locations are lower than high tide level. Similarly, low-lying

Table 18.4 Site-wise percentage of samples exceeding CPCB (24-h standards in the year 2008–2011 average)

Sr No.	Site	SO_2			NO_2			NH_3			SPM			Lead		
		08–09	09–10	10–11	08–09	09–10	10–11	08–09	09–10	10–11	08–09	09–10	10–11	08–09	09–10	10–11
1	Worli	0	0	0	46	6	7	0	0	0	41	45	36	0	1	0
2	Khar	0	0	1	47	9	2	1	0	0	59	60	50	0	0	0
3	Andheri	0	2	0	46	16	2	1	0	0	60	57	39	0	0	0
4	Bhandup	0	0	0	37	2	0	0	0	0	60	52	48	0	0	0
5	Borivali	0	0	0	2	0	0	0	0	0	9	12	11	0	0	0
6	Maravli	0	1	0	44	30	9	20	24	17	71	84	88	1	2	6

Source: Environmental Status of Brihanmumbai 2010–2011, MCGM

coastal edges and river floodplains are susceptible to flooding. Several areas around hill slopes in Greater Mumbai are prone to landslides. The risk is more during the monsoon. Areas around hill slopes in Ghatkopar, Bhandup and Kurla in the Eastern Suburbs are prone to landslides resulting in increased exposure of slopes to erosion and water infiltration. Slum populations residing on these hill slopes are at high risk.

Risks Due to Climate Change

Increased intensities of climatic events like increased rainfall, floods, unseasonal rain or drought, intense heat, sea level rise, cyclonic storm surges and increasing outbreaks of tropical diseases and epidemics are predicted outcomes of climate change and global warming. Greater Mumbai's coastal location and a large population living in close proximity to the coast render it highly vulnerable to many climate change effects, especially sea level rise and flooding. Since Mumbai is only a few metres above sea level and has four rivers flowing through it, it further increases its vulnerability to flooding.

Public health impacts in the city are largely attributed to environmental pollution; in particular, the illnesses and diseases spread as a result of the following environmental problems: poor air quality (pollutants from transportation, domestic and construction demolition activities), high levels of noise pollution, flooding during rains, poor quality of potable water, inadequate light and ventilation and inadequate sanitation facility.

Solid Waste Management

Improper solid waste management leads to aesthetic and environmental problems (Majumdar and Srivastava 2012). Emission of volatile organic compounds (VOCs) is one of the problems from uncontrolled dumpsite. VOCs are well known to be hazardous to human health and many of them are known or potential carcinogens. They also contribute to ozone formation at ground level and climate change as well. The recent study on VOCs emitting from two municipal waste (MSW) disposal sites in Mumbai, namely, Deonar and Malad, is a significant one. Air at dumpsites was sampled and analysed on gas chromatography-mass spectrometry (GC-MS) in accordance with the US Environmental Protection Agency (EPA) TO-17 compendium method for analysis of toxic compounds. As many as 64 VOCs were qualitatively identified, amongst which 13 are listed under hazardous air pollutants (HAPs). Study of environmental distribution of a few major VOCs indicates that although air is the principal compartment of residence, they also get considerably partitioned in soil and vegetation. The CO2 equivalent of target VOCs from the landfills in Malad and Deonar shows that the total yearly emissions are 7.89E + 03 and 8.08E + 02 kg, respectively. The total per hour ozone production from major

VOCs was found to be 5.34E-01 ppb in Deonar and 9.55E-02 ppb in Malad. The total carcinogenic risk for the workers in the dumpsite considering all target HAPs are calculated to be 275 persons in one million in Deonar and 139 persons in one million in Malad.

State's Efforts to Control Pollution

The Maharashtra Pollution Control Board (MPCB) is implementing various environmental legislations in the state of Maharashtra, mainly including Water (Prevention and Control of Pollution) Act, 1974; Air (Prevention and Control of Pollution) Act, 1981; Water (Cess) Act, 1977; some of the provisions under Environmental (Protection) Act, 1986; and the rules framed thereunder like Biomedical Waste (M&H) Rules, 1998; Hazardous Waste (M&H) Rules, 2000; Municipal Solid Waste Rules, 2000; and others. MPCB is functioning under the administrative control of Environment Department of Government of Maharashtra.

Various organisations are promoting good practices to curtail the pollutants in the environment and promoting the afforestation and solid waste management practices. One of the best practices is to monitor the air pollution (see Table 18.5) by the pollution control boards.

Efforts in Traffic Diversion

Though 2150 trains travel through the city, carrying millions of Mumbaikars to their destinations, for 75 lakh odd commuters, 'every day brings with it the challenge of searching for foot-space in a train that cannot hold a pebble more' (Mumbai Metro Rail Corporation LTD (2016) Available at: https://www.mmrcl.com/en/about-mmrc/know-your-metro).

Mumbai Metro Line-3 (MML-3) is one of such key projects to improve the transportation scenario in Mumbai. MML-3 project – a 33.5-km-long corridor running along Colaba-Bandra-SEEPZ – envisages to decongest the traffic situation in the city. It aims to provide a Mass Rapid Transit System that would supplement the inadequate suburban railway system of Mumbai by bringing metro closer to the doorsteps of the people.

By 2021, it aspires to bring reduction in vehicle trips/day by 456,771 and reduction in fuel consumptions – petrol and diesel – in litre/day by 243,390. The average daily money savings due to reduction in number of vehicle trips would be Rs. 158.14 lakhs, followed by 12,590 tonnes/year reduction in emission pollution (Mumbai Metro Rail Corporation LTD 2016).

The monorail is an efficient feeder transit system benefiting commuters and will offer efficient, safe, air-conditioned, comfortable and affordable public transport.

Table 18.5 Air pollution levels, type of area: industrial, residential, rural and other area

Parameters	Date	Time	Concentration	Unit	Concentration (previous 24 h)/ prescribed standard
Nitric oxide	19/01/2017	16:15:00	19.26	μg/m^3	21.70 μg/m^3
Nitrogen dioxide	19/01/2017	16:15:00	14.91	μg/m^3	15.19 μg/m^3 Prescribed standard: 80.00 μg/m^3
Oxides of nitrogen	19/01/2017	16:15:00	34.16	ppb	36.89 ppb
Sulphur dioxide	19/01/2017	16:15:00	16.54	μg/m^3	19.27 μg/m^3 Prescribed standard: 100.00 μg/m^3
Carbon monoxide	19/01/2017	16:15:00	0.97	mg/m^3	1.87 mg/m^3 Prescribed standard: 4.00 mg/m^3
Ozone	19/01/2017	16:15:00	93.59	μg/m^3	38.26 μg/m^3 Prescribed standard: 180.00 μg/m^3
PM10	19/01/2017	16:15:00	132.42	μg/m^3	258.64 μg/m^3
PM2.5	19/01/2017	16:15:00	63.59	μg/m^3	113.54 μg/m^3 Data under scrutiny Prescribed standard: 100.00 μg/m^3
Temperature	19/01/2017	16:15:00	34.00	°C	29.50 °C
Relative humidity	19/01/2017	16:15:00	48.12	%	59.44%
Wind speed	19/01/2017	16:15:00	0.08	m/s	0.63 m/s
Wind direction	19/01/2017	16:15:00	3.00	degree	164.62 degree
Vertical wind speed	19/01/2017	16:15:00	0.80	degree	0.75 degree
Solar radiation	19/01/2017	16:15:00	49.00	W/m^2	74.87 W/m^2
Barometric pressure	19/01/2017	16:15:00	766.29	mmHg	768.17 mmHg

Source: Central Pollution Control Board (CPCB)
*Prescribed standard for CO and ozone is one hourly average

Monorail carries 7500 commuters per hour per direction and has the capacity to carry 1.5–2 lakh commuters daily (MMRDA 2016).

In 2002, the state government, Indian railways and the MMRDA, with financial assistance from the World Bank, decided to undertake Mumbai Urban Transport

Project (MUTP) to find out long-term solution to city's transport and communication issues.

Besides these, other projects (MMRDA 2016) are also initiated to manage the vehicular traffic in the city.

- Eastern Freeway: This 16.8-km access-controlled freeway connects the Eastern Expressway at Ghatkopar with South Mumbai at P D'Mello Road. A 13.59-km stretch from Orange Gate on P D'Mello Road up to Panjarpol, near RK Studios in Chembur, is operational, reducing travel time from 90 min to a mere 15 min.
- Santa Cruz-Chembur Link Road: The 6.5-km double-deck flyover has reduced journey time from Santa Cruz to Chembur to 17 min.
- Andheri-Ghatkopar Link Road: The 7.9-km road connecting the Western Express Highway in Andheri to Ghatkopar via Saki Naka and Asalpha is almost fully operational.
- Sahar Elevated Access Road connecting to the international airport: This 2-km-long elevated road connects the Mumbai International Airport to the Western Express Highway.

With these projects, it is expected that the connectivity of various areas of Mumbai City is well networked and these measures greatly are contributing to combat air pollution in the region.

Suggestive Measures

To tackle further the issues related to environmental degradation, it is important to act at individual level as well as collectively. The idea of sustainable development needs to be imbibed and reflected in every developmental scheme, policy and project in the city. To specify a few such measures, the following needs emphasis:

For industries:

- Specific disincentives need to be imposed on the industries that use high-end energy, add-on to the emissions and carbon footprints.
- Machineries with anti-pollution techniques that would have zero emission effects need to be installed.
- Ensuring tall chimneys for the industries to avoid air pollution at the lower atmosphere.
- Providing incentives for small and medium towns to develop in the industrial sector. Necessary support of building the infrastructure like subsidised electricity, water and transport needs to be provided to such towns.
- Restricting the already industrialised areas from further installation of new industries to dilute the level of pollution.
- Tax holidays need to be facilitated.
- Separate space in the outskirts for recycling of the waste needs to be kept.

- Tying up with the organisations which have successfully combated the air pollution.

  Community level:

- Awareness programme on air pollution
- Promoting eco-friendly alternatives as a part of lifestyle
- Use of cycles and public transport as a green transport system for reducing congestion and pollution levels
- Collective action for conserving environment

Above all, the city dwellers have a major role to play, individually and collectively, in protecting the ecosystem of the city and to actively participate in anti-pollution measures which can go a long way in sustainable development of the city.

Acknowledgements The authors would like to thank Dr. B. Anil and Ms. Maya Pillai for their assistance in the preparation of the paper.

References

Arden Pope III C, Dockery DW (2012) Health effects of fine particulate air pollution: lines that connect. J Air Waste Manage Assoc 56, 2006 – Issue 6

Awasthi S, Glick HA, Fletcher RH, Ahmed N (1996) Ambient air pollution and respiratory symptoms complex in preschool children. Indian J Med Res 104:257–262

Bhagat RB, Jones GW (2013) Population change and migration in Mumbai metropolitan region: implications for planning and governance 201:1–25

Census (2011) Office of the registrar general of India. New Delhi

CPCB (2010) Air quality assessment, emissions inventory and source apportionment studies: Mumbai. Central Pollution Control Board. Available online at: http://cpcb.nic.in/Mumbai-report.pdf

CPCB (2015) National Air Quality Index. Central Pollution Control Board, Ministry of Environment, Forests and Climate Change, New Delhi

Davidson CI, Phalen RF, Solomon PA (2005) Airborne particulate matter and human health: a review. Aerosol Sci Technol 39(8):737–749. Available at: http://www.tandfonline.com/doi/abs/10.1080/02786820500191348

Gandhe HD, Kumar A (2016) Efficient resource recovery options from municipal solid waste: case study of Patna, India. Curr World Environ 11(1):72–76

Ghosh A, Mukherjee A (2010) Air pollution and child health in urban India. Indian Statistical Institute, New Delhi

Greenstone M et al (2015) Lower pollution, longer lives: life expectancy gains if India reduced particulate matter pollution. Econ Polit Wkly L(8)

Kamat SR (2000) Mumbai studies of urban air pollution and health resulting synergistic effects of SPM, S02 and NO, International conference on environmental and occupational respiratory diseases. Indian Toxicological Research Centre, Lucknow

Kumar R (1999) Effects of environmental pollution on the status of human health of Delhi residents. All India Institute of Medical Sciences, New Delhi

Majumdar D, Srivastava A (2012) Volatile organic compound emissions from municipal solid waste disposal sites: a case study of Mumbai, India. J Air Waste Manage Assoc 62(4):398–407. Available at: http://www.scopus.com/inward/record.url?eid=2-s2.0-84862265338&partnerID=tZOtx3y1

Mukhopadhyay T (2003) Urban commercial landscape- a case study of Mumbai. The City 1 (4):39–43
Mumbai City Development Plan 2005–2025. Available at: http://www.mcgm.gov.in/irj/go/km/docs/documents/MCGM%20Department%20List/City%20Engineer/Deputy%20City%20Engineer%20(Planning%20and%20Design)/City%20Development%20Plan/Vision%20Statement.pdf
Mumbai Metropolitan Region Development Authority (2016) Draft Mumbai metropolitan regional plan report 2016–2036. Available online at: https://mmrda.maharashtra.gov.in/regional-plan
NASA Global Climate Change. Available at: https://climate.nasa.gov/evidence/
Parikh JK, Hadker N (2003) Economic impacts of urban air pollution: valuation for Mumbai, India. Int J Environ Pollut 19:498–515
Rode Sanjay (2000) Waste Management in Municipal Corporation of Mumbai Metropolitan Region
Sharma VK, Tiwari P (2000) Environmental problems of coastal areas in India. Bookwell, Delhi, p 48
Sinha S, Bandhopadhyay TK (1998) Review of trace elements in air environment and its health impact in some Indian cities. J IPHE, India 1:35–48
Siva Raju S, Smita B (2016) Climate change and its vulnerability for the elderly in India. In: Akhtar R (ed) Climate change and human health scenario in south and Southeast Asia. Springer, New Delhi
Tata Institute of Social Sciences (TISS) (2015) Social economic conditions and vulnerabilities. A report of the baseline survey of M(East) ward, Mumbai. Available at: urk.tiss.edu/images/pdf/M-Ward-Report
TERI (2015) Air pollution and health. Discussion paper by The Energy and Resources Institute: New Delhi
United Nations Children's Fund, 2016. Available at: https://www.unicef.org/environment/index_60332.html
United Nations International Strategy for Disaster Reduction (UNISDR) (1999) Available at: http://www.unisdr.org/who-we-are/mandate
World Bank (2013) World development indicators. World Bank, Washington, DC
World Health Organisation (2016) Health topics: air pollution. Available at: http://www.who.int/topics/air_pollution/en/

S. Siva Raju is deputy director, Tata Institute of Social Sciences (Hyderabad Campus). His broad fields of interest for research projects are ageing, health and development. He has directed many research projects in these areas. He is a member of the Committee on Protection and Welfare of the Elderly Persons for the National Human Rights Commission; Expert Committee member on Ageing, Ministry of Social Justice and Empowerment; honorary director of International Longevity Centre, Pune; advisor to the Ministry of Social Security, Govt. of Mauritius; and a coordinator of the UNFPA Initiative on Building Knowledge Base on Population Ageing in India.

Khushboo Ahire is research scholar at the School of Development Studies, Tata Institute of Social Sciences, Mumbai, India. Currently, she is associated in a UNFPA sponsored project on Building Knowledge Base on Population Ageing in India as Project Officer. She is pursuing her research in the area of intergenerational bonds and quality of life of the elderly. Her research areas mainly include population ageing, environment and sustainability, corporate social responsibility (CSR), human development and research methodology. She has participated in various national and international conferences, workshops and seminars focusing on various social issues. She has to her credit publications in the areas of climate change and air pollution, wellbeing of the elderly and social impacts of CSR projects in the area. Khushboo has considerable experience in conducting social research projects.

Chapter 19
Climate Change and Air Pollution in East Asia: Taking Transboundary Air Pollution into Account

Ken Yamashita and Yasushi Honda

Abstract Co-benefit and co-control of SLCPs is the key concept to tackle simultaneously with problems of transboundary air pollution and climate change. Especially in East Asia, severe air pollution causing millions of premature mortality by $PM_{2.5}$ and ozone should be solved without delay as well as mitigation of global warming. Cost-benefit approach discussed in this chapter is one of the most effective and rational way to lead the feasible and appropriate policy for the challenge we need to do.

Keywords Risk assessment • Transboundary air pollution • Co-benefit/co-control • Human health • SLCPs

Introduction

The atmospheric environment is the critical issue in many regions of the world. The air pollution problems need to be assessed both in global and local scale due to its transboundary transportation and local effects. Hemispheric air pollution of ozone by intercontinental transportation, globally spread aerosols, and enormous nitrogen oxide emission from Asia is threatening human health, ecosystem, and also climate change (Akimoto 2003), and regional frameworks have been approaching the problems. In Europe, the atmospheric management has been tackled through the framework of the 1979 Convention on Long-Range Transboundary Air Pollution (CLRTAP) and European Union (EU) air pollution policy (Schroeder and Yocum 2006). Those regal institutions have the significant interlinkage between international, national, and local level. In Asia, the 2002 Agreement on Transboundary

K. Yamashita (✉)
Data Management Department, Asia Center for Air Pollution Research (ACAP), 1182 Sowa, Nishi-ku, Niigata-shi 950-2144, Japan
e-mail: yamashita.ken@home.nifty.jp

Y. Honda
Faculty of Health and Sport Sciences, University of Tsukuba, 1-1-1 Tennodai, Tsukuba 305-8577, Ibaraki, Japan

© Springer International Publishing AG 2018

R. Akhtar, C. Palagiano (eds.), *Climate Change and Air Pollution*, Springer Climate, DOI 10.1007/978-3-319-61346-8_19

Haze Pollution (Haze Agreement) by the Association of Southeast Asian Nations (ASEAN), Acid Deposition Monitoring Network in East Asia (EANET) which started in 1998, and Malé Declaration on Control and Prevention of Air Pollution and Its Likely Transboundary Effects for South Asia (Malé Declaration) which began in 1998 are the regimes for transboundary air pollution (Bergin et al. 2005). Haze Agreement is only for haze pollution from large and uncontrolled forest fire though it is the regal regime; EANET and Malé Declaration are non-regal regime and have limited activity scope. Recently co-benefit or co-control is the key concept to assess the air pollution and global warming problems simultaneously in terms of effective and efficient approach. Ozone and black carbon (BC) are called as short-lived climate pollutants (SLCPs) which cause global warming in relatively short time scale comparing with long-lived greenhouse gases (LLGHG) such as carbon dioxides (CO_2). The reduction of emission of SLCPs by 2050 as well as the mitigation of emission of LLGHG is indispensable for addressing near-term climate change and to keep the increase of the temperature within 2 °C in this century (UNEP 2011). The Climate and Clean Air Coalition (CCAC) was established in 2014 by governments, civil society, and private sectors to cope with this issue in the Asia and Pacific region, and the Asia Pacific Clean Air Partnership (APCAP) was also launched in 2014 by the United Nations Environmental Program (UNEP) in cooperation with governments and partners to tackle with air pollution problems in the region. We discuss here the adverse effect on human health by air pollution, especially transboundary air pollution of $PM_{2.5}$ and ozone in East Asia, and cost-benefit analysis of its control and related global warming issues on the health effects in terms of co-benefit/co-control approach. We then hope to provide the scientific evidence and suggestion for better achieving clean atmospheric environment.

Transboundary Air Pollution in East Asia

In East Asian region, air pollution problems are still big issue to be solved immediately in both of developed and developing countries. Due to continuing rapid economic growth and consequent enormous emission of air pollutants which are transported over countries (transboundary air pollution), the high concentration in atmospheric environment of nitrogen oxides (NOx), the sulfur oxides (SO_2), and the carbon monoxide (CO) is observed especially in megacities. Aerosol loadings and tropospheric ozone also has been increasing in past decades in Asia. The secondary particulate matter formed by air pollutant gases and aerosols through physical reaction as well as primary aerosol is closely associated with human activities, and emissions inside East Asia have the largest influence on East Asian ozone itself with 60% of East Asian surface ozone (TFRC 2015). Recently, ozone and fine particle ($PM_{2.5}$) which are typical transboundary air pollutants are focused on not only their adverse effect on human health such as respiratory and cardiac diseases but also their character as SLCPs. So in order to address regional air pollution and subsidiary global warming in East Asia, it is necessary to analyze

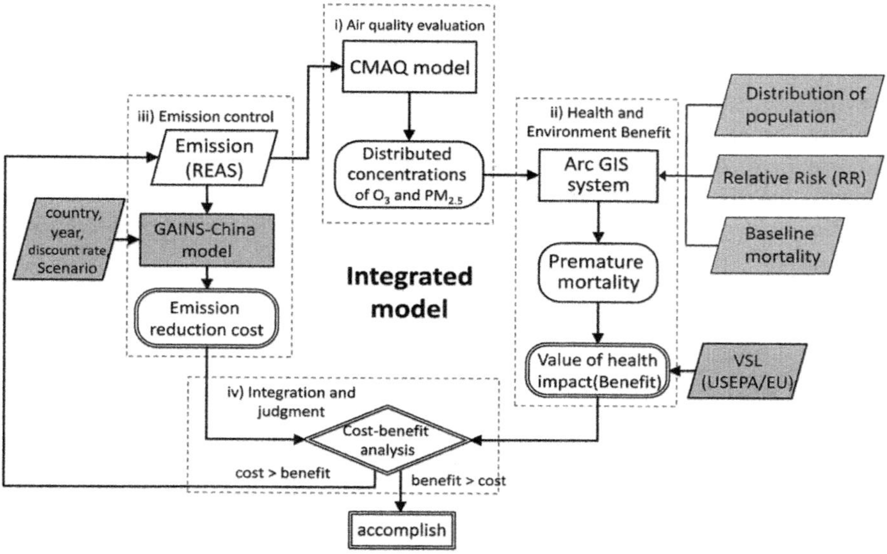

Fig. 19.1 Flow chart for the process of the cost and benefit estimation

the long-range transboundary transport of ozone and aerosols as well as their emission inventory and chemical reaction.

As to the estimation of health effect by air pollutants, WHO reports[1] that in 2012, around seven million people died as a result of air pollution exposure. Regionally, low- and middle-income countries in the WHO Southeast Asia and Western Pacific Regions had the largest air pollution-related burden in 2012, with a total of 3.3 million deaths linked to indoor air pollution and 2.6 million deaths related to outdoor air pollution. The 2010 Global Burden of Disease Study (GBD 2010) reported that approximately 2.0 million deaths were caused by air pollution in East Asia (Lim et al. 2013), which was the fourth highest risk factor, behind physiological risks, dietary risks, and high blood pressure. Premature mortality also leads to serious human and economic losses in environmental economics. Additionally, studies have shown that, in the field of pollutant reduction, more than 80 % of monetized benefits were attributed to reductions in premature mortality (Krupnick et al. 2002). Therefore, the cost-benefit analysis can provide essential information for prioritizing environmental policies.

In this chapter, we show an integrated approach to link the control cost of pollutants to air quality improvement and consequent health/environmental benefits (Fig. 19.1) (Nawahda et al. 2012; Chen et al. 2015). The framework includes the following elements: (1) Use CMAQ/REAS (the Models-3 Community Multiscale Air Quality Modeling System/the Regional Emission Inventory in Asia) to estimate

[1]http://www.who.int/mediacentre/news/releases/2014/air-pollution/en/

the emission and distributed concentration of ozone and $PM_{2.5}$ in East Asia; (2) a. evaluate premature mortalities in East Asia using the concentration-reaction function through geographic information system (GIS) and b. assess the benefit of saving lives based on value of statistical life (VSL); (3) use GAINS-China model (Amann et al. 2011) to evaluate the cost of reducing pollutant emissions; and (4) compare costs and benefits. If the benefit is larger than the cost, the emission-control scenario is beneficial for the welfare of society; otherwise the scenario is inefficient, and we need to either use new ozone and $PM_{2.5}$ emission scenarios or new technology of GAINS model.

Health Effects by Air Pollution (Ozone and $PM_{2.5}$)

Regional Chemical Transport Model and Emission Inventory

In this section, we show the health effects by air pollution: $PM_{2.5}$ and ozone in East Asia. The CMAQ/REAS modeling system is used to simulate the spatial distributions and temporal variations of $PM_{2.5}$ components and ozone in the East Asian region. The ozone and $PM_{2.5}$ concentrations were simulated by Uno et al. (2005) and Kurokawa et al. (2009) for the years 2000 and 2005 and by Yamaji et al. (2006, 2008) for the year 2020 scenarios using the three-dimensional regional-scale chemical transport model, based on the CMAQ ver. 4.4. This model is driven by the meteorological field simulated by the Regional Atmospheric Modeling System (RAMS) ver. 4.3 (for the year 2020) and ver. 4.4 (for the years 2000–2005). The grid resolution is 80 × 80 km, 14 layers for 23 km in the sigma-z coordinate system, and the height of the first layer is 150 m.

The CMAQ modeling system is coupled with REAS, which includes the following emissions: SO_2, NO_x, CO, NMVOC, black carbon (BC), and organic carbon (OC) from fuel combustion and industrial sources. REAS has three scenarios for China in 2020: PSC (policy success case), REF (reference case), and PFC (policy failure case). Regarding the emission of NO_x, in the 2020 PSC scenario, the NO_x emissions in China will have a slight decrease of 1% from 2000 to 2020. In the 2020 REF scenario, NO_x emissions in China (15.6 Tg) will increase by 40% from 2000 (11.2 Tg). In the 2020 PFC scenario, NO_x emissions emitted in China will increase by 128% from 2000. Regarding the emission of NMVOC, in the 2020 PSC scenario, the NMVOC emissions emitted in China will have a large increase of 97% from 2000. In the 2020 REF scenario, NMVOC emissions in China (35.1 Tg) will increase rapidly by 128% from 2000 (14.7 Tg). In the 2020 PFC scenario, NMVOC emissions in China will increase by 163% from 2000 (Ohara et al. 2007). The spatial distributions and annual variations of the annual mean $PM_{2.5}$ concentrations are calculated based on the annual mean concentrations of the following components: EC, OC, NO_3^-, SO_4^{2-}, and NH_4^+.

Though it is the best way to use the results of monitoring to estimate the amount of exposure by air pollutants, we usually use the results of the Regional Chemical

Transport Model (RCTM) such as CMAQ because the monitoring sites in East Asia are not deployed enough to cover the area concerned. The estimated exposure by RCTM, however, has some uncertainty. A study indicated the example of the uncertainty which has different results (premature mortality) by 2.5 times between using monitoring data and output of RCTM (Nawahda et al. 2013).

$PM_{2.5}$: Exposure and Premature Mortality Analysis

The distributed annual premature mortality rate in each grid cell is calculated as follows using Eq. 19.1 for $PM_{2.5}$ mean annual concentrations above 10 μgm^{-3} for the age group of 30 years and above:

$$\text{mortality}_{\text{PM2.5}}(i,j,t) = \text{pop}(i,j,t)M_b(i,j,t)\beta_{\text{PM2.5}}\triangle\text{PM}_{2.5}(i,j,t) \tag{19.1}$$

$$\beta = \ln(\text{RR})/\triangle C_{\text{PM2.5}} \tag{19.2}$$

where *mortality* indicates premature mortality, *i,j* specify the location of a grid cell within the simulation domain, *t* is the year of simulation, *pop* is the exposed population, M_b is the annual baseline mortality, β is the $PM_{2.5}$ CR coefficient, which can be calculated using Eq. (19.2), and $\triangle$C is the change in concentration. According to Pope III et al. (2002), an increase of 10 μgm^{-3} annual average of $PM_{2.5}$, within a range from around 7.5 to 30 μgm^{-3}, caused a 4% (95% confidence interval: 1.01–1.08) increase in mortality rate for the age group of 30 years and above. This gives β a value around 0.004 and $\triangle PM_{2.5}$ *(i,j,t)* is the change in the annual mean concentrations above 10 μgm^{-3}. We use the same β value also for mean annual concentrations above 30 μgm^{-3} similar to Cohen et al. (2005); they linearly extrapolated the $PM_{2.5}$ CR function to cover a wider range from 0 to 90 μgm^{-3}.

Ozone: Exposure and Premature Mortality Analysis

The distributed annual premature mortality rate based on a RR value of 1.003 (95% confidence interval (CI): 1.001–1.004) [0.3% increase in daily premature mortality caused by a 10 μgm^{-3} (~5 ppb) change in 8-h maximum mean concentration above 70 μgm^{-3} (~35 ppb)] at each grid cell is calculated by summing the daily premature mortality, which can be calculated using the following function (US-EPA 2006):

$$\text{mortality}_{O3}(i,j,n) = Y_o(i,j,n)\{1 - \exp[-\beta_{O3}\Delta O_3(i,j,n)]\} \tag{19.3}$$

where *n* is the calculation day and Y_o is the daily incidence of premature mortality at a certain ozone level where there is no clear health effect likely to occur. We estimate it in our study by multiplying the population of the age group of 30 years and above by the daily baseline mortality for this age group. β is estimated using Eq. (19.2) based on a RR value of 1.003, which gives β a value around 0.0003. ΔO_3 is the change in ozone concentration calculated based on the daily maximum 8-h mean concentrations above 35 ppb (or the value of the SOMO35 index of the day n) as follows:

$$\mathrm{SOMO35}(i,j,n) = [\max 8\text{ h mean} - 35]_n \tag{19.4}$$

The daily maximum 8-h mean concentration is the highest moving 8 h average to occur from 0:00 h to 23:00 h in a day.

Population Distribution

We obtain the population distribution in East Asia from the Gridded Population of the World (GPWv3) (CIESIN 2005); the size of the population grid cell is around 0.04167 degree. The total population within the simulation domain was about (1970) million in 2000 and (2057) million in 2005. In this study, we estimate the premature mortality rate for the age group of 30 years and above, which includes most of the working age groups in East Asia (WHO 2010a, b). According to the United Nations Department of Economic and Social Affairs/Population Division (2008), the fractions of population in East Asia that were 30 years and above for the years 2000 and 2005 were 51% and 55%, respectively. The distributed population for the year 2020 is estimated based on the population projections for the year 2015 by GPWv3 and the estimated growth rate of 0.42% in East Asia for the period from 2015 to 2020 by the United Nations Department of Economic and Social Affairs/Population Division (2008). However, there is no information about age-specific mortality rates for most of the countries in East Asia. Therefore, we estimate the baseline mortality for the age group of 30 years and above based on the WHO mortality database (WHO 2006) as shown in Table 19.1.

Premature Mortality

The premature mortality caused by exposure to both ozone and $PM_{2.5}$ in East Asia for the years 2000, 2005, and 2020 (PSC, REF, PFC) are estimated to be about (316), (520), (451), (649), and (1035) thousand, respectively (Nawahda et al. 2012). The estimated premature mortality of each country, caused by exposure to $PM_{2.5}$ annual mean concentrations above 10 μgm^{-3} and the daily maximum 8-h mean concentrations of ozone which are above 35 ppb for the age group of 30 years and above in East Asia for the years 2000, 2005, and 2020, is shown in Fig. 19.2.

Table 19.1 Population structure in East Asia from 2000 to 2020 and the corresponding baseline mortality

Year	2000	2005	2015	2020
Total population (thousand)	1,472,443	1,520,717	2,227,350	2,236,705
Population (+ 30) (thousand)	748,632	838,554	1,403,231	1,409,124
+30 years (%)	50.8	55.1	0.63	0.63
Total deaths (thousand)	10,063	10,063	[a]	[a]
Baseline mortality	0.0068	0.0066	[a]	[a]
+30 years baseline mortality	0.0103	0.0102	0.0102	0.0102

[a]No data

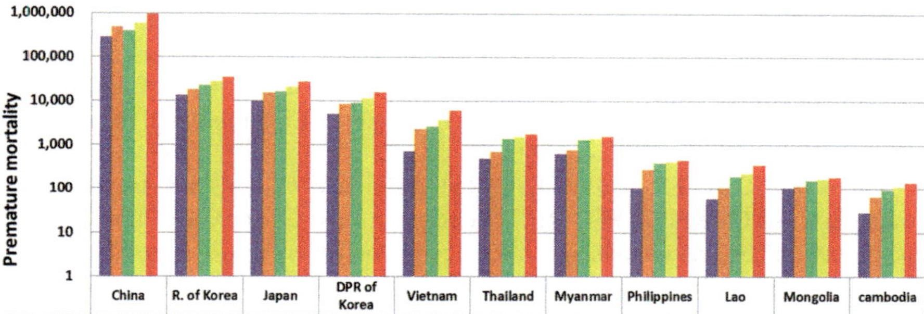

	China	R. of Korea	Japan	DPR of Korea	Vietnam	Thailand	Myanmar	Philippines	Lao	Mongolia	cambodia
2000	281,000	13,300	9,890	5,070	722	480	638	103	59	101	29
2005	467,000	17,800	15,300	8,350	2,290	704	773	271	102	112	65
2020 (PSC)	384,000	22,500	16,300	8,880	2,630	1,380	1,320	389	183	151	94
2020 (REF)	571,000	26,800	20,000	11,400	3,600	1,500	1,380	403	213	162	110
2020 (PFC)	935,000	34,400	26,900	15,500	6,170	1,790	1,550	437	340	180	137

Fig. 19.2 Estimated premature mortality affected by $PM_{2.5}$ and ozone in countries in East Asia in 2000, 2005, and 2020

Cost-Benefit Analysis and Policy Making

In this section, we compared the costs and benefits of reducing premature mortality caused by exposure to surface ozone and particulate matter in East Asia in 2020. The cost of ozone and $PM_{2.5}$ emission reduction is estimated using the Greenhouse Gas and Air Pollution Interactions and Synergies (GAINS)-China model. The benefit of reducing premature mortality caused by exposure to corresponding ozone and $PM_{2.5}$ emission is valued by the value of statistical life (VSL). The costs and benefits are evaluated for two emission reduction policies in 2020 with varying stringency in China.

The Process of the Estimation of Cost and Benefit

To calculate the benefits of saving lives, it is necessary to evaluate two scenarios: a baseline scenario, which describes the development without the implementation of environmental policy improvement, as well as the scenario including policy incentives. Then, the mortality change will show monetary benefits. In our study, we used the three REAS scenarios for China (PFC, REF, and PSC) developed by Ohara et al. (2007). The difference between the three REAS scenarios is the emissions generated in China in 2020 and consequent changes of the amount of pollutants in East Asian countries. Therefore we only valued the emission-control cost in China. The PFC is the baseline scenario, as it was realistic with high emission rates, increased energy consumption, and the slow deployment of new emission-control technologies. The PSC is the scenario with the implementation of advanced environmental policy, including energy efficiency measures and rapid deployment of new energy technologies and new emission-control technologies. REF represented the intermediate scenario between PFC and PSC.

We define Case FS as PFC-PSC (the difference of emission between two scenarios), and Case FR as PFC-REF. We then use Case FS and Case FR to estimate the reduction costs and benefits if the policy to control emissions becomes stricter. Figure 19.3 shows time series of NOx emissions and the relationships between the three scenarios in China. The data for 2000 and 2020 are from Ohara et al. (2007), and the data for 2005 are from Kurokawa et al. (2009) estimated using the same methodology as Ohara et al. (2007).

The Value of Statistical Life (VSL)

The value of statistical life (VSL) is used to estimate premature mortality in economic terms. There are three main methods to value VSL: labor market approach (hedonic wage studies), willingness to pay (WTP) approach, and human capital approach.

As the uncertainty and confidence intervals are quite high, we consider using a range to display the value of VSL as a reasonable choice. We used the data from OECD (2012) and defined the median VSL of environment category (3.0 million int. $, 2005) as the upper value while the median VSL of health category (1.1 million int. $, 2005) as the lower value,[2] with the standard deviation 1.5 and 0.45, respectively. As the VSL values are estimated from 24 countries, we consider the average GDP per capita by purchasing power parity (PPP) adjusted as the basic

[2]It is said on OECD (2012), "The distinction between the environment and health categories is not always obvious. In the classifications made here, the focus has been on whether or not an *explicit* reference to an environmental problem was made in the valuation-question posed to the sample. If that was not the case, the survey was classified as being "health-related". So we believe that both categories are related to our research.

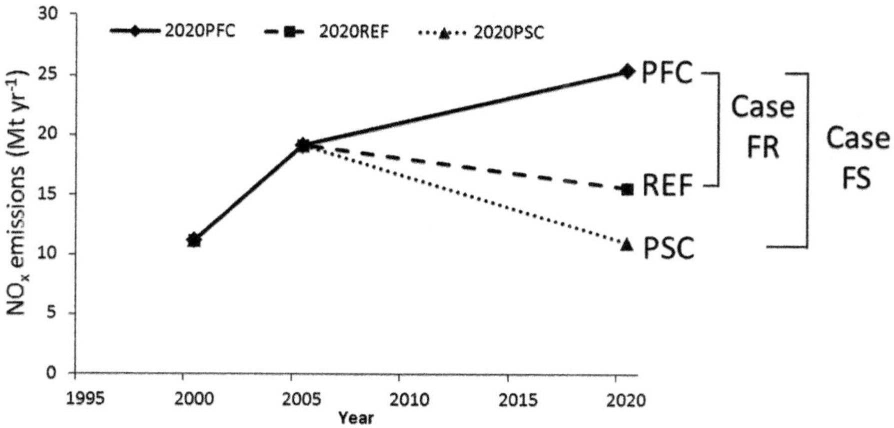

Fig. 19.3 Three scenarios of NO_x emissions in 2020

GDP per capita value. Then, we converted the upper and lower VSL into each East Asian country by GDP per capita of the country on PPP basis [World Development Indicators (WDI), Dec. 2013[3]]. Then, we calculated the economic value of health impacts (i.e., benefit) by the following function:

$$\text{Value of mortality change} = \text{Mortality Change of (FR/FS)} \times \text{VSL}$$

where mortality change is the amount of lives saved by Case FR and Case FS.

GAINS Model and Emission-Control Cost

The GAINS model is an integrated assessment model, which brings together information on future economic, energy, and agricultural development, emission-control potentials and costs, atmospheric dispersion, and environmental sensitivities toward air pollution (Amann et al. 2011). GAINS has been developed and applied for several world regions; here we use GAINS-China model.[4]

As we intended to use REAS as the emission inventory and GAINS-China model for deriving a cost function, the two models should correspond with each other. In order to perform the cost-benefit analysis, we used the applicable REAS emission and the cost curve of GAINS-China as in the benefit estimation. We assumed that the starting points of cost curve of REAS and GAINS-China for the reduction of emission were the points where no reduction technology is applied in each province (see Fig. 19.4). We also assumed that we can use the reduction methods/technology

[3]http://data.worldbank.org/data-catalog/world-development-indicators/wdi-2013

[4]GAINS-china online: http://gains.iiasa.ac.at/gains/EAN/index.login?logout=1

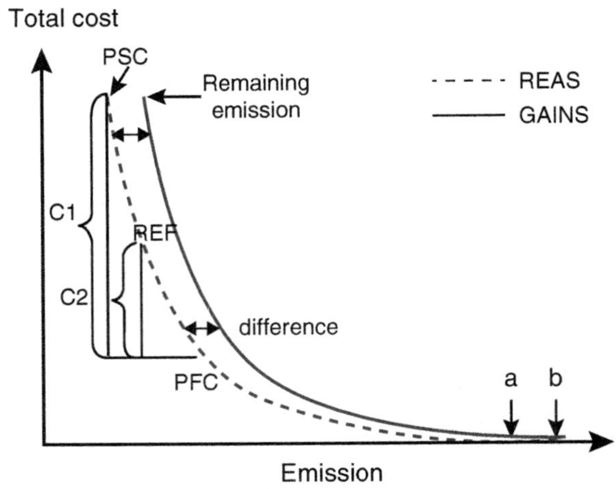

Fig. 19.4 Cost curve in GAINS model. Note: *a* reduction emission without technology in REAS; *b* reduction emission without technology in GAINS-China model; c_1 cost of Case FS, c_2 cost of Case FR

of GAINS-China with their unit cost to make the cost curve of REAS, and then the reduction methods/technologies should be applied one by one according to its marginal cost from the small marginal cost to large one. Because the starting points were different, the cost curves of GAINS-China and REAS were different though the shapes are similar. It means the parallel shift of cost curve in Fig. 19.3. Consequently, we made and used the new cost curve (dotted line in Fig. 19.4) followed by two reasons. Firstly, we calculated the difference ([remaining emission of GAINS-China-PSC]/emission of GAINS-China without reduction technology) of two emissions (REAS and GAINS-China), which was only 7.3%. The error is low enough. Secondly, there is no other cost function for China, thus far, that is available for use.

We choose baseline and current policy scenario[5] of GAINS-China model, and the cost curve covers overall emissions and technologies (energy projections updated by International Energy Agency [IEA] in September 2011; Birol 2011). The data from the GAINS-China website (GAINS-China online[6]) includes relative parameters that span 29 provinces of China (except for Chongqing city and Tibet), for the year 2020, and a discount rate of 10%. The GAINS-China model does not offer the ozone cost data directly. As ozone is mainly generated by NO_x and VOC, we took NO_x, VOC into account for the reduction cost of ozone. The difference between the three REAS scenarios is the emissions generated in China in 2020, so we only valued the emission-control cost in China.

The concentration of $PM_{2.5}$ in the CMAQ model includes primary and secondary particles; however, $PM_{2.5}$ emissions estimated from BC and OC of REAS only include primary particles. In the CMAQ model, atmospheric components of $PM_{2.5}$ include five components: EC, OC, SO_4^{2-}, NO_3^-, and NH_4^+. Thus, we calculated

[5]The scenaro is named “CP_WEO11_S10P50_v2”.

[6]http://gains.iiasa.ac.at/gains/EAN/index.login?logout=1

the ratios for each component in China in 2020, and the corresponding ratios are 2.1%, 7.1%, 62.5%, 9.4%, and 18.9%. Considering our interest in $PM_{2.5}$ and ozone, BC, OC, and NO_3^- are also significant though their ratios are not so high. Accordingly, we use the ratios mentioned above.

We considered the cost of emission reduction of NO_x, VOC, and $PM_{2.5}$. And in our study, we ignore the cost of VOC reduction and consider only the cost of ozone reduction coming from NO_x reduction. The costs of NO_x reduction are 32,800 and 8200 million (int. $, 2005) for Case FS and Case FR, respectively, the corresponding values for $PM_{2.5}$ are 3580 and 523 million (int. $, 2005), and the costs for the reduction of both ozone and $PM_{2.5}$ are 36,400 and 8720 million (int. $, 2005), respectively.

Benefits

Table 19.2 shows the VSL in East Asia adjusted by GDP per capita on PPP basis, and Fig. 19.5 shows the loss of VSL of countries.

Comparison of Cost and Benefit

The comparison between cost and benefit for the reduction of ozone, $PM_{2.5}$, and both of them is shown in Fig. 19.6 for Case FS and Fig. 19.7 for Case FR. The rectangles show the range for benefit, the error bars represent the 95% CI for benefit, and the lines show the cost of emission reduction. In Fig. 19.6 (Case FS), the cost line for ozone is a little lower than the benefit rectangle, indicating that the cost is a bit smaller than the lower benefit, and numerically, the ratio of benefit to cost (benefit/cost) is 1.1–3.0. The cost line of $PM_{2.5}$ is lower than the benefit rectangle in the total, indicating that the cost is lower than the benefit, and the ratio of benefit to cost is 82–220. For total ozone and $PM_{2.5}$, the ratio of benefit to cost is −9.0-25. In Fig. 19.7 (Case FR), the ratios of benefit to cost for the reduction ozone, $PM_{2.5}$, and both of them are 2.7–7.4, 370–1010, and 25–68, respectively. Specifically, in China, the benefits to cost ratios for ozone reduction are 1.0–2.7 and 2.4–6.6 in Case FS and Case FR, respectively. The corresponding ratios for $PM_{2.5}$ are 74–202 and 338–922, and for the reduction of both, they are 8.2–22 and 22–61. If we compare the benefit in Japan with the cost in China, the ratios for ozone, $PM_{2.5}$, and total are 0.07–0.18, 3.2–8.6, and 0.37–1.0 in Case FS and 0.15–0.43, 14–39, and 1.0–2.7 in Case FR. The reduction efficiency of $PM_{2.5}$ is substantially higher than O_3. It is possible that benefit to cost for ozone reduction is not so economically efficient if we consider only the case of ozone. However, when we consider the case of simultaneously reducing ozone and $PM_{2.5}$, the ratio of benefit to cost is quite beneficial (Table 19.3).

Table 19.2 Adjusted VSL in East Asia countries (int. $, 2005)

VSL	China	Japan	R. of Korea	Vietnam	Thailand	Myanmar	Philippines	Lao	Mongolia	Cambodia
Upper value	1,470,000	3,450,000	3,800,000	520,000	1,160,000	172,000	498,000	400,000	273,000	359,000
Lower value	541,000	1,270,000	1,390,000	191,000	426,000	63,200	183,000	147,000	100,000	132,000

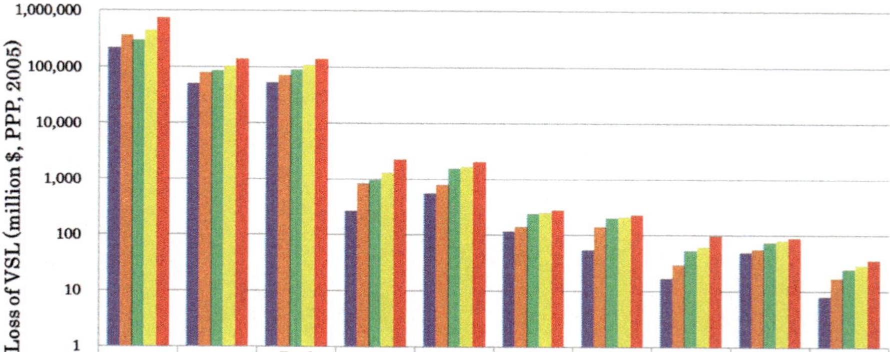

	China	Japan	R. of Korea	Vietnam	Thailand	Myanmar	Philippines	Lao	Mongolia	Cambodia
2000	221,000	52,300	54,100	271	557	118	55	17	50	8
2005	368,000	80,800	72,700	861	817	143	144	30	56	17
2020(PSC)	303,000	86,200	92,000	986	1,600	245	206	54	75	25
2020(REF)	450,000	106,000	110,000	1,350	1,750	256	214	62	81	29
2020(PFC)	737,000	142,000	140,000	2,320	2,080	285	232	100	90	36

Fig. 19.5 Loss of VSL of premature mortality by exposure of $PM_{2.5}$ and ozone in countries in East Asia in 2000, 2005, and 2020

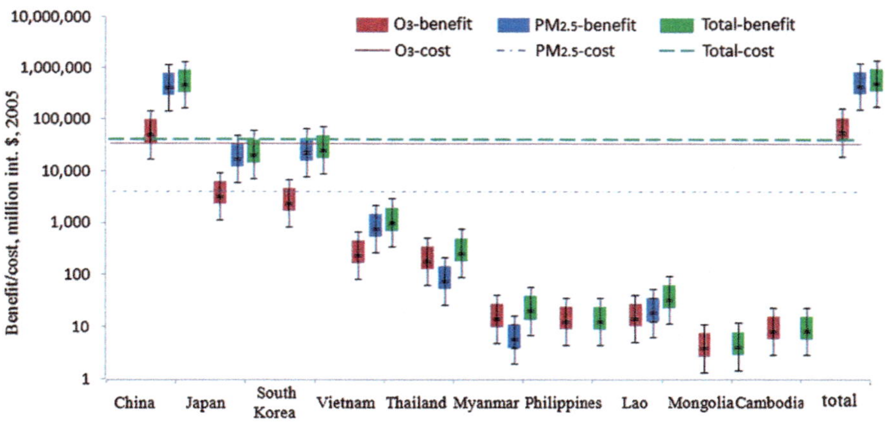

Fig. 19.6 Cost-benefit comparisons for Case FS in 2020

Climate Change and Transboundary Air Pollution in East Asia

As described in the introduction section, SLCPs are important contributors to the climate change, in addition to direct human threats (Smith et al. 2009). This implies that reducing emission of SLCPs would greatly contribute both to human health and mitigation. For this reason, East Asia is a very important region as shown below. In this section, some evidence about the local and transboundary air pollution in East Asia will be addressed.

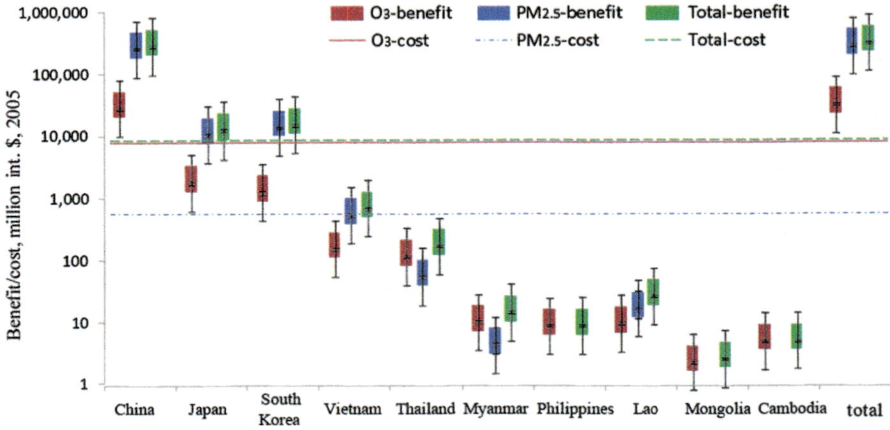

Fig. 19.7 Cost-benefit comparisons for Case FR in 2020

Table 19.3 Cost (in China) and benefit (in region) for reduction of $PM_{2.5}$ and ozone in 2020 (cases of FS and FR)

Case	Benefit		Cost		Benefit/Cost	
	FS	FR	FS	FR	FS	FR
Ozone	36,600–99,700	22,200–60,700	32,800	8200	1.1–3.0	2.7–7.4
$PM_{2.5}$	292,000–797,000	194,000–530,000	3580	523	82–220	370–1010
Total	329,000–897,000	216,000–591,000	36,400	8720	9.0–25	25–68

Unit: million int. $, 2005

In the 1970s, Japan suffered from severe air pollution from the heavy industrial area. Thanks to the legal actions taken and industrial transition, by which number of the polluting factories has become less in Japan and more in China or in other developing countries, Japan had been enjoying less polluted air. In recent years, however, this industrial transformation has created the present problems, i.e., heavy air pollution in many of the Chinese cities and transboundary air pollution from China to Korea and Japan.

Local Pollution

According to Kurokawa et al. (2013) (see Table 19.4), emission of PM_{10} and $PM_{2.5}$ in China were more than half of the whole Asia in 2008; the emission in Japan and Korea were less than 1/100 compared with China. In contrast, emission of CO_2, which can be regarded as an index of energy consumption, showed different relation; China occupied more than half of Asia, but only less than ten times of that in Japan and less than 20 times of that in Korea. These results suggest that developed countries' PM emission is much cleaner even when difference in energy

Table 19.4 Summary of national emissions in 2008

Country	SO_2	NO_x	PM_{10}	$PM_{2.5}$	CO_2
China	33,457	26,969	21,606	14,514	8814
Japan	761	2207	130	94	1192
Korea	417	1059	110	56	532
Asia	56,913	53,875	36,397	24,729	16,036

Extracted from Kurokawa et al. (2013)

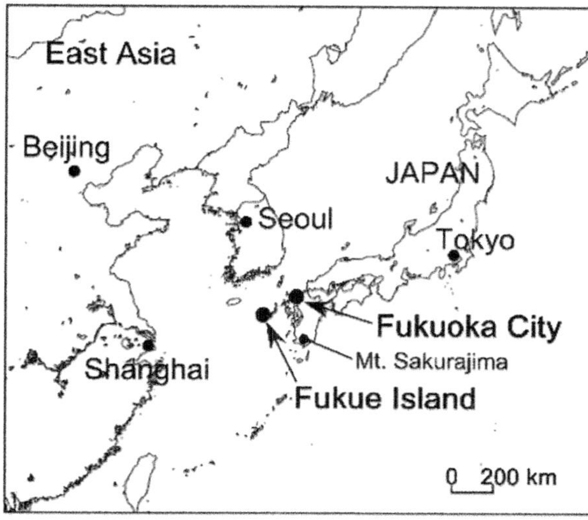

Fig. 19.8 Location of Seoul, Fukue Island, and Fukuoka City (Copied from Yoshino et al. 2016)

consumption is considered. In this regard, health and mitigation co-benefit through reduction of SLCPs is larger in developing countries.

Transboundary Air Pollution

Yoshino et al. (2016) reported the situation in two locations in southern Japan, i.e., Fukue Island, a rural island where the traffic is light and there is no local fixed source of air pollution, and Fukuoka City, a metropolitan city with the population of 1,5+ million in 2016 (Fig. 19.8). In Fukue Island, the chemical composition of $PM_{2.5}$ was dominated by sulfate and low-volatile oxygenated organic aerosols dominant for all of the $PM_{2.5}$ mass variations. In Fukuoka, sulfate was dominant when the $PM_{2.5}$ concentration was high, whereas organics and nitrate occupied a large fraction when the $PM_{2.5}$ concentration was low. Thus, they concluded that high $PM_{2.5}$ mass concentrations were attributed to the long-range transport of air pollution. They also reported that long-range transboundary air pollution was influential not only in winter-spring season but also in summer. Also in Korea, aerosols from China played a major role in the occurrence of severe air pollution episodes for 4+ days in cold seasons 2001–2013 in Seoul, Korea; the concentration of PM_{10} sometimes exceeded 100 μg/m^3 (Oha et al. 2015).

Since the global warming was first identified, the north-south problem was one of the toughest challenges we have had; developing countries which emitted little greenhouse gases suffer most, and developed countries which emitted a lot of greenhouse gases suffer less. As described above, however, massive emission of pollutants in China has been causing high PM concentration days not only domestically but also in neighboring countries such as Korea and Japan. This situation urges both developed and developing countries to solve this problem; developed countries should provide new technologies to reduce simultaneously greenhouse gas and pollutant emissions, especially SLCPs.

Summary and Conclusion

In this chapter, we showed an integrated approach to compare the reduction costs for ozone and $PM_{2.5}$ with the corresponding benefits of reducing the premature mortality rate due to exposure to ozone and $PM_{2.5}$ which are categorized as SLCPs. Assessing premature mortality risks caused by exposure to elevated concentrations of $PM_{2.5}$ and ozone in East Asia involves many uncertainties with regard to emission inventories, modeling systems, population distribution, and age-specific mortality rates. It is also recognized that statistically significant Asian epidemiological studies for ozone and $PM_{2.5}$ effects on human health are necessary. Based on the estimation of premature mortality, we see that the cost efficiency to reduce $PM_{2.5}$ is considerably higher than ozone, and reduction of both ozone and $PM_{2.5}$ is quite beneficial in either case of emission reduction policy. Also we introduced the relationship between SLCPs and transboundary air pollutants in East Asia.

Therefore, taking the crucial and complex aspects of the atmospheric environment into account, it is certainly pointed out that SLCPs are key factors, in terms of co-benefit/co-control, to approach the air pollution and climate change problems not only in East Asia but in the world. The further study should be enhanced to elucidate the uncertainty points mentioned above to provide the exact and persuadable scientific evidences to policy makers and related initiatives so that they can choose the most effective and efficient policy immediately.

References

Akimoto H (2003) Global air quality and pollution. Science 302:1716–1719

Amann M, Bertok I, Borken-Kleefeld J, Cofala J, Heyes C, Höglund-Isaksson L et al (2011) Cost-effective control of air quality and greenhouse gases in Europe: modeling and policy applications. Environ Model Softw 26(12):1489–1501

Bergin SM, West JJ, Keating JT, Russell GA (2005) Regional atmospheric pollution and transboundary air quality management. Annu Rev Environ Resour 30:1–37

Birol F (2011) World energy outlook 2011. International Energy Agency (IEA)

Chen F, Yamashita K, Kurokawa J, Klimont Z (2015) Cost-benefit analysis of reducing premature mortality caused by exposure to ozone and PM2.5 in East Asia in 2020. Water, Air and Soil Pollution

CIESIN (2005) Gridded population of the world, version 3 (GPWv3). Center for International Earth Science Information Network (CIESIN), Columbia University

Cohen A, Anderson H, Ostra B et al (2005) The global burden of disease due to outdoor air pollution. J Toxic Environ Health A 68:1–7

Krupnick A, Alberini A, Cropper M, Simon N, O'Brien B, Goeree R, Heintzelman M (2002) Age, health and the willingness to pay for mortality risk reductions: a contingent valuation survey of Ontario residents. J Risk Uncertain 24(2):161–186

Kurokawa J, Ohara T, Uno I, Hayasaki M, Tanimoto H (2009) Influence of meteorological variability on interannual variations of springtime boundary layer ozone over Japan during 1981–2005. Atmos Chem Phys 9(17):6287–6304

Kurokawa J, Ohara T, Morikawa T, Hanayama S, Janssens-Maenhout G, Fukui T, Kawashima K, Akimoto H (2013) Emissions of air pollutants and greenhouse gases over Asian regions during 2000–2008: regional emission inventory in Asia (REAS) version 2. Atmos Chem Phys 13 (21):11019–11058

Lim SS, Vos T, Flaxman AD, Danaei G, Shibuya K, Adair-Rohani H et al (2013) A comparative risk assessment of burden of disease and injury attributable to 67 risk factors and risk factor clusters in 21 regions, 1990–2010: a systematic analysis for the Global Burden of Disease Study 2010. Lancet 380(9859):2224–2260

Nawahda A, Yamashita K, Ohara T, Kurokawa J, Yamaji K (2012) Evaluation of premature mortality caused by exposure to $PM_{2.5}$ and ozone in East Asia: 2000, 2005, 2020. Water Air Soil Pollut 223(6):3445–3459

Nawahda A, Yamashita K, Ohara T, Kurokawa J, Ohizumi T, Chen F, Akimoto H (2013) Premature mortality in Japan due to ozone. Atmos Environ 81:538–545

OECD (2012) Mortality risk valuation in environment, health and transport policies. OECD Publishing. http://dx.doi.org/10.1787/9789264130807-en

Oha H, Hoa C, Kimb J, Chenc D, Leea S, Choid Y, Change L, Songe C (2015) Long-range transport of air pollutants originating in China: a possible major cause of multi-day high-PM10 episodes during cold season in Seoul, Korea. Atmos Environ 109:23–30

Ohara T, Akimoto H, Kurokawa JI, Horii N, Yamaji K, Yan X, Hayasaka T (2007) An Asian emission inventory of anthropogenic emission sources for the period 1980–2020. Atmos Chem Phys 7(16):4419–4444

Pope CA III, Burnett RT, Thun MJ et al (2002) Lung cancer, cardiopulmonary mortality, and long-term exposure to fine particulate air pollution. J Am Med Assoc 287(9):1132–1141

Schroeder H, Yocum D (2006) European institutions for controlling chemical air pollution: an analysis of CLRTAP-European Union interplay. In: Simeonov L, Chiria E (eds) Chemicals as intentional and accidental global environmental threats. Springer, p 321–336

Smith KR, Jerrett M, Anderson HR, Burnett RT, Stone V, Derwent R, Atkinson RW, Cohen A, Shonkoff SB, Krewski D, Pope CA 3rd, Thun MJ, Thurston G (2009) Public health benefits of strategies to reduce greenhouse-gas emissions: health implications of short-lived greenhouse pollutants. Lancet. Dec 19 374(9707):2091–2103. doi:10.1016/S0140-6736(09)61716-5

Task Force on Research Coordination (TFRC), Scientific Advisory Committee (SAC), Acid Deposition Monitoring Network in East Asia (EANET) (2015) Review on the state of air pollution in East Asia. EANET

United Nations Department of Economic and Social Affairs/Population Division (2008) World population prospects: the 2008 revision, http://esa.un.org/unpp/

United Nations Environment Programme (UNEP) (2011) Near-term climate protection and clean air benefits: actions for controlling short-lived climate forcers. UNEP

Uno I, Ohara T, Sugata S et al (2005) Development of the RAMS/CMAQ Asian scale chemical transport modeling system. J Jpn Soc Atmos Environ 40(4):148–164

US-EPA (2006) Ozone health risk assessment for selected urban areas. http://www.epa.gov/ttnnaaqs/standards/ozone/data/ozone_ra_2nd_draft_tsd_mainbody_7-18-2006.pdf

WHO (2006) Mortality database. World Health Organization. http://apps.who.int/whosis/database/mort/table1.cfm

WHO (2010a) Demographic tables. World Health Organization. http://www.wpro.who.int/

WHO (2010b) Mortality database. World Health Organization. http://www.who.int/

Yamaji K, Ohara T, Uno I, Tanimoto H, Kurokawa JI, Akimoto H (2006) Analysis of the seasonal variation of ozone in the boundary layer in East Asia using the Community multi-scale air quality model: what controls surface ozone levels over Japan? Atmos Environ 40(10):1856–1868

Yamaji K, Ohara T, Uno I, Kurokawa JI, Pochanart P, Akimoto H (2008) Future prediction of surface ozone over east Asia using models-3 Community multiscale air quality modeling system and regional emission inventory in Asia. J Geophys Res 113:D08306

Yoshino A, Takami A, Sato K, Shimizu A, Kaneyasu N, Hatakeyama S, Hara K, Hayashi M (2016) Influence of trans-boundary air pollution on the urban atmosphere in Fukuoka, Japan. Atmosphere 7(4):51. doi:10.3390/atmos7040051

Ken Yamashita PhD, head of Data Management Department, Asia Center for Air Pollution Research (ACAP), has more than 34 years of experience in research, administration and management and has coordinating experiences in the field of environmental sciences and policies with special emphasis on the risk assessment of atmospheric environment in East Asia.

Yasushi Honda MD, PhD, professor at the Faculty of Health and Sport Sciences, University of Tsukuba, has been working on climate change health impact, especially on heat-related mortality impact, and was a lead author of the Fourth (Asia) and Fifth (human health) Intergovernmental Panel on Climate Change Assessment Report and a convening lead author of SREX.

Chapter 20
Climate Change, Air Pollution and Health in South Africa

Eugene Cairncross, Aqiel Dalvie, Rico Euripidou, James Irlam, and Rajen Nithiseelan Naidoo

Abstract Climate change and air pollution pose significant short-term and long-term health risks to South Africans due to the carbon intensity of the national economy, the severe air pollution around coal mining and coal-fired power stations in many widespread populated areas and the particular vulnerability of many subgroups in a country burdened by extreme inequality and a severe quadruple epidemic of acute and chronic disease.

There are limited local studies on the respiratory, cardiovascular and other health risks of air pollution. Inadequate disease surveillance and air quality data pose a challenge for monitoring and research.

A number of interventions to mitigate or adapt to climate change with important co-benefits for air quality and public health are described for the following economic sectors: energy, industry, human settlements, transport, healthcare and business sector.

There is good policy commitment to address climate change and air pollution, but implementation needs to be drastically improved.

Keywords Air pollution • Climate change • Mitigation • Public health • Energy • Coal

E. Cairncross • J. Irlam (✉)
Primary Health Care Directorate, Faculty of Health Sciences, University of Cape Town, Observatory, Capetown 7925, South Africa
e-mail: cairncrosse@gmail.com; James.Irlam@uct.ac.za

A. Dalvie
School of Public Health & Family Medicine, Faculty of Health Sciences, University of Cape Town, Observatory, Capetown 7925, South Africa
e-mail: Aqiel.Dalvie@uct.ac.za

R. Euripidou
groundWork, Friends of the Earth South Africa, Pietermaritzburg, South Africa
e-mail: rico@groundwork.org.za

R.N. Naidoo
Discipline of Occupational and Environmental Health, University of KwaZulu-Natal, Durban, South Africa
e-mail: naidoon@ukzn.ac.za

© Springer International Publishing AG 2018

R. Akhtar, C. Palagiano (eds.), *Climate Change and Air Pollution*, Springer Climate, DOI 10.1007/978-3-319-61346-8_20

Key Points

- Climate change increases current exposures and health risks due to air pollution in South Africa.
- Interventions to mitigate or adapt to climate change can have important co-benefits to air quality and public health.
- South Africa needs to act with urgency and determination to mitigate and avoid further serious public health impacts from climate change and air pollution.

The South African Context

Carbon Intensity of the National Economy

South Africa has one of the most carbon intensive economies of middle-income countries in the world. In 2013, it emitted 0.71 kg CO_2/(2005US$GDP (PPP)),[1] ranking it within the top ten CO_2-emitting countries (International Energy Agency 2015). This is due to South Africa's dependence on coal-fired power (CFP) stations; the production of about 20% of its liquid transportation fuels using Sasol's energy-intensive coal-to-liquid (CTL) process (South African Petroleum Industries Association 2014); its heavy use of fossil fuels for energy-inefficient road freight and private commuter transport; the widespread domestic use of paraffin (kerosene), especially in low-income households; and many energy-intensive industries, such as mining and metal production.

In 2014/2015, the national power utility Eskom generated a calculated 94% of total electricity from 15 CFP stations, burning 119.2 million tons (Mt) of coal and emitting 223.4 Mt of CO_2, 1834 Mt of SO_2, 0.937 Mt of NO_x (NO plus NO_2) and 82,000 t of PM_{10} in the process (Eskom 2015). These CFP stations and their associated coal mining operations are major contributors to emissions in highly polluted priority areas.[2]

The Sasol Synfuels CTL plant, located in the HPA, is permitted to process about 35 Mt of coal per year at full production rates, equivalent to about 150,000 barrels per day of crude oil, to produce about 7 billion litres[3] of liquid transportation fuels, petrol and diesel (Synfuels 2014). The plant emits 48 Mt of CO_2-eq, 210,000 t of SO_2, 150,000 t of NO_x and 12,000 t of PM_{10} per year (Burger et al. 2014).

Road transport consumed 13.5 billion litres of diesel and 11.5 billion litres of petrol in 2015 (Dept. of Energy 2015), emitting a total of 63 Mt of CO_2.[4] Under-investment in rail compared with road infrastructure in recent years has

---

[1]GDP (PPP): Gross domestic product based on purchasing power parity.

[2]An area may be declared a priority area if the "Minister . . .. reasonably believes that... ambient air quality standards are being, or may be, exceeded in the area .." [National Environmental Management: Air Quality Act 39 of 2004, Chapter 4].

[3]Actual consumption and production data are not publicly available.

[4]Authors' estimate, using EPA emission factors for diesel and petrol from https://www.epa.gov/sites/production/files/2015–11/documents/emission-factors_nov_2015.pdf

compounded the problems of a poor commuter rail system countrywide, which drives the rapid increase in private commuter and road freight transport.

These three source categories (CFP stations, the Sasol CTL plant and road transport) together emitted about 334 Mt of CO_2 in 2013–2014 or 84% of South Africa's total annual CO_2 emissions of 397 Mt in 2014 (1.1% of global CO_2 emissions of 35,700 Mt).[5] In addition, household energy demand is relatively high due to the low penetration of solar water heaters (SWH), the low uptake of energy-efficient appliances (de la Rue du Can Stephane et al. 2013) and many extremely energy-inefficient mass housing developments.

South Africa launched the Renewable Energy Independent Power Producers Program (REIPPP) in 2011. Between 2012 and December 2015, 6300 MW of renewable energy, mainly wind and solar photovoltaic (PV), was procured under this programme. By December 2015 however, only 3920 MW of bids had achieved financial closure, which require Eskom's sign-off on power purchase agreements and grid connections (Dept. of Energy 2016). By mid-2016, only 2040 MW (supplying about 2.2% of system load) had been connected to the grid, with only 120 MW of capacity added during the first half of 2016. The remainder of the 6300 MW procured is scheduled for completion by 2018 (DoE 2016) but may be in jeopardy because Eskom appears to be delaying sign-off.[6] Bids for a further 1800 MW of RE were submitted in November 2015, but as of September 2016, the preferred bidders had not been announced (DoE 2016). Expansion of the REIPPP therefore appears to have stalled (Bofinger et al. 2016).

Despite excellent national RE potential, Eskom maintains a commitment to coal-based power, as exemplified by its building of two new CFP stations, Medupi and Kusile, scheduled for completion in 2021–2022 with a combined capacity of 9600 MW. South Africa has also committed to a further 2500 MW of coal-based power under its IPP (independent power producer) programme. These new coal plants will emit a further 85 Mt CO_2 per year when fully commissioned, which is an increase of 21% on current levels.

South Africa's Carbon Trajectory

South Africa's carbon trajectory is defined in the Climate Change Response White Paper (Department of Environmental Affairs 2011). Greenhouse gas (GHG) emissions are essentially allowed to peak between 2020 and 2025 in a range between 398 and 614 Mt CO_2-eq, to "plateau" in this range for up to 10 years and then to decline by 2050 to between 212 and 428 Mt CO_2-eq. South Africa's Intended

[5]EDGAR database: http://edgar.jrc.ec.europa.eu/overview.php?v=CO2ts1990-2014

[6]Fin24. Eskom review endangers biggest Africa renewable power plan. Jul 28 2016. http://www.fin24.com/Economy/Eskom/eskom-review-endangers-biggest-africa-renewable-power-plan-20160728

Nationally Determined Contributions (INDC) do not go beyond these inadequate climate change mitigation targets. The First Report (DEA 2016, Figure 8), which purports to show a substantial reduction in GHG emissions in the period through to 2012 relative to an implied "without measures" baseline, shows that absolute GHG emissions have continued to *increase* through to 2012, closely tracking the upper limit of the trajectory (DEA 2016, Figure 2). This implies that current annual emissions of about 583 Mt CO_2-eq will continue to rise to 614 Mt by 2025 and will remain there for a further 10 years before declining slowly to about 428 Mt per year by 2050.

Exposures to Air Pollution

CFP stations, coal mine operations and heavy industries using fossil fuels are not only major sources of greenhouse gas emissions (GHGs) in South Africa but also of the common air pollutants SO_2, NO_x, PM_{10} and the fine-fraction $PM_{2.5}$. In the three priority areas (Highveld (HPA), Vaal Triangle (VTPA) and Waterberg-Bojanala (W-BPA)), for example, coal combustion and mining activities are responsible for 96% of SO_2, 88% of NO_x and 78% of PM_{10} (Dept. of Environmental Affairs 2012a, b, 2013).

The concentrations of ambient $PM_{2.5}$, which consists of both directly emitted $PM_{2.5}$ and secondary $PM_{2.5}$ from the precursor gases SO_2 and NO_X, are high in each of the PAs. The annual average $PM_{2.5}$ concentrations for 2012–2015 are shown in Fig. 20.1 for towns in the HPA and in Fig. 20.2 for towns in the VTPA, most of them well in excess of the current South African National Ambient Air Quality Standard (SA NAAQS) of 20 $\mu g/m^3$ and the World Health Organization (WHO) guideline of 10 $\mu g/m^3$. In the adjacent densely populated metros of Tshwane and Johannesburg/Ekurhuleni, the annual average $PM_{2.5}$ concentrations of 39 $\mu g/m^3$ and 50 $\mu g/m^3$, respectively, in 2012 are also well in excess of air quality standards (Cairncross 2016).

In 2014, Eskom argued that it was unable to comply with air quality standards and that the health impacts of its emissions should be given less weight than the costs to comply with these standards. Eskom's non-compliance with numerous legislative requirements for its CFP stations[7] exacerbates these impacts and violates constitutional rights to an environment not harmful to health and well-being. Nevertheless both Eskom and Sasol were granted 5-year postponements from 2020 to 2025 to comply with more stringent emission standards.

[7]Department of Environmental Affairs' annual National Environmental Compliance and Enforcement Reports https://www.environment.gov.za/otherdocuments/reports

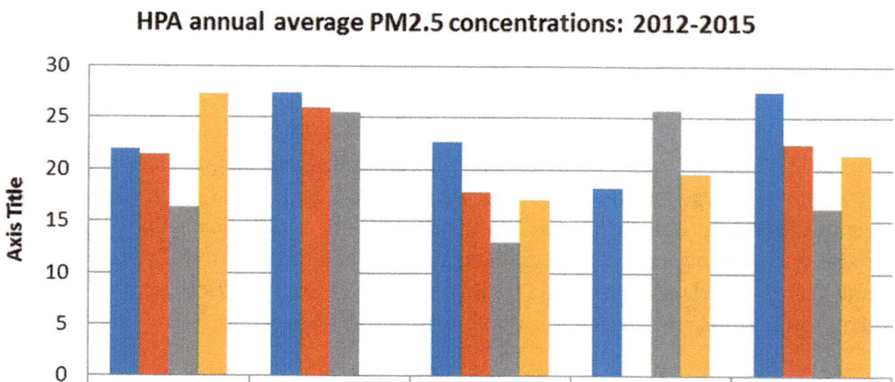

	Witbank	Secunda	Middleburg	Hendrina	Ermelo
2012	22	27	23	18	28
2013	21	26	18		22
2014	16	26	13	26	16
2015	27		17	20	21

Fig. 20.1 Annual average PM2.5 concentrations in the Highveld Priority Area 2012–2015

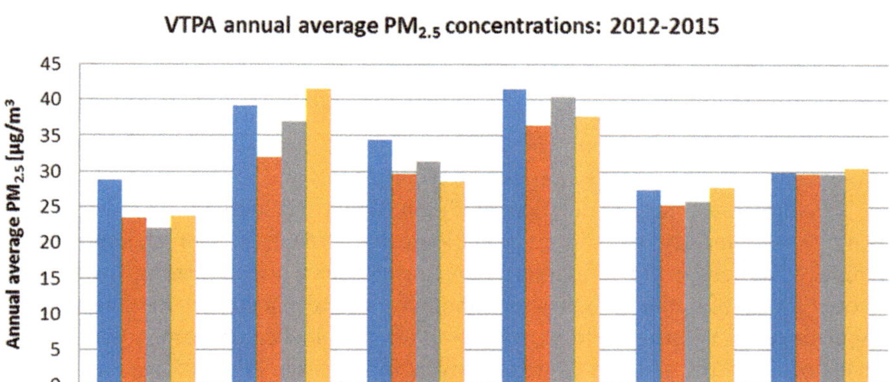

	Diepkloof	Klipriver	Sebokeng	Sharpville	Three Rivers	Zamdela
2012	29	39	34	41	27	30
2013	23	32	30	36	25	30
2014	22	37	31	40	26	30
2015	24	42	28	38	28	30

SA NAAQS ——— WHO guideline - - - - -

Fig. 20.2 Annual average PM2.5 concentrations in the Vaal Triangle Priority Area 2012–2015

Health Risks Due to Air Pollution in South Africa

South Africa carries a quadruple burden of disease, characterised by high morbidity and mortality from four broad groups of causes: injuries, non-communicable disease, HIV/AIDS and tuberculosis and communicable diseases, perinatal conditions, maternal causes and nutritional deficiencies. Significant subnational variations reflect deep and persistent health inequalities and indicate that population groups and provinces are at different stages of the health transition (Pillay-van Wyk et al. 2016). The health risks due to air pollution in South Africa are not insignificant however, as illustrated by a number of studies below.

Respiratory Conditions

The scientific literature over the past few decades provides substantial evidence for the association of air pollution with various respiratory outcomes, especially among children. These include the presentation of asthma symptoms (Mann et al. 2010; Zora et al. 2013), lung function impacts (Weinmayr et al. 2010) and visits to emergency departments (Nastos et al. 2010). Implicated pollutants include particulate matter, ozone, sulphur dioxide and oxides of nitrogen (Graveland et al. 2011; McConnell et al. 2010; Pan et al. 2010; Strickland et al. 2010). Proxy markers of air pollution, such as vehicle traffic (Jung et al. 2015) and industry (Rovira et al. 2014), have also been documented. Those at greatest risk include children, those with pre-existing respiratory diseases and the elderly.

The literature among populations in South Africa and southern Africa is limited and focused mostly on children. Ecological approaches have generally been employed as crude proxy markers of exposure, such as comparing towns with known levels of ambient exposure, although more recent studies have used more sophisticated exposure metrics. Studies have mostly been in areas with higher levels of industrial pollution, such as towns in the Vaal Triangle area, and in the south Durban and Cape Town metropoles.

Despite high levels of ambient and indoor pollution in low-income communities, early comparisons of exposed and less exposed communities provided limited evidence of pollutant-related respiratory outcomes among children. In a 1986 study of about 1000 schoolchildren from Sasolburg, site of the CTL plant, for example, little difference in reported symptoms was found compared to nearby less exposed communities, although there were small differences in lung function parameters (Coetzee et al. 1986). A more extensive study of about 10,000 schoolchildren within the Vaal Triangle area reported that 8–12-year-olds in communities without electricity had a 65% increased prevalence of upper airway symptoms and a 29% higher prevalence of lower respiratory tract illnesses. The risk of asthma was almost twofold higher among children from Vaal Triangle communities than among those from a less polluted town (Terblanche et al. 1992).

Similar ecologic studies have been done in the Highveld area of the north-eastern province of Mpumalanga, which is also home to several coal mines and CFP stations. A study of about 1000 children from higher exposed communities found an increased risk of respiratory symptoms (cough, wheeze and asthma) than among children from less exposed areas, although there was no difference in lung function measures (Zwi et al. 1990). A more recent study in the Highveld, using questionnaire-based exposure and outcome data, found a significantly increased risk of wheeze among schoolchildren due to environmental tobacco smoke, use of gas for indoor heating and the proximity to schools of heavy trucks (Shirinde et al. 2014).

Given the high burden of childhood infectious diseases in southern Africa, quantifying the additional risk from pollution is important for public health. Survival analysis of under-five mortality data from the World Health Survey of 2003, pooled for 16 African countries, showed a significant impact of indoor biomass fuels on acute lower respiratory tract infection mortality (adjusted HR = 2.35 (95% CI 1.22–4.52)) (Rehfuess et al. 2009). This finding has been replicated in studies in the Highveld area where the prevalence of respiratory symptoms among schoolchildren was substantially higher among those exposed to indoor biomass fuels (Albers et al. 2015). In other studies in South Africa, child tuberculosis has presented with an increased risk among those exposed to environmental tobacco smoke (du Preez et al. 2011) and to indoor air pollution (Jafta et al. 2015).

More recent respiratory health studies have developed more direct and sophisticated measures of air pollution exposure and have focused on short-term outcomes such as acute respiratory symptoms and measures of lung function. A study using a repeated measures panel design of young schoolchildren in the city of Durban, for example (Naidoo et al. 2013), enabled the direct associations between specific pollutants and short-term outcomes to be assessed. A previous study found high prevalence of asthma among children at primary schools in the industrially intense areas of south Durban, with short-term levels of PM_{10}, nitrogen dioxide (NO_2) and SO_2 significantly associated with increased respiratory symptoms and decrements of pulmonary function among asthmatic children (Kistnasamy et al. 2008). Another study in south Durban had found that acute respiratory outcomes, such as cough and wheeze, as well as daily lung function measures, were directly associated with short-term fluctuations in pollutants, particularly oxides of nitrogen and particulate matter (Naidoo et al. 2007). Naidoo et al. (2013) compared children in the southern areas and less industrialised northern areas of Durban and found a greater risk in the south of doctor-diagnosed asthma, persistent asthma, and airway hyper-reactivity. There was also a twofold increased risk for airway hyper-reactivity with SO_2 exposure (Naidoo et al. 2013).

More complex epidemiological studies in South Africa are likely to show more robust exposure-outcome relationships. Two birth cohort studies with a focus on environmental pollution and respiratory outcomes are underway in the city of Durban and in various settings in the Western Cape Province. The latter study has characterised indoor air pollution among the disadvantaged communities under study, with substantial use of biomass fuels resulting in high levels of benzene, carbon monoxide and oxides of nitrogen (Vanker et al. 2015). These birth cohort studies have used well-developed metrics of exposure and outcomes at various stages of childhood development, such as neonatal respiratory histories, increased frequency of infant respiratory infections, infant wheeze and lung function measures up to early childhood.

A particular concern in South Africa is exposure to asbestos, silica and other minerals from large mine dumps in areas generally remote from industrial centres. These dumps have been associated with both acute and chronic respiratory outcomes, including chronic obstructive lung disease, pneumoconiosis and cancers of the respiratory tract. The risk of living in close proximity to a mine dump has been associated with increased risks of asthma, chronic bronchitis, pneumonia and emphysema, as well as symptoms of chronic cough and wheeze among those above the age of 55 (Nkosi et al. 2015). The cancer-related risk in the study by Nkosi et al. (2015) is of interest, as cancer and pollution studies in southern Africa are almost non-existent. This is largely because of the absence of national population-based cancer registries and the lack of appropriate measures of cumulative pollution exposure within communities. Mzileni et al. (1999) showed an increased risk of lung cancer in men working in dusty environments and in men and women residing in asbestos-mining communities (Mzileni et al. 1999).

Faced with the challenges of conducting large epidemiological studies with robust measures of exposure, designs that incorporate burden of disease analyses or mortality databases become important approaches to understanding the relationships between pollution and respiratory health. The South African census data has therefore been used to determine the burden of respiratory disease in children and adults due to indoor and ambient pollution. Approximately 2500 excess deaths, or 0.5% of all deaths, and 60,934 disability-adjusted life years (DALYs) have been associated with exposure to indoor solid fuel use in South Africa (Norman et al. 2007a).

Indoor air pollution is now clearly understood to have the greatest burden of non-communicable disease globally causing approximately 3.5–4 million deaths per year (Gordon and et al. 2014).

Heart Disease

Cardiovascular diseases (CVD), such as heart valve problems, stroke, arrhythmia and "heart attacks", currently cause about a third of deaths globally and are likely increasing in both developing and developed countries (Deaton et al. 2011). The short- and long-term effects of air pollutants on CVD have been little studied in developing countries however.

The most recent burden of disease analysis of the South African census estimated that 3.7% of all cardiopulmonary deaths in South Africa are due to ambient air pollution (Norman et al. 2007b). The only completed South African epidemiological study used a case-crossover design for almost 150,000 cardiovascular and respiratory deaths in Cape Town, together with pollution data from the city's monitoring stations for 2001–2006 (Wichmann and Voyi 2012). It found a significant 3% increase in CVD mortality per interquartile increase in NO_2, PM_{10} and SO_2 (8 μg/m^3) in the warm season but found no effects in the winter season. Limitations of this study were the use of ecological measures of exposure (air pollution levels at one monitoring station representing levels in an area) and outcome (mortality in an area).

A cohort study of 600 adults in four informal settlements in the Western Cape is currently underway. CVD outcomes will be measured by questionnaire, and air pollution will be estimated by land-use regression modelling to produce spatially distributed air pollution levels (Dalvie et al. 2014).

Other Health Conditions

Cancers of the respiratory tract caused by air pollution accounted for 5.1% of all respiratory cancers and 1.1% of acute respiratory tract infection-related deaths in children under 5 years. These accounted for 0.9% of all deaths and 0.4% of all years of life lost (YLL) in South Africa (Norman et al. 2007b).

There is increasing evidence from epidemiologic and animal studies that air pollution might cause central nervous system (CNS) effects such as chronic brain inflammation, white matter abnormalities leading to increased risk for autism, lower IQ in children, behaviour problems and neurodegenerative diseases such as Parkinson's disease and Alzheimer's disease (Block et al. 2012). The mechanism of CNS effects is not well understood however; air pollutants either have direct effects on the CNS or else indirect effects via the cardiovascular system. Many air pollutants are associated with adverse effects on the CNS, including particulate matter, polycyclic aromatic hydrocarbons (PAHs), black carbon, heavy metals, volatile organic compounds (VOCs), environmental tobacco smoke (ETS), ozone and carbon monoxide (CO). The mechanisms by which outdoor pollutants could impact brain function include the indirect effects of peripheral inflammation, changes in the blood-brain barrier and direct neuronal and white matter injury. Neurotoxicity is likely to arise during periods of highest vulnerability (in utero, childhood and old age) and from lifetime exposure. Epidemiological studies have provided evidence that living in conditions with elevated air pollution is linked to decreased cognitive function, lower neurobehavioural testing scores in children, a decline in neuropsychological development in the first 4 years of life and neuropathology (Block et al. 2012). Genetic factors and epigenetic influences may modify CNS effects due to air pollution. Further research is required to establish CNS effects due to air pollution, as quantitative data from South Africa are lacking. A cohort study is being conducted on neurobehavioural effects due to pesticide drift among schoolchildren in the rural Western Cape (Baseera et al. 2016).

Current and Future Climate Hazards in South Africa

Global climate change is predicted to further the trends of marked temperature rise in South Africa, alongside increased rainfall variability, sea level rise and more extreme weather events (South Africa INDC 2015). Under a high-emission scenario (RCP8.5),[8] mean annual temperature is projected to rise by about 5.1 °C on average from 1990 to 2100 (and by 1.4 °C under the low scenario of RCP2.6) and the annual average of "heatwaves" (at least seven consecutive days with maximum temperatures above the 90th percentile threshold for that time of the year) from under 5 days to an average 145 days during the same period (or 25 days under RCP2.6). The longest dry spell is projected to increase by about 30 days to approximately 110 days in 2100 (or by less than 10 days under RCP2.6), with continuing large year-to-year variability (World Health Organisation 2015). The Mediterranean-type climatic region (the south-western Cape Province) is at particular risk of a drier climate (Dept. of Environmental Affairs 2016).

Without significant adaptation under a high-emission scenario, 13,900 people in South Africa per year on average may be affected by flooding due to sea level rise between 2070 and 2100. An additional 8500 people annually above the estimated affected population of 45,900 in 2010 may be at risk of inland river flooding by 2030 as a result of climate change.[9] No change in the number of days with very heavy precipitation (20 mm or more) is projected under either high- or low-emission scenarios [RCP 2.6], remaining around 6–7 days on average (World Health Organisation 2015).

Future Health Risks of Climate Change and Air Pollution

Increases in mean temperatures and prolonged heatwaves raise the risk of air pollution and consequent health impacts via several pathways: more ground-level ozone affects lung function and causes respiratory symptoms, eye irritation and broncho-constriction, as does smoke from more frequent and intense wildfires; and more aeroallergens (pollens, spores, moulds and allergenic plants) increase allergic reactions and asthma (Jonathan Patz and Frumkin 2016). Those with pre-existing respiratory and cardiovascular conditions, especially the elderly, are most vulnerable to heatwaves and episodes of poor air quality (Myers and Rother 2013).

Occupational groups with prolonged sun and heat exposure in South Africa, such as manual labourers in the agricultural, construction and mining sectors, are vulnerable to sunburn, sleeplessness, irritability and heat exhaustion (Mathee et al. 2010). Residents of poorly constructed and informal housing, which is highly

[8]Model projections from CMIP5 for RCP8.5 and RCP2.6.

[9]World Resources Institute Aqueduct Global Flood Analyzer, which assumes continued current socio-economic trends and a 25-year flood protection. http://www.wri.org

prevalent in cities and towns across South Africa, are most at risk during extreme events, such as heatwaves (Scovronick and Armstrong 2012), fires, floods and storms. In rural areas such as the northern Limpopo Province, future increases in temperature and declining rainfall under climate change scenarios up to 2050 may result in significantly more childhood diarrhoea and respiratory infections, which are currently most prevalent, and slight increases in the incidence of asthma, malaria and meningitis (Thompson et al. 2012).

A study in under-five children in the Cape Town Metropolitan Area (CTMA), based on diarrhoea incidence data from two peak periods in 2012–2013 and 2013–2014, found an association with rising minimum and maximum temperatures, which suggests the need for public early-warning systems when temperature changes are expected (Musengimana et al. 2016).

Opportunities for Climate-Health Co-benefits: Measures to Mitigate Air Pollutants and GHG Emissions

Although climate change has been recognised as a major threat to public health in the twenty-first century, it also presents a significant opportunity to improve public health by prioritising measures that can improve air quality in the short term and can mitigate GHG emissions and improve resilience in a changing climate (Watts et al. 2015). Key measures are described below with reference to contemporary data and examples from different sectors of the South African economy.

Energy Sector

South Africa has great potential for adding renewable energy to the electricity system: a large land area with a low population density so space is not a constraint, a widespread interconnected electricity system that enables spatial aggregation and minimal seasonality of solar and wind energy supply. A detailed analysis of national wind and solar resources concluded that more than 80% of South Africa's land mass has enough wind resource for economical wind farms with very high annual load factors of greater than 30%,[10] that up to 65% of electricity supply can be achieved from a combined wind and solar PV fleet without any significant excess energy and that low seasonality in both wind and solar PV

[10]Another study calculated that in order to generate enough electricity to meet current South African demand (approx. 250 TWh/year), 0.6% of available South African land mass would need to be dedicated to wind farms with an installed capacity of approx. 75 GW (Energy Centre 2016).

supplies makes the integration easier, because no seasonal storage is required to balance fluctuations (Energy Centre 2016).

A more recent analysis calculated that a "re-optimised" mix of solar PV, wind and natural gas providing 70% of national power by 2040 would be almost R90 billion per year cheaper than a "business-as-usual" scenario by then, even before factoring in the cost savings of an estimated 60% reduction in CO_2 emissions.

South Africa therefore has the potential to replace a large proportion, if not all, of its coal-based power with renewable wind and solar energy, at negative carbon-avoidance cost and stimulate a local solar PV and wind manufacturing industry (Wright et al. 2016).

The total phase-out of coal-based power by 2050, the decommissioning of the CTL plant by 2040, and reducing transport emissions could enable South Africa to meet its low-carbon target of 214 Mt CO_2-eq per year by 2050 and achieve a carbon budget 40–60% lower than the current trajectory. Co-benefits would include the reduction of air and water pollution from coal mining and combustion and job creation and re-industrialisation in the renewable energy sector.[11]

Industrial Sector

The industrial sector is the biggest user of energy in South Africa, accounting for approximately half of national electricity consumption. Consumption in the mining sector is primarily for ore processing, pumping and heating, ventilation and cooling systems, for iron and steel production furnaces and for electrochemical processes in the non-ferrous metals subsector. Concerted efforts to improve both technological and process efficiencies in these energy-intensive subsectors would yield large savings in electricity consumption and significant decreases in air pollution.

The case study below on the Multi-Point Plan for reduction of SO_2 in the South Durban Basin illustrates the importance too of stakeholder engagement and oversight in significantly reducing chronic air pollution in a large metropolitan industrial area.

Human Settlements

The 18 major metropoles in South Africa, home to 46% of the total population but occupying only 4.6% of land space, consume, with a number of secondary cities, about 37% of national energy, 46% of national electricity consumption, 52% of petrol and diesel consumption, 32% of energy-related GHG emissions and 70% of wealth production (Wolpe and Reddy 2015).

[11]Coal power plants consume 7–8% of the power that they produce, which will reduce national power demand.

The state's Free Basic Alternative Energy (FBAE) programme aims to provide alternative energy free of charge to indigent households in non-electrified areas to support their basic needs. Pilot projects have included installation of SWH and geysers, clean cooking fuels (methanol), insulated ceilings, hot water boxes, biogel lighting and more efficient cooking stoves and lighting. The FBAE is poorly implemented however, and existing projects are benefiting only a limited number of households (Mohlakoana 2014).

Climate mitigation measures in some South African cities include RE from landfill gas, sewerage methane, micro-hydro on water distribution systems and solar PV on rooftops, provision of energy subsidies to the poor, promotion of EE, reduction of water leakage distribution systems and waste recycling (Department of Environmental Affairs 2015).

Transport Sector

The transport sector is the main consumer of energy in most South African cities. Effective management of transport supply and demand to reduce transport emissions can have significant co-benefits for public health and quality of life due to less air pollution and increased physical activity (Woodcock et al. 2009). Measures recommended by national and municipal transport policies in South Africa to help promote public and non-motorised transport modes include urban densification, better public transport and better infrastructure for cycling and walking, and these are being implemented by means of integrated transport plans in metropolitan areas. Transport emissions could also be reduced by requiring all new public transport vehicles to be low carbon, by shifting road freight to rail where possible and by promoting greater fuel efficiency, driving efficiency and system efficiency (Department of Transport 2006). Nevertheless, emission reductions in the transport sector have been limited, and traffic congestion and private commuter vehicles remain the norm. It is clear that a range of strategies are required to shift behavioural dynamics, using both incentive and disincentive schemes.

Healthcare Sector

Climate change and air pollution have important implications for the healthcare sector. Extreme weather events, such as flooding and heatwaves, may directly impact health system infrastructure in the form of structural damage and power outages at times when health centres may be struggling already with the health impact of such events. Direct and indirect health impacts of climate change may also weaken an already overburdened health system by adding to staff workload and overwhelming emergency response capacity to disease outbreaks (Myers and Rother 2013).

The Global Green and Healthy Hospitals Network (GGHHN) of Health Care Without Harm (HCWH)[12] challenges health institutions to reduce their considerable carbon footprint, to strengthen their resilience to the growing health impacts of extreme weather events and to show leadership in educating staff, raising public awareness and promoting policies to protect public health from climate change. A number of hospitals in South Africa have recently joined the GGHHN, including Groote Schuur Hospital in Cape Town (site of the world's first human heart transplant in 1967), which has almost halved its electricity and coal consumption in recent years by improving system efficiency.[13]

Business Sector

There are a number of current initiatives in South Africa with experience in improving energy efficiency (EE) and mitigating GHGs in the private business sector. The National Business Initiative (NBI)[14] uses advocacy, collective action and partnership approaches to improve EE in commercial and industrial companies.

The Green Building Council SA[15] provides the tools, training, knowledge and networks to promote green building practices in the South African property and construction industry through market-based solutions. It supports government to legislate and facilitate the adoption of green building practices and rewards industry leaders who achieve green building excellence.

Recommendations to Protect Public Health from Climate Change and Air Pollution

Policy and Governance

The Department of Environmental Affairs (DEA), through its Climate Change and Air Quality (CCAQ) Unit, is responsible to improve air and atmospheric quality and to *ensure that reasonable legislative and other measures are developed, implemented and maintained in such a way as to protect and defend the right of all to air and atmospheric quality that is not harmful to health and well-being.*[16]

[12]Health Care Without Harm https://noharm-global.org/

[13]Personal communication with Prof Edda Weimann, GSH Climate Change Management Team.

[14]The National Business Initiative (NBI) in South Africa is a voluntary coalition of South African and multinational companies since 1995 undertaking business action for sustainable growth www.nbi.org.za

[15]The Green Building Council SA is a non-profit company formed in 2007 to lead the greening of South Africa's commercial property sector.

[16]Department of Environmental Affairs (DEA) Climate Change and Air Quality Unit https://www.environment.gov.za/branches/climatechange_airquality

The draft report (September 2016) of the South Africa National Adaptation Strategy (NAS), which seeks to link climate adaptation efforts more coherently to South Africa's national developmental goals, proposes priorities for a number of key economic sectors: energy, water, health, disaster risk reduction, transport, human settlements, biodiversity, agriculture and mining. Air quality receives relatively little attention in the strategy, leading to call for it to be included as one of the key sectors. The NAS does however recommend an increased budget for monitoring air pollution, GHG emissions and climate parameters such as ambient air temperatures. It also recognises the need for improved capacity within DEA to provide mechanisms and oversight for measuring, reporting and verifying sectoral emissions (Dept. of Environmental Affairs 2016).

Public Health Advocacy

The public health community in South Africa needs to advocate more strongly for action on climate change and air pollution in several areas:

- Scaling up the renewable energy programme to replace South Africa's heavy dependence on coal-based power;
- Greening the health sector by means of greater energy efficiency, water efficiency, and waste reduction measures, as well as reduced use of transport and greener procurement of goods and services;
- Public health promotion to minimise the health risks of climate change and air pollution, such as effective early warning systems about heatwaves, high pollen counts and air pollution levels, especially for the most vulnerable (children, the elderly, people with chronic respiratory and cardiovascular conditions, and those in heat-exposed occupations);
- Better monitoring of air quality and enforcement of air pollution legislation;
- Stronger programmes for surveillance of key health impacts of climate change and pollution based on reliable and valid mortality and morbidity data (cancers, respiratory conditions, cardiovascular diseases etc.);
- Funding and training for developing greater research capacity.

Conclusions

The carbon intensity of the South African economy makes the country one of the primary contributors to climate change worldwide, which is increasing the health risks from air pollution in the short term and from extreme weather events and indirect climate impacts in the longer term. South Africa is a very unequal country with many groups especially vulnerable to these risks, such as children, people living with chronic respiratory and cardiovascular diseases, those living in informal settlements, and those working in heat-exposed environments. South Africa therefore needs to act with greater urgency and commitment to mitigate emissions of

GHGs and related pollutants and to adapt to projected climate change impacts across all economic sectors. The public health community has an important role to play in urging further action and research at the national, provincial and local levels.

Case Study: The Multi-Point Plan for Reduction of SO_2 in the South Durban Basin

The *eThekwini* Metropolitan area, which includes the port of Durban, the largest on the African continent, and is the centre of the petrochemical industry, has a long history of high levels of industrial ambient pollution, especially south of the city. Ambient SO_2 concentrations in the early 2000s were approximately 42,000 tons per annum and were driven by two oil refineries and a pulp and paper plant, responsible for approximately 80% of the SO_2 pollution load.[17]

The Multi-Point Plan (MPP) for the South Durban Basin was announced by the Environment Minister in 2007 to control and reduce ambient pollution by means of an air quality management system backed by a state-of-the-art air quality monitoring network. The MPP included two key oversight structures, the Stakeholders Consultative Forum (SCF) and the Inter-Governmental Co-ordinating Committee (IGCC) (DEA 2007). At each stage of the project, there was strong, informed participation from all stakeholders, particularly the affected communities and industries.

The municipality's air quality management system and its directive to industry to phase out dirty fuels and reduce emissions soon resulted in positive air quality impacts, especially a marked decrease in ambient SO_2 concentrations (Fig. 20.3) and immediate decreases in the number of 10-min average SO_2 guideline exceedances (Fig. 20.4).

By 2005, the Engen Environmental Improvement Program had resulted in a 65% reduction in SO_2 emissions (their permit was reduced from 72 to 25 tpd), a 70% reduction in particulate matter emissions and in major reductions in VOC emissions, NOx emissions and flaring. The SO_2 emission permit of the South African Petroleum Refinery (SAPREF), the largest crude oil refinery in Southern Africa, was reduced from 50 to 20 tpd from 2004 onwards, although actual emissions declined from 52 tpd in 1995 to 11 tpd in 2006, representing a 79% reduction with fewer 10-min average SO_2 exceedances[18] (www.sapref.com/initiatives). The installation of a SO_2 scrubber at Mondi reduced their SO_2 emissions by 50% with a co-benefit of particulate matter removal.[19]

---

[17]eThekwini Health and Norwegian Institute for Air Research, 2007: Air Quality Management Plan for eThekwini Municipality.

[18]South African Petroleum Refinery (Pty) Ltd. http://www.sapref.com/

[19]DEA (2007) South Durban Basin Multi-Point Plan Case Study Report: Governance Information. Publication Series C Book 12. Output A2: DEAT AQA Implementation: Air Quality Management Planning

Authors: Lisa Guastella*, Svein Knudsen^ October 2007. *Zanokuhle Environmental Services (ZES)

^ Norwegian Institute for Air Research (NILU)

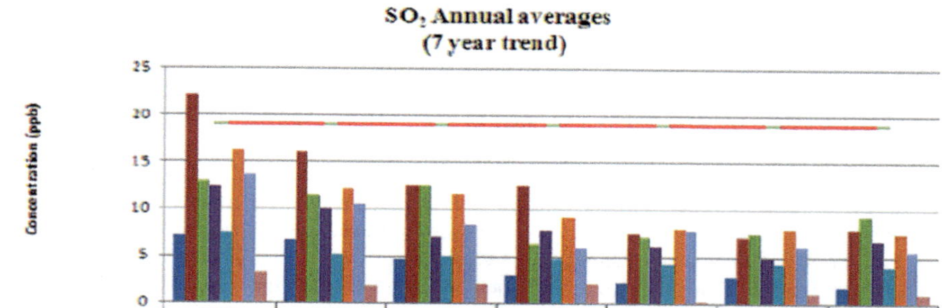

	2004	2005	2006	2007	2008	2009	2010
Prospecton	7.2	6.7	4.6	2.9	2.2	2.8	1.7
Southern Works	22.2	16.2	12.5	12.5	7.5	7	7.9
Settlers	13	11.4	12.6	6.3	7.0	7.5	9.3
Ganges	12.4	10	7	7.8	6.0	4.9	6.7
Grosvenor	7.5	5.1	5	4.8	4.3	4.2	3.9
Wentworth	16.3	12.2	11.7	9.1	7.9	8	7.5
Jacobs (Balfour)	13.7	10.5	8.3	5.9	7.8	6	5.5
Ferndale	3.2	1.8	2.1	2	0.1	1.1	0.9
GLV	19	19	19	19	19	19	19

Fig. 20.3 Ambient SO_2 concentrations, 2004–2010 (eThekwini health department: Pollution Control Support. eThekwini air quality monitoring network: Annual report 2010)

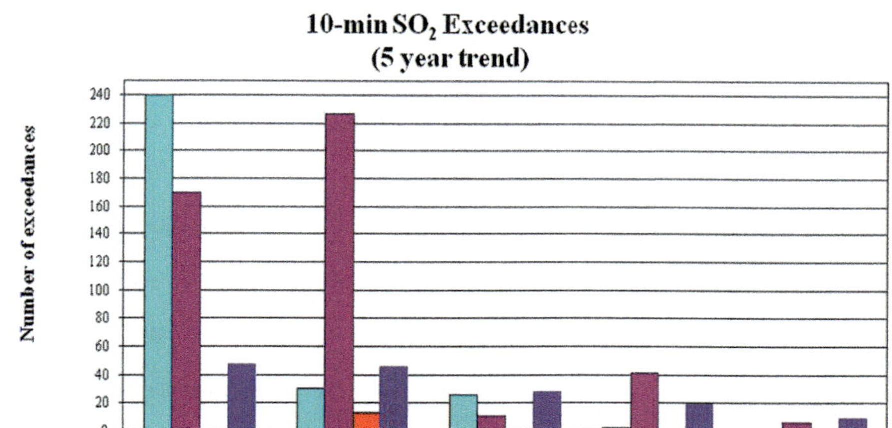

	2005	2006	2007	2008	2009
Southern Works	240	30	26	2	0
Settlers	170	227	11	41	6
Ganges	0	13	1	0	0
Wentworth	48	46	28	20	10

Southern Works Settlers Ganges Wentworth

Fig. 20.4 10-min average SO_2 guideline exceedances, 2005–2009 (eThekwini health department: Pollution Control Support. eThekwini air quality monitoring network: Annual report 2009)

References

Albers PN, Wright CY, Voyi KV, Mathee A (2015) Household fuel use and child respiratory ill health in two towns in Mpumalanga, South Africa. S Afr Med J 105(7):573–577. doi:10.7196/SAMJnew.7934

Baseera W, Mhlanga S, Roosli M, Dalvie M (2016) An epidemiological cohort study of school-going children investigating reproductive and neurobehavioral effects due to environmental pesticide and cell phone use exposure in the Western Cape, South Africa (Research proposal). Faculty of Health Sciences, University of Cape Town

Block ML, Elder A, Auten RL, Bilbo SD, Chen H, Chen J-C,. .. Dorman DC (2012) The outdoor air pollution and brain health workshop. Neurotoxicology 33(5):972–984

Bofinger S, Zimmermann B, Gerlach-Fraunhofer A, Bischof-Niemz T, Mushwana C (2016) Wind and solar PV resource aggregation study for South Africa Pretoria

Burger L, von Gruenewaldt R, Bird T (2014) Atmospheric impact report: Sasol Secunda facility. Retrieved from http://www.srk.co.za/files/File/South-Africa/publicDocuments/Sasol_Postponement/Nitro/Annexure_A_Nitro_Atomospheric_Impact_Report.pdf

Cairncross E (2016) The State of South Africa's air quality monitoring network and its air quality. Paper presented at the National Association for Clean Air

Coetzee AM, Smith FC, van der Merwe CA, Dreyer RJ (1986) The influence of air pollution on health in the Sasolburg area. S Afr Med J 70(6):339–343

CSIR Energy Centre (2016) Statistics of utility-scale solar PV and wind in South Africa in the first half of 2016. Retrieved from Pretoria

Dalvie M, Jeebhay M, Naidoo R (2014) An epidemiological cohort study of school children investigating asthma and of adults investigating cardiopulmonary outcomes (Research proposal). Department of Environmental Affairs and Developmental Planning. University of Cape Town

de la Rue du Can Stephane LV, Leventis Greg, Berkeley Lawrence, Covary Theo, Professor Xia (2013) Energy efficiency country study: Republic of South Africa. Retrieved from https://ies.lbl.gov/sites/all/files/south_africa_country_study_lbnl_report_final_0.pdf

Deaton C, Froelicher ES, Wu LH, Ho C, Shishani K, Jaarsma T (2011) The global burden of cardiovascular disease. Eur J Cardiovasc Nurs 10(Suppl 2):S5–13. doi:10.1016/S1474-5151(11)00111-3

Department of Environmental Affairs (2011) National climate change response white paper. Republic of South Africa, Pretoria

Department of Environmental Affairs (2015) SA 2050 carbon calculator retrieved from http://2050.lateral.co.za/. http://2050.lateral.co.za/

Department of Transport (2006) National Land Transport Strategic Framework (NLTSF). Retrieved from http://us-cdn.creamermedia.co.za/assets/articles/attachments/03496_notice1468.pdf

Dept. of Energy (2015) Annual fuel sales volumes pretoria. Retrieved from http://www.energy.gov.za/files/media/media_SAVolumes.html

Dept. of Energy (2016) Independent power producers program. Retrieved from http://www.energy.gov.za/files/media/media_SAVolumes.html

Dept. of Environmental Affairs (2012a) Highveld priority area: air quality management plan. Government Gazette No. 35072, 2 Mar 2012. Pretoria

Dept. of Environmental Affairs (2012b) Waterberg-Bojanala priority area air quality management plan

Dept. of Environmental Affairs (2013) The medium term review of the 2009 vaal triangle priority area: air quality management plan final review report. Retrieved from

Dept. of Environmental Affairs (2016) South Africa National Adaptation Strategy- DRAFT (Sept 2016). Retrieved from

du Preez K, Mandalakas AM, Kirchner HL, Grewal HM, Schaaf HS, van Wyk SS, Hesseling AC (2011) Environmental tobacco smoke exposure increases Mycobacterium tuberculosis infection risk in children. Int J Tuberc Lung Dis 15(11):1490–1496, i. doi:10.5588/ijtld.10.0759

Eskom (2015) Factsheets 2015. Statistical table 1A: ten year technical statistics. Retrieved from http://www.eskom.co.za/IR2015/Documents/Eskom_fact_sheets_2015.pdf

Gordon et al (2014) Respiratory risks from household air pollution in low and middle income countries. Lancet Respir Med Comm 2(10):823–860

Graveland H, Van Roosbroeck SA, Rensen WM, Brunekreef B, Gehring U (2011) Air pollution and exhaled nitric oxide in Dutch schoolchildren. Occup Environ Med 68(8):551–556. doi:10.1136/oem.2010.056812

International Energy Agency (2015) CO2 emissions from fuel combustion highlights. Retrieved from www.iea.org

Jafta N, Jeena PM, Barregard L, Naidoo RN (2015) Childhood tuberculosis and exposure to indoor air pollution: a systematic review and meta-analysis. Int J Tuberc Lung Dis 19(5):596–602. doi:10.5588/ijtld.14.0686

Jung DY, Leem JH, Kim HC, Kim JH, Hwang SS, Lee JY, Kim BJ, Hong YC, Hong SJ, Kwon HJ (2015) Effect of traffic-related air pollution on allergic disease: results of the Children's health and environmental research. Allergy, Asthma Immunol Res 7(4):359–366. doi:10.4168/aair.2015.7.4.359

Kistnasamy EJ, Robins TG, Naidoo N (2008) The relationship between asthma and ambient air pollutants among primary school students in Durban, South Africa. Int J Environment and Health 2:365–385

Mann JK, Balmes JR, Bruckner TA, Mortimer KM, Margolis HG, Pratt B, Hammond SK, Lurmann FW, Tager IB (2010) Short-term effects of air pollution on wheeze in asthmatic children in Fresno, California. Environ Health Perspect 118(10):1497–1502. doi:10.1289/ehp.0901292

Mathee A, Oba J, Rose A (2010) Climate change impacts on working people (the HOTHAPS initiative): findings of the South African pilot study. Global Health Action, 3

McConnell R, Islam T, Shankardass K, Jerrett M, Lurmann F, Gilliland F, Gauderman J, Avol E, Künzli N, Yao L, Peters J, Berhane K (2010) Childhood incident asthma and traffic-related air pollution at home and school. Environ Health Perspect 118(7):1021–1026. doi:10.1289/ehp.0901232

Mohlakoana N (2014) Implementing the south african free basic alternative energy policy. (Ph.D.), University of Twente. Retrieved from http://doc.utwente.nl/93045/1/thesis_N_Mohlakoana.pdf

Musengimana G, Mukinda FK, Machekano R, Mahomed H (2016) Temperature variability and occurrence of diarrhoea in children under five-years-old in Cape town metropolitan sub-districts. Int J Environ Res Public Health 13(9):859

Myers JE, Rother H (2013) Public health impact and response to climate change in South Africa. In: Padarath A, English R (eds) South African health review 2012/13. Health Systems Trust, Durban

Mzileni O, Sitas F, Steyn K, Carrara H, Bekker P (1999) Lung cancer, tobacco, and environmental factors in the African population of the Northern Province, South Africa. Tob Control 8(4):398–401. Retrieved from http://tobaccocontrol.bmj.com/content/8/4/398.full.pdf

Naidoo R, Gqaleni N., Batterman S, Robins T (2007) South Durban health study. Final Project Report. Center for Occupational and Environmental Health, University of KwaZulu-Natal, Durban, South Africa. Retrieved from http://doeh.ukzn.ac.za/Libraries/Documents/SDHS_FINAL_Report_revision_February_2007.sflb.ashx

Naidoo RN, Robins TG, Batterman S, Mentz G, Jack C (2013) Ambient pollution and respiratory outcomes among schoolchildren in Durban, South Africa. Sajch 7(4):127–134. doi:10.7196/sajch.598

Nastos PT, Paliatsos AG, Anthracopoulos MB, Roma ES, Priftis KN (2010) Outdoor particulate matter and childhood asthma admissions in Athens, Greece: a time-series study. Environ Health 9:45. doi:10.1186/1476-069X-9-45

Nkosi V, Wichmann J, Voyi K (2015) Chronic respiratory disease among the elderly in South Africa: any association with proximity to mine dumps? Environ Health 14:33. doi:10.1186/s12940-015-0018-7

Norman R, Barnes B, Mathee A, Bradshaw D (2007a) Estimating the burden of disease attributable to indoor air pollution from household use of solid fuels in South Africa in 2000. S Afr Med J 97(8 Pt 2):764–771

Norman R, Cairncross E, Witi J, Bradshaw D (2007b) Estimating the burden of disease attributable to urban outdoor air pollution in South Africa in 2000. S Afr Med J 97(8 Pt 2):782–790

Pan G, Zhang S, Feng Y, Takahashi K, Kagawa J, Yu L et al (2010) Air pollution and children's respiratory symptoms in six cities of Northern China. Respir Med 104(12):1903–1911. doi:10.1016/j.rmed.2010.07.018

Pillay-van Wyk V, Msemburi W, Laubscher R, Dorrington RE, Groenewald P, Glass T et al (2016) Mortality trends and differentials in South Africa from 1997 to 2012: second National Burden of Disease Study. Lancet Glob Health 4(9):e642–e653. Retrieved from http://ac.els-cdn.com/S2214109X16301139/1-s2.0-S2214109X16301139-main.pdf?_tid=7f53138e-855f-11e6-b1c8-00000aab0f6c&acdnat=1475055768_86aaba8344268e04584729b710d57390

Rehfuess EA, Tzala L, Best N, Briggs DJ, Joffe M (2009) Solid fuel use and cooking practices as a major risk factor for ALRI mortality among African children. J Epidemiol Community Health 63(11):887–892. doi:10.1136/jech.2008.082685

Rovira E, Cuadras A, Aguilar X, Esteban L, Borras-Santos A, Zock JP, Sunyer J (2014) Asthma, respiratory symptoms and lung function in children living near a petrochemical site. Environ Res 133:156–163. doi:10.1016/j.envres.2014.05.022

Scovronick N, Armstrong B (2012) The impact of housing type on temperature-related mortality in South Africa, 1996–2015. Environ Res 113:46–51. Retrieved from http://ac.els-cdn.com/S0013935112000072/1-s2.0-S0013935112000072-main.pdf?_tid=78c1d89e-8563-11e6-806c-00000aab0f6b&acdnat=1475057475_3e4fcf5dd1f15c684303fee5e73ee603

Shirinde J, Wichmann J, Voyi K (2014) Association between wheeze and selected air pollution sources in an air pollution priority area in South Africa: a cross-sectional study. Environ Health 13(1):32. doi:10.1186/1476-069x-13-32

South African Petroleum Industries Association. (2014). Annu Rep. Retrieved from www.sapia.org.za/

Strickland MJ, Darrow LA, Klein M, Flanders WD, Sarnat JA, Waller LA, Sarnat SE, Mulholland JA, Tolbert PE (2010) Short-term associations between ambient air pollutants and pediatric asthma emergency department visits. Am J Respir Crit Care Med 182(3):307–316. doi:10.1164/rccm.200908-1201OC

Synfuels (2014) Synfuels AIR2014

Terblanche AP, Opperman L, Nel CM, Reinach SG, Tosen G, Cadman A (1992) Preliminary results of exposure measurements and health effects of the Vaal Triangle Air Pollution Health Study. S Afr Med J 81(11):550–556

Thompson AA, Matamale L, Kharidza SD (2012) Impact of climate change on children's health in Limpopo Province, South Africa. Int J Environ Res Public Health 9(3):831–854. Retrieved from https://www.ncbi.nlm.nih.gov/pmc/articles/PMC3367281/pdf/ijerph-09-00831.pdf

Vanker A, Barnett W, Nduru PM, Gie RP, Sly PD, Zar HJ (2015) Home environment and indoor air pollution exposure in an African birth cohort study. Sci Total Environ 536:362–367. doi:10.1016/j.scitotenv.2015.06.136

Watts N, Adger WN, Agnolucci P, Blackstock J, Byass P, Cai W,. .. Cooper A (2015) Health and climate change: policy responses to protect public health. Lancet 386(10006):1861–1914

Weinmayr G, Romeo E, De Sario M, Weiland SK, Forastiere F (2010) Short-term effects of PM10 and NO2 on respiratory health among children with asthma or asthma-like symptoms: a systematic review and meta-analysis. Environ Health Perspect 118(4):449–457. doi:10.1289/ehp.0900844

Wichmann J, Voyi K (2012) Ambient air pollution exposure and respiratory, cardiovascular and cerebrovascular mortality in Cape Town, South Africa: 2001–2006. Int J Environ Res Public Health 9(11):3978–4016. doi:10.3390/ijerph9113978

Wolpe P, Reddy Y (2015) The contribution of low-carbon cities to South Africa's greenhouse gas emissions reduction goals: briefing on urban energy use and greenhouse gas emissions. Stockholm Environment Institute, Stockholm

Woodcock J, Edwards P, Tonne C, Armstrong BG, Ashiru O, Banister D, Roberts I (2009) Public health benefits of strategies to reduce greenhouse-gas emissions: urban land transport. Lancet 374(9705):1930–1943

World Health Organisation (2015) Climate and health country profile 2015: South Africa Retrieved from http://www.who.int/globalchange/resources/country-profiles/en/#S

Wright J, Bischof-Niemz T, Calitz J, Mushwana C (2016, November) Least-cost electricity mix for South Africa by 2040: scenarios for South Africa's future electricity mix. Paper presented at the Windaba, Cape Town

Zora JE, Sarnat SE, Raysoni AU, Johnson BA, Li WW, Greenwald R, Holguin F, Stock TH, Sarnat JA (2013) Associations between urban air pollution and pediatric asthma control in El Paso, Texas. Sci Total Environ 448:56–65. doi:10.1016/j.scitotenv.2012.11.067

Zwi S, Davies JC, Becklake MR, Goldman HI, Reinach SG, Kallenbach JM (1990) Respiratory health status of children in the eastern Transvaal highveld. S Afr Med J 78(11):647–653

Eugene Cairncross has a chemical engineering degree and a PhD. After many years working in a variety of chemical industries, followed by a similar period in academia, he now focuses on air pollution, coal power emissions, the burden of disease attributable to air pollution, on climate change, marathon running, swimming and two grandchildren.

Aqiel Dalvie has a PhD in Public Health and is an academic at the University of Cape Town's Faculty of Health Sciences within the Centre for Environmental and Occupational Health Research in the School of Public Health and Family Medicine. His research interest include health impacts due to endocrine disruptors, pesticides, air pollution, water pollution and climate change.

Rico Euripidou manages an environmental health campaign for groundWork, Friends of the Earth, South Africa. He trained as an environmental epidemiologist at the London School of Hygiene and Tropical Medicine. Previously, Rico worked for the National Poisons Information Service and the University of Witwatersrand. Rico's interests lie in energy policy, climate change and public health, all of course interrelated.

James Irlam is an academic at the University of Cape Town Faculty of Health Sciences within the Primary Health Care (PHC) Directorate. He believes that the PHC vision of Health for All depends on a healthy environment. He is interested in the health impacts of climate change and in a socially just transition towards renewable energy in South Africa.

Rajen Nithiseelan Naidoo, an associate professor and head of the Department in Occupational and Environmental Health at the University of KwaZulu-Natal, has been in academic research for over 25 years. His areas of research focus on occupational and environmental respiratory diseases, with projects funded by major national and international agencies. He has over 50 peer-reviewed publications, with presentations at several international conferences.

Chapter 21
The Impact of Climate Change and Air Pollution on the Caribbean

Muge Akpinar-Elci and Olaniyi Olayinka

Abstract A review of air pollution, the impact of climate change on air pollution, and the population health impacts of these in the Caribbean region are discussed. Air quality standards are not usually enforced in many Caribbean countries thereby increasing the risks of morbidity and mortality from exposure to air pollutants. Among people living in the Caribbean, an increase in respiratory diseases such as asthma has been linked to exposure to air pollutants resulting from natural events and especially human activities. Unfortunately, dependence on fossil fuels (regionally and globally), poor land use and waste management, and industrialization all contribute to poor air quality in the Caribbean. In addition, climate change is predicted to exacerbate air pollution and its negative health effects in a region considered to be one of the most vulnerable to global climate change. Key drivers of air pollution in the region are discussed, and recommendations on climate change adaptation and mitigation strategies are highlighted.

Keywords Air pollution • Caribbean • Particulate matter • Air quality • Climate change • Health impacts

Introduction

History is replete with the negative human impacts of air pollution (WHO 2008). Although it is hard to find historical data on air pollution in the Caribbean, there are reports suggesting a long history of air quality issues in the region (de Koning et al. 1985; Romieu et al. 1989; Sanhueza et al. 1982). For example, a 1996 World Bank report on global air pollution from automobiles showed that one of the most industrialized countries in the Caribbean produced leaded gasoline for local use while exporting unleaded fuel (The World Bank 1996, p. 226); by the mid-1990s, the use of leaded gasoline had significantly declined in many developed countries due to public health safety concerns (Nriagu 1990). In a 2005 review of the public

---

M. Akpinar-Elci (✉) • O. Olayinka (✉)
School of Community & Environmental Health and Center for Global Health, Old Dominion University, Norfolk, VA 23529, USA
e-mail: makpinar@odu.edu; oolayink@odu.edu

© Springer International Publishing AG 2018
R. Akhtar, C. Palagiano (eds.), *Climate Change and Air Pollution*, Springer Climate, DOI 10.1007/978-3-319-61346-8_21

health impacts of urban air quality in Latin America and the Caribbean, Cifuentes and his colleagues suggested that exposure to particulate matter in 26 cities across the region is "more than twice the US standard"; while ground-level ozone might be a problem in the region, the lack of data made it difficult for the authors to conduct ozone exposure-impact analysis (Cifuentes et al. 2005). Although countries in WHO's Southeast Asia and Western Pacific regions are the hardest hit, a couple of population-based studies across the Caribbean suggests that significant air quality problem still exists in the region (Akpinar-Elci et al. 2015, 2015; Amadeo et al. 2015; Bautista et al. 2009; Brauer et al. 2015; Chafe et al. 2015; PAHO-WHO 2005).

Clean air is considered a fundamental human right globally; unfortunately, air pollution remains a major contributor to morbidity and mortality, especially in developing countries (including Caribbean countries) due to the general lack of air quality regulations and enforcement coupled with socioeconomic, geographic, and climatological factors (Amadeo et al. 2015; Jessamy 2016; Krzyzanowski and Cohen 2008; Macpherson and Akpinar-Elci 2015; Schwindt et al. 2010; Segal and Nilsson 2015; Tanveer et al. 2014). According to WHO, the attributable mortality and disability adjusted life years (DALYs) due to outdoor air pollution in the Americas subregion B (which include states and territories in the Caribbean) were 30 deaths and 307 DALYS per 1000 population; these values exceed the attributable mortality and DALYs (28 deaths and 200 DALYs per 1000 population) reported from their more developed neighbors (the Americas subregion A including Canada and the United States) (Ostro 2004). These statistics are not surprising as air pollution is considered the largest environmental health risk factor globally. In fact, the World Health Organization (WHO) estimated seven million deaths were linked to air pollution in 2012. During the same year, outdoor air pollution accounted for 3.7 million deaths globally (WHO 2014a). It is projected that deaths from air pollution will increase in the future as air quality deteriorates in major cities of low- and middle-income countries. Globally, carbon dioxide (CO2), ground-level ozone, nitrogen dioxide, particulate matter, and sulfur dioxide remain the major air pollutants (Jacobson 2009).

In general, maintaining ambient air quality standards remain a challenge in many parts of the Caribbean (Cifuentes et al. 2005; Jessamy 2016; Prospero et al. 2014). This is likely to be compounded by climate change given that meteorological and climatological factors (including local temperature, wind speed, wind direction, poor air circulation, precipitation, and level of humidity) significantly impact air quality (Jacob and Winner 2009; UNEP 2005, p. 3). Additionally, scientific evidence has emerged suggesting a relationship between long-term weather patterns (the climate) and human activities (IPCC 2007). For example, a change in the climate favoring a rise in atmospheric temperature (either from natural or human activities) is likely to increase the demand for air conditioning especially in tropical climates where the mean daily minimum temperature is typically above 180 °C (Trewin 2014). This invariably increases energy consumption in residential and commercial buildings. Because energy production is largely dependent on the burning of fossil fuels, the downstream effects are an increase in the atmospheric concentration of air pollutants (e.g., particulate matter such as black carbon) and greenhouse gases (GHG). The long-term cumulative effects of GHG include global warming, an important indicator of climate change (IPCC 2007). Also, there is scientific evidence that a changing climate will alter the concentration of airborne respiratory allergens due to the effect of CO_2 and temperature on plant growth and the health burden of meteorological events such as windblown dust and mold (Gennaro et al. 2014; Gyan et al. 2005; Jacob and Winner 2009; Monteil 2008). Since the Intergovernmental Panel on Climate Change (IPCC) was established to assess the evidence on climate change in 1988, studies on the link between air pollution and climate change have been widely investigated. Similarly, the scientific community ramped up efforts to address air pollution-climate-sensitive health issues. In this chapter, we will review the relationship between air pollution and climate change and their impacts on the health of people in the Caribbean. Small Island Developing States (SIDS) communities constitute around 5% of the global population (AOSIS 2015). Caribbean states are developing economies and represent about half of the Alliance of Small Island States (AOSIS 2015; UN 2012; UNEP et al. 2004). We especially focused on the Caribbean in this chapter due to their large coastal areas and relatively small economies, which makes the region highly vulnerable to the impact of climate change despite contributing little to global greenhouse gas emission (GHG) (CDKN and ODI 2014).

Climate change and air pollution impact a range of health indicators in Small Island Developing States (SIDS) raising problems for economies and national security. While a comprehensive presentation of the scientific evidence is beyond the scope of this chapter, we have tried to highlight some of the key relationships between climate change and air pollution. Although historical events are alluded to in this chapter, our assessment of the air quality issues facing people in the Caribbean (Fig. 21.1.) is based on a review of epidemiologic studies, anecdotal reports, and evidence presented in the 2014 IPCC Fifth Assessment Report. These are followed by suggestions for mitigation and adaptation strategies to combat the negative impacts of climate change.

Sources of Air Pollutants in the Caribbean

Generally, the main cause of air pollution in the Caribbean is human activities including those related to the use of fossil fuels (Akpinar-Elci and Sealy 2014; CDKN and ODI 2014; IPCC 2014). Some air pollutants, particularly GHGs, alter the composition of the atmosphere and worsen the health impact of air pollution on the Caribbean people despite the region contributing relatively little to global GHG emissions (Akpinar-Elci and Sealy 2014; CDKN and ODI 2014; Dodman 2009). As an indicator of urban air quality, the majority of the Caribbean countries reference the WHO Air Quality Guidelines (AQG) for ambient PM2.5 (i.e., 10 μg/m3 annual mean and 25 μg/m3 24-h mean) and PM10 (i.e., 20 μg/m3 annual mean and 50 μg/m3 24-h mean) (Cifuentes et al. 2005; Krzyzanowski and Cohen 2008). However,

Fig. 21.1 Islands in the Caribbean region

air quality data from the Caribbean are sparse; hence, we have to rely on pockets of scientific evidence suggesting that air pollution is still a problem in the region (Amadeo et al. 2015; Bautista et al. 2009; Cifuentes et al. 2005; Gyan et al. 2005; Matthew et al. 2009).

Other than a couple of volcanic air pollution, the process of burning fossil and biomass fuels to generate electricity, and for heating, cooking, and transportation, especially leads to the emission of major air pollutants (including PM2.5, PM10, carbon monoxide, nitrogen dioxide, lead, sulfur dioxide, ground-level ozone, and CO2 in the Caribbean) (Akpinar-Elci et al. 2015; Akpinar-Elci and Sealy 2014; Amadeo et al. 2015; Bautista et al. 2009; Cadelis et al. 2013; Cifuentes et al. 2005; Han and Naeher 2006; Macpherson and Akpinar-Elci 2015; Monteil et al. 2004; UNEP 1998). The sources of these pollutants largely fall into one or more of the fuel types listed in the 2006 IPCC Guidelines which include crude oil and petroleum products (e.g., gasoline), coal and coal products, natural gas, peat, biomass (e.g., wood/wood waste, charcoal, and the biomass fraction of municipal wastes), and other fossil fuels (e.g., municipal waste, industrial wastes, and waste oils) (IPCC 2006).

In the Caribbean, the CO2 emission and contribution to air pollution and climate change of each member state vary widely. For example, Grenada has a small population and economy (population 104,000; gross national income per capita US$ 8430); Barbados is a midsized country (population 256,000; gross national income per capita US$ 18,240); and Trinidad and Tobago is a larger, wealthier, and more industrialized country (population 1,339,000; gross national income per capita US$ 24,240) (The World Bank 2016). United Nations data show that in 2011, Trinidad and Tobago emitted significantly more CO2 per capita than the United States (37.2 and 16.8 metric tons of CO2 per capita, respectively), while emissions in Barbados and Grenada were significantly lower (5.6 and 2.4 metric tons of CO2 per capita, respectively) (The United Nations 2015). Therefore, air pollution is a huge public health concern in the highly industrialized Trinidad and Tobago. Because of the close proximity of the Caribbean islands, air pollutants from one island travel around the whole region, hence impacting the health of people at distant sites.

According to a recent report, the energy and transportation sectors are responsible for most of the air pollution in Trinidad and Tobago (UNFCCC 2013). In fact, Trinidad, along with the Bahamas and Saint Kitts and Nevis, has one of the highest registered vehicles rate per 1000 population in the Caribbean (WHO 2013). The preponderance of older cars on many islands (Jacobson estimates that 1000 old cars without emission controls produce as much pollution as 100,000 new cars), along with the fact that many of these idyllic places burn sugarcane, winds up causing pollution (Jacobson 2009; The World Bank 1996). Additionally, unhealthy practices such as sugarcane harvesting burning practices and the uncontrolled burning of forest and bushes are not uncommon in the country and in other parts of the Caribbean (Akpinar-Elci, Coomansingh et al. 2015; EMA 2001; Macpherson and Akpinar-Elci 2015). Recent population-based studies and focus group discussion conducted in Grenada found domestic bush burning is a common practice on the island (Akpinar-Elci et al. 2015; Macpherson and Akpinar-Elci 2015). In addition to CO_2 emission, vehicle emissions and ash from bush/forest burning generate a significant amount of fine particles (i.e., $PM_{2.5}$).

Air quality is also impacted by pollutants from natural sources including windblown dust, wildfires, and gases and PM emitted during volcanic eruptions. Of note, air pollutants can originate from a local/regional source or from a distant/global source. Some natural events, such as the transportation of volcanic ash and dust across long distances, have been shown to contribute to air pollution and respiratory diseases in some Caribbean countries. In a 2015 study of air pollution and respiratory health among elementary school children in Guadeloupe, the authors found that the mean PM10 levels in over 70% of the schools exceeded the WHO AQG (Amadeo et al. 2015). There is a high index of suspicion that Saharan dust is responsible for the high PM10 levels in Guadeloupe. Similarly, climate-driven humidity interacting with dust from the Sahara has been shown to produce PM in Barbados, Grenada, Trinidad and Tobago, and US Virgin Islands, hence increasing visits to the emergency department due to exacerbated asthma in the Caribbean (Akpinar-Elci et al. 2015; Garrison et al. 2014; Gyan et al. 2005; Monteil 2008). Furthermore, ash from the Soufriere volcano in Montserrat was linked to an increase in asthma admissions in Guadeloupe after it erupted in 2010 (Cadelis et al. 2013). It is worth noting that the particle size of Saharan dust varies from less than 5 μm (as reported in studies from Barbados and Bermuda) to between 5 and 30 μm (Goudie and Middleton 2001). Similarly, studies have shown the particle size of fine volcanic ash/dust (an admixture of PM, toxic gases like sulfur dioxide, and water vapor) to vary up to less than 60 μm (Lowe and Hunt 2001).

Air pollutants that are released directly into the atmosphere are classified as "primary pollutants" and are a source of indoor and outdoor air pollution in parts of the Caribbean (PAHO-WHO 2005). Fine particulate matter (e.g., particles less than 2.5 μm [$PM_{2.5}$]) has been reported to occur from indoor activities such as smoking, "cooking, cleaning, and other general activities involving either combustion (e.g., candles) or resuspension (e.g., any physical movement such as walking, dusting, vacuuming, etc.)" (Long et al. 2000). Direct exposure to $PM_{2.5}$ from cooking stove, for instance, is particularly common among low-income populations, as was found in a 2009 study of children in parts of the Dominican Republic (Bautista et al. 2009).

On the other hand, secondary pollutants are formed in the atmosphere following a series of photochemical reactions. Although studies suggest that the atmospheric concentration of secondary pollutants (especially ground-level ozone) in the Caribbean is less compared with developed countries, industrialization and increase in fossil fuel-powered vehicles in countries like Trinidad and Tobago may reverse this trend (Amadeo et al. 2015). Both short- and long-term exposures to ozone increase the risk of morbidity and mortality from cardiovascular and respiratory diseases (Bell et al. 2005).

Overall, domestic and commercial activities including the use of fossil fuels are likely to contribute more to air quality problems in the Caribbean, especially as the demand for energy increases as population grows. However, if Caribbean countries and the global community adopt the "stringent mitigation scenario," in addition to effective adaptation strategies, air quality in the region is likely to improve in the near future (Akpinar-Elci and Sealy 2014; IPCC 2014).

Air Pollution, Climate Change, and Health Effects

The human health impact of air pollution on the Caribbean people is well documented. According to a USAID 2009 report: "The burden of disease associated with non-communicable chronic diseases (NCDs) is greater than the burden of disease associated with communicable diseases or injuries in Latin America and the Caribbean (LAC); however, much less attention has been given to NCDs" (Anderson et al. 2009). Current literature reports smoking, allergy, infection, tropical climate, diesel exposure, charcoal smoke, mite, and Sahara dust as risk factors for asthma in the Caribbean (Bautista et al. 2009; Calo et al. 2009; Ivey et al. 2003; Matthew et al. 2009; Milián and Díaz 2004; Monteil 2008; Monteil et al. 2004). Outdoor air pollution is particularly a major public health concern in the Caribbean with a 2014 ambient air pollution data from the WHO showing the annual mean concentrations of $PM_{2.5}$ in some Caribbean countries were above the recommended annual mean of 10 μg/m3 (WHO 2014b) (Fig. 21.2.).

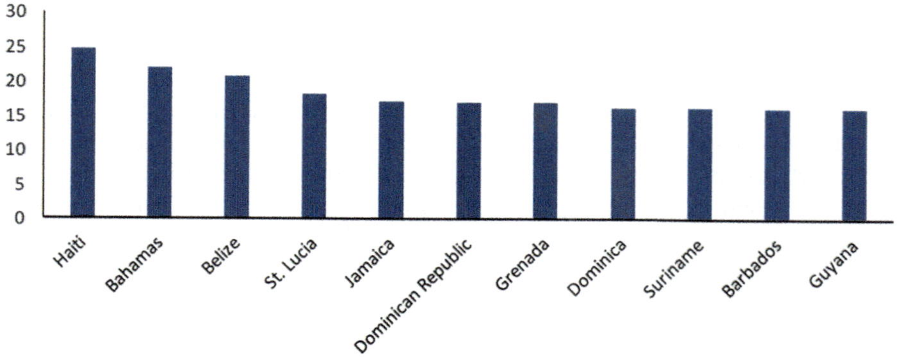

Fig. 21.2 Caribbean countries with annual mean concentrations of $PM_{2.5}$ in urban areas exceeding the WHO recommendation of 10 μg/m^3 (Source of data: WHO http://gamapserver.who.int/gho/interactive_charts/phe/oap_exposure/atlas.html)

There is a growing concern that climate change will exacerbate the human health impacts of air pollution among the Caribbean people (Macpherson and Akpinar-Elci 2015). Climate change is predicted to impact air quality by altering the concentration and distribution of major air pollutants particularly CO_2, ozone, fine particulate matter, and aeroallergens. For example, extreme weather events (including hurricanes, heavy precipitation, and flooding) in the Caribbean create environments conducive for mold, mildew, and other bioaerosols (Ivey et al. 2003; Milián and Díaz 2004). The complex relationship between air-polluting GHGs, climate change, and health is another public health issue. Based on evidence presented in the 2014 IPCC Fifth Assessment Report, the global impact of climate change over the last few decades is significant. According to the report, there is high confidence that climate change will have a major impact on terrestrial ecosystem (i.e., forests) of small islands, hence increasing atmospheric carbon concentration via a reduction in natural carbon sinks. This scenario is likely to be exacerbated by poor land use management, indiscriminate forest and bush burning practices, urbanization and industrialization, rapid population growth, and an increase in energy demand by the Caribbean people and tourists.

In the 2014 Office of Evaluation and Oversight of the Inter-American Development Bank (OVE) evaluation of climate change in nine Caribbean countries (including the Bahamas, Barbados, Belize, Dominican Republic, Guyana, Haiti, Jamaica, Suriname, and Trinidad and Tobago), OVE found that the use of fossil fuels for the production of electricity accounts for 60% of GHG emissions in these countries (OVE 2014). In addition, the report found that 90% of the power plants in the nine countries depend on fossil fuels making electric power generation the largest contributor to air pollution in the Caribbean. The process of burning fossil fuels to generate electric power leads to the release of CO_2 (a major GHG and that is also essential for plant growth), sulfur dioxide, and nitrogen oxides (a precursor of ozone, an air pollutant that affects cardiovascular and respiratory health) (Elenikova et al. 2008).

Extrapolating from studies conducted in other parts of the world, climate change is predicted to affect the respiratory and cardiovascular health of populations across the Caribbean. The impact on the population's health will result from increases in environmental exposure to PM (e.g., black carbon, soot, and Saharan dust), pollens, mold, other bioaerosols, and ground-level ozone. $PM_{2.5}$, for instance, has been proposed to induce and worsen inflammation and oxidative stress in both the pulmonary and cardiovascular systems (Brook et al. 2010). Aeroallergens also affect respiratory health by inducing inflammatory reaction in the respiratory airway. Studies suggest that increased atmospheric CO_2 levels is associated with an increase in ragweed, an allergenic and immunogenic weed that flourishes in tropical and subtropical climates and native to Guadeloupe, Jamaica, and Martinique (CABI 2016; Ziska et al. 2011). Unfortunately, aeroallergens from pollen-producing plants are expected to rise in the future (Richter et al. 2013).

Adaptation Strategies to Climate Change

According to IPCC, an integrated approach to climate adaptation and mitigation is the best way to combat climate change (IPCC 2014). With regard to air pollution, atmospheric pollutants in most Caribbean countries are either generated locally (e.g., from automobiles), while most result from activities at distant sites (e.g., Sahara dust, GHGs emitted by "heavy polluters," and volcanic eruption). Considering the relatively lower socioeconomic and political status and the low carbon footprint of Caribbean countries in general, the Caribbean people need the collaboration of the global community in implementing climate mitigation and adaptation strategies in the region. We believe these strategies should include (1) the enactment of laws and regulations targeted at reducing uncontrolled forest, bush, and trash burning [e.g., sustainable municipal waste management, improved land use management, and agricultural practices such as reforestation], (2) investment in sustainable and green technologies that reduce dependence on fossil fuels, (3) strengthening of public health infrastructure and surveillance systems, and (4) education of the population on the health risks of air pollution and climate change.

The Nairobi Work Programme of the United Nations Framework Convention on Climate Change (UNFCCC) also recommends a number of good practices in the adaptation planning process. These include engaging members of the community in the development of a structured and iterative knowledge base, establishing monitoring systems that are participatory to provide a consistent and reliable source of information, leveraging technology to increase the capacity of the health sector to respond to climate change variability, and raising public awareness of the potential health risks under a changing climate and the need for taking action to address these risks (UNFCCC 2015).

Conclusions

In summary, the burden of air pollution on the Caribbean people will increase with climate change, unless stringent measures are taken at the community, country/ government, and global levels. Particularly, given the established human health effects of air pollutants such as ozone, environmental surveillance of these pollutants and longitudinal studies of their impact on the health of populations across the Caribbean are recommended. Finally, how climate change is likely to influence the effects of air pollution on states and territories in the region should be considered.

References

Akpinar-Elci M, Sealy H (2014) Climate change and public health in small island states and Caribbean countries. In: Global climate change and public health. Springer, New York, pp 279–292

Akpinar-Elci M, Coomansingh K, Blando J, Mark L (2015a) Household bush burning practice and related respiratory symptoms in Grenada, the Caribbean. J Air Waste Manage Assoc 65 (9):1148–1152

Akpinar-Elci M, Martin FE, Behr JG, Diaz R (2015b) Saharan dust, climate variability, and asthma in Grenada, the Caribbean. Int J Biometeorol 59(11):1667–1671

Amadeo B, Robert C, Rondeau V, Mounouchy M-A, Cordeau L, Birembaux X, Marcin G (2015) Impact of close-proximity air pollution on lung function in schoolchildren in the French West Indies. BMC Public Health 15(1):1

Anderson GF, Waters H, Pittman P, Herbert R, Chu E, Das K (2009) Non-communicable chronic diseases in Latin America and the Caribbean. Retrieved September, 2016. http://www.archive.healthycaribbean.org/publications/documents/NCD-in-LAC-USAID.pdf

AOSIS (2015) The alliance of small island states. Retrieved October, 2016, from http://aosis.org/about-aosis/

Bautista LE, Correa A, Baumgartner J, Breysse P, Matanoski GM (2009) Indoor charcoal smoke and acute respiratory infections in young children in the Dominican Republic. Am J Epidemiol 169(5):572–580

Bell ML, Dominici F, Samet JM (2005) A meta-analysis of time-series studies of ozone and mortality with comparison to the national morbidity, mortality, and air pollution study. Epidemiology 16(4):436

Brauer M, Freedman G, Frostad J, Van Donkelaar A, Martin RV, Dentener F, Apte JS (2015) Ambient air pollution exposure estimation for the global burden of disease 2013. Environ Sci Technol 50(1):79–88

Brook RD, Rajagopalan S, Pope CA, Brook JR, Bhatnagar A, Diez-Roux AV, Mittleman MA (2010) Particulate matter air pollution and cardiovascular disease an update to the scientific statement from the American Heart Association. Circulation 121(21):2331–2378

CABI (2016) Ambrosia artemisiifolia (common ragweed). distribution table. Retrieved September 2016, from http://www.cabi.org/isc/datasheet/4691

Cadelis G, Tourres R, Molinie J, Petit RH (2013) Exacerbations of asthma in Guadeloupe (French West Indies) and volcanic eruption in Montserrat (70 km from Guadeloupe). Rev Mal Respir 30(3):203–214. doi:10.1016/j.rmr.2012.11.002

Calo WA, Quintana R, Catoni I, Valle Y, Alvarez JJ, Colon WM,... Kallis M (2009) Clinical conditions associated with environmental exposures: an epidemiologic study in two communities in Juana Diaz, Puerto Rico. P R Health Sci J, 28(2):126.

CDKN, ODI (2014) The IPCC's fifth assessment report| what's in it for small island developing states? Climate & Development Knowledge Network (CDKN) and Overseas Development Institute (ODI). Retrieved August, 2016, from http://cdkn.org/wp-content/uploads/2014/08/CDKN_IPCC_Whats_in_it_for_SIDS.pdf

Chafe ZA, Brauer M, Klimont Z, Van Dingenen R, Mehta S, Rao S,... Smith KR (2015) Household cooking with solid fuels contributes to ambient PM2.5 air pollution and the burden of disease. University of British Columbia

Cifuentes LA, Krupnick AJ, O'Ryan R, Toman M (2005) Urban air quality and human health in Latin America and the Caribbean: Inter-American Development Bank

de Koning HW, Smith K, Last J (1985) Biomass fuel combustion and health. Bull World Health Organ 63(1):11

Dodman D (2009) Blaming cities for climate change? An analysis of urban greenhouse gas emissions inventories. Environ Urban 21(1):185–201

Elenikova Z, Goossens E, Adams A (2008) Air pollution from electricity-generating large combustion plants: An assessment of the theoretical emission reduction of SO2 and NOX through implementation of BAT as set in the BREFs

EMA (2001) Trinidad and Tobago State of the Environment Report 2000. Retrieved September, 2016, from http://apps.unep.org/redirect.php?file=/publications/pmtdocuments/-Trinidad%20and%20Tornado%20State%20of%20the%20Environment%20Report%202000%20-%20Air%20Pollution-2000Trinidad%20and%20Tobago%20EMA%20SOE%20Report%202000.pdf

Garrison VH, Majewski MS, Foreman WT, Genualdi SA, Mohammed A, Simonich SM (2014) Persistent organic contaminants in Saharan dust air masses in West Africa, Cape Verde and the eastern Caribbean. Sci Total Environ 468:530–543

Gennaro DA, Karl CB, Lorenzo C, Isabella A-M, Sanduzzi Zamparelli A, Gennaro L,... Stanziola AA (2014) Climate change and air pollution effects on pollen allergy and other allergic respiratory diseases. Allergo J, 23(1):570–576.

Goudie A, Middleton N (2001) Saharan dust storms: nature and consequences. Earth Sci Rev 56 (1):179–204

Gyan K, Henry W, Lacaille S, Laloo A, Lamsee-Ebanks C, McKay S, Antoine RM, Monteil MA (2005) African dust clouds are associated with increased paediatric asthma accident and emergency admissions on the Caribbean island of Trinidad. Int J Biometeorol 49 (6):371–376. doi:10.1007/s00484-005-0257-3

Han X, Naeher LP (2006) A review of traffic-related air pollution exposure assessment studies in the developing world. Environ Int 32(1):106–120

IPCC (2006) Guidelines for national greenhouse gas inventories. Retrieved September, 2016, from http://www.ipccnggip.iges.or.jp/public/2006gl/pdf/2_Volume2/V2_1_Ch1_Introduction.pdf

IPCC (2007) Climate Change 2007: synthesis report. contribution of Working Groups I, II and III to the fifth assessment report of the Intergovernmental Panel on Climate Change [Core Writing Team, Pachauri RK, Reisinger A (eds)]. IPCC, Geneva, 104 pp

IPCC (2014) Climate Change 2014: synthesis report. contribution of Working Groups I, II and III to the fifth assessment report of the Intergovernmental Panel on Climate Change [Core Writing Team, Pachauri RK, Reisinger A (eds)]. IPCC, Geneva, 151 pp

Ivey M, Simeon D, Monteil MA (2003) Climatic variables are associated with seasonal acute asthma admissions to accident and emergency room facilities in Trinidad, West Indies. Clin Exp Allergy 33(11):1526–1530

Jacob DJ, Winner DA (2009) Effect of climate change on air quality. Atmos Environ 43(1):51–63

Jacobson MZ (2009) Review of solutions to global warming, air pollution, and energy security. Energy Environ Sci 2(2):148–173

Jessamy V (2016) A potentially explosive health crisis in the Caribbean. Newsday. Retrieved from http://www.newsday.co.tt/commentary/0,138724.html

Krzyzanowski M, Cohen A (2008) Update of WHO air quality guidelines. Air Qual Atmos Health 1(1):7–13

Long CM, Suh HH, Koutrakis P (2000) Characterization of indoor particle sources using continuous mass and size monitors. J Air Waste Manage Assoc 50(7):1236–1250
Lowe DJ, Hunt JB (2001) A summary of terminology used in tephra-related studies
Macpherson CC, Akpinar-Elci M (2015) Caribbean heat threatens health, well-being and the future of humanity. Pub Health Ethics 8:196–208. phv008
Matthew J, Pereira LMP, Pappas TE, Swenson CA, Grindle KA, Roberg KA, Gern JE (2009) Distribution and seasonality of rhinovirus and other respiratory viruses in a cross-section of asthmatic children in Trinidad, West Indies. Ital J Pediatr 35(1):1
Milián E, Díaz AM (2004) Allergy to house dust mites and asthma. P R Health Sci J 23(1)
Monteil MA (2008) Saharan dust clouds and human health in the English-speaking Caribbean: what we know and don't know. Environ Geochem Health 30(4):339–343
Monteil MA, Joseph G, Chang Kit C, Wheeler G, Antoine RM (2004) Smoking at home is strongly associated with symptoms of asthma and rhinitis in children of primary school age in Trinidad and Tobago. Rev Panam Salud Publica 16(3):193–198
Nriagu JO (1990) The rise and fall of leaded gasoline. Sci Total Environ 92:13–28
Ostro B (2004) Outdoor air pollution. WHO Environ Burden Dis Ser 5
OVE (2014) Thematic evaluation. Climate change at the IDB: building resilience and reducing emissions. Office of Evaluation and Oversight of the Inter-American Development Bank. Retrieved September, 2016, from https://publications.iadb.org/bitstream/handle/11319/6692/CC_EnglishBRIK.pdf?sequence=1
PAHO-WHO (2005) An assessment of health effects of ambient air pollution in Latin America and the Caribbean. Retrieved September, 2016, from http://www.bvsde.paho.org/bvsea/fulltext/assess.pdf
Prospero JM, Collard FX, Molinié J, Jeannot A (2014) Characterizing the annual cycle of African dust transport to the Caribbean Basin and South America and its impact on the environment and air quality. Glob Biogeochem Cycles 28(7):757–773
Richter R, Berger UE, Dullinger S, Essl F, Leitner M, Smith M, Vogl G (2013) Spread of invasive ragweed: climate change, management and how to reduce allergy costs. J Appl Ecol 50 (6):1422–1430
Romieu I, Weitzenfeld H, Finkelman J (1989) Urban air pollution in Latin America and the Caribbean: health perspectives. World Health Stat Q. Rapport trimestriel de statistiques sanitaires mondiales 43(3):153–167
Sanhueza E, Africano M, Romero J (1982) Air pollution in tropical areas. Sci Total Environ 23:3–10
Schwindt CD, Bacchus H, Leu S-Y, Tjoa T, Klebanova Y, Delfino RJ (2010) Characterization of the asthmatic population of St. Vincent and the Grenadines: asthma severity levels and atopic sensitization. J Asthma 47(8):871–877
Segal C, Nilsson A (2015). The use of human rights in the fight against air pollution.
Tanveer Z, Nadeem I, Raza S, Akram N (2014) Impact of global air pollution on human health: an empirical investigation. J Glob Sci Issues 2(4)
The United Nations (2015) Carbon dioxide emissions (CO2), metric tons of CO2 per capita (CDIAC). Retrieved October, 2016, from http://mdgs.un.org/unsd/mdg/SeriesDetail.aspx?srid=751
The World Bank (1996) Air pollution from motor vehicles: standards and technologies for controlling emissions. Retrieved September, 2016, from http://www.un.org/esa/gite/iandm/faizpaper.pdf
The World Bank (2016) Latin America & Caribbean. Retrieved October, 2016, from http://data.worldbank.org/region/latin-america-and-caribbean?view=chart
Trewin B (2014) The climates of the tropics and how they are changing. State Trop:39–52
UN (2012) World economic situation and prospects 2012. Retrieved September, 2016, from http://www.un.org/en/development/desa/policy/wesp/wesp_current/2012country_class.pdf
UNEP (1998) Management of wastes in small island developing states. Retrieved September, 2016, from http://islands.unep.ch/dd98-7a2.htm

UNEP (2005) Urban air quality management toolbook. Retrieved September, 2016, from http://www.unep.org/urban_environment/PDFs/toolkit.pdf

UNEP, Villasol A, Beltrán J (2004) Caribbean Islands, GIWA Regional assessment 4. Fortnam M, Blime P (eds) University of Kalmar, Kalmar, Sweden. Retrieved September, 2016, from http://www.unep.org/dewa/giwa/areas/reports/r4/giwa_regional_assessment_r4.pdf

UNFCCC (2013) Trinidad and Tobago. Intended nationally determined contribution under the United Nations framework convention on climate change. Retrieved September, 2016, from http://www4.unfccc.int/Submissions/INDC/Published%20Documents/Trinidad%20and%20Tobago/1/Trinidad%20and%20Tobago%20Final%20INDC.pdf

UNFCCC (2015) Good practices and lessons learned in adaptation planning processes addressing ecosystems, human settlements, water resources and health, and in processes and structures for linking national and local adaptation planning: a synthesis of case studies. Retrieved September, 2016, from http://unfccc.int/resource/docs/2015/sbsta/eng/04.pdf

WHO (2008) Outdoor air pollution. Children's health and the environment. Retrieved August, 2016, from http://www.who.int/ceh/capacity/Outdoor_air_pollution.pdf

WHO (2013) Road safety facts in the region of the Americas, 2013. Retrieved September, 2016, from http://www.who.int/violence_injury_prevention/road_safety_status/2013/report/factsheet_paho_en.pdf

WHO (2014a) 7 million premature deaths annually linked to air pollution. Retrieved August, 2016, from http://www.who.int/mediacentre/news/releases/2014/air-pollution/en/

WHO (2014b) Public health and environment: ambient air pollution. Annual mean concentrations of fine particulate matter (PM2.5) in urban areas, 2014. Retrieved September, 2016, from http://gamapserver.who.int/gho/interactive_charts/phe/oap_exposure/atlas.html

Ziska L, Knowlton K, Rogers C, Dalan D, Tierney N, Elder MA, ... Hedberg C (2011). Recent warming by latitude associated with increased length of ragweed pollen season in central North America. Proc Natl Acad Sci, 108(10):4248–4251

Muge Akpinar-Elci, MD, MPH, worked for NIH and CDC/NIOSH. She was also a director of World Health Organization (WHO) Collaborating Center for Environmental and Occupational Health in Grenada. She joined Old Dominion University as a director of Center for Global Health and became a chair of the School of Community and Environmental Health. Her clinical and field research experiences include pulmonary medicine and environmental global health.

Olaniyi Olayinka, MBChB, MPH, is an assistant professor at the Center for Global Health at Old Dominion University. His research experience and interests include studies of the human health impacts of environmental and occupational hazards. Current research interests include environmental epidemiology, disaster science, community resilience and preparedness activities.

Chapter 22
Compounding Factors: Air Pollution and Climate Variability in Mexico City

María Eugenia Ibarrarán, Iván Islas, and José Abraham Ortínez

Abstract In early 2016, Mexico City suffered from repeated severe episodes of high ozone concentrations. Tropospheric ozone is a secondary compound produced by precursors such as nitrogen oxides and volatile organic compounds. However, other conditions such as cloud coverage, solar radiation, humidity, wind speed, and temperature play a significant role on the rate at which ground-level ozone forms. During periods of low precipitation, that is, March through May 2016, Mexico City Metropolitan Area (MCMA) witnessed high concentrations of tropospheric ozone. We look at the correlation between the occurrence of El Niño events, meteorological conditions, and ground concentration of ozone. We also describe other features of MCMA that can contribute to explain this deterioration of air quality as well as discuss health and economic costs this may entail. We finally address some public policies that may help reduce low air quality in this and other metropolitan areas.

Keywords Mexico City • Air pollution • Climate variability • Ozone peaks • Atmospheric stability • Supreme Court rulings

Introduction

In the spring of 2016, Mexico City faced several air pollution events that led to implementing harsh restrictions on the population to improve air quality. There are several reasons why pollution levels met contingency level concentrations, some being a Supreme Court ruling allowing all private passenger vehicles to circulate every day, no matter model year, as long as they approve the inspection and maintenance test;

M.E. Ibarrarán (✉)
Universidad Iberoamericana Puebla, Blvd. del Niño Poblano 2901, Reserva Territorial Atlixcáyotl, San Andrés Cholula, Puebla 72810, Mexico
e-mail: mariaeugenia.ibarraran@iberopuebla.mx

I. Islas
Mexico Low Emissions Development Program, Mexico City, Mexico

J.A. Ortínez
Instituto Nacional de Ecología y Cambio Climático and Centro de Ciencias de la Atmósfera, UNAM, México City, México

© Springer International Publishing AG 2018

R. Akhtar, C. Palagiano (eds.), *Climate Change and Air Pollution*, Springer Climate, DOI 10.1007/978-3-319-61346-8_22

lowering of the threshold to call upon a contingency; and great atmospheric stability probably linked to climate change. This article is divided into three parts. First we describe the recent trends in pollutants and the regulations that have shaped air quality in the city. The second part describes the recent evolution of air quality and the feedbacks that contributed to this, namely, climate variability and its impact on meteorological conditions and ultimately air quality. Finally, we suggest some policy recommendations that go beyond the usual regulations used to reduce emissions from the private transport sector only but take into account other sources and that can significantly improve air quality and reduce carbon emissions further.

Context and Background

Mexico City faces a wide array of challenges, one being air quality. In the spring of 2016, pollution levels led to a partial shutdown of the city and to the upscaling of prohibitions of the Hoy No Circula Program (HNC). This program implies that, on average, based on plate terminations, one day a week each car is prohibited from running in the larger Mexico City Metropolitan Area. Only recent year models that have better technology and therefore produce less emissions can run daily. To identify these vehicles, they were granted a zero or double zero hologram during the verification process that is to be held generally twice a year, depending of the year model of the car. Older cars are also expected to be idle on Saturdays. Hologram 1 is for cars that have to remain idle for one weekday and two Saturdays a month; these are the cars with electronic injection. Hologram 2 is for cars that have to remain idle once a week and all Saturdays, and these have mechanical injection. Cars with plates from outside the city have to observe these same regulations plus they are banned from 5 to 11 am every day, and from 9 pm to midnight, unless they hold a zero (or double zero) hologram.

HNC has been in place since 1986, but the point at which additional constraints kick in has become more stringent, and therefore circulation prohibitions have occurred more often, making more cars idle. Contingencies are announced when ozone concentrations go beyond levels that may harm human health. Contingency measures in Phase I include recommendations to restrict outdoor exercise, limit activities that increase congestion and the use of chemicals without filters, prevent fires, and avoid any activities that may use chemicals that are precursors to ozone. Vehicles with holograms 1 and 2, depending on if they have even or odd termination on their license plate, may also have to stop from circulating. Phase II stops 50% of the vehicle fleet from running, including federal public transport; schools may have to stop, as well as museums and parks; gas stations cannot operate; food preparation using coal or wood for cooking is prohibited; highly polluting cars are stopped; and industrial facilities have to reduce emissions by 60%. Finally, it gives discretionary power to the environmental authorities to implement other measures they see fit.

Calling for contingency actions has led to the misperception that air quality has not improved regardless the many years of HNC because contingencies are still called upon. Actually, in 2009 the activation values went from 166 IMECAS to

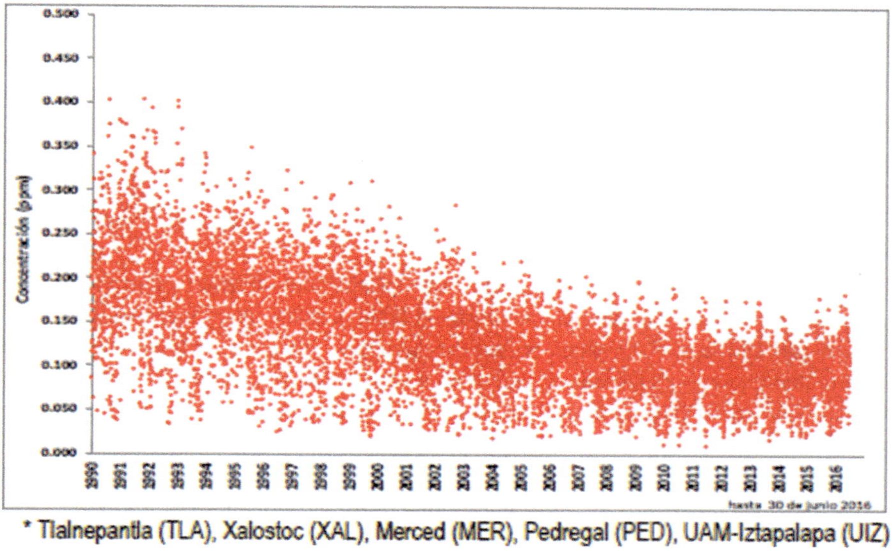

Fig. 22.1 Ozone maximum daily concentration trends in ppm from 1990 to 2016 at five historical stations of Mexico City Air Monitoring Network* (Source data: SEDEMA 2016; INECC 2016)

161, in 2010 to 156 and finally to 151 in 2011.[1] This in itself shows that the air quality has improved.

Pollution concentrations were on the right track, decreasing due to HNC and other regulations implemented during the last 25 years. These trends are shown in Fig. 22.1. However, in late 2015, a Supreme Court ruling declared that exempting a car of the HNC program based on the age of the car rather than on emission levels violated the rights to no discrimination and to equality (SCJN 2015). This obliged environmental authorities in the city to allow cars of older age to attain the zero hologram and run daily regardless their age or injection system, as long as they complied with the vehicle verification limits. Corruption played a great role into granting many more hologram zero stickers to older cars that did not meet the verification standards. This ruling, in turn, increased the number of cars on the streets on a daily basis in about 1.7 million, causing increased traffic problems in the city, increased perceived congestion, presumably lower speeds, and most likely emissions. At this point, there does not seem to exist actual estimates of these changes (INECC 2016).

[1]IMECAS stands for the Metropolitan Index of Air Quality and compares absolute values to the norm set by the WHO. Values equal to the norm are represented as 100. Values above the norm are above 100.

Peaks and Feedbacks

In mid-March of 2016, the highest ozone concentration episode of the last 14 years took place, and Phase I of an environmental contingency was called upon. Several factors played a role for this to happen. As Fig. 22.2 shows, there seems to be an inverse relation between ozone concentration levels and wind speed. Reduced wind speeds come from changes in meteorological patterns probably fostered by climate change.

Since, due to climatological conditions, additional high-concentration level events were expected to happen, a revamped HNC was designed that would operate from April 1 to June 30. This increased the days in which each car had to be idle and eliminated exceptions for newer cars with a zero hologram. Now all cars had to be idle for two Saturdays a month as well. New standards were set for the vehicle verification program, giving holograms an exemption from the HNC program based on emissions rather than on the year model of the car. Restrictions on circulation, even under the presence of a zero hologram, were reestablished for all cars.

Mexico City faced 80 atmospheric contingencies between March and June of 2016. Phase I contingencies became active and lasted anywhere from a few hours to 3 days. Most lasted for 1 day only. The day after the contingency was called upon and emission control actions were implemented, maximum ozone concentration decreased from 23% to 37% (INECC 2016). However, in one case, in May 2–5, even though ozone concentrations reduced the next day, the day later it climbed back up, maybe due to atmospheric stability in this central part of the country that inhibited dispersion of pollutants. On the other hand, CO concentrations went down anywhere from 11% to 47% and NOx from 5% to 46% after doubling up HNC. This

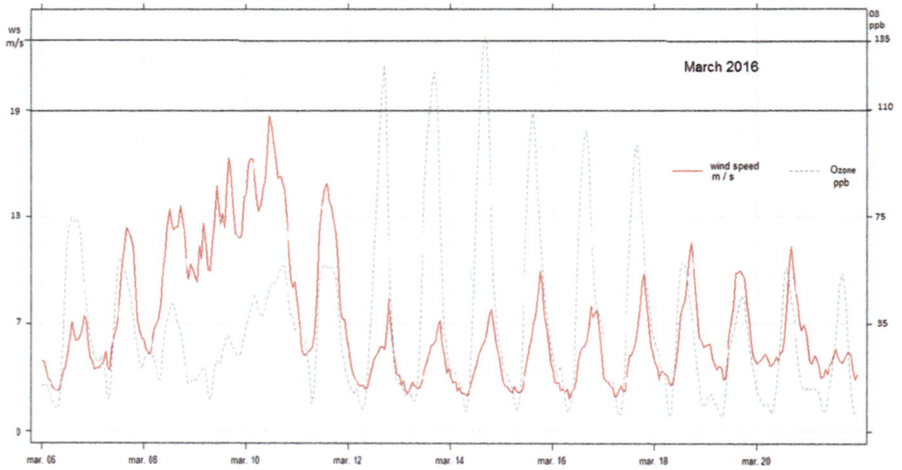

Fig. 22.2 Ozone and wind speed during the high pollution concentration episode (Source data: SEDEMA 2016)

undoubtedly led to lower health impacts on the population, but they have not been measured.

This reloaded HNC program ended on June 30, 2016, and no contingencies were called upon for the rest of the year. Upscaling HNC and the beginning of the rainy period have contributed to a better air quality, but the cost has been significant to citizens. Among these costs, there was a significant increase in transport tariffs, such as those of Uber, that due to their dynamic prices, increases up to 9.9 times during contingency days. This was because they had to attend about 64% more rides with 40% less of their vehicle fleet. Since then, they have made agreements with the government of Mexico City to control the increase in tariffs during contingency days. These price increases, however, are a good example of the shadow costs of such contingencies.

In a longer-run perspective, several analyses have found that even though HNC had some effects when perceived as a short-term program, once it became permanent, it only gave way to more cars being bought to compensate for the car that had to be left idle (Margolis 1991). Usually, the second car that was bought was old, and therefore pollution increased per household because the older car was also used the other days of the week that it was allowed. Authorities knew this had occurred at the early stages of HNC and did not want to give signs that this newer version of HNC was permanent to avoid motivating the purchase of yet another (and older) car fleet, so they announced that the program was temporary and stopped it as soon as climatological conditions, such as rain, changed.

Atmospheric Background

The positive radiative forcing of the long-lived greenhouse gases and of short-lived climate pollutants impacts directly on the general equilibrium balance of temperature and therefore on climate change. The incoming solar radiation is mainly absorbed by gases such as ozone, carbon dioxide, methane, and nitrous oxide, as well as by tropospheric particle matter that includes black carbon aerosols and other co-pollutants, both organic and inorganic, like sulfates that are light scattering in many global climate models. The understanding of both heating and cooling atmospheric processes is currently being explored, and the temperature modeling results are quite uncertain. Thus, the effects of global climate change on air quality are still unknown.

From an air quality standpoint, there seems to be a better grasp of the effect of changes in climate, known as climatic variability due to the time scale, on ambient quality, but global models need to make further assessments. However, from a meteorological perspective, climate variability is a new normal at the larger scale. This in itself is an impact of climate change that may play a role on air quality. Pollution concentration levels are affected by perturbing ventilation rates, e.g., wind speed and convection, and other physical and chemical atmospheric processes (Jacob and Winner 2009). For instance, in cities such as Mexico City, local weather conditions have fostered the formation of tropospheric ozone and secondary particle matter, which together with atmospherics conditions like high-pressure systems that are dry and free of clouds create the conditions to increase the reactive and formation of chemical pollutants.

Thus, evaluating the effects of variations of weather conditions on air quality requires an improvement in temporal and spatial resolutions in the air quality models to align them with the global models. This also entails improving emission inventories, since often the analysis is limited by the availability information, particularly emission sources, land use, and meteorological data. However, it is possible to evaluate air quality conditions with acceptable uncertainty for short-term periods implementing weather forecasting models coupled with chemical models. Nevertheless, it is important to highlight that the uncertainties involved in modeling climate and air quality are carried into determining the feedbacks between climate change and air quality.

In any case, there is some evidence of the probable impacts of climate perturbation on regional- and local-scale atmospheric processes. In the Fourth Assessment Report (AR4) of the Intergovernmental Panel on Climate Change (2007), climate change is defined as the modifications of the mean or variability of climate properties, e.g., the increment of the global surface temperature by about 0.2 °C/decade in the past 30 years, for example (Hansen et al. 2006). If temperature increases and there is more variability, the rate of transport of pollutants from urban and regional scale to global scale could increase, and the chemical composition of the atmosphere may in turn cause a feedback effect on local weather, affecting temperature, precipitation, cloud formation, wind speed, and wind direction (Bernard et al. 2001). This may, in turn, affect anthropogenic and natural emission such as biogenic VOC releases.

Composition of Air Pollution in Mexico City

In addition to the expanded number of cars because of the Supreme Court ruling, the corruption it promoted, and the lowering of the threshold for calling upon a contingency, there are atmospheric conditions that exacerbated the effect of higher emission levels and contributed to the buildup of higher concentration of pollutants. On the one hand, ozone formations are used to respond to nitrogen oxide (NOx) concentrations, but in recent years, it was more related to concentrations of volatile organic compounds (VOC) (Molina et al. 2010; Zavala et al. 2009). This itself has significant implications that call for different policies, with a closer focus on controlling VOC to a larger extent than before. This, however, has not been turned into actual policy, e.g., the VOCs are used to manufacture goods and come in many industrial products such as paint, aerosols, and thinners, or in rugs, also mostly of these organic compounds, for instance; benzene, toluene, and formaldehyde are

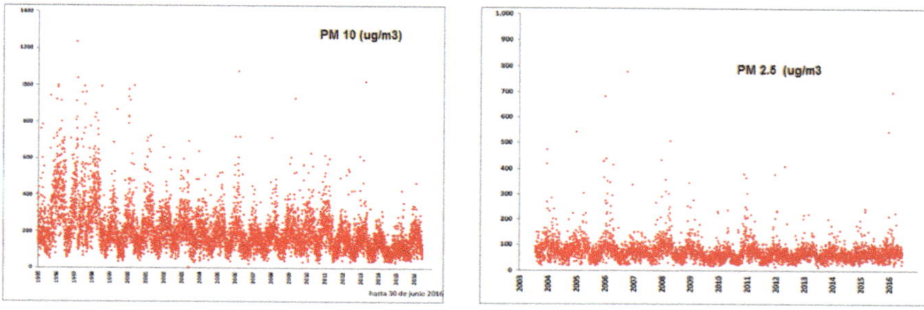

A) * Merced (MER), Tlalnepantla (TLA), Xalostoc (XAL), Pedregal (PED), Tláhuac(TAH)

B) * Merced (MER), Tlalnepantla (TLA), Coyoacán (COY), UAM-Iztapalapa (UIZ), San Juan de Aragón (SJA), San Agustín (SAG), Netzahualcóyotl (NEZ) y Camarones (CAM)

Fig. 22.3 Particle matter maximum daily concentrations trends in μg/m3. (**a**) PM_{10} at five historical stations from 1990 to 2016 and (**b**) $PM_{2.5}$ at eight historical stations of Mexico City Air Monitoring Network (Source data: SEDEMA 2016; INECC 2016)

released from fossil fuel combustion (Bravo et al. 2002). Moreover, the concentration of air pollutants has decreased in the last two decades; particularly those of NOx, SOx, and CO are now below the norm. However, particulate matter (PM_{10} and $PM_{2.5}$) (Fig. 22.3) and ozone still exceed the local regulations and those of the World Health Organisation (WHO 2016).

Concentration of pollutants respond to climatology and this is seasonal throughout the year. Ozone concentrations tend to be higher between February and June, peaking in May, when days turn longer, solar radiation increases, and lack of clouds and wind turn the lower atmosphere very stable. Figure 22.4 shows how ozone concentration lowers as winds have greater speed.

Changing Meteorological Conditions

During the low-humidity period of 2016, ozone levels and those of its precursors have been above average, compared to previous years (INECC 2016). This has been compounded by global circulation patterns causing El Niño effect. El Niño generates anomalies in Mexico's climatic conditions, reducing rain in the spring-summer period and increasing temperature, thus setting the conditions for drought. During strong El Niño events in 1982–1983 and 1997–1998, drought and high temperatures led to significant forest fires, particularly in the center of the country because of the delay in the rainfall season. Even higher temperatures have been recorded for 2016, and this in turn may increase forest fires throughout the country and therefore more emissions and VOCs. For Mexico, the maximum temperature recorded in March 2016 was 0.8 °C higher than for the 1981–2010 average, and most of the country faced maximum temperatures between 25 and 30 °C.

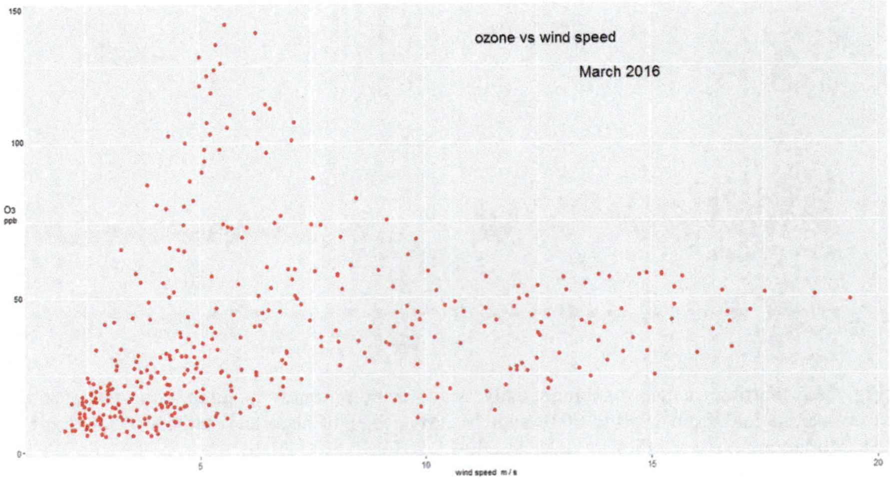

Fig. 22.4 Behavior of high concentration of ozone and low wind speed (Source data: SEDEMA 2016)

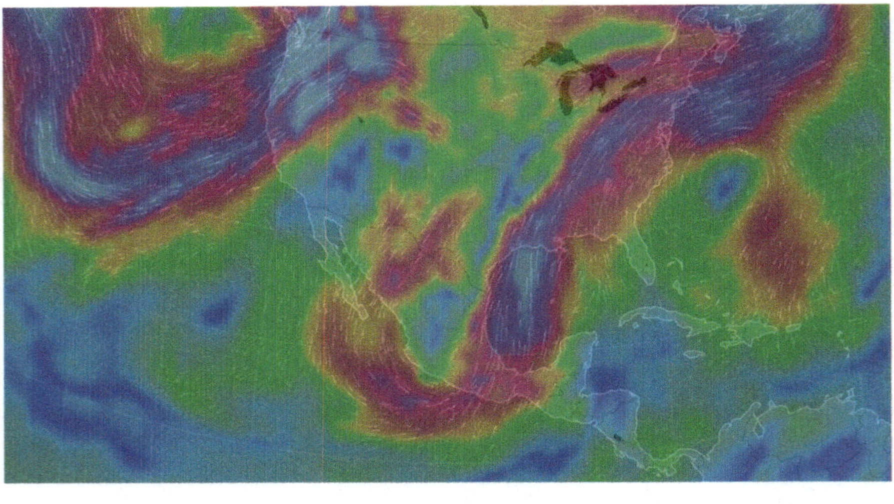

Fig. 22.5 Wind currents in March 2016. Note: Flow wind current @ 700 hPa and wind speed of 19.1 m/s, with cyclonic circulation over the central region of México, March 10 @ 18 UTC. 2016. (Source: www.windytv.com)

Meteorological conditions may have played an important role in increasing air pollution in 2016. Early in the year, wind currents covered most of the country, as shown in Fig. 22.5. During that period, the anticyclonic perturbations covered the central region of Mexico, where Mexico City is located. However, in April, the current moved northward and by May winds were located in the north, and the central part had

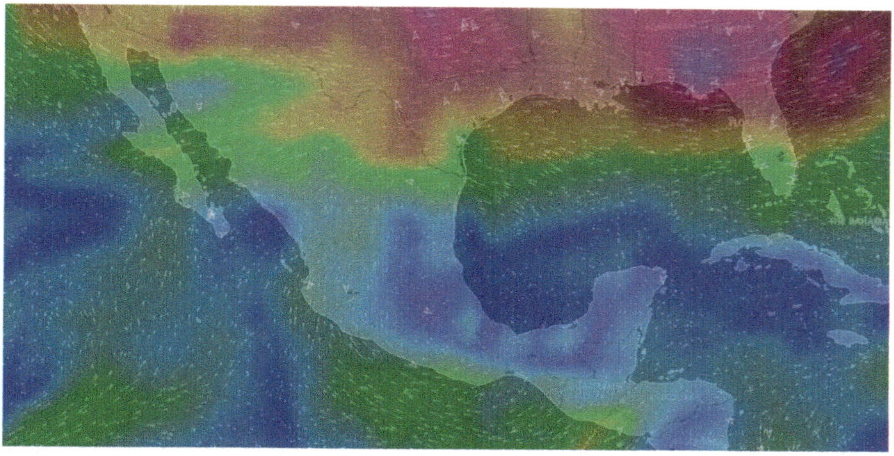

Fig. 22.6 Wind currents in May 2016 (Note: Flow wind current @ 700 hPa and wind speed of 0.3 m/s, with anticyclonic circulation over the central region of México March 14 @ 18 UTC. 2016. Source: www.windytv.com)

very week wind circulation, as seen in Fig. 22.6. This in turn created stability in the atmosphere and, colloquially, less movement of particles, air pollution included.

Clearly meteorological conditions seem to create the circumstances for pollutants to concentrate. Such conditions are attributable to climate change, and they hint at the relationship between climate change and the worsening of air pollution in the city.

Recommendations on Public Policies

As it has been argued in the paragraphs above, poor air quality interacts with climate issues with negative impacts on human health. These impacts might worsen in time as we continue to experience changes in weather patterns as a consequence of climate change. In spite of 30 years of public policies to tackle air pollution, Mexico City still faces a severe air quality problem. Although pollutants have changed, being peaks in ozone now related to VOCs the threat, the solutions remain the same. Public policies aim to change technologies in the private vehicle fleet with new ways of testing, trying to create incentives for new cleaner technology cars. This end-of-the-pipe policy might be necessary but not sufficient to tackle the entire air pollution problem.

On July 1, 2016, the Mexican Official Emergency Standard (NOM-EM-167-SEMARNAT-2016) came temporarily into force. It establishes the testing methods and emission levels of pollutants for motor vehicles circulating in Mexico City, State of Mexico, Hidalgo, Morelos, Puebla, and Tlaxcala. The new regulation seeks to solve two problems related to the current mandatory vehicle emissions testing

(PVVO). First, it addresses the technological and regulatory backwardness of the testing centers, updating its procedures by making use of diagnostic systems on board the vehicle for model year 2006 and later model year, and reduces the maximum allowable emission limits. Second, it aims to tackle corruption in testing centers through a centralized system of processing and storage of data and by on-the-road monitoring using remote sensing. These measures are undoubtedly an improvement in the PVVO, and despite not being infallible, they ensure that vehicles on the road are appropriately tested lowering the probability of gaming the system.

However, public policy must be evaluated in terms of its effectiveness to solve a problem, in this case an acute air pollution suffered by Mexico City Metropolitan Area, and its efficiency, that is, cost relative to other measures that curb air pollutants. To evaluate public policies, there must be a set of measurable indicators and targets they should achieve. Regarding the Mexican Official Emergency Standard, environmental authorities have not set a verifiable goal and a quantitative indicator to measure the potential success of this new emergency standard. Even more, the results will have to be assessed in similar weather conditions without other affecting factors, such as the rainy season.

A possible indirect indicator could be the number of vehicles verified, approved, and rejected, and compare them to those of previous semesters. The new standard could result in fewer vehicles obtaining hologram zero that allows them to circulate every day. After 1 year, this measure would probably reduce the number of vehicles on the road since most likely not all the extra 1.7 million vehicles would be able to get hologram zero again. If we assume that the measure is effective in restricting the holograms, it should be expected that at least part of that universe will return to hologram 1. This in turn should be related to a more direct indicator, the number of environmental contingencies enacted in 1 year compared to the previous one.

If there are less environmental contingencies in future years relative to 2016, remains to be seen. However, this might not be the case, as there are two factors that the new standard is not tackling: driving activity, measured in kilometers driven annually by private passenger cars, and other sources of pollution. The standard addresses only part of a component of the problem: anti-pollution vehicle technologies. This measure sets aside vehicle activity. Restricting the use of the vehicle once a week does not translate directly into lower vehicle travel or less emissions, since generally the substitute for that vehicle is another motor vehicle with the same or older technology. While the program has encouraged a newer and cleaner fleet in the Valley of Mexico, it has been unable to reduce the volume of the on-road fleet or the driving behavior of the population as there are few other quality options to commute in the city. Kilometers traveled by car increase year after year, causing more pollution and serious congestion problems that affect the economy of the city.

The second problem not addressed by the current policy is fixed sources of pollution, which include area sources. They are the first source of emissions of VOCs and the other important precursor of O_3.

Economic Instruments Toward a Change in the Energy Matrix

Environmental authorities stated that the new regulation is only one of several measures that are being taken to strengthen the system of air quality monitoring. Other mechanisms are the use of economic instruments to finance a megalopolitan fund to improve public transport and to build infrastructure for other non-motorized transport means. Other policies include better standards for fixed and mobile sources. This means that measures that aim to address the background environmental problem and its long-term effects are yet to be announced and that without them, the new vehicle verification standard will do little to mitigate air pollution, becoming, at best, a necessary but insufficient measure.

If authorities want to send the right signals of the social costs of fossil fuels use, not only those of mobile sources, it is important to attack directly the pollutant emissions coming from those sources or the use of fossil fuels as inputs of other activities. The goal of an economic instrument is to explicitly set this social external cost and internalize it. There are two ways it can be done, either by setting a cap on emissions or by imposing a price through a tax to pollutant emissions. The more general and directed to the pollutants, the more effective such taxes will be to curb emissions.

Economic instruments serve as incentives to influence the behavior of individuals. Contrary to regulatory instruments, economic instruments provide greater freedom to people to make decisions on energy use, for example, so they are more efficient in reducing the social impact. In addition, economic instruments can contribute to strengthen pollution control by generating tax revenue. It is important to mention that economic instruments do not replace but complement and reinforce regulatory approaches. Economic instruments therefore must be considered as important components of the mixtures of policies and not as independent policy packages (GTZ 2010).

Mexico already has a carbon tax at the national level. The carbon tax is part of the economic package of fiscal year 2014. This tax covers approximately 40% of total GHG emissions at the national level. It is not a tax on the total carbon content of fuels but rather additional emissions compared to those of natural gas, which is not subject to the carbon tax. The tax rate varies between US$ 1 and US$ 4/tCO_2, depending on the type of fuel and with a limit of 3% of the sale price of the fuel. The tax is paid at the time of importation or production and can be credited, except for the final sale (similar to VAT). According to the Federal Revenue Act for fiscal year 2014, the federation would receive tax revenues representing 0.328% of the federal government's total revenues. By 2015, they represented 0.210% of total revenues. So far, revenues from this tax are not labeled to direct investment in environmental measures.

There are three drawbacks that stop the Mexican carbon tax from becoming a true Pigouvian tax controlling global and local pollution. The first one is that the tax is not based on the social cost of carbon. Even more, it is too low to change investment decisions and does not creates incentives to shift to clean technologies across different sectors. It has become only another source of fiscal revenue. The second one is that it does not tax gas, a fossil fuel producing methane fugitive emissions at the source of extraction and in its transportation. At the local level, its combustion in fixed and mobile sources produces important local pollutants that impact human health. The third one is that it misses the opportunity of a double dividend by not directly recycling tax revenues either by a reduction on income taxes or lowering any other tax that imposes a cost on economic activities (Landa et al. 2016).

Conclusions

In sum, regardless the long-run efforts put into reducing air pollution in Mexico City, most of the policies have concentrated on vehicle emissions. This has proved not to be enough given the increase in vehicles circulating in the city and the poor public transport options that have not kept pace with demand for mobility. A reduction in urban local pollutants and greenhouse gases that may reduce air pollution and mitigate climate change will only come from a true change in the energy matrix. Such a change may only be produced in the medium run by the use of economic incentives to deter the use of highly polluting fuels and to embark into long-term investments that will need less and cleaner energy sources.

References

Bernard SM, Samet JM, Grambsch A, Ebi KL, Romieu I (2001) The potential impacts of climate variability and change on air pollution-related health effects in the United States. Environ Health Perspect 109(Supplement 2):199–209

Bravo H, Sosa R, Sánchez P, Bueno E, González L (2002) Concentrations of benzene and toluene in the atmosphere of the southwestern area at the Mexico City Metropolitan Zone, Atmos Environ 36(23):3843–3849. ISSN 1352-2310, http://dx.doi.org/10.1016/S1352-2310(02)00292-3

Deutsche Gesellschaft für Technische Zusammenarbeit (GTZ) (2010) Economic Instruments in the waste management sector. Experiences from OECD and Latin American Countries. Report prepared by Green Budget Germany on behalf of GmbH. Berlin. Disponible en: http://www.foes.de/pdf/2010%20FOES%20Economic%20Instruments%20Waste%20Management%20final.pdf

Hansen J, Sato M, Ruedy R, Lo K, Lea DW, Medina-Elizade M (2006) Global temperature change. Proc Natl Acad Sci U S A 103(39):14288–14293

INECC (2016) Evolución de la Calidad del Aire en la ZMVM y episodios de ozono durante la temporada seca-caliente 2016. Informe Técnico, SEMARNAT

IPCC (2007) Climate Change 2007: synthesis report. contribution of Working Groups I, II and III to the fourth assessment report of the Intergovernmental ¨Panel on Climate Change [Core Writing Team, Pachauri RK, Resinger A (eds)]. IPCC, Geneva, 104 pp

Jacob DJ, Winner DA (2009) Effect of climate change on air quality. Atmos Environ 43(1):51–63. http://dx.doi.org/10.1016/j.atmosenv.2008.09.051

Landa G, Reynès F, Islas I, Bellocq F, Grazi F (2016) Towards a low carbon growth in Mexico: is a double dividend possible? A dynamic general equilibrium assessment. Energy Policy 96 (2016):314–327. www.elsevier.com/locate/enpol

Margolis S (1991) Back-of-the-envelope estimates of environmental damage costs in Mexico. Working paper IDP104. The World Bank, Washington, DC

Molina LT, Madronich S, Gaffney JS, Apel E, de Foy B, Fast J, Ferrare R, Herndon S, Jimenez JL, Lamb B, Osornio-Vargas AR, Russell P, Schauer JJ, Stevens PS, Volkamer R, Zavala M (2010) An overview of the MILAGRO2006 campaign: Mexico City emissions and their transport and transformation. Atmos Chem Phys 10(8697–8760):2010. doi:10.5194/acp-10-8697-2010

SCJN (2015) Gaceta del Semanario de la Suprema Corte de la Nación, No. 25894. México D.F. http://sjf.scjn.gob.mx/SJFSist/paginas/DetalleGeneralScroll.aspx?id=25894&Clase=DetalleTesisEjecutorias&IdTe=2010225

SEDEMA (2016), http://www.aire.cdmx.gob.mx/estadisticas-consultas/concentraciones/index.php

WHO World Health Organization, Air quality guidelines global update. http://www.who.int/mediacentre/factsheets/fs313/en/

Zavala M, Lei W, Molina MJ, Molina LT (2009) Modeled and observed ozone sensitivity to mobile-source emissions in Mexico City. Atmos Chem Phys 9:39–55

María Eugenia Ibarrarán is director of the Research Institute on Environment, Xabier Gorostiaga S. J., at Universidad Iberoamericana Puebla. She earned a PhD in geography and an MA in energy and environmental studies from Boston University and a BA in economics from Instituto Tecnológico Autónomo de México (ITAM). Her research interests are development policies, energy and environmental economics modelling and policy-making, as well as mitigation and adaptation to climate change.

Iván Islas holds a degree in economics from Universidad de las Americas Puebla and a master's degree in environmental economics and natural resources from University College London. He was the director of Environmental Economics at the National Institute of Ecology and Climate Change (2005–2015), where he specialized in economic analysis of energy and climate policies. He is now the climate policy lead for the Mexico Low Emissions Development Program.

José Abraham Ortínez was the former deputy director of the Environmental Modelling Division at INECC. He received his BSc in chemical engineering and MSc in atmospheric physics from the UNAM and currently is finishing his PhD in the field of emission inventories and modelling and black carbon measurements. His main area of experience is air quality modelling, emission inventories, air pollution monitoring and policy-making.

Chapter 23
Air Pollution, Climate Change, and Human Health in Brazil

Júlia Alves Menezes, Carina Margonari, Rhavena Barbosa Santos, and Ulisses Confalonieri

Abstract Air pollution, especially after the industrial revolution, has adversely affected human health both in Brazil and worldwide. In Brazil, the most common pollutants are associated with biomass burning and the energy sector (transport) and include aldehydes, sulfur dioxide nitrogen dioxide, hydrocarbons (methane and non-methane), particulate matter, and ozone. These gases accumulate in the stratosphere and may influence both directly and indirectly the greenhouse effect which, in turn, impacts the climate and human health. The combination of changes in precipitation and temperature patterns coupled with increased pollution may intensify problems related to infectious diseases, coronary-respiratory diseases, cancer, and premature death, among other health issues. Surveys designed locally may reveal where the data is insufficient and what information on climate risks and associated health conditions need to be better understood. This may provide accurate information on national policies and support the most urgent adaptation actions to the populations at risk.

Keywords Air pollution • Pollution in Brazil • Human health • Climate change • Particulate matter • Ozone

Introduction

Atmospheric pollutant is any form of matter or energy with intensity and in quantity, concentration, time, or characteristics not in accordance with established levels and which render or may render the air inappropriate, harmful, or offensive to health; inconvenient to public welfare; harmful to materials, fauna, and flora; and detrimental to the safety, usage, and enjoyment of property and the normal activities of the community (CONAMA Resolution 003 1990).

J.A. Menezes (✉) • C. Margonari • R.B. Santos • U. Confalonieri
René Rachou Institute, FIOCRUZ, Belo Horizonte, Brazil
e-mail: julia.menezes@cpqrr.fiocruz.br; margonari@cpqrr.fiocruz.br; rhavena.santos@cpqrr.fiocruz.br; uconfalonieri@cpqrr.fiocruz.br

© Springer International Publishing AG 2018
R. Akhtar, C. Palagiano (eds.), *Climate Change and Air Pollution*, Springer Climate, DOI 10.1007/978-3-319-61346-8_23

A milestone for the air pollution background was the measurement of carbon monoxide (CO) concentration over Asia, Africa, and South America, in 1981. Performed by the Columbia space shuttle, it was the first time that pollution was perceived as an international problem. The images showed that, in addition to the burning of fossil fuels, the biomass burning (forest fires and burning of agricultural residues, among others) could affect regional and global air quality (Akimoto 2003).

In general, air quality is the result of the interaction of a complex set of factors, such as the magnitude of the emissions, topography, and meteorological conditions of the region, which may be favorable or not to the dispersion of the pollutants. Regarding the magnitude of emissions, anthropic activities related to industrial processes and power generation, motor vehicles, and forest fires are considered the major causes of the introduction of polluting substances into the atmosphere. The pollutants emitted by these activities are diverse and comprise an important group due to the frequency of occurrence and adverse effects to the environment and health, namely, aldehydes (RCHO), sulfur dioxide (SO_2), hydrocarbons (methane and non-methane hydrocarbons), total suspended particles (TSP) and inhalable particles (particulate matter, PM), carbon monoxide (CO), photochemical oxidants expressed as ozone (O_3), and nitrogen oxides (NOx) (Brazilian Ministry of Health 2013).

Some environmental and human health damage from the most important air pollutants are shown in Table 23.1.

Atmospheric Pollution: A Brief Brazilian Policy Scenario

The Brazilian government established air quality patterns for some of these pollutants through the Resolution 003/90 of the National Council of Environment (CONAMA), and there is a National Air Quality Control Program, implemented in 1989 (CONAMA resolution n° 005 1989), and its goal is to fix parameters to the emission of gaseous pollutants and particulate matter by stationary sources. The mean values were established for the following pollutants: total suspended particles (TSP), smoke, inhalable particles (PM_{10}), nitrogen dioxide (NO_2), sulfur dioxide (SO_2), carbon monoxide (CO), and ozone (O_3), as shown in Table 23.2. This resolution also defined two types of air quality pattern, a primary one, in which overcoming the established threshold may impact health, and a secondary one, where the concentration causes minimum adverse effect on the human well-being. Although $PM_{2.5}$ is of great relevance to the air pollution and public health issue, the country does not yet have any regulations for the concentration of this pollutant. The World Health Organization (WHO) recognizes countries' autonomy in regulating their air quality parameters rather than following global standardizations, once important conditions such as health risks, technical viability, and economic factors are locally defined (WHO 2006).

Table 23.1 Pollutants, their origins, and effects on health and the environment

Pollutant	Source	Health damages	Environmental damages
Carbon monoxide (CO)	Incomplete combustion of materials containing carbon such as petroleum and coal	Causes respiratory distress and suffocation. It is dangerous for those who have heart and lung problems	–
Ozone (O_3)	It is not a pollutant emitted directly by anthropic sources but formed in the atmosphere through the reaction between the volatile organic compounds and nitrogen oxides in the presence of sunlight	Irritation in the eyes and respiratory tract, aggravating preexisting diseases such as asthma and bronchitis, reduced lung functions	Damage to crops, natural vegetation, and ornamental plants. It can damage materials due to its high oxidizing power
Nitrogen oxides (NOx)	Burning of fuels at high temperatures in vehicles, airplanes, and incinerators	They act on the respiratory system and may cause irritation and respiratory problems or pulmonary edema at high concentrations	NO_2 can lead to the formation of photochemical smog and acid rain and has effects on global climate change
Sulfur dioxide (SO_2)	Burning of fossil fuels containing sulfur, such as fuel oil, coal, and diesel. Natural sources, such as volcanoes, also contribute to the increase of SO_2 concentrations in the environment. It can react with other compounds in the atmosphere to form particulate material of reduced diameter	Irritating action in the respiratory tract, which causes coughing and even shortness of breath. It aggravates the symptoms of asthma and chronic bronchitis and, still, other sensory organs	May react with water in the atmosphere forming acid rain
Suspended particles – size <100 microns	Incomplete combustion from industry, combustion engines, fires, and dust	Interferes in the respiratory system and can affect the lungs and the whole organism	Damage to vegetation, reduced visibility, and soil contamination
Total hydrocarbons	Industrial and natural processes. In urban centers the main sources of emissions are cars, buses, and trucks, in the processes of burning and evaporation of fuels	–	They are precursors for the formation of tropospheric ozone and present potential greenhouse effect (methane)
Inhalable particles – size <10 micron	Combustion processes (industries and automotive vehicles) and secondary aerosol (formed in the atmosphere). In nature, they can originate from pollen, marine aerosol, and soil	Respiratory cancer, arteriosclerosis, lung inflammation, worsening of asthma symptoms, increased hospitalizations, and death	Damage to vegetation, reduced visibility, and soil contamination

Source: Brazilian Ministry of Health (2016), State Foundation for Environmental Protection Henrique Luiz Roessler

Table 23.2 Air quality standards adopted in Brazil

Pollutant	Mean time sampling	Concentration (annual violations allowed) Primary standard	Secondary standard
TSP (μg.m^{-3})	24 h	240 (1)	150 (1)
	Annual (geometric mean)	80	60
Smoke (μg.m^{-3})	24 h	150 (1)	100 (1)
	Annual	60	40
Inhalable particles – PM_{10} (μg.m^{-3})	24 h	150 (1)	Equal to the primary standard
	Annual	50	
SO_2 (μg.m^{-3})	24 h	365 (1)	100 (1)
	Annual	80	40
CO (μg.m^{-3} – ppm)	1 h	40.000–35 (1)	Equal to the primary standard
	8 h	10.000–9 (1)	
O_3 (μg.m^{-3})	1 h	160 (1)	Equal to the primary standard
NO_2 (μg.m^{-3})	1 h	320	190
	Annual	100	Equal to the primary standard

Source: Brazilian Ministry of Environment (2016); CONAMA Resolution n° 003 (1990)

Regarding the impacts of pollutant on health, Brazil presents a surveillance in environmental health related to the Air Quality Program ("Vigiar") to promote the health of populations exposed to factors related to air pollutants, either from metropolitan regions, industrial plants, and areas under the impact of mining or the influence of biomass burning (Freitas et al. 2013). Vigiar's strategies to achieve the goal of health promotion are focused on the situation diagnosis. Freitas et al. (2013) argue that these strategies include both the prioritization of municipalities with greater population at risk to atmospheric pollution, the so-called risk identification instrument, and the mapping of critic areas related to air quality that might be of interest to the health issue. In addition, Vigiar proposes health impact assessment strategies, such as knowing the health situation of populations regarding air pollution, risk assessments related to disease emergence by exposure to air contaminants, and the deployment of sentinel units in priority areas (Freitas et al. 2013).

Meteorology and Pollution

The impacts of air pollution on health of urban populations may vary depending on the characteristics of the pollutants present and of their concentration in the atmosphere. On the other hand, the concentration of pollutants is capable of determining an increased likelihood pathological effects, such as allergies, although some meteorological and climatic aspects may also influence on the permanence and generation of pollutants, contributing to the higher or lower incidence of diseases.

The concentration of atmospheric pollutants is a result of interactions between the local climate patterns, the atmospheric circulation characteristics, the wind, topography, human activities (transportation, energy generation), and human response to climate change, among other factors (Ebi and McGregor 2008). Air pollution events are frequently associated to some phenomena such as (1) anticyclone or stationary high-pressure systems that reduce the dispersion, diffusion, and deposition of pollutants; (2) physical characteristics of the wind, like turbulence and temperature; and (3) meteorological conditions that influence chemical and physical processes involved in the formation of secondary pollutants, such as ozone (Arya 2000; Nilsson et al. 2001; Rao et al. 2013).

Although the meteorological conditions may act dissipating or concentrating pollutants, some of them present a straight relationship to climate, impacting it either at the regional or global levels. The report *Integrated assessment of black carbon and tropospheric ozone: summary for decision makers*, published by the United Nations Environment Programme and the World Meteorological Association, highlights the impacts of black carbon, a component of particulate matter, and ozone as pollutants affecting the climate dynamics, since they are capable of disturbing tropical rainfall and regional circulation patterns (UNEP and WMO 2011). Currently, there are only expectations about climate change effects on air pollution concentrations, although the human-induced climate change is known to influence some meteorological factors, such as temperature, precipitation, and solar radiation, which directly affect the concentration of some pollutants (Kinney 2008). These climate drivers of the pollutant cycle are modified as climate changes, and this in turn is expected to affect air quality (Giorgi and Meleux 2007).

In general, regional and global climate models have shown that global warming may result in a worsening of air quality in urban centers, including increased levels of ozone, particulate matter (PM), and pollens, among others (Harlan and Ruddell 2011; Jacob and Winner 2009; Kinney 2008). In the case of PM, for example, studies pointed out that climate change may reduce or increase its concentration due to regional differences in rainfall or temperature (Heald et al. 2008; Jacob and Winner 2009). The studies for tropospheric ozone are quite conflicting, since some of them have shown a decrease in its concentration due to humidity and temperature rise, whereas others have observed an increase related to warmer temperatures (Aw and Kleeman 2003; Girogi and Meleux 2007; Sillmaan and Samson 1995). Some mechanisms by which the climate change may affect air quality, either local or regionally, are changes in rates of chemical reactions and the height of the atmospheric layers that are closest to the ground, influencing the vertical mixture of pollutants. The modification of human behavior or changes in the levels of biogenic emissions – some vegetation species naturally produce ozone precursors, but in larger quantities at higher temperatures – is considered an indirect effect that may increase or decrease anthropogenic emissions (Ebi and McGregor 2008).

Factors like temperature, wind speed, and precipitation may influence air quality and climate. Although the studies are conflicting, there is evidence that atmospheric pollutants, such as ozone and fine particulate matter, interact with temperature by raising heat-related mortality, even in milder climates (Nawrot et al. 2007; Ren et al. 2008). Other studies have also shown that the pollution-climate interaction presents distinct effects in each locality evaluated; both the combined effect and the individual contribution of each of the two factors may change according to the local profile, generating different mortality risks (Filleul et al. 2006). While many researches have associated air pollution to increased mortality, what is more prominent for acute episodes like London in 1952, there is evidence that even low concentrations of pollutants can raise mortality due to decreased lung function, respiratory symptoms, asthma, chronic bronchitis, and cardiovascular disease (Brabin et al. 1994; Gonçalves et al. 2005; Logan 1953; Pope et al. 1992, 1999; White et al. 1994). In this sense, as argued by Kinney (2008), future control of levels of key health-relevant pollutants, like ozone and fine particles, should incorporate assessment of potential future climate conditions and their possible influence on the attainment of air quality objectives.

Among the atmospheric pollutants, those that cause the greatest public health concern are the particulate matter and ozone. These pollutants have been showing consistent associations with certain health conditions, both internationally and locally. The contribution of these pollutants to climate and public health is addressed in the following topics.

Particulate Matter (PM_{10} and $PM_{2.5}$) and Meteorological Factors

Particulate matter is closely linked to anthropic activities, and its main source is the burning of fossil fuels, whether from automotive vehicles or from industrial plants, energy, or biomass burning. In Brazil, pollutant emissions in urban areas play an important role in the local climate, with vehicles being considered the main emission source of these compounds. Although the country has a very distinct emission profile, given that its energy matrix is mostly hydroelectric and the light vehicular fleet makes massive use of alcohol, transport planning is mainly based on diesel-powered heavy-duty vehicles, one of the main sources of inhalable particulate matter and other pollutants (Miranda et al. 2012).

The particulate matter presents itself in aerosol form and may vary in size, number, shape, surface area, chemical composition, solubility, and origin. The distribution of these suspended particles is trimodal comprising coarse (PM_{10}), fine ($PM_{2.5}$), and ultrafine particles (Fig. 23.1), which are especially classified for their relevance in causing health damage. The thick particles, PM_{10}, are derived mainly from the suspension or resuspension of dust, soil, and other materials such as asphalt, sea salt, and pollen, among others (Pope III and Dockery 2006). The fine

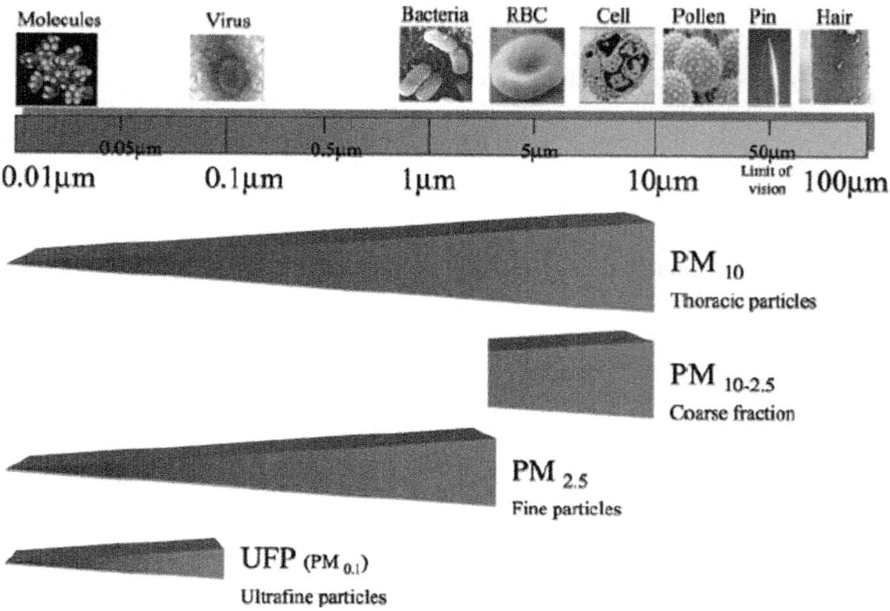

Fig. 23.1 Size distribution of the polluting particulate matter (Source: Brook et al. 2004)

particles, $PM_{2.5}$, originate directly from combustion processes (gasoline or diesel automotive vehicles), biomass burning, and coal and industrial processes – foundries, steelworks, cement, etc. (Pope III and Dockery 2006). Particularly, $PM_{2.5}$ is the most studied air pollutant and is commonly used as a proxy for exposure to air pollutants in general.

Among the urban atmospheric pollutants, sulfur dioxide, ammonia, and nitrogen oxides act as precursors of other compounds, such as sulfuric acid and ammonium derivatives, which constitute important fractions of PM_{10} and $PM_{2.5}$ (Miranda et al. 2012). These particles are efficiently eliminated by precipitation, which works as the main sink, causing a reduced availability of PM in the atmosphere. In general, this occurs in a few days' time, in the boundary layer of the troposphere, to a few weeks, in the free troposphere (Jacob and Winner 2009). For this reason, as far as air quality is concerned, the concentration of PM has local rather than global relevance, since precipitation inhibits the transfer of the particles by continental air masses, as with other pollutants. Exceptions are plumes from large dust storms and forest fires, which can be transported on intercontinental scales (Jacob and Winner 2009).

Although the influence of aerosols on the global climate is well studied but not fully understood, the relationship between PM and some meteorological variables is still poorly demonstrated in the scientific literature. Jacob and Winner (2009) compiled studies of climate models and atmospheric pollution and observed that some studies were able to establish a positive relationship between regional atmospheric stagnation and PM concentration and a negative relationship between relative humidity and PM. The precipitation, as the main dissipating mechanism, presents a tendency to decrease the concentration of particles, wherein the frequency of these rains is a determinant factor (Balkanski et al. 1993; Dawson et al. 2007). A study conducted in six Brazilian capitals showed that there are differences in PM concentration due to some meteorological factors (Miranda et al. 2012). In the cities of São Paulo, Rio de Janeiro, Belo Horizonte, and Curitiba, there was a strong negative correlation between $PM_{2.5}$ and accumulated precipitation. However, the wind speed was not associated with the concentration of these particles in any of the cities studied.

An important aspect related to PM is its composition – if derived from sulfuric acid, there is a tendency to increase the concentration along with the temperature. Yet, if derived from nitrates, compounds that experience conversion from particle to gas with increased temperatures, the tendency is of PM reduction (Dawson et al. 2007; Tsigaridis and Kanakidou 2007). Miranda et al. (2012) observed that the concentrations of SO_4^{2-} ions were the highest among several types of cations and anions measured in some Brazilian cities. These results demonstrate that climate change may significantly influence the air quality in the country, especially in urban centers. Moreover, in large cities, a fraction of the fine particulate matter produced by vehicular combustion engines has the property of strongly absorb radiation, the so-called black carbon. This compound, which is also widely produced in forest fires, is able to interfere on climate in three different ways: (1) directly absorbing the radiation, (2) reducing the albedo of snow and ice by deposition, and (3) interacting with clouds, due to its aerosol nature (Costa and Pauliquevis 2014). Due to its ability to raise the atmospheric temperature, black carbon also plays an important role in global climate change. Jacobson (2001) suggests both that atmospheric warming due to black carbon-type aerosols could balance the cooling effect associated with other types (sulfates) and that its direct radiative forcing may exceed that associated to CH_4. Thus, aerosol particles, a product of incomplete combustion processes, would be second only to CO_2 in the radiative contribution to the warming of the atmosphere (Freitas et al. 2005).

Although not conclusive, some climate models examined the impacts of climate change on air pollution and pointed out that (1) PM may reduce in some regions and increase in others, mainly due to differences in precipitation regime, and (2) there may be a positive response from PM to the expected temperature raise for the next decades, especially in already polluted areas (Heald et al. 2008; Jacob and Winner 2009). Other indirect processes of climate change may also be responsible for raising the concentration of PM, as is the case of forest fires becoming more frequent in a drier climate. In this sense, air pollution maps, produced by the WHO for the 5-year period 2008–2013, show for Brazil an annual estimate for $PM_{2.5}$ of at least 11–15 $\mu g.m^{-3}$ and for the region known as "arc of deforestation," and with high forest fires' frequency of fires, this value rises to 16–25 $\mu g.m^{-3}$ (WHO 2016).

Ozone and Climate

Ozone (O_3) may occur naturally or be formed in a secondary process by the photochemical oxidation of carbon dioxide, methane, and other volatile organic compounds (VOCs) under conditions of intense radiation and high temperatures. Biogenic emissions of O_3 occur mainly by vegetation that produces VOCs, one of its precursors, but this pattern has been changing in the last century due to changes in land use associated with biomass burning, urbanization, and massive use of petroleum-powered automotive vehicles, which produce nitrogen oxides and VOCs in large quantities (Ebi and McGregor 2008).

Unlike particulate matter, O_3 may remain in the atmosphere for days to weeks and is liable to be carried out from the continents to very distant locations when available in the free troposphere. This allows high concentrations of this gas to be extended for thousands of miles, including rural or nonexposed areas far from the emission source. In Brazil, a study carried out in Cubatão, a state of São Paulo, showed that only a small portion of the observed pollutants were from local sources; the rest were due to mass transport-high-pollutant concentrations associated with north-northeast (land breeze) and south-southeast (sea breeze) air flow (Silva 2013). Historically, atmospheric concentrations of O_3 have increased in both polluted and remote regions, having since the industrial age doubled due to the anthropogenic emissions of its major precursors, methane and nitrogen oxides (Brook et al. 2004; Wang and Jacob 1998).

Atmospheric O_3 is considered a greenhouse gas, with two roles in the thermal balance of the planet: (1) to absorb ultraviolet radiation, warming the stratosphere, and (2) to absorb infrared radiation that is reflected by the earth's surface, trapping heat in the troposphere. The influence of O_3 concentration on climate, then, depends on the altitude at which these processes occur. Thus, although industrial ozone-depleting gases such as chlorine and bromine may have a cooling effect on the stratosphere, other O_3 precursor gases produced in anthropic processes remain in the troposphere, leading to surface warming.

Meteorological and chemical factors, such as temperature, humidity, winds, and the presence of other gases, influence O_3 formation, and the formed O_3 affects other components of the atmosphere. The higher solar radiation during the summer months, for example, is related to increase O_3 concentrations (Ebi and McGregor 2008; Nilsson et al. 2001). In general, the temperature increase can accelerate the reaction rates, with a strong correlation between high levels of O_3 and very warm days. Several studies have shown that high concentrations of O_3 may be due to higher biogenic hydrocarbon production, higher anthropogenic VOC production, and stagnation of atmospheric circulation, all of which are influenced by temperature (Sillmaan and Samson 1995; Lamb et al. 1987). In urban areas, the ozone formation peak is quite extensive and tends to last between the late morning and late afternoon, when the radiation is at its maximum. However, meteorological processes, such as thermal inversion, wind direction, and velocity, and the presence of other precursor compounds, may affect this pattern, causing peaks to occur at any time between morning and afternoon (Brook et al. 2004).

Regarding the urban impacts associated with the combination of air pollution and climate change, regional climate models have shown that, for the twenty-first century, the correlations found in the present between ozone and meteorological variables are sustained in the long-term projections (Jacob and Winner 2009). Furthermore, the ozone concentrations observed in the modeling were reasonably consistent with the current surface ozone measurements (West et al. 2007). Changes in ozone concentrations projected by future emission scenarios have been developed for various regions of the world, as well detailed by Ebi and McGregor (2008). The global maximum ozone concentration measured at 8 h is projected to increase by 9.4 parts per billion per volume (ppbv) compared to a concentration simulation in the year 2000, with the highest increases over South Asia (almost 15 ppbv) and with remarkable increases for the Middle East, Southeast Asia, Latin America, and East Asia (West et al. 2007).

The CONAMA, through resolution 003/90, states that the mean concentration of O_3 per hour cannot exceed 160 $\mu g.m^{-3}$. However, studies in the two Brazilian megacities, São Paulo and Rio de Janeiro, have shown a different scenario. In the year 2015 for São Paulo, for example, the national limit was exceeded by 80 days (CETESB 2016). For both cities, the phenomenon of higher concentration of O_3 during weekends was observed, precisely when there is less vehicle circulation. This phenomenon was first reported in the United States, in 1970, and it is common to large centers, presenting several explanations that relate to the availability of other O_3 precursor compounds. In general, the VOC/NOx ratio defines O_3-forming process in which one of the possible paths is the high VOC/NOx ratios favoring reactions with OH radicals, which increases ozone formation (Martins et al. 2015). This was the case of Rio de Janeiro, where the highest O_3 concentrations at weekends were controlled by VOC. The VOC/NOx ratio was high during weekends because the NOx reductions were more significant, which increased ozone formation in the period of the study (Martins et al. 2015).

Impacts on Health of Particulate Matter and Ozone in Brazil

The climate change perspective presents challenges to the issue of urban air pollution and its impacts on health, as the pollutants can either exacerbate some climatic parameters as be influenced by them. In regard to health, the particulate matter is associated with a range of acute and chronic diseases mainly related to the respiratory tract and the cardiovascular system. A publication of the Organization for Economic Cooperation and Development (OECD) estimates that more than 3.5 million people die prematurely because of atmospheric particulate matter concentration and that air pollution is expected to become the main environmental cause of mortality in the world by 2050 (OECD 2014).

Several studies have demonstrated the relationship between the high concentration of PM and cardiovascular or respiratory diseases worldwide (Gouveia and Fletcher 2000; Pope et al. 1992; Peng et al. 2005; Orsini et al. 1986). Several epidemiological studies have evidenced associations of particulate matter with the incidence of respiratory diseases in Brazil (Braga et al. 1998; Gouveia and Fletcher 2000; Gouveia et al. 2006; Miranda et al. 2012; Nardocci et al. 2013; Romieu et al. 2012; Saldiva et al. 1994). Gouveia et al. (2006) identified an association of inhalable particulate matter with increases of 4.6% in hospitalizations for asthma in children and 4.3% for chronic obstructive pulmonary disease and 1.5% for ischemic heart disease in the elderly. In fact, the population at greater risk are the elderly, children, those with chronic lung diseases or coronary disease, and patients with diabetes (Ribeiro 2008). The large Brazilian cities have shown higher levels than those established by the WHO for both pollutants, PM_{10} and $PM_{2.5}$, with an estimated excess of deaths associated with these concentrations of materials (Miranda et al. 2012; Orsini et al. 1986). Miranda et al. (2012) observed an excess mortality risk of more than 13,000 deaths per year associated with $PM_{2.5}$ concentrations above that recommended by the WHO for several Brazilian capitals.

Regarding ozone, it is one of the pollutants that contributes the most to the degradation of air quality in large urban centers. Exposure to high concentrations is associated with increased hospital admissions for pneumonia, chronic obstructive pulmonary disease, asthma, bronchitis, allergic rhinitis, and other respiratory diseases, as well as premature mortality (Aris et al. 1993; Bell 2005; Ebi and McGregor 2008; Frampton et al. 1999; Gryparis et al. 2004; Ito et al. 2005). A study conducted in nine megacities of Latin America examined the association between exposure to air pollution and mortality. It was observed that, in São Paulo and Rio de Janeiro, besides all-cause mortality being significantly associated with ozone, there was also an estimated higher risk of death for the summer (Romieu et al. 2012). For both cities, the higher risk of ozone-related mortality was associated with respiratory causes, especially in the low and high socioeconomic status groups. Although the impacts in the respiratory system are more common, Nardocci et al. (2013) observed, in addition to the association between O_3 and respiratory diseases in children under 5 years, also an association between this pollutant and cardiovascular diseases in adults above 39 years old in the city of Cubatão, São Paulo, a Brazilian city known for industrial pollution.

The Case of Megacities

The rapid growth of the world's population, especially in developing countries, coupled with the processes of continuous industrialization and migration to urban centers, has transformed megacities into important sources of pollutant emissions. While some health impacts on the inhabitants of these large centers also have a social character, environmental consequences can be noticed at regional and global scales. Therefore, air quality at these various scales and their related problems and impacts, including climate change, should be addressed in an integrated approach (Akimoto 2003).

However, there are several megacities yet understudied around the globe, especially in Africa and Asia. Thus, monitoring data are not readily available to these and other regions, making comparative studies difficult. A 2010 study assessed the health risks in megacities in terms of mortality and morbidity due to air pollution (Gurjar et al. 2010). From the WHO standardization of atmospheric pollutants SO_2, NO_2, and total suspended particles, mortality and morbidity risk due to atmospheric pollution were calculated. The findings showed that some cities such as Los Angeles, New York, Osaka, Kobe, São Paulo, and Tokyo presented low mortality rate from these pollutants. On the other hand, high numbers of deaths (15,000 a year) and elevated TSP concentration (~ 670 $\mu g.m^{-3}$) were observed in Karachi, Pakistan. The research points out to the importance of calculating types of risk estimate and the need to do so in parallel with the development of air pollution monitoring networks to obtain a more realistic basis for the consequences of air pollution (Gurjar et al. 2010). Besides that, there is a need to solve uncertainties among different monitoring methodologies, which will assist in estimating the pollution health effects and the projections of future changes (Marlier et al. 2016).

Molina and Molina (2004) argue that there is no single strategy to address the problems of air pollution in megacities. But a possible strategy based on experiences, successful or not, in many cities, is the integrated approach that considers scientific, technical, infrastructure, economic, social, and political aspects.

Highlights for São Paulo Capital

São Paulo is considered one of the most polluted cities in the world occupying the sixth position along with Mexico City; it is behind only to Beijing (China), Cairo (Egypt), Jakarta (Indonesia), Los Angeles (USA), and Moscow (Russia). The polluted air of the city of São Paulo is considered a public health problem by several researchers (Böhm et al. 1989; Saldiva et al. 1994 1995; Coelho et al. 2010). Thus, the city suffers from the worsening of pulmonary diseases and clinical condition of the patients with cardiac diseases, as well as neonatal deaths and hematological, ophthalmological, neurological, and dermatological problems, among others (Imai et al. 1985; Saldiva et al. 1994, 1995; Braga 1998; Braga et al. 2002; Gonçalves et al. 2005).

The first studies relating air pollution and population health in Brazil were developed by Ribeiro (1971). In the region of Santo André, a state of São Paulo, the author observed an association between the number of visits for upper respiratory infection and asthmatic bronchitis in children under 12 years old and the monthly rates of sulfate and suspended dust. Later, Mendes and Wakanatsu (1976) observed, for the first time, the acute effects of three intense episodes of air pollution in São Caetano do Sul, a city in the state of São Paulo. The review of 8000 medical records occurred in June 1979, showed morbidity peaks overlapping pollution peaks of particulate material and SO_2. The authors also verified an increase in the number of cases of respiratory and cardiovascular diseases surpassing the increase of attendances by other causes. Soon afterwards, Ribeiro et al. (1976) compared, through respiratory function tests, the conditions of 2000 schoolchildren aged 7–12 years living in two distinct areas of Greater São Paulo, one industrialized and the other semirural. The results showed lower rates of ventilatory capacity and symptoms of chronic lung diseases in children of the industrial region, even after controlling for socioeconomic variables.

Regarding at-risk age groups, studies have shown that children and the elderly are the most affected by air pollution, both in Brazil and internationally (Barbosa et al. 2015; Braga et al. 1999, 2001; Martins et al. 2002a, b, Rodrigues et al. 2010; Romão et al. 2013; Saldiva et al. 1994, 1995). In Brazil, Barbosa et al. (2015) observed a significant association between visits of children and adolescents with sickle cell anemia to the pediatric emergency room in São Paulo and the variation (increase) of PM_{10}, NO_2, SO_2, CO, and O_3. Another survey studied the association of respiratory morbidity in children under 13 years old to thermal comfort, air pollutants, and meteorological variations in the city of São Paulo (Coelho et al. 2010). The analysis performed showed that the air pollutants were statistically correlated with (a) hospitalizations for upper respiratory tract infections and other diseases of the respiratory tract, (b) respiratory infections of the lower respiratory tract, and (c) infections caused by influenza and pneumonia. Despite these positive results, it is known that health depends not only on environmental factors but also on results from hereditary, nutritional, and economic factors.

The WHO sets safe limits for annual mean concentration of air pollutants: 20 $\mu g.m^{-3}$ for PM_{10} and 10 $\mu g.m^{-3}$ for $PM_{2.5}$. Brazil has a national air quality standard that specifies limits for the availability of inhaling thick particles (150 $\mu g.m^{-3}$/day), but makes no mention to finer particles, $PM_{2.5}$, which are able to penetrate the respiratory tract in more depth and are associated with significant health conditions (Saldiva et al. 1994; Lanki et al. 2006; Stölzel et al. 2007). Previous research has shown that impacts relapse in a more adversely way upon the extremes of the age spectrum due to physiological and sensitivity conditions. Gouveia and Fletcher (2000) found an increase in mortality due to respiratory diseases of 6% together with increased fine particulate matter and sulfur dioxide concentrations – an even higher mortality risk for the population over 65 years old was observed in the São Paulo city. The trend of higher mortality risk for the elderly population was also confirmed in other studies, whereas the same was observed for children in Brazil, who presented an increase in hospital respiratory admission of 12% when considered PM_{10} (Braga et al. 1999; Saldiva et al. 1995).

Reviewing air pollution and pregnancy problems, various degrees of association between air pollution and problems in intrauterine growth have been found: low birth weight, conception problems, premature birth, and death from respiratory diseases due to exposure to particulate matter in the postnatal period. Romão et al. (2013) developed a study in Santo André, a state of São Paulo, a municipality heavily affected by traffic and pollution. A significant association was found between the risk of being born with low weight and exposure to PM_{10} between the first and second trimester of pregnancy. Santos et al. (2016) observed similar results regarding maternal exposure in the first and third trimester of gestation to air pollution in the city of São José dos Campos, São Paulo, with effects on weight of newborns.

About the elderly, Martins et al. (2002a) verified the effect of air pollution on the attendance of this group due to pneumonia and influenza in São Paulo city. The ecological study encompassed the time series from 1996 to 1998 and used descriptive statistics of the following atmospheric pollutants: particulate matter (PM), carbon monoxide (CO), sulfur dioxide (SO_2), nitrogen dioxide (NO_2), and ozone (O_3). The number of visits for pneumonia and influenza had a significant positive correlation with CO, SO_2, and PM_{10}. In São Paulo, studies have shown an increase of 18% in hospitalizations for chronic obstructive pulmonary disease and of 14% for asthma among the elderly. This increase was associated with daily variations in ozone concentrations up to 35.87 $\mu g.m^{-3}$ (Braga et al. 2001; Martins et al. 2002b).

Biological studies, developed in the city of São Paulo, also demonstrate the consequences that pollution might bring to the human/animal organism. de Brito et al. (2014) observed that mice exposed to concentrated atmospheric particles (CAPs) presented lung inflammation with increased neutrophils and macrophages. Mice exposed in the cold/dry period presented the most prominent inflammations. This was due to the difficulty of dispersing pollutants in the cold/dry season, which aggravates air quality in large urban centers (Albuquerque et al. 2012; Matsumoto et al. 2010). In the short term, the findings demonstrated that exposure to low concentrations of CAPs caused significant pulmonary inflammation and, to a lesser extent, changes in blood parameters. In addition, the data suggest that changes in climate may slightly alter the toxicity of CAPs in the cold/dry period and may produce a more exacerbated response.

Nationally, there is a huge contribution of the biomass burning to the concentration of particulate matter in the air, whether due to forest fires – mainly in the northern region of Brazil – or by sugarcane burning, a common procedure in the southeast region. Studies conducted in Araraquara and Piracicaba, located in the state of São Paulo, which produces 60% of Brazil's sugarcane, found a positive association between the number of daily therapeutic inhalations in health services and the concentration of particulate matter generated by sugarcane burning (Arbex et al. 2000; Cançado et al. 2006). The annual mean PM_{10} was 56 $\mu g.m^{-3}$, the same as that of the city of São Paulo in 1997, with variations between 88 and 29 $\mu g.m^{-3}$, corresponding to the harvest and inter-harvest periods, respectively. These studies raise an interesting point in demonstrating that the sugarcane straw burning emits pollutants that lead to an increase in respiratory morbidity like the pollution produced by fossil fuels in large urban centers (Abex et al. 2000; Cançado et al. 2006). In addition to worsening local air quality, pollution from this type of biomass burning may extend miles away, reaching populations far from the emission source.

Recently, there has been an effort by Brazilian researchers to understand and demonstrate the atmospheric pollution effects, especially in the state of São Paulo. A review carried out in 2015 by Pereira and Limonge showed that among the studies selected for analysis, 76% were developed in the state of São Paulo

(Table 23.3). According to the results presented, the inhalable fraction of PM_{10} was positively associated with health outcomes in 62.5% of the evaluated surveys, even though it was below the daily and annual limits recommended by CONAMA. This result points out to two evidences. The first is related to the particulate matter comprising the air pollution indicator mostly used in the monitoring of air quality. The other evidence reveals the necessity to revise national parameters of particulate matter and the inclusion of $PM_{2.5}$ fraction in the national environmental legislation (Andrade-Filho et al. 2013; Mascarenhas et al. 2008; Ignotti et al. 2010a, b).

Forest Fires and Health in Northern Brazil

Burnings in the Brazilian rain forests of the northern region, where the Amazon biome is located, are related to the human occupation of the territory, which has been occurring in migratory pulses with a focus on mining and/or the opening of agricultural frontiers (Ribeiro and Asunção 2002). Biomass burning has become a common practice in the Amazon and, in the last decades, has been mainly related to agricultural production and pasture formation. According to Ribeiro and Assunção (2002), the practice consists of incomplete combustion in the open air and depends on the type of biomass being burned and its density, humidity, and environmental conditions, especially wind speed. In this process, the resulting emissions initially comprise of carbon monoxide (CO) and particulate matter (soot), as well as simple and complex organic compounds represented by hydrocarbons (HC) and other volatile and semi-volatile organic compounds, which are of great interest in terms of public health due to high toxicity characteristics. In addition to direct emissions, atmosphere reactions between these pollutants and several other compounds present in the air occur, such as photochemical reactions with important participation of the sun's ultraviolet radiation, resulting in compounds that may be more toxic than their precursors, namely, ozone (O_3), peroxyacyl nitrates (PAN), and aldehydes (Ribeiro and Assunção 2002; Artaxo et al. 2005).

One of the most important episodes recorded in the northern region was the 1998 fire in the state of Roraima, where burnings used to clear pastures and remnants of forest escaped human control and destroyed an area of around 40,000 km^2 – about 20% of the state (Ribeiro and Asunção 2002). The effects on the environment were severe; however, those related to human health did not present great magnitude because of the low population density of the state and the northern region. In spite of this demographic factor, the risk of occurrence of similar events is constant for the region, since the same situation observed in Roraima is reproduced along the "arc of deforestation" to the south of the Amazon, comprising part of the states of Rondônia, Acre, Amazonas, Pará, Mato Grosso, Tocantins, and Maranhão (Nascimento et al. 2000; Ribeiro and Assunção 2002). According to the National Institute of Space Research (INPE), the number of forest fires accumulated in Brazil between 2012 and 2016 was 149,385 (INPE 2016). In relation to the Legal Amazon, made up of nine Brazilian states, in the same period, the region accumulated 72% of

Table 23.3 Characterization of the studies evaluated for the year of publication, period evaluated, population studied, type of pollutant evaluated, positive associations, and location, per each reference

Reference	Period evaluated	Population studied	Pollutants evaluated	Pollutants positively associated	Location
Rumel et al. (1993)	1989–1991	Total	CO	CO	São Paulo-SP
Saldiva et al. (1994)	1990–1991	Under 5 years	SO_2, CO, NO_X, PM_{10}, O_3	NO_X	São Paulo-SP
Saldiva et al. (1995)	1990–1991	Over 65 years old	SO_2, CO, NO_X, PM_{10}, O_3	SO_2, CO, NO_X, PM_{10}	São Paulo-SP
Lin et al. (1999)	1991–1993	Under 13 years	SO_2, CO, NO_X, PM_{10}, O_3	SO_2, CO, PM_{10}	São Paulo-SP
Pereira et al. (1998)	1991–1992	Fetuses up to 28 weeks	SO_2, CO, NO_2, PM_{10}, O	SO_2, CO, NO_2	São Paulo-SP
Gouveia and Fletcher (2000)	1991–1993	Under 5 years and over 65 years old	SO_2, CO, NO_2, PM_{10}, O_3	SO_2, CO, PM_{10}, O_3	São Paulo-SP
Botter et al. (2002)	1991–1993	Over 65 years old	SO_2, CO, NO_2, PTS, O_3	SO_2	São Paulo-SP
Gouveia and Fletcher (2000)	1992–1994	Under 5 years	SO_2, CO, NO_2, PM_{10}, O_3	NO_2, PM_{10}, O_3	São Paulo-SP
Gonçalves et al. (2005)	1992–1994	Under 13 years	SO_2, PM_{10}, O_3	O_3	São Paulo-SP
Kishi and Saldiva (1998)	1992–1993	Under 5 years	SO_2, CO, NO_2, PM_{10}, O_3	CO, PM_{10}, O_3	São Paulo-SP
Freitas et al. (2004)	1993–1997	Hospitalizations in children under 15 years and mortality in patients older than 65 years	CO, PM_{10}, O_3	CO, PM_{10}, O_3	São Paulo-SP
Braga et al. (2001)	1993–1997	Under 19 years old	SO_2, CO, NO_2, PM_{10}, O_3	CO, PM_{10}	São Paulo-SP
Conceição et al. (2001)	1994–1997	Under 5 years	SO_2, CO, PM_{10}, O_3	SO_2, CO, PM_{10},	São Paulo-SP
Lin et al. (2003)	1994–1995	People between 45 and 80 years	SO_2, CO, PM_{10}, O_3	SO_2, CO, PM_{10}, O_3	São Paulo-SP
Arbex et al. (2000)	1995	Total	Mass of sedimented material	Mass of sedimented material	Araraquara-SP

(continued)

Table 23.3 (continued)

Reference	Period evaluated	Population studied	Pollutants evaluated	Pollutants positively associated	Location
Martins et al. (2002a, b)	1996–1998	Over 64 years old	SO_2, CO, NO_2, PM_{10}, O_3	SO_2, O_3	São Paulo-SP
Sharovsky et al. (2004)	1996–1998	People between 35 and 109 years	SO_2, CO, PM_{10}	SO_2	São Paulo-SP
Gouveia et al. (2006)	1996–2000	Children under 5 years and over 65 years	SO_2, CO, NO_2, PM_{10}, O_3	SO_2, CO, NO_2, PM_{10}	São Paulo-SP
Martins et al. (2002)	1996–1998	Over 64 years old	SO_2, CO, NO_2, PM_{10}, O_3	SO_2, O_3	São Paulo-SP
Farhat et al. (2005)	1996–1997	Children under 13	SO_2, CO, NO_2, PM_{10}, O_3	NO_2	São Paulo-SP
Martins et al. (2006)	1996–2001	Over 64 years old	SO_2, CO, NO_2, PM_{10}, O_3	SO_2, CO, NO_2, PM_{10}, O_3	São Paulo-SP
Jasinski et al. (2011)	1997–2004	Under 19 years old	SO_2, NO_2, PM_{10}, O_3	PM_{10}, O_3	Cubatão-SP
Gouveia et al. (2004)	1997	Born in 1997	SO_2, CO, NO_2, PM_{10}, O_3	CO	São Paulo-SP
Martins et al. (2004)	1997–2000	Over 65 years old	PM_{10}	PM_{10}	São Paulo-SP
Yanagi et al. (2012)	1997–2005	Total	PM_{10}	PM_{10}	São Paulo-SP
Lin et al. (2004)	1998–2000	Children under 28 days	SO_2, CO, NO_2, PM_{10}, O_3	SO_2, PM_{10}	São Paulo-SP
Medeiros and Gouveia (2005)	1998–2000	Total	SO_2, CO, NO_2, PM_{10}, O_3	CO, NO_2, PM_{10}	São Paulo-SP
Cendon et al. (2006)	1998–2000	Over 64 years old	SO_2, CO, NO_2, PM_{10}, O_3	SO_2, CO, NO_2, PM_{10}, O_3	São Paulo-SP
Santos et al. (2008)	1998–2000	Over 17 years old	SO_2, CO, NO_2, PM_{10}, O_3	CO, NO_2, PM_{10}	São Paulo-SP
Nishioka et al. (2000)	1998	Born in 1998	SO_2, CO, NO_2, PM_{10}, O_3	SO_2, NO_2, PM_{10}, O_3	São Paulo-SP
Nascimento et al. (2006)	2000–2001	Under 10 years old	SO_2, PM_{10}, O_3	SO_2, PM_{10}, O_3	São José dos Campos-SP

(continued)

Table 23.3 (continued)

Reference	Period evaluated	Population studied	Pollutants evaluated	Pollutants positively associated	Location
Romão et al. (2013)	2000–2006	Born between 2000 and 2006	PM_{10}	PM_{10}	São Bernardo do Campo-SP
Vidotto et al. (2012)	2000–2007	Under 19 years old	SO_2, CO, NO_2, PM_{10}, O_3	SO_2, CO, NO_2, PM_{10}	São Paulo-SP
Negrete et al. (2010)	2000–2007	Over 35 years old	PM_{10}	PM_{10}	Santo André-SP
Pereira et al. (2008)	2001–2003	Older than 18 years	SO_2, CO, NO_2, PM_{10}, O_3	SO_2, CO, NO_2	São Paulo-SP
Nascimento and Moreira (2009)	2001	Mothers aged 20–34	SO_2, PM_{10}, O_3	SO_2, O_3	São José dos Campos-SP
Arbex et al. (2009)	2002–2003	Over 40 years old	SO_2, CO, NO_2, PM_{10}, O_3	SO_2, PM_{10}	São Paulo-SP
Arbex et al. (2007)	2003–2004	Total	PTS	PTS	Araraquara-SP
Arbex et al. (2009)	2003–2004	Total	PTS	PTS	Araraquara-SP
Amancio and Nascimento (2012)	2004–2005	Under 10 years old	SO_2, PM_{10}, O_3	SO_2, PM_{10}	São José dos Campos-SP
Nascimento (2011)	2006	Over 60 years old	SO_2, PM_{10}, O_3	PM_{10}	São José dos Campos-SP
Nascimento and Francisco (2013)	2007–2010	Total	SO_2, PM_{10}, O_3	PM_{10}	São José dos Campos-SP
Nascimento et al. (2012)	2007–2008	Total	SO_2, PM_{10}, O_3	PM_{10}	São José dos Campos-SP

Source: Fonte: Pereira and Limongi (2015)

forest fire episodes registered in the country. In general, the northern region is responsible for 62% of the fires occurred in Brazil during the dry season. The climate issue has also been preponderant to determine the highest frequency of forest fires in the Brazilian Amazon. In 2005, the region experienced a prolonged drought and recorded numerous outbreaks of forest fires – estimates were of over

400,000 people affected by smoke, with a total area of over 300,000 ha of devastated forests and about $50 million direct financial losses (Brown et al. 2006).

Biomass burning is often adopted by the local population due to its low cost, causing serious damage to the environment (biodiversity loss, destruction of forest ecosystems), to human health (increase in respiratory diseases, problems in newborns, ocular discomfort, discomfort caused by soot), to air quality (increased emissions of greenhouse gases and air pollution), as well as economic losses (closing of airports and traffic accidents, among others) (Silva 2005). Despite the known impacts, studies on the effects of burnings on human health are very scarce, both in Brazil and abroad, although the deleterious effects of biomass burning on human health are reported in the scientific literature (Ribeiro and Assunção 2002).

In terms of damage to human health associated with exposure to biomass-burning pollutants, studies have shown the increased air pollution levels associated with an increase in the number of respiratory disease hospitalizations (Arbex et al. 2000; Braga et al. 2001; Cançado et al. 2006; Ignotti et al. 2010a, b). It is also known that children, the elderly, and individuals with cardiorespiratory diseases, including asthmatics, are the most susceptible to the effects of air pollution exposure. According to Gonçalves et al. (2012), most of the infant vulnerability is due to factors such as increased growth rate, increased heat loss area per unit weight, and high rates of metabolism at rest and oxygen consumption, which facilitates the entry of chemical agents in the airways. In the elderly, factors related to low immunity and reduction of bronchial ciliary function contribute to increased vulnerability to respiratory illness related to air pollutants (Gonçalves et al. 2012). In rural or remote areas, gaseous pollutants and fine particulate matter have direct effects on the respiratory system, especially for the most sensitive groups (Carmo et al. 2010). Gonçalves et al. (2012) performed a nonsystematic review of epidemiological studies linking air pollution arising from burning and respiratory illness in the Brazilian Amazon, and the results showed increased involvement of children and the elderly by the presence of atmospheric particulate matter. The surveys compiled by these authors and other recent studies are summarized in Table 23.4.

The seriousness of the issue becomes relevant when it is observed that about 60% of the particulate matter emitted in the region comes from the burning, which contributes significantly to changing the chemical composition of the Amazon atmosphere, with important implications at the local, regional, and global level. In some cases, the values exceed the limits observed in many urban centers (Artaxo et al. 2002). In addition to the burning effects to the Amazon ecosystem, pollutant emissions contribute to increased respiratory morbidity in the municipalities of the Amazon "arc of deforestation" (Carmo et al. 2010; Mascarenhas et al. 2008). According Carmo et al. (2010), forest fires in the region have the characteristic of exposing the population to a high magnitude of pollutants during an annual mean of 3–5 months, combined with low rainfall, which is different from the exposure profile observed in urban centers. During this period, concentrations of particulate matter less than 10 μm reach up to 400 $\mu g.m^{-3}$ (Artaxo et al. 2002). The study on the concentration of particulate matter in Tangara da Serra, a state of Mato Grosso, corroborates these findings, since the PM_{10} concentrations found were only high in

Table 23.4 Main studies developed for the Brazilian Amazon region

Estudo	População e local	Resultados
Mascarenhas et al. (2008)	All ages Rio Branco, Acre	Higher incidence of respiratory system diseases in children <10 years; positive correlation between the concentration of $PM_{2.5}$ and visits for asthma
Souza (2008)	Children <4 years and elderly over 65 years Rio Branco, Acre	Relationship between the increase in the forest fires outbreaks and hospital admissions for respiratory system diseases
Rosa et al. (2008)	Children >15 years Tangará da Serra, Mato Grosso	Increase in hospital admissions for respiratory diseases in the forest fire season (dry season)
Saldanha and Botelho (2008)	Children with asthma <5 years Cuiabá, Mato Grosso	Relationship between asthma and hot spots
Castro et al. (2009)	Elderly <65 years. Rondônia	Relationship between mortality from respiratory diseases and chronic obstructive pulmonary disease and the number of hot spots
Ignotti et al. (2010b)	All ages Microregions of the Brazilian Amazon	Relationship between $PM_{2.5}$, rate of hospitalizations due to respiratory diseases, and complications at childbirth
Carmo et al. (2010)	All ages Alta Floresta, State of Mato Grosso	Relationship between $PM_{2.5}$ and outpatient care for respiratory diseases in children and the elderly
Rodrigues et al. (2010)	Asthma in the elderly All states of the Legal Amazon	Hospitalizations tripled in the dry season when compared to the rainy season, with higher rates in Rondônia and Mato Grosso states
Silva (2010)	All ages Cuiabá, Mato Grosso	Relationship between $PM_{2.5}$ and hospitalization rate for respiratory diseases in children and the elderly
Andrade (2011)	Children with respiratory diseases Manaus, Amazonas	Relationship between $PM_{2.5}$ and hospitalization rate for respiratory diseases in children
Oliveira (2011)	Children between 6 and 14 years old Tangará da Serra, Mato Grosso	During the dry season, exposure to $PM_{2.5}$ levels posed a toxicological risk for children aged 6–14 residing in biomass-burning areas
Silva et al. (2013)	Children <5 years and elderly ≥65 years Cuiabá, Mato Grosso	Influence of $PM_{2.5}$ on the occurrence of hospitalizations due to respiratory diseases in children <5 years
Barros et al. (2014)	Children between 29 days and 12 years old Porto Velho, Rondônia	Increase in hospital readmissions for respiratory diseases in the dry season, together with an increase in the number of hot spots. Pneumonia accounted for 54% of the causes of rehospitalization

Adapted from: Gonçalves et al. (2012)

the months of August, September, and October, just when the largest forest fire numbers occurred in the state – between 2008 and 2009 (Moreira et al. 2014). Similarly, Santiago et al. (2015), to characterize the present particulate matter in Cuiabá, state of Mato Grosso, found the highest concentration of suspended particulate matter in September – 306 $\mu g.m^{-3}$ after a long dry period, which exceeds the primary limit set by CONAMA.

Conclusions

In the Brazilian scenario, two major factors influence the patterns of emission and the air quality associated: the economic development model based on commodities, which puts great demands on natural resources and is linked to some poor technological practices such as slash-and-burn agriculture and deforestation for livestock expansion, and the traffic in large urban centers. This is characterized by intense vehicle flows, absence of urban and traffic planning, and the massive usage of diesel-powered vehicle fleet. As demonstrated in here, the result has been the overcoming of the national and international thresholds of emission of important pollutants, such as PM and ozone, in several Brazilian cities, with relevant consequences to the health of the population. In the near future, the climate change represents a threat to the maintenance of the basic air quality patterns, since there is a consistent relationship between the availability of some pollutants in the atmosphere and alterations in the regional dynamic of climate, since the generation and dispersion of air pollutants may also be influenced by certain meteorological and climatic factors. Therefore, to maintain the air quality in satisfactory levels, considering the prospects of climate change, it is necessary both to improve the national patterns, once not even $PM_{2.5}$, recognized for its ability to cause significant harm to human health, is parameterized by Brazilian standards, and to promote the use of renewable energy and less aggressive economic practices to the environment aiming to mitigate the climate change impacts. In addition, a more effective monitoring network to assess the emissions and types of pollutants present in all parts of the country would greatly contribute to a better understanding of the association between health problems and pollution in the various regions of Brazil, and not only in the southeastern region, the most populated and polluted region of Brazil.

References

Akimoto H (2003) Global air quality and pollution. Science 302(5651):1716–1719
Albuquerque TT, Andrade MF, Ynoue RY (2012) Characterization of atmospheric aerosols in the city of São Paulo, Brazil: comparisons between polluted and unpolluted periods. Environ Monit Assess 184(2):969–984

Amâncio CT, Nascimento LFC (2012) Asma e poluentes ambientais: um estudo de séries temporais. Rev Assoc Med Bras 58(3):302–307

Andrade Filho VSD (2011) Estudo da associação entre material particulado emitido em queimadas e doenças respiratórias no município de Manaus, AM. Master dissertation. National Institute of Research of Amazon

Andrade Filho VSD, Artaxo P, Hacon S, Carmo CND, Cirino G (2013) Aerossóis de queimadas e doenças respiratórias em crianças, Manaus, Brasil. Rev Saude Publica 47(2):239–247

Arbex MA, Böhm GM, Saldiva PH, Conceiçao GM III, A. C. P, Braga AL (2000) Assessment of the effects of sugar cane plantation burning on daily counts of inhalation therapy. J Air Waste Manage Assoc 50(10):1745–1749

Arbex MA, Martins LC, de Oliveira RC, Pereira LAA, Arbex FF, Cançado JED, . . ., Braga ALF (2007) Air pollution from biomass burning and asthma hospital admissions in a sugar cane plantation area in Brazil. J Epidemiol Community Health 61(5):395–400

Arbex MA, de Souza Conceição GM, Cendon SP, Arbex FF, Lopes AC, Moyses EP, . . ., Braga ALF (2009) Urban air pollution and chronic obstructive pulmonary disease-related emergency department visits. J Epidemiol Community Health 63(10):777–783

Aris RM, Christian D, Hearne PQ, Kerr K, Finkbeiner WE, Balmes JR (1993) Ozone-induced airway inflammation in human subjects as determined by airway lavage and biopsy. Am Rev Respir Dis 148(5):1363–1372

Artaxo P, Martins JV, Yamasoe MA, Procópio AS, Pauliquevis TM, Andreae MO, . . ., Leal AMC (2002) Physical and chemical properties of aerosols in the wet and dry seasons in Rondonia, Amazonia. J Geophys Res: Atmos 107(D20)

Artaxo P, Gatti LV, Leal AMC, Longo KM, Freitas SRD, Lara LL, . . ., Rizzo LV (2005) Química atmosférica na Amazônia: a floresta e as emissões de queimadas controlando a composição da atmosfera amazônica. Acta Amazon 35(2):185–196

Arya SP (2000) Air pollution meteorology and dispersion. Bound Lay Meteorol 94:171–172

Aw J, Kleeman MJ (2003) Evaluating the first-order effect of intraannual temperature variability on urban air pollution. J Geophys Res Atmos 108(D12):4365

Balkanski YJ, Jacob DJ, Gardner GM, Graustein WC, Turekian KK (1993) Transport and residence times of tropospheric aerosols inferred from a global three-dimensional simulation of 210Pb. J Geophys Res Atmos 98(D11):20573–20586

Barbosa SMM, Farhat SCL, Martins LC, Pereira LAA, Saldiva PHN, Zanobetti A, Braga ALF (2015) Air pollution and children's health: sickle cell disease. Cad Saude Publica 31 (2):265–275

Barros MMA, Gonçalves KDS, Hacon SDS (2014) Reinternações hospitalares por doenças respiratórias em menores de 12 anos e as queimadas em Porto Velho–RO, Amazônia Ocidental. Rev Saúde, Meio Ambiente Sustentabilidade 8(3)

Bell ML, Dominici F, Samet JM (2005) Meta-analysis of ozone and mortality. Epidemiology 16 (5):S35

Böhm GM, Saldiva PHN, Pasqualucci CAG, Massad E, de Arruda Martins M, Zin WA., . . ., Negri EM (1989) Biological effects of air pollution in São Paulo and Cubatão. Environ Res 49 (2):208–216

Botter DA, Jorgensen B, Peres AA (2002) A longitudinal study of mortality and air pollution for São Paulo, Brazil. J Expo Anal Environ Epidemiol 12(5):335–343

Brabin B, Smith M, Milligan P, Benjamin C, Dunne E, Pearson M (1994) Respiratory morbidity in Merseyside schoolchildren exposed to coal dust and air pollution. Arch Dis Child 70 (4):305–312

Braga ALF (1998). Quantificação dos efeitos da poluição do ar sobre a saúde da população pediátrica da cidade de São Paulo e proposta de monitorização. Doctoral thesis, São Paulo University, School of Medicine

Braga AL, Conceição G, Pereira LA, Kishi HS, Pereira JC, Andrade MF, . . ., Latorre MR (1999) Air pollution and pediatric respiratory hospital admissions in São Paulo, Brazil. J Environ Med 1(2):95–102

Braga AL, Saldiva PH, Pereira LA, Menezes JJ, Conceição GM, Lin CA,. . ., Dockery DW (2001) Health effects of air pollution exposure on children and adolescents in São Paulo, Brazil. Pediatr Pulmonol 31(2):106–113

Braga ALF, Zanobetti A, Schwartz J (2002) The effect of weather on respiratory and cardiovascular deaths in 12 US cities. Environ Health Perspect 110(9):859

de Brito JM, Macchione M, Yoshizaki K, Toledo-Arruda AC, Saraiva-Romanholo BM, Andrade MdeF, Mauad T, Rivero DHRF, Saldiva PHN (2014) Acute cardiopulmonary effects induced by the inhalation of concentrated ambient particles during seasonal variation in the city of Sao Paulo. J Appl Physiol 117(5):492–499

Brazilian Ministry of Environment (2016) Poluentes Atmosféricos. Available at: http://www.mma.gov.br/cidades-sustentaveis/qualidade-do-ar/poluentes-atmosf%C3%A9ricos. Accessed 3 Nov 2016

Brazilian Ministry of Health (2013) Department of Health Surveillance. Surveillance in Environmental Health related to the Air Quality Program – VIGIAR. Instrument of Identification of Municipalities of Risk, 2012: Situation of the Brazilian states. Brasilia

Brook RD, Franklin B, Cascio W, Hong Y, Howard G, Lipsett M, ... Tager I (2004) Air pollution and cardiovascular disease. A statement for healthcare professionals from the expert panel on population and prevention Science of the American Heart Association. Circulation 109 (21):2655–2671

Brown IF, Schroeder W, Setzer A, Maldonado R, Pantoja N, Duarte A, Marengo J (2006) Monitoring fires in southwestern Amazonia rain forests. Eos 87(26):253–264

Cançado JE, Saldiva PH, Pereira LA, Lara LB, Artaxo P, Martinelli LA, . . ., Braga AL (2006) The impact of sugar cane-burning emissions on the respiratory system of children and the elderly. Environ Health Perspect 725–729

Carmo CN, Hacon S, Longo KM, Freitas S, Ignotti E, Ponce de Leon A, Artaxo P (2010) Associação entre material particulado de queimadas e doenças respiratórias na região sul da Amazônia brasileira. Rev Panam Salud Publica 27(1):10–16

Castro HA, Gonçalves KS, Hacon SS (2009) Tendência da mortalidade por doenças respiratórias em idosos e as queimadas no estado de Rondônia/Brasil – período entre 1998 e 2005. Cien Saude Colet 14(6):2083–2090

Cendon S, Pereira LA, Braga AL, Conceição G, Cury Junior A, Romaldini H, . . ., Saldiva PH (2006) Air pollution effects on myocardial infarction. Rev Saude Publica 40(3):414–419

Coelho MSZS, Gonçalves FLT, Latorre MRDO (2010) Statistical analysis aiming at predicting respiratory tract disease hospital admissions from environmental variables in the city of São Paulo. J Environ Public Health 2010:209270

Companhia de Tecnologia de Saneamento Ambiental do estado de São Paulo – CETESB (2016) Qualidade do ar no estado de São Paulo, 2015.CETESB, 167 pp. Available at: http://ar.cetesb.sp.gov.br/wp-content/uploads/sites/37/2013/12/RQAR-2015.pdf. Accessed on 30 Oct 2016

Conceicao GM, Miraglia SG, Kishi HS, Saldiva PH, Singer JM (2001) Air pollution and child mortality: a time-series study in São Paulo, Brazil. Environ Health Perspect 109(Suppl 3):347

Conselho Nacional do Meio Ambiente – CONAMA. Resolution N°005 of June 15 (1989) Official Journal of the Federative Republic of Brazil. Institutes the National Air Quality Control Programee – PRONAR, Brasília. 1989

Conselho Nacional do Meio Ambiente – CONAMA. Resolution N° 003 of June 28 (1990) Official Journal of the Federative Republic of Brazil. Sets standards of primary and secondary air quality and even the criteria for acute episodes of air pollution, Brasilia, Brazil, 1990

Costa AA, Pauliquevis Junior TM (2014) Aerossóis atmosféricos e nuvens. In: Ambrizzi T, Araujo M (eds) Base científica das mudanças climáticas. Contribuição do Grupo de Trabalho 1 do Painel Brasileiro de Mudanças Climáticas ao Primeiro Relatório da Avaliação Nacional sobre Mudanças Climáticas, Chap. 6. COPPE. Universidade Federal do Rio de Janeiro, Rio de Janeiro, pp 209–236

Dawson JP, Adams PJ, Pandis SN (2007) Sensitivity of PM 2.5 to climate in the Eastern US: a modeling case study. Atmos Chem Phys 7(16):4295–4309

Ebi KL, McGregor G (2008) Climate change, tropospheric ozone and particulate matter, and health impacts. Environ Health Perspect 116(11):1449

Farhat SCL, Paulo RLP, Shimoda TM, Conceição GMS, Lin CA, Braga ALF, . . ., Saldiva PHN (2005) Effect of air pollution on pediatric respiratory emergency room visits and hospital admissions. Braz J Med Biol Res 38(2):227–235

Filleul L, Cassadou S, Médina S, Fabres P, Lefranc A, Eilstein D, . . ., Declercq C (2006) The relation between temperature, ozone, and mortality in nine French cities during the heat wave of 2003. Environ Health Perspect 1344–1347

Frampton MW, Pryor WA, Cueto R, Cox C, Morrow PE, Utell MJ (1999) Ozone exposure increases aldehydes in epithelial lining fluid in human lung. Am J Respir Crit Care Med 159 (4):1134–1137

Freitas C, Bremner SA, Gouveia N, Pereira LA, Saldiva PH (2004) Internações e óbitos e sua relação com a poluição atmosférica em São Paulo, 1993 a 1997. Rev Saude Publica 38 (6):751–757

Freitas SR, Longo KM, Dias MAF, Dias PL (2005) Emissões de queimadas em ecossistemas da América do Sul. Estudos Avançados 19(53):167–185

Freitas CUD, Junger W, Leon APD, Grimaldi R, Silva MAFR, Gouveia N (2013) Poluição do ar em cidades brasileiras: selecionando indicadores de impacto na saúde para fins de vigilância. Epidemiologia e Serviços de Saúde 22(3):445–454

Giorgi F, Meleux F (2007) Modelling the regional effects of climate change on air quality. Compt Rendus Geosci 339(11):721–733

Gonçalves FL, Carvalho LMV, Conde FC, Latorre MRDO, Saldiva PHN, Braga ALF (2005) The effects of air pollution and meteorological parameters on respiratory morbidity during the summer in Sao Paulo City. Environ Int 31(3):343–349

Gonçalves KDS, Castro HAD, Souza Hacon SD (2012) As queimadas na região amazônica e o adoecimento respiratório. Revista Ciência & Saúde Coletiva 17(9):1523–1532

Gouveia N, Fletcher T (2000) Time series analysis of air pollution and mortality: effects by cause, age and socioeconomic status. J Epidemiol Community Health 54(10):750–755

Gouveia N, Bremner SA, Novaes HMD (2004) Association between ambient air pollution and birth weight in São Paulo, Brazil. J Epidemiol Community Health 58(1):11–17

Gouveia N, Freitas CUD, Martins LC, Marcilio IO (2006) Respiratory and cardiovascular hospitalizations associated with air pollution in the city of São Paulo, Brazil. Cad Saude Publica 22 (12):2669–2677

Gryparis A, Forsberg B, Katsouyanni K, Analitis A, Touloumi G, Schwartz J, . . . Wichmann HE (2004) Acute effects of ozone on mortality from the "air pollution and health: a European approach" project. Am J Respir Crit Care Med 170(10):1080–1087

Gurjar BR, Jain A, Sharma A, Agarwal A, Gupta P, Nagpure AS, Lelieveld J (2010) Human health risks in megacities due to air pollution. Atmos Environ 44(36):4606–4613

Harlan SL, Ruddell DM (2011) Climate change and health in cities: impacts of heat and air pollution and potential co-benefits from mitigation and adaptation. Curr Opin Environ Sustain 3(3):126–134

Heald CL, Henze DK, Horowitz LW, Feddema J, Lamarque JF, Guenther A, . . ., Fung I (2008) Predicted change in global secondary organic aerosol concentrations in response to future climate, emissions, and land use change. J Geophys Res: Atmos 113(D5)

Ignotti E, Hacon SDS, Junger WL, Mourão D, Longo K, Freitas S, . . ., Leon ACMPD (2010a) Air pollution and hospital admissions for respiratory diseases in the subequatorial Amazon: a time series approach. Cad Saúde Pública 26(4):747–761

Ignotti E, Valente JG, Longo KM, Freitas SR, Hacon SDS, Artaxo Netto P (2010b) Impact on human health of particulate matter emitted from burnings in the Brazilian Amazon region. Rev Saude Publica 44(1):121–130

Imai M, Yoshida K, Kotchmar DJ, Lee SD (1985) A survey of health effects studies of photochemical air pollution in Japan. J Air Pollut Control Assoc 35(2):103–108

Ito K, De Leon SF, Lippmann M (2005) Associations between ozone and daily mortality: analysis and meta-analysis. Epidemiology 16(4):446–457

Jacob DJ, Winner DA (2009) Effect of climate change on air quality. Atmos Environ 43(1):51–63

Jacobson MZ (2001) Strong radiative heating due to the mixing state of black carbon in atmospheric aerosols. Nature 409(6821):695–697

Jasinski R, Pereira LAA, Braga ALF (2011) Poluição atmosférica e internações hospitalares por doenças respiratórias em crianças e adolescentes em Cubatão, São Paulo, Brasil, entre 1997 e 2004. Cad Saúde Pública 27:2242–2252

Kinney PL (2008) Climate change, air quality, and human health. Am J Prev Med 35(5):459–467

Kishi HS, Saldiva PHN (1998) Associação entre poluição atmosferica e internacoes pediatricas por causas respiratorias na cidade de São Paulo. Rev med (São Paulo) 77(1):2–10

Lamb B, Guenther A, Gay D, Westberg H (1987) A national inventory of biogenic hydrocarbon emissions. Atmos Environ (1967) 21(8):1695–1705

Lanki T, Pekkanen J, Aalto P, Elosua R, Berglind N, D'Ippoliti D, …, Salomaa V (2006) Associations of traffic related air pollutants with hospitalisation for first acute myocardial infarction: the HEAPSS study. Occup Environ Med, 63(12):844–851

Lin CA, Martins MA, Farhat SC, Pope 3rd, CA, Conceicao GM, Anastácio VM, …, Saldiva PH (1999) Air pollution and respiratory illness of children in Sao Paulo, Brazil. Paediatr Perinat Epidemiol 13(4):475–488

Lin CA, Pereira LAA, de Souza Conceição GM, Kishi HS, Milani R, Braga ALF, Saldiva PHN (2003) Association between air pollution and ischemic cardiovascular emergency room visits. Environ Res 92(1):57–63

Lin CA, Pereira LAA, Nishioka DC, Conceição GMS, Braga ALF, Saldiva PHN (2004) Air pollution and neonatal deaths in Sao Paulo, Brazil. Braz J Med Biol Res 37(5):765–770

Logan WPD (1953) Mortality in the London fog incident, 1952. Lancet 261(6755):336–338

Marlier ME, Jina AS, Kinney PL, DeFries RS (2016) Extreme air pollution in global megacities. Curr Clim Change Rep 2(1):15–27

Martins LC, Latorre MRD, Saldiva PHN, Braga ALF (2002a) Air pollution and emergency room visits due to chronic lower respiratory diseases in the elderly: an ecological time-series study in Sao Paulo, Brazil. J Occup Environ Med 44(7):622–627

Martins LC, Latorre MRD, Cardoso MRA, Gonçalves FLT, Saldiva PHN, Braga ALF (2002b) Air pollution and emergency room visits due to pneumonia and influenza in Sao Paulo, Brazil. Rev Saude Publica 36(1):88–94

Martins MCH, Fatigati FL, Vespoli TC, Martins LC, Pereira LAA, Martins MA, Saldiva PHN, Braga ALF (2004) Influence of socioeconomic conditions on air pollution adverse health effect in elderly people: an analysis of six region in São Paulo, Brasil. J Epidemiol Community Health 58:41–46

Martins LC, Pereira LA, Lin CA, Santos UP, Prioli G, Luiz ODC, …, Braga ALF (2006) The effects of air pollution on cardiovascular diseases: lag structures. Rev Saúde Pública 40 (4):677–683

Martins EM, Nunes AC, Corrêa S (2015) Understanding ozone concentrations during weekdays and weekends in the urban area of the city of Rio de Janeiro. J Braz Chem Soc 26 (10):1967–1975

Mascarenhas MDM, Vieira LC, Lanzieri TM, Leal APPR, Duarte AF, Hatch DL (2008) Poluição atmosférica devido à queima de biomassa florestal e atendimentos de emergência por doença respiratória em Rio Branco, Brasil-setembro, 2005. J Bras Pneumol 34(1):42–46

Matsumoto G, Nakagawa NK, Vieira RP, Mauad T, Silva LF, André CD …, Garcia ML (2010) The time course of vasoconstriction and endothelin receptor a expression in pulmonary arterioles of mice continuously exposed to ambient urban levels of air pollution. Environ Res 110(3):237–243

Medeiros A, Gouveia N (2005) Relação entre baixo peso ao nascer e a poluição do ar no Município de São Paulo. Rev Saude Publica 39(6):965–972

Mendes R, Wakanatsu CT (1976) Avaliação dos efeitos agudos sobre a saúde através do estudo da morbidade diária em São Paulo. Cetesb (Mineo)

Miranda RM, de Fatima Andrade M, Fornaro A, Astolfo R, de Andre PA, Saldiva P (2012) Urban air pollution: a representative survey of PM2.5 mass concentrations in six Brazilian cities. Air Qual Atmos Health 5(1):63–77

Molina MJ, Molina LT (2004) Megacities and atmospheric pollution. J Air Waste Manage Assoc 54(6):644–680

Moreira PSP, Dallacort R, de Fatima Lima I, Tieppo RC, Santos C (2014) Concentração de Material Particulado (PM10) na Região de Tangará da Serra-MT, Sul da Amazônia Legal. Rev Bras Geogr Fís 7(6):1145–1152

Nardocci AC, Freitas CU, Leon ACMP, Junger WL, Gouveia NC (2013) Poluição do ar e doenças respiratórias e cardiovasculares: estudo de séries temporais em Cubatão, São Paulo, Brasil. Cad Saude Publica 29(9):1867–1876

Nascimento LFC (2011) Air pollution and cardiovascular hospital admissions in a medium-sized city in São Paulo state, Brazil. Braz J Med Biol Res 44(7):720–724

Nascimento LFC, Francisco JB (2013) Particulate matter and hospital admission due to arterial hypertension in a medium-sized Brazilian city. Cad Saude Publica 29(8):1565–1571

Nascimento LFC, Moreira DA (2009) Are environmental pollutants risk factors for low birth weight? Cad Saude Publica 25(8):1791–1796

Nascimento MT, Felfiti JM, Oliveira FAT, Leite Fontes MA; França JT, Hay J, Gribel R (2000) Efeitos do fogo nas florestas. Ciência Hoje 27(157) SBPC

Nascimento LFC, Pereira LAA, Braga AL, Módolo MCC, Carvalho JA Jr (2006) Effects of air pollution on children's health in a city in southeastern Brazil. Rev Saude Publica 40(1):77–82

Nascimento LFC, Francisco JB, Patto MBR, Antunes AM (2012) Environmental pollutants and stroke-related hospital admissions. Cad Saude Publica 28(7):1319–1324

National Institute of Space Research – INPE (2016) Programa queimadas. Monitoramento por satélite. Relatório Diário Automático 27/Nov/2016. Available at: https://prodwww-queimadas.dgi.inpe.br/cadastro/relatorio-diario-automatico/relatorio-diario-automatico.pdf. Accessed on 10 Oct 2016

Nawrot TS, Torfs R, Fierens F, De Henauw S, Hoet PH, Van Kersschaever G, …, Nemery B (2007). Stronger associations between daily mortality and fine particulate air pollution in summer than in winter: evidence from a heavily polluted region in western Europe. J Epidemiol Community Health 61(2):146–149

Negrete BR, Rosa CDC, Ikeuti DH, Delena PJ, Borba TMD, Braga ALF (2010) Poluição atmosférica e internações por insuficiência cardíaca congestiva em adultos e idosos em Santo André (SP). Arq Bras Ciências Saúde 35(3):208–212

Nilsson ED, Rannik Ü, Kulmala M, Buzorius G, O'dowd CD (2001) Effects of continental boundary layer evolution, convection, turbulence and entrainment, on aerosol formation. Tellus B 53(4):441–461

Nishioka DC, Coura FLB, Pereira LAA, de Souza Conceição GM (2000) Estudo dos efeitos da poluição atmosferica na mortalidade neonatal e fetal na cidade de São Paulo, Brasil. Rev Med 79(2–4):81–89

Oliveira BFA (2011) Avaliação do risco a saúde infantil relativa à exposição ao PM2,5 proveniente da queima de biomassa no município de Tangará da Serra – MT. Master dissertation, Oswaldo Cruz Foundation

Organization for Economic Co-operation and Development – OECD (2014) The cost of air pollution: health impacts of road transport. OECD Publishing, Paris. http://dx.doi.org/10.1787/9789264210448-en

Orsini CQ, Tabacniks MH, Artaxo P, Andrade MDF, Kerr AS (1986) Characteristics of fine and coarse particles of natural and urban aerosols of Brazil. Atmos Environ (1967) 20 (11):2259–2269

Peng RD, Dominici F, Pastor-Barriuso R, Zeger SL, Samet JM (2005) Seasonal analyses of air pollution and mortality in 100 US cities. Am J Epidemiol 161(6):585–594

Pereira Filho MA, Pereira LAA, Arbex FF, Arbex M, Conceição GM, Santos UP, . . ., Cendon S (2008) Effect of air pollution on diabetes and cardiovascular diseases in São Paulo, Brazil. Braz J Med Biol Res 41(6):526–532

Pereira BB, Limongi JE (2015) Epidemiology of human health outcomes related to air pollution in Brazil: a systematic review. Cad Saúde Coletiva 23(2):91–100

Pereira LA, Loomis D, Conceicao GM, Braga AL, Arcas RM, Kishi HS, . . ., Saldiva PH (1998) Association between air pollution and intrauterine mortality in Sao Paulo, Brazil. Environ Health Perspect 106(6):325

Pope CA III, Dockery DW (2006) Health effects of fine particulate air pollution: lines that connect. J Air Waste Manage Assoc 56(6):709–742

Pope CA III, Schwartz J, Ransom MR (1992) Daily mortality and PM10 pollution in Utah Valley. Arch Environ Health Int J 47(3):211–217

Pope CA III, Dockery DW, Kanner RE, Villegas GM, Schwartz J (1999) Oxygen saturation, pulse rate, and particulate air pollution: a daily time-series panel study. Am J Respir Crit Care Med 159(2):365–372

Rao S, Pachauri S, Dentener F, Kinney P, Klimont Z, Riahi K, Schoepp W (2013) Better air for better health: forging synergies in policies for energy access, climate change and air pollution. Glob Environ Chang 23(5):1122–1130

Ren C, Williams GM, Morawska L, Mengersen K, Tong S (2008) Ozone modifies associations between temperature and cardiovascular mortality: analysis of the NMMAPS data. Occup Environ Med 65(4):255–260

Ribeiro HP (1971) Estudo das correlações entre infecções das vias aéreas superiores, bronquite asmatiforme e poluição do ar em menores de 12 anos em Santo André. Pediatr Prat 42:9

Ribeiro H (2008) Queimadas de cana-de-açúcar no Brasil: efeitos à saúde respiratória. Rev Saúde Pública 42(2):370–376

Ribeiro H, Assunção JVD (2002) Efeitos das queimadas na saúde humana. Estud Avançados 16 (44):125–148

Ribeiro HP, Nogueira DP, Bongiovanni CAT, Souza JMP, Koga RK, Nefussi N, Guimarães FA, Oliveira WE (1976) Estudo da Função Ventilatória em Escolares Vivendo em Áreas com Diferentes Níveis de Poluição do Ar. São Paulo University, Public Health School, São Paulo

Rodrigues PCO, Ignotti E, Rosa AM, Hacon SDS (2010) Spatial distribution of asthma-related hospitalizations of the elderly in the Brazilian Amazon. Rev Bras Epidemiol 13(3):523–532

Romão R, Pereira LAA, Saldiva PHN, Pinheiro PM, Braga ALF, Martins LC (2013) The relationship between low birth weight and exposure to inhalable particulate matter. Cad Saude Publica 29(6):1101–1108

Romieu I, Gouveia N, Cifuentes LA, de Leon AP, Junger W, Vera J, . . ., Carbajal-Arroyo L (2012) Multicity study of air pollution and mortality in Latin America (the ESCALA study). Res Rep (Health Effects Institute) (171):5–86

Rosa AM, Ignotti E, Hacon SS, Castro HA (2008) Análise das internações por doenças respiratórias em Tangará da Serra – Amazônia Brasileira. J Bras Pneumol 34(8):575–582

Rumel D, Riedel LF, Maria do Rosario DO, Duncan BB (1993) Infarto do miocárdio e acidente vascular cerebral associados à alta temperatura e monóxido de carbono em área metropolitana do sudeste do Brasil. Rev Saude Publica 27(1):15–22

Saldanha CT, Botelho C (2008) Queimadas e suas influências em crianças asmáticas menores de cinco anosatendidas em um hospital público. Rev Bras Alerg Imunopatol 31(3):108–112

Saldiva PH, Lichtenfels AJFC, Paiva PSO, Barone IA, Martins MA, Massad E, . . ., Bohm GM (1994) Association between air pollution and mortality due to respiratory diseases in children in São Paulo, Brazil: a preliminary report. Environ Res 65(2):218–225

Saldiva PH, Pope III CA, Schwartz J, Dockery DW, Lichtenfels AJ, Salge JM, . . ., Bohm GM (1995) Air pollution and mortality in elderly people: a time-series study in Sao Paulo, Brazil. Arch Environ Health Int J 50(2):159–163

Santiago A, Prado RJ, Filho PM, Alonso RV (2015) Caracterização do material particulado suspenso no ar de Cuiabá-MT no período de queimadas. Rev Mater 11595:273–283

Santos UP, Terra-Filho M, Lin CA, Pereira LAA, Vieira TCB, Saldiva PHN, Braga ALF (2008) Cardiac arrhythmia emergency room visits and environmental air pollution in Sao Paulo, Brazil. J Epidemiol Community Health 62(3):267–272

Santos VDP, Medeiros APPD, Lima TACD, Nascimento LFC (2016) Air pollutants associated with insufficient birth weight. Rev Bras Epidemiol 19(1):89–99

Sharovsky R, César LAM, Ramires JAF (2004) Temperature, air pollution, and mortality from myocardial infarction in Sao Paulo, Brazil. Braz J Med Biol Res 37(11):1651–1657

Sillman S, Samson PJ (1995) Impact of temperature on oxidant photochemistry in urban, polluted rural and remote environments. J Geophys Res Atmos 100(D6):11497–11508

Silva RG (2005) Disposição a pagar para evitar danos à saúde oriundos das queimadas: uma aplicação do método de valoração contingente no estado do Acre. Doctoral thesis. Viçosa University, Postgraduate program in applied economics

Silva AMC (2010) Quantificação dos efeitos na saúde da exposição à queima de biomassa: uma contribuição ao entendimento dos efeitos da exposição ao material particulado(PM2,5) em grupos populacionais sensíveis na Amazônia Legal. Doctoral thesis, Oswaldo Cruz Foundation

Silva MFD (2013) Reatividade fotoquímica da atmosfera de Cubatão e a influência de fontes exógenas. Doctoral thesis, São Paulo University, Public Health Faculty. 256 p

Silva AMCD, Mattos IE, Ignotti E, Hacon SDS (2013) Material particulado originário de queimadas e doenças respiratórias. Rev Saude Publica 47(2):345–352

Souza LSN (2008). Análise de Impactos das Queimadas sobrea Saúde Humana: Um estudo de caso do Município de Rio Branco, Acre. Master dissertation. National School of Public Healt (ENSP), Rio de Janeiro

State Foundation for Environmental Protection Henrique Luiz Roessler ("FEPAM"). Secretariat of state of the environment of the state of Rio Grande do Sul. Qualidade ambiental: poluentes e seus efeitos. Available at: http://www.fepam.rs.gov.br/qualidade/poluentes.asp#

Stölzel M, Breitner S, Cyrys J, Pitz M, Wölke G, Kreyling W, . . ., Peters A (2007) Daily mortality and particulate matter in different size classes in Erfurt, Germany. J Expo Sci Environ Epidemiol 17(5):458–467

Tsigaridis K, Kanakidou M (2007) Secondary organic aerosol importance in the future atmosphere. Atmos Environ 41(22):4682–4692

United Nations Environment Programme, World Meteorological Organization (2011) Integrated assessment of black carbon and tropospheric ozone: summary for decision makers. Nairobi, United Nations Environment Programme. Availabe at: http://www.unep.org/dewa/Portals/67/pdf/BlackCarbon_SDM.pdf. Accessed 26 Mar 2013

Vidotto JP, Pereira LAA, Braga ALF, Silva CA, Sallum AM, Campos LM, . . ., Farhat SCL (2012) Atmospheric pollution: influence on hospital admissions in paediatric rheumatic diseases. Lupus 21(5):526–533

Wang Y, Jacob DJ (1998) Anthropogenic forcing on tropospheric ozone and OH since preindustrial times. J Geophys Res 103(D23):31123

West JJ, Szopa S, Hauglustaine DA (2007) Human mortality effects of future concentrations of tropospheric ozone. Compt Rendus Geosci 339(11):775–783

White MC, Etzel RA, Wilcox WD, Lloyd C (1994) Exacerbations of childhood asthma and ozone pollution in Atlanta. Environ Res 65(1):56–68

World Health Organization (2006) Air quality guidelines: global update 2005: particulate matter, ozone, nitrogen dioxide, and sulfur dioxide. World Health Organization. Available at: http://www.euro.who.int/__data/assets/pdf_file/0005/78638/E90038.pdf?ua=1

World Health Organization (2016) Global ambient air pollution database (update 2016). Available at: http://maps.who.int/airpollution/. Accessed on 4 Nov 2016

Yanagi Y, Assunção JVD, Barrozo LV (2012) The impact of atmospheric particulate matter on cancer incidence and mortality in the city of Sao Paulo. Cad Saúde, Brazil

Júlia Alves Menezes is a biologist, holds a master's degree in health sciences and studies the epidemiological aspects of infectious diseases and public health, especially leishmaniasis. Recently, she has been dedicated to studies regarding human and socio-environmental vulnerability to climate and the impacts of global environmental changes on health and human systems at the René Rachou Institute, Oswaldo Cruz Foundation, Brazil.

Carina Margonari is a biologist with a master's in molecular and cell biology and a PhD in parasitic biology. As a researcher at René Rachou Institute, Oswaldo Cruz Foundation, Brazil, her work is focused on parasitology and molecular biology, with emphasis in entomology and epidemiology, mainly in subjects related to tropical diseases, such as leishmaniasis transmitted by female phlebotomine sandflies, and public health, and collaborates with studies related to human vulnerability to climate change.

Rhavena Barbosa Santos has a degree in nursing and a master's in public health and studies vulnerability to climate focusing on the human and social aspects of environmental change. Recently, Rhavena has been dedicated to the study of a metric of vulnerability related to hydrometeorological disasters at the René Rachou Institute, Oswaldo Cruz Foundation, Brazil.

Ulisses Confalonieri is a physician, holds a master's in parasitology and a PhD in science and is senior researcher at the René Rachou Institute, Oswaldo Cruz Foundation, Brazil. Works with epidemiology; the ecology of infectious processes, especially emerging diseases; and the impacts of global environmental changes on health, particularly the dynamics of infectious diseases. Coordinated working groups at the Intergovernmental Panel on Climate Change (IPCC) and the Millennium Ecosystem Assessment.

Chapter 24
Climate Change, Air Pollution, and Infectious Diseases: A New Epidemiological Scenario in Argentina

Daniel Oscar Lipp

Abstract Over the past 50 years, human activity, in particular the consumption of fossil fuels, has released quantities of CO_2 and other greenhouse gases sufficient to retain more heat in the lower layers of the atmosphere and to alter global climate. Sea level is increasing, glaciers are melting, and rainfall regimes are changing. Extreme weather events are becoming more intense and frequent. On the other hand, it is estimated that by 2030, climate change will increase the risk of some health parameters to double. Health effects related to climate change can be either direct, as heat waves, or indirect, through changes in vectors, water quality, and food, which favors the onset of diseases. Our intention is to provide the reader with what is being done in Argentina about these diseases provoked and increased by climate change. Of course, when answering questions like these, we should limit ourselves to making a report of each particular noxa, despite the obvious importance of it, and to stop in those with the greatest impact in the country.

Keywords Air pollution-climate change in Argentina • Infectious diseases • Emerging diseases and climate change • Global warming • Climate variability • Climate change and health

Climate Change and Infectious Diseases

In Argentina, there are very limited studies that anticipate epidemiological consequences due to climate change. From the outset, since this phenomenon had an impact in Argentina, it did carry out how many speculations occurred to health specialists without a firm basis in their determinations. It is not my way to proceed with such a current issue, such as climate change and its potential health impacts. Many people expect reliable reports of this very specific, changing, and

D.O. Lipp (✉)
Catholic University Argentina, Buenos Aires, Argentina

Buenos Aires' University, Buenos Aires, Argentina
e-mail: Daniellipp@arnet.com.ar

© Springer International Publishing AG 2018
R. Akhtar, C. Palagiano (eds.), *Climate Change and Air Pollution*, Springer Climate, DOI 10.1007/978-3-319-61346-8_24

controversial field. Therefore I will be careful in this study and will limit myself to the best medical information available in Argentina.

Until now it has not been possible to prove conclusively and emphatically that climate change experienced in recent decades has increased the overall risk of transmission of insect-borne diseases, but there is enough scientific evidence to suspect it. In addition to climate change, there are many factors that can influence the epidemiology of vector diseases, such as atmospheric composition, urbanization, economic and social development, international trade, human migration, industrial development, land use, irrigation, and agricultural development. The recent resurgence of many of these diseases in the world could be attributed more to political, economic, and human activity changes rather than climate change. Therefore, climate alone is not a sufficient cause for the establishment of endemic foci in Argentina, although a requirement. The latter must be clear.

The direct effects of climate change on health include all those diseases caused by direct exposure to meteorological variables. Among these are diseases caused by extremes of heat and cold, such as heat waves and cold waves that raise rates of death, especially among older people with chronic pathologies linked to the heart and lungs. The elderly and the sick are therefore at greater risk of contracting them. The well-known phenomenon of the urban heat island can increase the negative effects of these impacts. Another effect on health is given by air pollution. Climate change can affect the ozone concentration at ground level by increasing the number of respiratory diseases caused by this gas. Another direct effect on health is given by the accumulation of powder in suspension. This dust appears as a contamination of the particulate material of different granulometry that can be transported by the winds through great distances causing serious respiratory-like ailments. Research in Argentina on the dangers of these diseases is very limited. There is a clear need to expand knowledge about this issue in Argentina because it will allow us to act quickly, safely, and firmly in the face of the forthcoming climate change. The indirect effects of climate change on health are probably the most important. Changes in climatic conditions affect health indirectly, particularly through changes in the biological and ecological processes that influence the transmission of some diseases, especially infectious diseases. They have a strong character of being influenced by global warming (Flannery 2006-38).

In general terms, infectious diseases can be classified into two broad categories according to their mode of transmission. On the one hand, if they are transmitted directly from person to person through direct contact or if they are indirectly propagated through a vector or host such as mosquitoes and ticks or a non-biological physical element such as water and the ground. In general, diseases that are transmitted by direct contact, or from person to person, are much less influenced by climatic factors as the disease agent spends very little time outside the human host (measles, tuberculosis, and transmitted infections such as HIV, herpes, and syphilis). In contrast, cycles of transmission requiring, for example, a nonhuman vector or host are more susceptible to external environmental influences than those diseases which include only the pathogen and the human (Perczyk et al. 2004-7).

The Diseases of Major Impact

A recent report by the Intergovernmental Panel on Climate Change (IPCC) argues that Argentina will face during this century the dramatic increase in some infectious diseases such as Chagas, dengue, and malaria, three pathologies that currently prevail the tropical and subtropical regions. They would find a more favorable climate for their expansion and would be favored by the possible new conditions of humidity and heat. Climatic conditions have a strong influence on insect-borne diseases or other intermediaries. Climate changes are likely to prolong transmission stations of important vector-borne diseases and alter their geographical distribution. In this sense, Argentina is expected to expand considerably in areas affected by dengue fever and Chagas' disease.

Dengue is a growing problem for global public health due to several factors such as climate change, increasing world population in urban areas in an accelerated and unplanned manner, inadequate collection of waste, and accumulation of containers that favor breeding of mosquitoes. These factors are compounded by the risk of travel and migrations to endemic areas and insufficient control of vectors, all elements that have an impact on the spread of this disease. The disease is caused by a virus that is transmitted through the bite of infected mosquitoes, mainly of the species "*Aedes aegypti*," which makes the control of the vector a fundamental tool for the prevention of the disease. It does not spread from person to person, through objects, or orally, respiratory, or sexual. "*Aedes aegypti*" is a small, house-eating insect.

The bite of this mosquito also transmits the virus Zika. Climate change also favors the proliferation of Zika virus and other mosquito-borne viruses. The increase in temperatures has contributed to expand the habitat of these vectors by increasing the incidence of the disease in areas that until then were free of the flagellum. The heat and humidity, associated with climate change, create the ideal conditions for the procreation of mosquitoes. Regions that were earlier drier and colder now experience temperature rise higher and more rainfall, which causes mosquitoes to expand their breeding grounds, which increases the number of populations at risk. At least 22 Latin American countries have reported cases of Zika virus, among which Brazil is the most affected. It is believed that the Zika virus could have come to Brazil through Asian tourists. In October 2015, Brazil's health authorities confirmed an increase in the prevalence of microcephaly in the northeast of the country, which coincided with an outbreak of the Zika virus. Later, there were described other congenital anomalies, placental insufficiency, late fetal growth, and fetal death associated with the infection of Zika virus during the pregnancy. The latter event I lead that the 1° of February, 2016, the World Health Organization (WHO) was declaring an emergency of public health of international importance. In Argentina, the first outbreak was identified this year by virus of the Zika of vectorial transmission in Tucumán province. In the same one, 25 autochthonous cases were confirmed, three of which corresponded to pregnant women. In the course of 2016, other 266 cases were notified studies for Zika in the frame of the integrated vigilance of arbovirus.

Malaria is another evil that threatens to spread due to climate change. Of parasitic origin, this is produced by protozoa of the genus *Plasmodium* and transmitted to man through the bite of hematophagous Diptera of the genus *Anopheles*. The parasite that causes the disease, which can be fatal if not treated in a timely manner, reproduces in the liver of the person who contracts and then infects the red blood cells. This disease is preventable and curable by treatment with medication. In Argentina, the main risk zone is the north of the province of Salta, especially the rural area of the departments San Martín and Oran. In the last 3 years, no cases of the disease have been registered, so the country is in the process of declaring itself free of indigenous cases of malaria. In 2010, the last reported autochthonous cases were recorded in the border area. In 2011, 2012, and 2013, respectively, 18, 4, and 2 cases were checked, all imported. Through the years, with qualified technical staff and distributed in different operational bases and through a unified methodology, consisting of the development of epidemiological surveillance actions, search for febrile patients, timely diagnosis, supervised treatment, and spraying of patient housing and neighboring areas, a significant reduction of the vector transmission surface was achieved, which currently reaches an area of 28,000 km^2 (Ministry of Health, Presidency of the Nation 2016a, b, c, d).

Chagas Disease in Argentina

Chagas disease, on the other hand, is one of the most widespread diseases in Latin America. It is a life-threatening disease caused by the protozoan parasite *Trypanosoma cruzi*. The most frequent form of contagion is by the bite of the vinchuca. The latest case estimates indicate that in Argentina, there would be 7,300,000 people exposed, 1,505,235 infected, of whom 376,309 would present Chagasic heart diseases. This constitutes the disease as one of the main public health problems. There are people with Chagas in the whole country because in addition to the vectorial transmission, human migrations and the existence of other transmission routes spread the disease throughout the whole territory. Chagas disease is a current topic of study in Argentina as it constitutes a real threat due to climate change. Its prevention is one of the most significant points of health authorities because it prevents the occurrence of evil and spread throughout the region. Extreme personal and environmental hygiene measures are carried out, and disinfection campaigns are carried out in the most affected areas. This disease is notifiable. Regarding their vectorial transmission, the Argentine provinces are classified as high, medium, and low risk of transmission of the parasite. There are also some so-called safe areas due to the magnitude of the number of existing vectors. In what areas of the country does the disease exist? The Chagas is found in those areas of our territory where there are vinchucas, although migratory movements have generated an increase of infection in places where the insect is not found. That is why there are only vinchucas in some provinces, but Chagas disease exists throughout the country (Ministry of Health, Presidency of the Nation, Diagnosis of Situation 2015).

Currently, the national scenario for Chagas disease is as follows (Fig. 24.1):

- High-risk situation for vector transmission: Chaco, Catamarca, Formosa, Santiago del Estero, San Juan, and Mendoza provinces present a reemergence of Chagas vector transmission due to an increase in home infestation and a high seroprevalence in vulnerable groups.
- Moderate-risk situation for vector transmission: The provinces of Córdoba, Corrientes, La Rioja, Salta, and Tucumán show an intermediate-risk situation with a reinfestation rate greater than 5% in some departments and insufficient surveillance coverage in some cases.
- Situation of low risk for the vectorial transmission: In May of 2015, the province of San Luis managed to certify the interruption of the vectorial transmission of *Trypanosoma cruzi* by *Trypanosoma infestans*. In 2012, they were able to certify

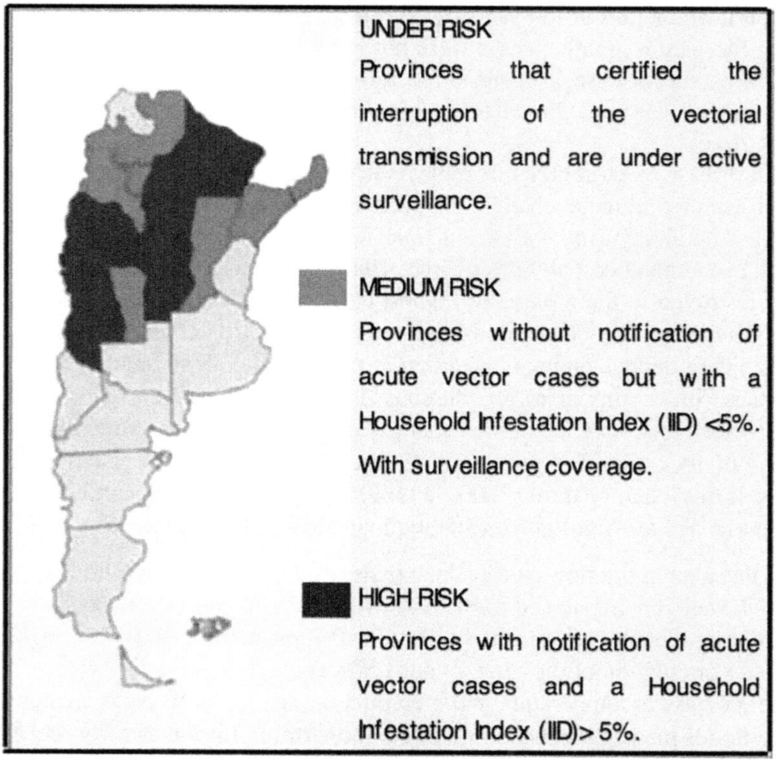

Fig. 24.1 Mal De Chagas in Argentina. Risk areas (Source: Argentina. Epidemiology Department. Ministry of Health of the Nation)

the provinces of Misiones and Santa Fe, along with six departments in the south of Santiago del Estero (Aguirre, Miter, Rivadavia, Belgrano, Quebracho, and Ojo de Agua). The provinces of Entre Ríos, Jujuy, La Pampa, Neuquén, and Río Negro managed to recertify the interruption of vector transmission (Ministry of Health, Presidency of the Nation, Chagas in the country and Latin America 2016).

Situation of the Dengue in Argentina

For the scientific community of our country, the progressive increase of temperature is a fact that does not admit discrepancies, so the health authorities have proposed prevention and treatment programs to be carried out in areas affected by dengue, pathology which today affects more than 38,000 people in Argentina. In Argentina, the behavior of dengue has so far been epidemic. Outbreaks began with the introduction of the virus by travelers to countries with viral circulation. During the winter months, cases were not recorded between 1 year and the next, reemerging the disease in some areas during the months of high temperatures (Ministry of Health, Presidency of the Nation, Information for Health Teams 2016).

This disease is currently the subject of special attention in Argentina due to several factors: climate change, in particular, increased travel and migration, inadequate collection of waste, and inadequate provision of drinking water for storage in containers usually discovered. The occurrence of dengue cases in Argentina is restricted to the months of highest temperature (November to May) and is directly linked to the occurrence of outbreaks in bordering countries. During 2009, the first major dengue outbreak occurred in Argentina, with autochthonous cases of the disease in 11 provincial jurisdictions. In contrast, in the current season, outbreaks of dengue prevailed at the usual period of its beginning, affecting a greater number of localities and provinces. Figure 24.2 shows the registered cases of dengue in the country during 2009 and 2016. Among the causes that motivated these outbreaks are highlighted (Stamboulian Health Services 2016):

- An increase in the flow of travelers, mainly due to the summer holiday season, which were directed to and from areas with viral circulation in the country and in bordering countries (especially Brazil, Paraguay, and Bolivia), favoring a greater circulation of the virus in our territory
- The increase in temperature and precipitation due to the El Niño phenomenon
- The floods mainly produced in the provinces of the Litoral as a consequence of this phenomenon

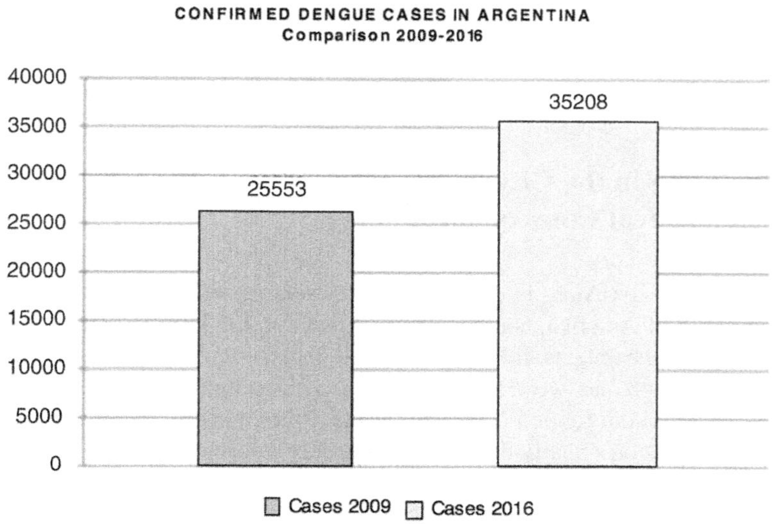

Fig. 24.2 Confirmed dengue cases in Argentina. Comparison 2009–2016 (Source: Argentina. Stamboulian. Health Services 2016)

Other Pathologies

However, other pathologies are also expected in the country, such as cardiovascular stress diseases, cancer oncology, diarrhea and acute respiratory infections, particularly in the malnourished, etc. An investigation by specialists from the National University of the Northeast (UNNE) found that diarrhea and acute respiratory infections suffered a significant increase, accompanied by a rise in the average minimum temperature and humidity, in a vulnerable ecological region of the province of Corrientes. The study, conducted by researchers of the Department of Infectology of the Faculty of Medicine and Institute of Regional Medicine of the UNNE, showed results that establish the growth of different diseases in the town of Ituzaingó, measuring health and environmental profiles in the period between the 2001 and 2006, and compare them with those obtained between 1994 and 2000 in the same locality. Twelve thousand eight hundred cases were analyzed (Ministry of Health, Presidency of the Nation, Country profile on climate change and health 2014). After 7 years, the observation showed that the data of diarrheas and acute respiratory infections suffered an increase of remarkable magnitude accompanying a rise in the average minimum temperature and minimum relative humidity, probably due to ecological instability in the area of impact, environmental or by the impact of global warming, which is a worrying indicator. The presence or combination of pathologies such as those mentioned are significantly influenced by global warming, which undoubtedly constitutes a factor of undoubted significance. But it is not the only cause that determines the health commitment in all its complexity, other components that have an impact on the magnitude of the problem and are largely responsible for the resulting environmental impact must therefore be taken into account. Epidemiological multifactority.

Air Pollution in the City of Buenos Aires and Its Epidemiological Consequences

The city of Buenos Aires is capital of the Argentine territory, is located in the extreme south of America, and counts on an area of 203 km^2 and a population of almost 3,000,000 habitants. It limits to the east with the River of the Silver, whereas by the north, south, and west, it is surrounded by urban municipalities conforming the call Metropolitan Region of Buenos Aires. From a geomorphological point of view, the city sits on a nearly flat surface, is widely spread, and has no geographical features that cause an accumulation of gases caused by the automobile transport and the industries that reside in the place. However, despite this, pollution is high because of the innumerable urban canyons that the city holds. Winds blow generally from the northeast, in winter, and frontal systems can be broken in from the south, whereas between autumn and spring, intense winds of the southeast can occur that cause great floods and floods in the waterfront.

In Buenos Aires, the effects of climate change are a major concern because they are beginning to be noticed. An increase in minimum temperatures, changes in the length of the seasons, an increase in precipitation averages, and a tendency to increase extreme events have been observed in the region. During the twentieth century, it has been noted that the average level of the Río de la Plata increased by about 17 cm and that change would be associated with the increase of mean sea level. These trends in climate dynamics have led to visible consequences in the region such as floods, heat waves, forest fires, and rangelands (Environmental Protection Agency, Buenos Aires 2012).

In the city of Buenos Aires, the concern on the part of the authorities to keep the atmosphere clean is very reduced. Compared to cities like Mexico or Santiago de Chile, whose topographic conditions do not favor the cleaning of atmospheric pollutants, there is an environmental perception of "clean city" in the city of Buenos Aires, which is not, and should be reviewed because concentration levels of particulate matter in suspension and nitrogen oxides exceed the permissible marks indicated by the World Health Organization. Among urban air pollutants, particulate matter is a serious threat to health. Its greatest danger is related to its entry into the lungs, staying there and damaging the tissues involved in gas exchange. It is also associated with a series of acute and chronic diseases mainly related to the respiratory tract and cardiovascular system. A publication by the Organization for Economic Cooperation and Development (OECD) estimates that more than 3.5 million people die prematurely due to the concentration of atmospheric particles and that air pollution will become the main environmental cause of death in the world in 2050 (OECD 2014). Several studies have also shown the

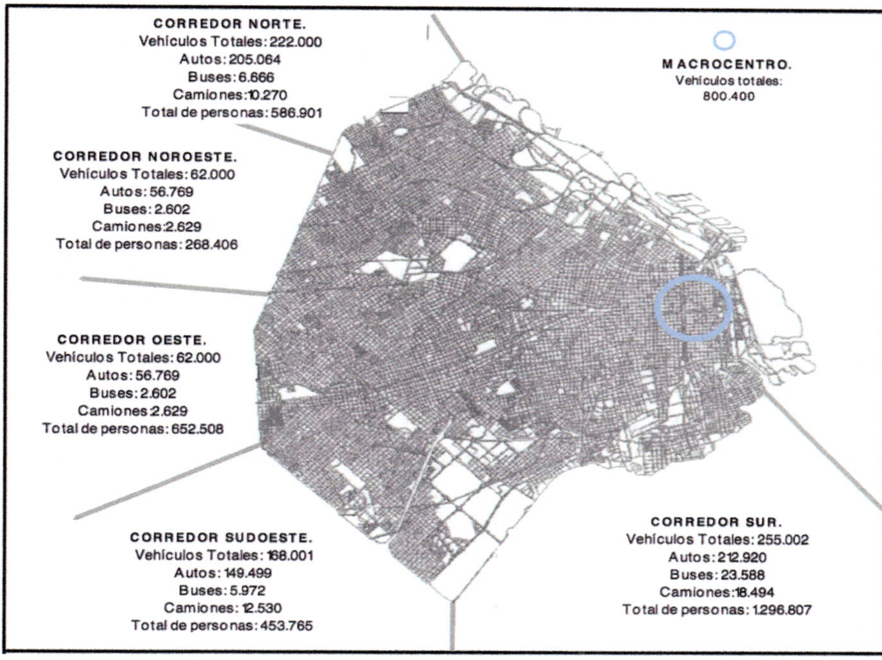

Fig. 24.3 Daily total revenue of vehicles and people to the city of Buenos Aires (Source: calculations based on information from various agencies. Transit Department. Undersecretary of Traffic and Transportation)

relationship between high particle concentration and cardiovascular or respiratory disease worldwide (Gouveia and Fletcher 2000; Pope et al. 1992; Peng et al. 2005; Orsini et al. 1986). In fact, the populations most at risk are the elderly, children, those with chronic lung disease or coronary heart disease, and patients with diabetes (Ribeiro 2008). Other effects of particulate matter in suspension are related to reduced visibility, increased dispersion, and/or absorption of solar radiation affecting short-wave radiation and increasing the number of condensation nuclei in the atmosphere. Aerosols or particulate matter, reflecting sunlight, can produce local and temporary cooling that could partly compensate for global warming caused by greenhouse gases, but since aerosols have a very short life in the atmosphere, they cannot make up for it forever. Also, there is evidence of damage caused by the deposit of particulate material on buildings and monuments.

In the city of Buenos Aires, on the other hand, given the remarkable increase that the car has acquired, there are areas in the downtown area with critical pollution problems. It is possible to determine two types of air pollution suffered by Buenos Aires, whose automotive traffic is practically uncontrollable today (Fig. 24.3): one caused by the carbon monoxide, of very long data, being a poison that we all consume when to cross the city. The phenomenon is, of course, very serious in the central areas, especially when they are very congested by vehicles. The increase in vehicular traffic in the center of large cities has increased the traffic to unexpected levels. However, the pollutant activity of man does not end with this primary pollutant, insidiously toxic and even more deadly. There is a second type of pollution caused by the car: photochemical pollution. When the sun illuminates in the morning, the gases released by vehicles, especially NO_x and hydrocarbons, react photochemically and generate, among other things, ozone (Venegas et al. 2003). This substance, in almost infinitesimal concentrations, is very irritating to our mucous membranes. In addition, its effects on the respiratory system, and particularly on the pulmonary parenchyma, have also been recognized and documented. Ozone is one of the pollutants that contribute most to the degradation of air quality in large urban centers. Exposure to high concentrations is associated with increased hospital admissions for pneumonia, chronic obstructive pulmonary disease, asthma, bronchitis, allergic rhinitis and other respiratory diseases, as well as premature mortality (Aris et al. 1993; Bell and McGregor 2008; Frampton et al. 1999; Gryparis et al. 2004; Ito et al. 2005). But the increased risk of ozone-related mortality is associated with respiratory causes, especially in low and high socio-economic status groups. Although the impacts on the respiratory system are more common, Nardocci et al. (2013) observed, in addition to the association between O_3 and respiratory diseases in children under 5 years of age, an association between this contaminant and cardiovascular diseases in adults older than 39 years. Due to the toxic nature of this gas and the potential risk it poses to human health, its permitted levels have already been carefully established by institutions such as the United States Environmental Protection Agency (EPA).

However, a phenomenon that is likely to be a major concern for large cities if they do not fit into a preventive action on air pollution is the so-called ozone weekend effect whose sole and exclusive responsibility is attributed to the car. The "ozone weekend effect" refers to the curious finding in certain metropolis of high concentrations of ozone during the weekends compared to other days of the week. This is very striking because the higher emissions of ozone-producing compounds usually occur on weekdays rather than on weekends (Lipp et al. 2010).

Research has been carried out in Buenos Aires for the damages caused by atmospheric pollution. The SAEMC through a study showed a clear correlation between the variation of air compounds measured in the city and mortality. Data on the concentration of carbon monoxide and nitrogen oxides were used as these are the compounds measured by the monitoring network of the Government of the city of Buenos Aires.

SAEMC verified a 3.6% increase in daily deaths the following day to a rise in 1 ppm (part per million) of atmospheric CO. Analysis of nitrogen oxides (NO_x) also shows a significant correlation with daily mortality, particularly due to respiratory causes. On the same day that the NO_x level in air increases by 10 ppb (parts per billion), mortality from this cause increases by 0.7% and cardiovascular mortality by 0.4%. The results of this work express strongly that even in a city that generally does not surpass the concentration levels of contaminants established in local regulations, pollution causes an effect on health that in extreme cases leads to death. There is no further study (Abrutzky et al. 2014). The epidemiological information in the city is extremely poor. On the other hand, the data obtained from studies carried out in cities of developed countries are not totally extrapolable to our environment, since the susceptibility of children and adults to the effects of air pollutants is potentially greater in the region due to poverty, malnutrition, immunological deficiencies, and poor living conditions. In addition, many of the health effects are not easy to isolate technically since they may be obeying also, or jointly, to other causes.

Actions in the Field of Climate Change in the Country

Given the state of concern about climate change, many countries, including Argentina, have emphasized reducing greenhouse gas emission levels while improving local air pollution conditions. In this direction, the Ministry of Environment and Sustainable Development of the Nation, today elevated to the rank of Ministry, has been working in a systematic way in order to contribute to the development of policies to avoid the increase of its emissions. Within the United Nations, the United Nations Framework Convention on Climate Change (UNFCCC) was created in 1992 to ensure that the concentration of greenhouse gases in the atmosphere does not continue to increase, that is to say, to stabilize to a level that prevents dangerous interferences with the climatic system. Our country signed and ratified this international agreement in 1992 and 1993, respectively. However, for developing countries, there are no quantifiable targets for reducing their GHG emissions, but there are particular commitments, including an inventory of emissions by sources and removals by GHG sinks, an overview of the steps taken or to be taken to implement the convention and any other information deemed relevant to achieve the objective of the convention. Our country complied with this commitment by submitting this first communication to the UNFCCC, according to the methodology established by the IPCC in June 1997 (Secretariat of Environment and Sustainable Development of the Nation 2012).

Since that date, our country has published several emission inventories, the last of which was presented in 2000. In addition, through the Argentine Carbon Fund program, promotion and technical assistance are offered to proponents of potential GHG emission reduction projects, in order to assess whether they comply with the requirements of the Clean Development Mechanism. At present, the Argentine Carbon Fund has a portfolio of more than 340 projects in different degrees of maturity and of different sectors, among which the most numerous correspond to the category of energy, waste management, and forestry. On the other hand, a carbon footprint calculator has been developed at the municipal level that allows the identification of the emission sources that are under the administration of a municipal government, such as offices and municipal buildings, hospitals, schools, bus terminals, vehicle fleets, among others, and determines their temporal evolution as a basis for the development of mitigation strategies. Currently this tool is under review and adjustment.

Conclusion

Argentine experts on climate change say that global warming will affect the country if the concentration of carbon dioxide is almost double the current. In this sense, however, the country has advanced in the development of environmental strategies with a particular imprint in which interinstitutional cooperation between national, provincial, and municipal levels is combined, as well as intersectoral articulation, integrating actors, and organizations of society civil. On the other hand, the establishment of institutional and legal frameworks such as the enactment of the Law of Forests and the Law of Glaciers, the ratification of international conventions, and the nationalization of strategic natural resources. According to experts, since 1995, a number of technologies have been developed which have made it possible to moderate this phenomenon in particular, such as the construction of low emission engines and turbines, the techniques used by some metallurgical industries and chemical products to reduce the emission of gases and, mainly in some areas of the country, the substitution of coal, oil, and its derivatives by other nonpolluting energy sources such as wind, solar, and nuclear energy. With regard to urban transport, which we have been discussing here for being responsible for the enormous quantities of gases that are sent to the atmosphere, there have also been notable and valuable advances in the field of climate change in the last decade. In the above line, there is an incipient tendency of the Argentine automotive industry to offer less polluting public and private transport units, based on both the technical improvements of its engines and the type of fuel used. However, these technical changes would not have immediate effects on mitigating the problem, due to the slow replacement of public and private transport fleets. Currently, the Congress of the Argentine Nation is debating how to progress toward the fulfillment of the goals established to generate energy from renewable sources. The development of renewable energies requires the absolute attention of the state, since they can contribute not only to the global climatic situation but also bring economic benefits for Argentina that allow to recover the energy self-sufficiency. Argentina depends on 87% of fossil fuels to generate its energy. In 2013, only 1.4% of electricity came from renewable sources, despite having a law that establishes that this contribution should reach 8% in 2016. The new findings of the IPCC show that this is only possible if investments and subsidies for the development of renewable energies and energy efficiency are reoriented.

References

Abrutzky R, Dawidowski L, Murgida A, Natenzon CE (2014) Contaminación del aire en la Ciudad Autónoma de Buenos Aires: El riesgo de hoy o el cambio climático futuro, una falsa opción. Ciência Saúde Coletiva 19(9):3763–3773. Associação Brasileira de Pós-Graduação em Saúde Coletiva Rio de Janeiro, Brasil

Aris RM, Christian D, Hearne PQ, Kerr K, Finkbeiner WE, Balmes JR (1993) Ozone-induced airway inflammation in human subjects as determined by airway lavage and biopsy. Am Rev Respir Dis 148(5):1363–1372

BUENOS AIRES (ARGENTINA). Agencia de Protección Ambiental. Plan Estratégico 2008–2012. Gobierno de la ciudad de Buenos Aires

Flannery TF (2006) la amenaza del cambio climático: historia y futuro. Madrid, Ed. Santillana, 2006. 393p

Frampton MW, Pryor WA, Cueto R, Cox C, Morrow PE, Utell MJ (1999) Ozone exposure increases aldehydes in epithelial lining fluid in human lung. Am J Respir Crit Care Med 159 (4):1134–1137

Gouveia N, Fletcher T (2000) Time series analysis of air pollution and mortality: effects by cause, age and socioeconomic status. J Epidemiol Community Health 54(10):750–755

Gryparis A, Forsberg B, Katsouyanni K, Analitis A, Touloumi G, Schwartz J, ... Wichmann HE (2004) Acute effects of ozone on mortality from the "air pollution and health: a European approach" project. Am J Respir Crit Care Med 170(10):1080–1087

Hernández Nava E (2014) Cambio Climático, Salud Humana y Enfermedades emergentes. Universidad de Cantabria, Facultad de Filosofía y Letras, Cantabria. Curso 20–2014

Ito K, De Leon SF, Lippmann M (2005) Associations between ozone and daily mortality: analysis and meta-analysis. Epidemiology 16(4):446–457

Lipp DO, Gassmann MI (2010) Modelling the weekend effect in Buenos Aires City. Turin, Italia: Italia. Congreso. 31st NATO/SPS international technical meeting on air pollution modelling and its application. NATO

Ministerio de Salud. Presidencia de la Nación (2014) Perfil de país sobre cambio climático y salud. http://www.msal.gob.ar/. Argentina

Ministerio de Salud. Presidencia de la Nación (2015) Diagnóstico de Situación. http://www.msal.gob.ar/chagas/index.php/institucional/diagnostico Argentina

Ministerio de Salud. Presidencia de la Nación (2016a) http://www.msal.gob.ar/chagas/. Argentina

Ministerio de Salud. Presidencia de la Nación (2016b) http://www.msal.gob.ar/index.php/component/content/article/48/98-paludismo-o-malaria. Argentina

Ministerio de Salud. Presidencia de la Nación (2016c) El Chagas en el país y América Latina. http://www.msal.gob.ar/chagas/index.php/informacion-para-ciudadanos/el-chagas-en-el-pais-y-america-latina Argentina

Ministerio de Salud. Presidencia de la Nación (2016d) Información para Equipos de Salud. http://www.msal.gov.ar/dengue/informacion-para-equipos/. Argentina

Nardocci AC, Freitas CU, Leon ACMP, Junger WL, Gouveia NC (2013) Poluição do ar e doenças respiratórias e cardiovasculares: estudo de séries temporais em Cubatão, São Paulo, Brasil. Cad Saude Publica 29(9):1867–1876

Organization for Economic Co-operation and Development – OECD (2014) The cost of air pollution: health impacts of road transport. OECD Publishing, Paris. DOI: http://dx.doi.org/10.1787/9789264210448-en

Orsini CQ, Tabacniks MH, Artaxo P, Andrade MDF, Kerr AS (1986) Characteristics of fine and coarse particles of natural and urban aerosols of Brazil. Atmos Environ (1967) 20 (11):2259–2269

Peng RD, Dominici F, Pastor-Barriuso R, Zeger SL, Samet JM (2005) Seasonal analyses of air pollution and mortality in 100 US cities. Am J Epidemiol 161(6):585–594

Perczyk D, Bromioli M, Carlino H, González MP, Andelman M (2004) Para Entender el Cambio Climático. Ministerio de Salud y Ambiente de la Nación, Buenos Aires. 2004

Pope CA III, Schwartz J, Ransom MR (1992) Daily mortality and PM10 pollution in Utah Valley. Arch Environ Health: Int J 47(3):211–217

ArgentinaSecretaria de Ambiente y Desarrollo Sustentable (2009) El cambio climático en Argentina. SA y DS, Buenos Aires

Secretaría de Ambiente y Desarrollo Sustentable de la Nación (2012) Informe sobre el Estado del Ambiente Año 2012. Buenos Aires

Stamboulian. Servicios de Salud (2016) http://www.stamboulian.com.ar/viajeros.aspx?idContenido=450. Argentina

Venegas LE, Mazzeo NA (2003) Calidad del aire en una zona de la Ciudad de Buenos Aires (Argentina). III Congresso Interamericano de Qualidade do Ar, ULBRA, Canoas, RS, Brasil. CD

Daniel Oscar Lipp is doctor in geography at the University of Salvador (Argentina) and has earned a master's in environmental sciences (natural resources) from the University of Buenos Aires and worked on issues related to the environmental area, more specifically the study of air pollution. He has worked as a researcher and teacher at the Catholic University of Salta (Argentina) and is the author of scientific publications (with reference) of national and international circulation in environmental issues. He is currently a member of the Argentine Society of Geographical Studies (GAEA) and participates in its activities.

Part IV
Conclusion

Chapter 25
Summary and Conclusion

Rais Akhtar and Cosimo Palagiano

Keywords Urban air quality • Hippocrates • Industrial revolution • Great Smog • Ozone pollution

Concern about air pollution has been known for thousands of years. "Complaints about its effects on human health and the built environment were first voiced by the citizens of ancient Athens and Rome. Urban air quality, however, worsened during the Industrial Revolution, as the widespread use of coal in factories in Britain, Germany, the United States and other nations ushered in an 'age of smoke'" (Mosley 2014). As urban areas developed, pollution sources, such as chimneys and industrial processes, were concentrated, leading to visible and damaging pollution dominated by smoke. The harmful effects of air pollution were recognized by Hippocrates in his fifth-century treatise Air, Water and Places; Hippocrates noted that people's health could be affected by the air they breathe and that quality of the air differed by area (cited in Adams 1891).

Air pollution disasters such as London's sulphur-laden "Great Smog" in 1952 that killed an estimated 4000 people demonstrated conclusively the damage it caused to human health and instigate parliament to enact the 1956 Clean Air Act to reduce coal burning and begin serious air pollution reform in England.

In the United States, concern for the air quality in and around large cities was increasing during the latter 1800s and resulted in local laws and regulations followed ultimately by federal air pollution control regulations. A degree of particulate air pollution in Australia before colonization is likely to have been frequent, due to the widespread indigenous practice of deliberately lighting fires to manage their landscape, a process today called "fire-stick farming" (Gammage 2011, Jones 2012 cited in Butler and Whelan's chapter in this book on Australia).

R. Akhtar (✉)
International Institute of Health Management and Research (IIHMR), New Delhi, India
e-mail: raisakhtar@gmail.com

C. Palagiano
Dipartimento Di Scienze Documentarie, Linguistico-Filologiche e Geografiche, Sapienza University of Rome, Rome, Italy
e-mail: cosimo.palagiano@uniroma1.it

© Springer International Publishing AG 2018

R. Akhtar, C. Palagiano (eds.), *Climate Change and Air Pollution*,
Springer Climate, DOI 10.1007/978-3-319-61346-8_25

Thus climate change represents a range of environmental hazards including air pollution and will affect populations wherever the current burden of climate-sensitive disease is high – such as the urban poor in low- and middle-income countries. Understanding the current impact of weather, climate and air pollution variability on the health of populations is the first step towards assessing future impacts.

About 54% of the world's population lives in urban areas, a proportion that is expected to increase to 66% by 2050. Projections show that urbanization combined with the overall growth of the world's population could add another 2.5 billion people to urban populations by 2050, with close to 90% of the increase concentrated in Asia and Africa, according to a new United Nations report launched today.

We are aware about the scientific explanations that climate change occurs because excessive amount of greenhouse gases were emitted into the atmosphere due to human activity. Human influence on the climate system is clear, and recent anthropogenic emissions of greenhouse gases are the highest in history. The Earth's atmosphere has already warmed by 0.85 °C from 1880 to 2012. Recent climate changes have had widespread impacts on human and natural systems (IPCC 2014).

Having said that, "Weather and climate play important roles in determining patterns of air quality over multiple scales in time and space, owing to the fact that emissions, transport, dilution, chemical transformation and eventual deposition of air pollutants all can be influenced by meteorological variables such as temperature, humidity, wind speed and direction and mixing height. There is growing recognition that development of optimal control strategies for key pollutants like ozone and fine particles now requires assessment of potential future climate conditions and their influence on the attainment of air quality objectives. In addition, other air contaminants of relevance to human health, including smoke from wildfires and airborne pollens and moulds, may be influenced by climate change" (Kinney 2008). In the study by Kinney, the focus was on the ways in which human health-relevant measures of air quality, including ozone, particulate matter and aeroallergens, may be influenced by climate variability and change.

Focusing on climate change impacts on air pollution, particularly ozone pollution, IPCC has also clearly stressed that "pollen, smoke and ozone levels likely to increase in warming world, affecting health of residents in major cities. Rising temperatures will worsen air quality through a combination of more ozone in cities, bigger wild fires and worse pollen outbreaks, according to a major UN climate report. It is formed by the reaction with sunlight (photochemical reaction) of pollutants such as nitrogen oxides (NO_2)" (Wynn 2014). Frequent forest fires in certain regions in Australia and in the state of California are examples of such events. World Meteorological Organization (WMO) has now certified that 2016 was the warmest year.

In Italy and in many Mediterranean regions, ozone is particularly dangerous, because of the solar radiation, which captures the oxide from the chemical compounds such as the sulphur and nitrogen dioxide and binds them to oxygen free in the air. So ozone rises producing many health problems to humans. This comes true in the cities, where the car traffic is very congested.

Rome is the Italian city with the much congested traffic. But the car traffic problems are severe in many cities both of developed and developing countries. We can give as example Beijing and Bangkok, where the car traffic is very intense, with many air pollution and health impacts.

With reference to human health implications, the air pollution is currently the leading environmental cause of premature deaths. The findings of the World Health Organization contend that air pollution is the world's biggest environmental health risk, killing seven million people in 2012 (in comparison to four million deaths due to malaria and 3.1 million deaths of children under five due to malnutrition). Deteriorating air quality will mostly affect the elderly, children, people with chronic ill-health and expectant mothers, with growing population in urban areas in the coming decades and the rise in vulnerable population.

The present book comprises studies on developed and developing countries. The book is aimed to present a regional analysis pertaining to climate change, air pollution and human health, focusing on climate change, air pollution and adaptation strategies in geographically and socio-economically varied countries of the world.

In the context of developed countries, for instance, Australia, there also needs to be a much greater appreciation of the health and economic costs of air pollution and climate change. It is enormously misleading to claim that coal-fired electricity is "cheap". Coal mining, coal combustion and coal export cause significant health costs, in the past, present and future. Furthermore, the price of alternatives such as wind and solar continues to fall. Reducing emissions from the burning of wood and the combustion of vehicular fuel is more challenging, but much can also be accomplished in these spheres too, including electric vehicles, public transport, and, in the foreseeable future, domestic production and consumption of solar energy, incorporating batteries.

In developing countries, for instance, Mexico, most of the policies have concentrated on vehicle emissions. This has proved not to be enough given the increase in vehicles circulating in the city and the poor public transport options that have not kept pace with demand for mobility. A reduction in urban local pollutants and greenhouse gases that may reduce air pollution and mitigate climate change will only come from a true change in the energy matrix. Such a change may only be produced in the medium run by the use of economic incentives to deter the use of highly polluting fuels and to embark into long-term investments that will need less and cleaner energy sources.

In the Caribbean, the research indicates that the burden of air pollution on the people will increase with climate change, unless stringent measures are taken at the community, country/government and global levels. Particularly, given the established human health effects of air pollutants such as ozone, environmental surveillance of these pollutants and longitudinal studies of their impact on the health of populations across the Caribbean are recommended. Finally, how climate change is likely to influence the effects of air pollution on states and territories in the region should be considered.

Wildland fire is an important component to ecological health in California forests. Wildland fire smoke is a risk factor to human health. Exposure to smoke from fire cannot be eliminated, but managed fire in a fire-prone ecosystem for forest health and resiliency allows exposure to be mitigated while promoting other ecosystem services that benefit people. California's Sierra Nevada is a paragon of land management policy in a fire-prone natural system. Past fire suppression has led to extreme fuel loading where extreme fire events are much more likely, particularly with climate change increasing the length of fire season and the probability of extreme weather. California's Sierra Nevada is an example to showcase the clash of increased development and urbanization, past land management policy, future scenarios including climate change and the intertwining of ecological health and human health. Fire suppression to avoid smoke impact has proven to be an unreliable way to decrease smoke-related health impacts. Instead, ecological beneficial fires should be employed, and their management should be based on smoke impacts at monitors, making air monitoring the foundation of fire management actions giving greater flexibility for managing fires. Tolerance of smoke impacts from restoration fire that is best for forest health and resiliency, as well as for human health, is paramount and preferred over the political expediency of reducing smoke impacts today that ignores that we are mortgaging these impacts to future generations.

Another example from developing countries comes from South Africa. Climate change and air pollution pose significant short-term and long-term health risks to South Africans due to the carbon intensity of the national economy, severe air pollution around coal mining and coal-fired power stations and the vulnerability of many subgroups in a nation burdened by extreme inequality and severe acute and chronic diseases.

There are limited local studies on the respiratory, cardiovascular and other health risks of air pollution. Inadequate disease surveillance and air quality data pose a challenge for air pollution monitoring and research to its health impacts.

Key measures suggested to mitigate emissions are concerned with the energy, industry, human settlements, transport, health care and business sectors. The public health community has an important role to play in urging further action and research at the national, provincial and local levels.

Mitigating air pollution as well as greenhouse gases in Delhi, one of the most polluted cities in the world, without adversely impacting development remains a crucial goal. Further, climate change has profound impacts that Delhi must adapt to. From a health perspective, in addition to health impacts of pollution, addressing health impacts of climate change such as heat waves is important.

This study on Delhi understands the transitions of key drivers of energy use such as population, vehicle use and per-capita incomes that in turn drive emissions of pollutants and greenhouse gases. It provides estimates of greenhouse gas and pollutant emissions from Delhi. It estimates pollution as well as future heat-related mortality for Delhi. Finally, it argues that policies for GHG as well as pollutant mitigation require to be better aligned. This will ensure that health co-benefits are accrued for Delhi.

In case of developed region, Hong Kong is one of the densest cities on the planet and has adopted a stronger (60–65% carbon reduction by 2030) mitigation plan for combating climate change. Although it may not be significant from a global perspective, it shows a strong commitment as a global citizen. The major reduction of GHGs in Hong Kong, is focused on the energy sector, where changing carbon-intensity fossil fuel (i.e. coal) into less intense fuel such as natural gas, or nuclear, reducing building-related energy usage and adopting more green transportation. These mitigation plans have moved Hong Kong towards becoming a healthier city. In terms of air pollution, these mitigation plans carry some co-benefit on local air quality, where reduction of coal/gasoline burning would reduce $PM_{2.5}$ and NO_x emitted into both roadside and ambient environments. Under the future emission projections (IPCC AR5), $PM_{2.5}$ air quality for Hong Kong in 2050 would be improved under RCP2.6, 4.5 and 8.5 due to the reduction of primary PM and its precursors, while it is increased under RCP6.0. In terms of ozone, less exceedance of ozone (based on MDA8) is projected in 2050 under RCP2.6, 4.5 and 8.5 in PRD area.

To sum up, the analysis of different chapters regionally diversified did reveal the association between climate change and air pollution and impacts on human health. The discussion also highlights the mitigation strategies to be adopted to combat climate change and minimize GHG emissions in both developed and developing countries.

References

Adams F (1891) The genuine works of Hippocrates. William Wood and Company, New York
IPCC (2014) Climate change: impact, adaptation and mitigation, WG II, AR5, chapter 24 Asia. Cambridge University Press, Cambridge
Kinney PL (2008) Climate change, air quality, and human health. Am J Prev Med 35(5):459–467
Mosley S (2014) Environmental history of air pollution and protection. In: World environmental history, encyclopedia of life support systems (EOLSS), Paris. http://www.eolss.net
Wynn, G. (2014) Climate change will hike air pollution deaths says UN study, Climate Home, March 28

Rais Akhtar is presently adjunct professor of the International Institute of Health Management Research, New Delhi.

Cosimo Palagiano is emeritus professor in geography at the Department of Documentary, Linguistic-Philological and Geographical Sciences, Sapienza University of Rome.

Index

© Springer International Publishing AG 2018

R. Akhtar, C. Palagiano (eds.), *Climate Change and Air Pollution*,
Springer Climate, DOI 10.1007/978-3-319-61346-8

Printed by Printforce, the Netherlands